GIS for Environmental Applications

GIS for Environmental Applications provides a practical introduction to the principles, methods, techniques and tools of GIS (geographical information system or systems) for spatial data management, analysis, modelling and visualisation, and their applications in environmental problem solving and decision making. It covers the fundamental concepts, principles and techniques in spatial data, spatial data management, spatial analysis and modelling, spatial visualisation, spatial interpolation, spatial statistics and remote sensing data analysis, as well as demonstrating the typical environmental applications of GIS, including terrain analysis, hydrological modelling, land use analysis and modelling, ecological modelling and ecosystem service valuation. Case studies are used in the text to contextualise these subjects in the real world, and examples and detailed tutorials are provided in each chapter to show how the GIS techniques and tools introduced in the chapter can be implemented using ESRI ArcGIS (a popular GIS software system for environmental applications) and other third-party extensions to ArcGIS.

The emphasis is placed on how to apply or implement the concepts and techniques of GIS through illustrative examples with step-by-step instructions and numerous annotated screenshots. The features include:

- over 350 figures and tables illustrating how to apply or implement the concepts and techniques of GIS;
- learning objectives along with the end-of-chapter review questions;
- authoritative references at the end of each chapter;
- GIS data files for all examples, as well as PowerPoint presentations for each chapter downloadable from the companion website.

GIS for Environmental Applications weaves theory and practice together, assimilates the most current GIS knowledge and tools relevant to environmental research, management and planning, and provides step-by-step tutorials with practical applications. This volume will be an indispensable resource for any students taking a module on GIS for the environment.

Xuan Zhu is Senior Lecturer in the School of Earth, Atmosphere and Environment, Monash University, Australia.

In memory of my father, Baiming Zhu (1925–2009)

GIS for Environmental Applications

A practical approach

Xuan Zhu

Routledge
Taylor & Francis Group

LONDON AND NEW YORK

First published 2016
by Routledge
2 Park Square, Milton Park, Abingdon, Oxon OX14 4RN

and by Routledge
711 Third Avenue, New York, NY 10017

Routledge is an imprint of the Taylor & Francis Group, an informa business

© 2016 Xuan Zhu

British Library Cataloguing in Publication Data
A catalogue record for this book is available from the British Library

Library of Congress Cataloging-in-Publication Data
Names: Zhu, Xuan (Earth scientist), author.
Title: GIS for environmental applications : a practical approach / Xuan Zhu.
Other titles: Geographic information systems for environmental applications
Description: Abingdon, Oxon ; New York, NY : Routledge, 2016. | Includes bibliographical references and index.
Identifiers: LCCN 2015044508 | ISBN 9780415829069 (hardback : alk. paper) | ISBN 9780415829076 (pbk. : alk. paper) | ISBN 9780203383124 (ebook)
Subjects: LCSH: Environmental monitoring—Geographic information systems. | Environmental geography.
Classification: LCC G70.212 .Z5386 2016 | DDC 363.7/0630285—dc23
LC record available at http://lccn.loc.gov/2015044508

ISBN: 978-0-415-82906-9 (hbk)
ISBN: 978-0-415-82907-6 (pbk)
ISBN: 978-0-203-38312-4 (ebk)

Typeset in Minion Pro
by Swales & Willis Ltd, Exeter, Devon, UK

Contents

Figures and tables

FIGURES

TABLES

Preface

Geographical information systems (GIS – the same acronym can serve as either the singular or plural form) are effective tools for reporting on the environment, modelling environmental interactions, assessing environmental impacts, evaluating the effectiveness of environmental policies and actions, and dissimilating environmental information. This book introduces the concepts, methods, techniques and tools involved in using GIS for environmental analysis, modelling and decision making. It weaves theory and practice together, assimilates the most current GIS knowledge and tools relevant to environmental studies, and provides detailed examples with practical applications.

Many existing GIS textbooks provide theoretical understandings of geographical information and GIS operations, some of which are well-written and comprehensive. However, they provide little guidance on how to implement GIS tools and techniques to master their applications in environmental studies and provide little real hands-on understanding of GIS. A handful of GIS books with an environmental focus largely provide case studies or edited research chapters, which either do not provide enough coverage of GIS principles and techniques or are not accessible to students. This book aims to provide detailed and comprehensive coverage of GIS principles and techniques, and in the meantime to provide a practical guide to the implementation of these principles and techniques in environmental research, management and planning through a large number of carefully chosen examples, practicals and case studies to help readers get a real grasp of knowledge and skills. It brings the knowledge, tools and practices into one cohesive, comprehensive, concise and self-contained book accessible not only to students, but also to environmental scientists and practitioners.

For most of us, it is through examples and guided practicals or experiments that we really learn. Accordingly, the book takes a practical approach and uses examples and hands-on practicals as a means to quickly test, verify and experiment with GIS principles and techniques in an instructive and interactive way. The book comes complete with a companion website containing self-assessment questions and data for working out all the examples and practicals. It is intended for upper-division undergraduates and postgraduates in environmental science, geography, Earth science, atmospheric science, ecology and biological science. It can also be used as a reference book for researchers and practitioners in these fields who wish to use GIS as a tool for environmental analysis, modelling and decision making. In addition, it is written for instructors looking to provide a hands-on and structured course on GIS.

The book is organised into ten chapters. Chapters 1–9 cover the fundamental concepts, principles and techniques in spatial data, spatial data management, spatial analysis, spatial statistics, remote sensing data analysis, terrain analysis, spatial visualisation, and spatial decision analysis and modelling. Each of these chapters includes examples and practicals that show how the GIS principles, techniques and tools introduced can be implemented using ESRI ArcGIS (a popular GIS software system for environmental applications) and other third-party extensions to ArcGIS to address environmental issues. Although case studies are provided in the first nine chapters, Chapter 10 contains a collection of more detailed case studies organised under the themes of hydrological modelling, land use analysis and modelling, atmospheric modelling, ecological modelling and ecosystem service valuation. These case studies, although they do not cover all areas of environmental research and practice, show some typical environmental applications of GIS with an emphasis on how GIS has been used in the analysis, modelling or solution processes. They are fully cross-referenced to relevant sections in the text and provide additional case studies to those sections.

Each chapter includes the following learning aids:

- a list of learning objectives stated at the beginning;
- full-colour maps, diagrams, tables and photos;
- boxes providing tutorial-style practical guidance on how to apply or implement the concepts and techniques of GIS with step-by-step instructions;
- a summary at the end covering the main concepts introduced;
- a set of end-of-chapter review questions, the answers to which can be found through careful reading of the chapter;
- a list of references including all publications cited in the text, serving as both source material for the chapter and a starting point for further investigation.

The content of the book is fairly modular and sufficiently flexible to support a variety of course needs. These needs are determined by the backgrounds of students, the course length, the course position in a curriculum and the course objectives. For example,

Chapter 9, 'Spatial decision analysis and modelling', is an advanced topic which is suitable for postgraduates, and can be excluded from an undergraduate course. Case studies in Chapter 10 can be used as selective reading for students according to their backgrounds.

In writing this book, I have many people to thank. First, I thank the many students who have taken my GIS courses over the past several years. Their feedback has been instrumental in organising the ideas presented in the text. My heartfelt thanks are due to Katherine Yu, who produced many of the diagrams, proofread chapters and provided constant support and frequent understanding over the years. I wish to thank the anonymous reviewers of the book for carefully reading and critiquing the work. Their input resulted in significant improvements to the quality of the final output. I would also like to thank Andrew Mould, Sarah Gilkes and Egle Zigaite, the editors, for their help, guidance, encouragement and professionalism throughout the years. Last and not least, thanks go to my family, who over the years have been neglected during my deepest concentration on writing.

Xuan Zhu
Monash University
Melbourne

Companion website and GIS software

The website that accompanies this book is available at:

www.routledge.com/cw/zhu

The resources on the site include:

- GIS data files for practicals – most of the practicals in the book use a GIS database created for a hypothetical virtual catchment; all the data files are compressed in **GIS4EnvSci.zip**, which you should download, then extract all the GIS data files to the **C:** directory on your hard drive before starting practicals;
- updates on changes to ArcGIS as new versions are released – every new release of ArcGIS software may involve addition of new functions, re-grouping and renaming of functions, and modification of user interfaces; the practicals will be updated with these changes and made available on the website;
- online quizzes for each chapter;
- PowerPoint presentations for each chapter;
- useful links to GIS-related professional organisations, journals, magazines, data sources and software.

ESRI ArcGIS for Desktop is required in order to carry out the majority of the practicals in the book. You will need access to a current ArcGIS for Desktop licence. ESRI (http://www.esri.com/) offers a free twelve-month ArcGIS for Desktop software licence to students at colleges and universities which have signed up to the ESRI Education Site License Program. Students can also purchase a twelve-month ArcGIS for Desktop licence for $100.

CHAPTER ONE

GIS overview

1

This chapter provides an overview of GIS for environmental applications. It introduces the history, the general framework and the functionality of GIS for environmental problem solving and decision making, discusses the roles of GIS in environmental research, management and planning, presents a brief introduction to several popular GIS software packages, and provides a guided tour of ArcGIS – the GIS software used in the examples throughout the book.

LEARNING OBJECTIVES

After studying this chapter, you should be able to:

- distinguish GIS from other types of computerised systems;
- understand the purposes for which GIS may be applicable;
- discuss the roles of GIS in environmental research, management and planning;
- describe the process of environmental problem solving with GIS;
- outline the key features of a range of proprietary and open source GIS software currently available for environmental studies;
- become familiar with the ArcGIS software environment.

1.1 INTRODUCTION TO GIS

The fundamental goal of environmental research is to understand and solve environmental problems, focusing on how humans use natural resources and how human activities change the environment. For example,

an ecologist might like to study the consequences of landscape fragmentation caused by agricultural intensification and habitat loss for wildlife. A wildlife biologist might want to analyse the movement patterns of migratory animals. A geographer might be interested in land use changes and their environmental impacts in a region. A climatologist might want to predict the responses of ecosystems to climate change in order to provide options for sustainable landscape management. A geologist might like to assess landslide susceptibility according to past history, slope steepness, bedrock and other physical factors and identify areas of differing landslide potential. A hydrologist might be interested in modelling hydrological responses to the urbanisation of areas where there are significant groundwater–surface water interactions and in understanding urban hydrology and contaminant flow pathways, which will inform water-sensitive urban design. All disciplines of environmental science work with data which are largely spatial in nature – data related to places on the Earth's surface. Such data are called spatial data. They describe where things are or where they were or will be on the surface of the Earth. Box 1.1 shows a typical example of species distribution modelling relying on the use of spatial data. GIS provides a technology and suite of tools for collecting, organising, searching, analysing and visualising spatial data.

What is GIS?

The acronym GIS stands for geographical information system(s). Different from other types of computerised systems such as spreadsheets, word processors and database management systems, GIS processes and manages spatial data. While word processors and spreadsheets are computer tools for dealing with text and numbers respectively, GIS handles maps, images and other types

Box 1.1   Modelling the distribution of Sambar deer (*Cervus unicolor*) in Victoria, Australia   **CASE STUDY**

Species distribution modelling aims to understand where a species occurs or likely occurs, how the occurrence of a species is related to the environment, and how a species might respond to changes in its environment. It is based on the assumption that each species has its own environmental niche – living in areas within a range of environmental factors such as temperature, rainfall and vegetation type. Environmental scientists from the Arthur Rylah Institute (ARI) for Environmental Research in Melbourne combined field surveys with biophysical data to estimate the current and potential distributions of Sambar deer (*Cervus unicolor*) in Victoria. Sambar deer were introduced to Victoria in the nineteenth century from Sri Lanka, India and the Philippines. They have subsequently expanded their geographical range in southeastern Australia, and their continued expansion could have negative impacts on native biodiversity and water quality. Understanding the current and potential distributions of Sambar deer will allow wildlife managers to concentrate their management efforts more effectively.

The research team from ARI used faecal pellet counts to index Sambar deer abundances. They first collected the data on the presence of Sambar deer by counting the number of Sambar deer faecal pellets along 100 randomly located transects (150m long) in the field. The presence data are essentially a sample of locations with known presence of Sambar deer, which describe the total number of intact pellets counted on each transect (called the faecal-pellet index, or FPI) and the bearing and geographical locations of the transects. Then biophysical variables around each transect were estimated, including elevation, aspect, slope, proximity to water, vegetation type, solar radiation and proximity to roads. All these variables vary from transect to transect – in other words, they are location-specific. Based on the presence and biophysical data, several hypotheses about how the biophysical variables might affect the abundance of Sambar deer were formulated. Models were built to test the hypotheses, and finally the best model was selected to predict the spatial variations in Sambar deer abundance. The result was a habitat suitability model and a map of predicted Sambar deer abundance measured in FPI. This spatially explicit habitat model can help identify locations where Sambar deer might be found and predict how their geographical range might be expanded due to environmental change or other causes.

In this case study, the presence data for Sambar deer, the biophysical data and the predicted Sambar deer abundance data are all spatial data. These spatial data can be handled effectively in GIS. The 'where' and 'what' questions addressed in species distribution modelling – such as where a species occurs or likely occurs, where occupied or unoccupied habitats of a species are and what biophysical factors are associated with the occurrence of a species – are typical questions that a GIS is designed to help answer.

Source: Forsyth et al. (2007).

of spatial data with specific references to locations on the Earth's surface.

Generally speaking, spatial data describe a set of observations and measures of their attributes or properties in geographical space at a particular time or in a certain time period. A geographical space is a space in which the observations correspond to locations on the Earth's surface defined with a spatial measurement framework that captures their spatial relationships (such as proximity, connectivity and direction) in the real world. Here a spatial measurement framework is a coordinate

system which defines spherical or planimetric coordinates, height, orientation and other space measurement properties in ways that take the Earth's complex shape into account. Observations in geographical space are made on spatially distributed features such as plants, wildlife animals, streams and rivers, which are also often referred to as geographical features, spatial objects or spatial entities (Gatrell 1991; Longley et al. 2011). They are defined as a set of locations in a geographical space together with a set of attributes characterising those locations. For example, a sighting of Sambar deer is an observation, which may be defined with a pair of latitude and longitude (spherical coordinates in the geographical coordinate system, discussed in Chapter 2) and a number of attributes attached to the location, including its elevation, aspect, slope, vegetation type and solar radiation.

A map is a traditional means of representing a geographical space. Therefore, spatial data can be linked to a map or represented in map form (Figure 1.1). Spatial data are also called geographically referenced data, georeferenced data or geospatial data.

Non-spatial data: a summary of the landslides occurred in Australia since 1900

TYPES OF LANDSLIDES	NO. OF OCCURRENCE
debris avalanche	2
debris fall	3
debris flow	26
debris slide	23
debris slump	14
debris topple	1
dry sand flow	1
earth block slide	4
earth fall	3
earth flow	27
earth slump	22
rock fall	111
rock slide	14
rock slump	8
rock topple	4
slump-earth flow	8
slump and topple	1
soil creep	3
wet sand or silt flow	1

Spatial data: locations of the landslides occurred in Australia since 1900

DATE	LOCATION	TYPE	LONGITUDE	LATITUDE
23/03/1997	Trinity Beach, Cairns	debris fall	145.70	-16.79
4/02/2000	O'Connor, Kalgoorlie-Boulder area, WA	debris fall	121.48	-30.76
1/04/1911	Nisbet Range, 5 km E of Cairns	debris flow	145.80	-16.90
1/01/1925	Mt Ainslie, Canberra, ACT	debris flow	149.17	-35.27
1/08/1970	Plenty River, 10km west of New Norfolk	debris flow	146.94	-42.80
30/04/1988	Coledale	debris flow	150.93	-34.28
22/09/1989	Sulphur Creek, NW Tasmania	debris flow	146.02	-41.10
1/05/1996	Mt Mitchell, near Cunningham's Gap, Qld	debris flow	152.40	-28.07
19/01/1997	Killaloe, near Mossman, Qld	debris flow	145.38	-16.47
30/07/1997	Thredbo, Snowy Mountains	debris flow	148.31	-36.51
10/01/1998	Mt. Stuart, Townsville, Queensland	debris flow	146.78	-19.35
11/01/1998	Near Kelly St, Nelly Bay, Qld	debris flow	146.84	-19.16
7/08/1998	Below Seaforth Crescent, Sydney	debris flow	151.25	-33.80
17/08/1998	Wollongong	debris flow	150.90	-34.43
11/02/1999	Kuranda Range Road, Kuranda, Qld	debris flow	145.64	-16.82
11/02/1999	Barron Gorge Road, Cairns, Qld	debris flow	145.67	-16.87
3/03/1999	Great Western Tiers, Meander, Tasmania	debris flow	146.62	-41.75
23/10/1999	Nareena Hills, Wollongong	debris flow	150.83	-34.43
...	...	...	...	...

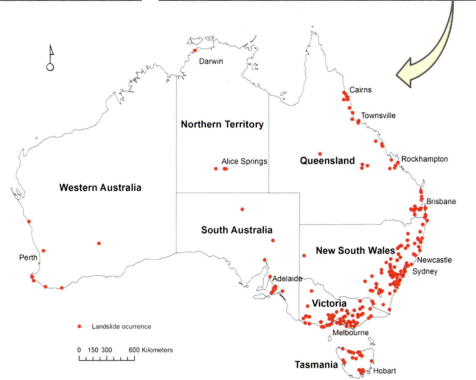

Figure 1.1 Spatial and non-spatial data (data source: Geoscience Australia).

The concepts of the modern GIS were first articulated in the late 1970s. Dueker (1979) defined GIS as:

> a special case of information systems where the database consists of observations on spatially distributed features, activities or events, which are definable in space as points, lines, or areas. A geographic information system manipulates data about these points, lines, and areas to retrieve data for ad hoc queries and analyses.

In the 1980s and 1990s, many new GIS definitions emerged. For example, Marble et al. (1984) defined GIS as 'a set of tools for the input, storage and retrieval, manipulation and analysis, and output of spatial data'. Cowen (1988) contended that the ultimate goal of GIS is to support decision making processes, and therefore defined GIS 'as a decision support system involving the integration of spatially referenced data in a problem solving environment'. Chrisman (1999) argued that GIS is not purely a technology, but a set of activities carried out by people within human organisations. He defined GIS as 'organized activity by which people measure and represent geographic phenomena then transform these representations into other forms while interacting with social structures'.

GIS draws on concepts and ideas from many disciplines, including cartography, geography, computer science, surveying, photogrammetry, remote sensing and statistics, to name but a few. They gradually converged to form a new field of scientific study called geographical information science (GIScience). Goodchild (1992) defined GIScience as 'the generic issues that surround the use of GIS technology, impede its successful implementation, or emerge from an understanding of its potential capabilities'. Therefore, it involves the scientific study of the fundamental issues arising from the creation, processing, management and use of spatial data and information, as well as the impacts of GIS on society and the influence of society on GIS. GIScience is the science behind GIS technology, not intended to replace GIS in terminology and in practice.

It should be noted that GIS is content-free expression. It means different things to different people. There is no universally accepted definition of GIS. To environmental researchers and practitioners, GIS is a computerised tool that assists them in managing, integrating, presenting and distributing environmental data and information; building, analysing and utilising environmental models; and facilitating and improving environmental problem solving and decision making processes. Hence, a working definition of GIS is provided in this book, which characterises GIS as a toolkit in terms of its capabilities:

> A GIS is a computerised system for capturing, managing, manipulating, integrating, analysing, visualising and disseminating spatial data.

The chapters of this book are structured around this order of capabilities.

Components of GIS

A general view is that a GIS is primarily comprised of hardware, software, data, people and procedures (or methods) (Dueker and Kjerne 1989; Longley et al. 2011). Hardware is the device on which a GIS operates or is operated. Today, GIS runs on a wide range of devices, from centralised computer servers to desktop computers, laptops, tablets, in-vehicle devices and mobile phones. Software is a set of computer programs, algorithms and related data that drive GIS operations. Every GIS operation involves data. A GIS operation may take a map (a dataset) as an input and produce a new map (a new dataset) as an output. Data are prepared and entered into a GIS for specific problem solving or research purposes. People are users of GIS. They may be system developers who design and develop the GIS system, or system administrators who manage and maintain the system, or end users who use it as a tool to perform spatial analysis and interpret results. When a GIS is operated in an organisational context, procedures (including institutional arrangements, implementation plans and protocols) need to be established to ensure that the GIS system supports organisational activities efficiently and effectively, and is properly integrated into the entire business strategy and operation. Formal methodologies have been developed for GIS design and implementation within an organisation (see Harmon and Anderson 2003; Sugarbaker 2005). These are beyond the scope of this book. Here, we focus on the functional components of a GIS.

A GIS typically consists of six functional components:

1. spatial data input;
2. spatial data storage and management;
3. spatial data manipulation;
4. spatial data analysis;
5. spatial data output;
6. interfaces.

The components are organised into three tiers: the interface tier, tools tier and data management tier (Figure 1.2).

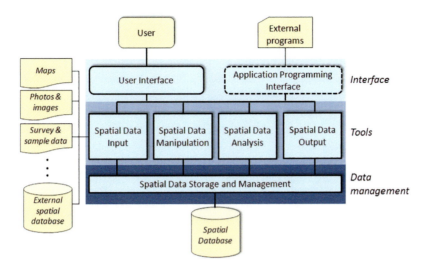

Figure 1.2 GIS functional components.

Spatial data input

The spatial data input component provides facilities to capture spatial data from existing analogue maps, aerial photos and satellite images and convert them into digital form, or to import existing digital spatial data from other spatial databases, or to enter spatial data captured through surveys or sampling or from other data sources.

Spatial data storage and management

This component includes a spatial database or several spatial databases and performs spatial data management tasks, including spatial data storage, retrieval and editing. After spatial data are entered, they are stored and organised in a spatial database in such a way that they can be easily accessed and quickly retrieved. The component provides specific editing tools for updating and editing spatial data – for example, adding new incidental sightings of wildlife into the spatial database, deleting a forest stand that has been cleared from a forest map, changing the shape of a lake based on new surveyed data, or detecting and correcting errors that have occurred during data capturing.

Spatial data manipulation

The spatial data manipulation component provides tools for manipulating and transforming spatial data – for example, converting the coordinate system used for recording the locations in a spatial dataset, merging adjacent map sheets into a single map, subsetting spatial data

and so on. Chapters 2 and 3 provide detailed discussions of spatial data input, spatial data management and manipulation functions.

Spatial data analysis

This component provides a GIS with analysis capabilities. It contains a collection of spatial data analysis tools or functions which allow users to explore patterns, associations and interactions of environmental phenomena as well as evidence of their changes through time and over space. These tools can also be used to estimate parameters and define constraints for various environmental models. Typical spatial data analysis functions in GIS include spatial query, geometric measurement, buffering and proximity measurement, overlay analysis, surface analysis, network analysis, image analysis and spatial statistics.

Spatial query functions are used to retrieve or extract data from a spatial database based on locations, spatial relations (for example, adjacency, connectivity, contiguity, direction and containment) or non-spatial attributes. They are useful for examining and reviewing data and results and making thematic maps to show the distribution or pattern of particular geographical features. For example, we can use spatial query functions to find all buildings within a 100-year flood plain, or to search for all septic tanks within 50m from rivers.

Geometric measurement functions calculate geometric properties of geographical features, such as length, perimeter, area, shape, orientation, centroid and

(a)

(b)

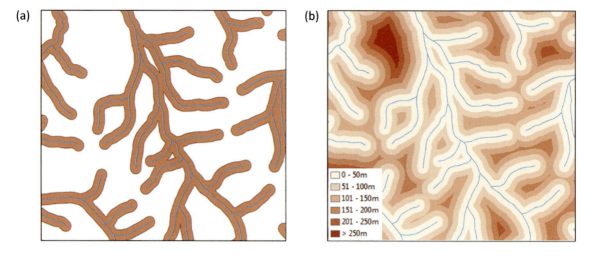

Figure 1.3 Buffering and proximity measurement: (a) 50m buffers around streams and (b) proximity to streams.

distance. These are particularly useful for calculating landscape metrics and indices, such as patch size, nearest neighbour distance, shape index, shape compactness and landscape fragmentation, and for characterising landscape structures and various ecological processes at the landscape or ecosystem level.

Buffering and proximity measurement functions are used to delineate physical zones of a certain distance around features called buffers and to calculate distances from a feature or features to all points in the surrounding area. For example, we can use the buffering function to create 50m buffers around streams and designate the buffer areas as riparian vegetation protection zones (Figure 1.3a), or apply the proximity measurement function to calculate the distance from every location to streams and produce a proximity map (Figure 1.3b).

Overlay analysis functions allow users to stack multiple maps of different themes on the basis of their common geography to examine the correspondence between these themes, to identify spatial associations among them, or to conduct environmental modelling using the themes as input parameters. For example, we may superimpose a land cover map upon a soil map through the overlay operation to analyse how land cover types correspond to soil types (Figure 1.4), combine incidental sightings of rare plant species with a number of environmental factors via overlay analysis to establish spatial associations between the occurrence of the plant species and each of the environmental factors, and apply a soil erosion model to soils, vegetation and slope through the overlay operation to derive a soil erosion potential map. Therefore,

overlay operations can be used with models for complex environmental modelling. In addition, overlay analysis can be used to combine a time series of maps of the same theme to detect and measure temporal changes in the phenomena, such as land use changes over time.

Surface analysis functions are used to estimate, characterise and analyse environmental variables whose values change continuously across the geographical space, such as atmospheric temperature, air pollutants, soil moisture and terrain. This group of functions provides operations for deriving surface features such as slope, aspect and curvature, computing volume, cut-and-fill, profile and intervisibility. They are fundamental tools for terrain, atmospheric and hydrological modelling. Figure 1.5 provides an example of terrain surface and the surface's slope and aspect maps.

Network analysis functions are used to analyse the movement of goods, services, energy, information and many other materials, or even people, throughout networks such as water supply pipes, stream and street networks. They can be applied to find the optimal path that minimises flow travel time, distance, cost or other objective functions (for example, finding the optimal path for distribution of water resources and determining the quickest route for dispatching of emergency services), assess accessibility to certain resources across a network, or to determine which facilities, such as water storage facilities in a water supply network, are to operate based on their potential interaction with demand points, such as the water consumers (for example, individual houses, industrial or commercial establishments) and other water

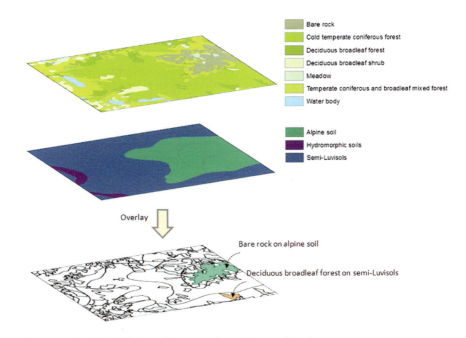

Figure 1.4 Overlay analysis: identifying soil types under each type of land cover.

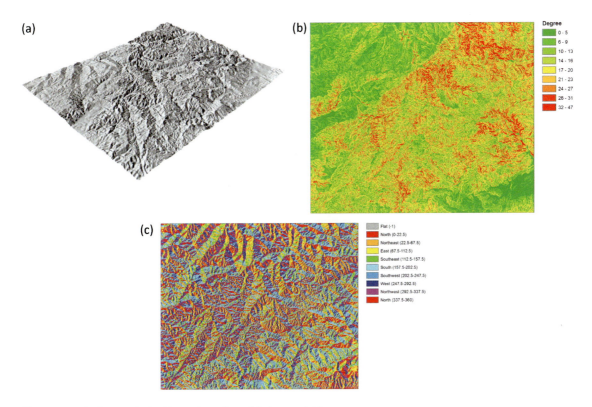

Figure 1.5 Surface analysis: (a) terrain surface, (b) slope and (c) aspect.

usage points (such as fire hydrants). Figure 1.6 shows two examples of network analysis.

Image analysis functions are applied to extract biophysical information from remote sensing imagery, including aerial photos and satellite images, which are among the primary sources of data for environmental studies. Three groups of image analysis functions are available. Image preprocessing functions aim to correct distorted or degraded images to create more accurate representations of the original scenes. They are often required prior to the main image analysis and extraction of information. Image enhancement functions are used to improve the appearance of an image in order to assist in visual interpretation and analysis. They involve techniques for increasing the visual distinctions between features in an image, thus increasing the amount of information that can be visually interpreted from the image. Image classification functions are used to automatically classify pixels in an image into land cover classes or themes (for example, water, coniferous forest, deciduous forest, corn, wheat and so on) and generate a thematic map of the original image (Figure 1.7).

Spatial statistics provide tools for measuring spatial distributions and associations, identifying spatial patterns and detecting spatial clusters. They can be used to analyse the variations in the mean value of an event or a process over space, and explore the spatial correlation or dependence in the event or process. For example, spatial statistics can be used to answer the questions like: Do cancer cases show a surprising tendency to cluster together? Does the cancer incidence rate vary spatially? Is the cancer incidence rate elevated near a particular location? Are cancer incidence rates spatially associated with concentration of airborne pollutants? All such questions can be expressed as hypotheses to be tested using spatial statistics.

The spatial data analysis functions mentioned above can be used as model building blocks to construct environmental models. For instance, in the example described in Box 1.1, a predictive model was developed to predict the Sambar deer abundance, which is expressed as:

$$\log (Y) = 16.14 - 0.005 \times \text{Distance\_to\_water} \\ - 0.024 \times \text{Elevation} + 0.022 \times \\ \cos(90 - \text{Aspect}) - 0.007 \times \\ \sin(90 - \text{Aspect}) \qquad (1.1)$$

where Y is the predicted count of Sambar deer faecal pellets (Forsyth et al. 2007). This model can be built using proximity measurement and surface analysis functions as building blocks, and implemented via map algebra.

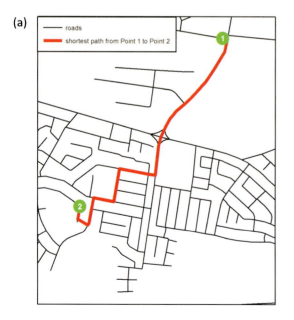

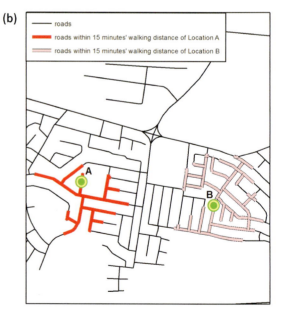

Figure 1.6 Network analysis: (a) routing – finding optimal path and (b) district allocation – finding nearest network links to a centre.

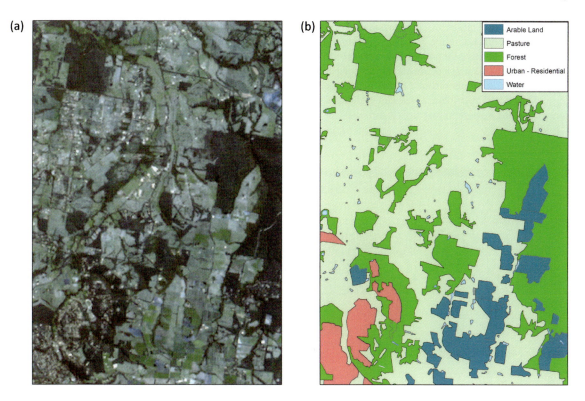

Figure 1.7 Image classification: (a) Landsat satellite image and (b) land cover map derived from (a) (image source: US Geological Survey).

Map algebra is a unique and powerful GIS tool for using maps as variables and algebraic operations (for example, add, subtract, exponentiate, log, root, cosine) to form mathematical equations similar to conventional algebra. It provides a major means of environmental modelling in GIS. We will discuss spatial data analysis and modelling in detail in Chapters 4, 5, 6, 7 and 9.

Spatial data output

The spatial data output component contains visualisation tools for presenting and disseminating spatial data from the spatial database or outcomes from spatial data manipulation, analysis and modelling in various map, graph and tabular forms. Modern GIS systems provide not only mapping tools for designing two-dimensional (2D) and 2.5-dimensional (2.5D) map presentations with traditional cartographic approaches, but also incorporate advanced visualisation techniques for making three-dimensional (3D), multimedia, animated and interactive maps, and allowing users to navigate through the real world in virtual reality. In addition, tools are available for designing maps which can be presented and disseminated via different types of media (such as hard copy papers, computer screens and hand-held devices like mobile phones). In particular, maps can be designed in such a way that they can be published from GIS as map services on the Internet through Google Earth or Google Maps or other map servers, which greatly facilitates the dissemination of environmental data and information to a wide user community. Chapter 8 focuses on this component.

Interfaces

Every GIS system has a user interface. The user interface covers all aspects of communication between a user and a GIS application. It provides the user with access to all other GIS components. For example, it allows the user to enter and edit spatial data, select and run spatial data analysis tools, and display and view analysis results. Some GIS systems provide command-line user interfaces

(Figure 1.8a). Most modern GIS systems have graphical user interfaces (GUIs), which are able to accommodate the user with a variety of input options, such as entering data using forms, issuing a command by selecting a tool from a toolbar or an item from a menu, and formulating a spatial model by combining relevant spatial analysis functions using a diagramming tool (Figure 1.8b).

In addition, many GIS systems have application programming interfaces, which enable users to create custom tools and domain-specific models for environmental analysis and modelling using computer programming languages, such as Visual Basic (VB), C++, C#, Java, Python, VBScript and JavaScript, and to integrate the tools or models directly into the GIS system.

(a)

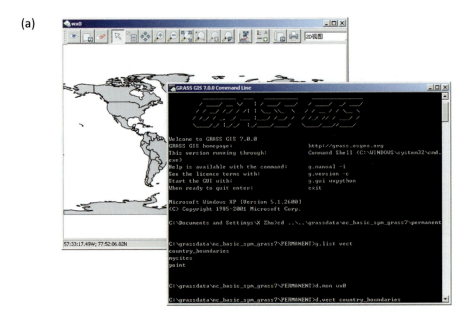

(b)

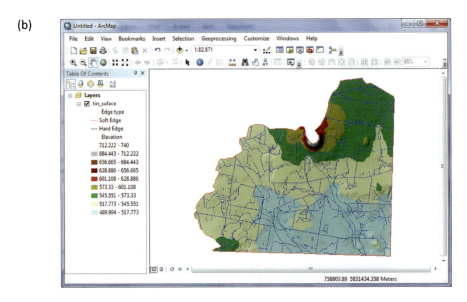

Figure 1.8 GIS interfaces: (a) command-line and (b) GUI.

1.2 GIS FOR ENVIRONMENTAL APPLICATIONS IN A HISTORICAL PERSPECTIVE

Computerised GIS became practical with the development of microcomputers, timeshare operating systems and distributed computing. The history of the implementation of computerised GIS systems begins in the mid-1960s, when the first real GIS system – the Canada Geographic Information System, or CGIS – was developed. CGIS aimed to assist the Canadian Federal Government to locate and quantify the natural resource and land capability of rural Canada (Tomlinson 1967). Since then, GIS technology has evolved and new GIS applications have been developed and studied. The evolution of GIS technology is characterised as a series of innovations and developments in spatial visualisation, spatial data management, spatial analysis and spatial data collection techniques as well as GIS computing platforms, as shown in Figure 1.9.

GIS is rooted in computer mapping. Maps have served as a storage medium for information to meet society's needs for centuries. They utilise symbols to represent geographical distributions of environmental phenomena based on a particular georeferencing system (discussed in Chapter 2) and according to a set of map

design principles (discussed in Chapter 8). Prior to the computer revolution in cartography, maps were mainly produced by hand and photo-chemical procedures, called photo-mechanical map production (Robinson et al. 1995). This was an expensive and lengthy process, involving data collection, compilation, drafting and printing, and it was technically demanding. By the time a map was published, some of its contents might have become out of date. Environmental analysis and monitoring, such as monitoring of earthquakes, air pollution and forest fires, often require maps to be updated frequently. Some monitoring tasks may need maps updated every day or even every hour. Obviously photo-mechanical map production cannot meet such a demand. In addition, although printed maps provided a major source of information to help us understand the spatial patterns and relationships of environmental phenomena, using them for quantitative analysis was difficult. Moreover, environmental problems are complex. Many factors need to be considered in finding solutions. For example, site selection for a landfill is based on the integrated analysis of soil, land use, hydrology, population, transport and other environmental and socio-economic factors. A commonly used approach in the past to such integrated analyses was map overlay. It involved manually tracing the printed maps of various factors and then superimposing the tracings

	1960s	1970s	1980s	1990s	2000s–present
Spatial data collection	Manual coding	Manual digitising	Scanning, digital remote sensing, digital photogrammetry	GPS and GLONASS	BeiDou, Very high resolution remote sensing, Internet map-based crowdsourcing
Spatial data management		Simple data retrieval	Spatial data editing, spatial query, DBMS for non-spatial data	DBMS with spatial extensions	All major DBMS systems supporting spatial data types and operations
Spatial analysis	Simple map manipulation	Map overlay, line-of-sight analysis	Fundamental operations for spatial analysis, map algebra	Coupled with spatial analytical and modelling, and decision analysis tools	Expanded and enhanced spatial data exploration and analytical capabilities
Spatial visualisation	Mapping with line-printers	Mapping with pen plotters	Mapping with inkjet printers and on computer screens	Multimedia and Internet mapping	Virtual reality and virtual globes
Computing platform	Mainframe GIS	Mainframe GIS	Workstation GIS	Desktop GIS, Internet map server	Distributed GIS, Cloud GIS

Figure 1.9 Evolution of GIS technology.

to identify areas where conditions overlapped. Manual overlay analysis is time-consuming and becomes impractical when a large number of maps need to be combined. Computer mapping offered an alternative approach.

In the 1960s and 1970s, GIS systems were virtually computer mapping systems which aimed to automate the process of map production, and facilitate quantitative analysis and integration of map data. Spatial data were entered by manual encoding or by manually digitising printed maps, aerial photos or images using a digitising tablet – an electronic equivalent of the drafting table. Standard line printers were used as mapping devices, which produced very crude and poor-quality maps using characters. Pen plotters were later used to draw maps of improved quality. Several cartographic data structures were developed for encoding map data (Peuquet 1984). For example, the DIME (Dual Independent Map Encoding) data structure was devised in the late 1960s by the US Census Bureau to encode streets and their topological relationships for referencing and aggregation of census records. The ODYSSEY GIS system, developed in the 1970s by Harvard University's Laboratory for Computer Graphics and Spatial Analysis, pioneered the arc/node data structure for digitally encoding line and area features on maps (Peucker and Chrisman 1975). These cartographic data structures provided foundations for the development of spatial data models and data structures for spatial data management in modern GIS, which will be discussed in Chapter 2. However, the GIS systems during this period were simply 'map in–map out' systems. They had inflexible spatial data input, poor spatial data management, limited cartographic display, and simple map manipulation and analysis capabilities. In addition, they generally ran on large mainframe computers, which usually operated in batch mode. The early GIS systems were mainly applied to land and natural resource inventory, such as CGIS, MLMIS (the Minnesota Land Management System) and LUNR (the Land Use and Natural Resources Inventory System) (Coppock and Rhind 1991).

The first microcomputer for home consumer use, the IBM PC, was introduced in 1981. It reduced the cost of computing hardware for GIS operation and started the era of workstation and desktop GIS. It was in the 1980s when GIS really began to take off. As the computing power of computers increased, fundamental spatial data management and analysis functions were developed, which combined computer mapping capabilities with database management and some analytical capabilities. GIS started to mature as a spatial data processing, management, analysis and mapping technology. The first generation of GUIs, including X-Windows,

Microsoft Windows and Apple's Macintosh, made the GIS software much easier to use. A significant development in the 1980s was the launch of ARC/INFO, the first commercial GIS software package designed for microcomputers. ARC/INFO was composed of two software components: ARC, a set of spatial data input, processing and output tools, and INFO, a relational database management system (DBMS) for managing non-spatial data. It was the starting point when conventional non-spatial DBMS technology was used in GIS for database building and management. Along with this, spatial query functions were developed which allow users to retrieve information from the GIS database based on geographical locations, or to search for geographical entities in the database to make maps. Forestry companies and natural resource agencies were among the first to adopt GIS as a tool to build spatial databases for monitoring and regulating the uses of forest and other natural resources. The 1980s also experienced a rapid demand for spatial data, which led to the emergence of a spatial data industry and a marketplace for digital spatial data. New technologies were employed for spatial data collection, including automated scanner and remote sensing technology. Moderate-resolution digital remote sensing, particularly the Landsat Earth resource satellite programme, was becoming a new source of digital spatial data for GIS. In addition to increased data availability and improved spatial data management, spatial data manipulation and analysis functions became available in most GIS systems, such as digitising error detection and correction, map generalisation (for example, simplification of geographical features), map sheet manipulation (for example, scale change), geometric measurement, buffer generation, overlay analysis and digital terrain analysis (Berry 1987; Dangermond 1983). In particular, the concept and techniques of map algebra were developed, which enable the manipulation of maps as variables with mathematical operations. Implementation of map algebra facilitates environmental modelling in the GIS environment. However, GIS systems in the 1980s lacked spatial analytical and modelling capabilities (Goodchild 1990).

The 1990s saw significant growth in both GIS technology and application. Desktop GIS became popular, new spatial analysis tools were added, the Internet emerged, giving rise to distributed and Web GIS and mapping systems, and GIS applications expanded from computer mapping and natural resource inventory to environmental analysis, modelling and decision making. The capabilities of GIS for environmental analysis were mainly based on the spatial relationship principles of connectivity (for example, to measure how connected or spatially continuous wildlife habitat corridors are), contiguity (for example, to identify the types of land use/land cover neighbouring a habitat), proximity (for example, to create buffer zones to define areas with good accessibility to water resources) and overlay methods (for example, to identify potential sites for nuclear waste disposal through overlay analysis with a number of environmental and socio-economic factors). But GIS systems still lacked the kinds of spatial analysis and modelling features required for complex environmental modelling and decision making. In particular, GIS is not well designed for handling either time or the interactions of continuously changing variables, which environmental models can handle effectively. This promoted the integration of GIS and environmental modelling. Some GIS software packages increased the scope of spatial analysis operations to support environmental modelling, such as IDRISI. Links between some statistical packages (S-Plus) and GIS (such as ArcView) were built to increase the power of GIS for statistical modelling. A stand-alone spatial data analysis package, the Geographical Analysis Machine (GAM), was developed to support spatial statistical analysis, including spatial pattern analysis and clustering (Openshaw 1998). More remarkably, the potential of GIS for spatial analysis and modelling had attracted environmental scientists around the world. The integration of GIS and environmental modelling was actively pursued during this period by coupling GIS with a variety of environmental models, such as land surface/subsurface (Wilson 1996), hydrological (Sui and Maggio 1999), atmospheric (Pielke et al. 1996) and ecological models (Hunsaker et al. 1993), to address contemporary environmental issues ranging from global change research to land and water resource management and environmental risk assessment. Development of spatial decision support systems to support the environmental decision making process by integrating GIS and domain-specific environmental models and decision analysis techniques (such as multi-criteria decision making and multi-objective optimisation) also gained momentum in the 1990s (Densham 1991; Zhu 1997; Zhu and Dale 2000). Therefore, the 1990s was a period of exploring and extending the spatial analysis and modelling capabilities of GIS for environmental modelling.

In the meantime, the American Global Positioning System (GPS) was complete, with a network of twenty-four Navstar satellites in 1993, and started to be used for data collection for GIS. GPS was the first global navigation satellite system (GNSS) in the world. Russia completed the world's second GNSS system, GLONASS, in 1995. A Chinese experimental system, BeiDou-1, was established in 2003 to offer services covering China and neighbouring regions. A Chinese full-scale GNSS system,

BeiDou-2, became operational in 2011, began offering services for the Asia-Pacific region in 2012, and will start to serve global customers upon its completion in 2020. A European system, GALILEO, will come into service by 2016, and its full completion is expected by 2020. A GNSS provides three dimensional positions – latitude, longitude and altitude – on the Earth's surface as well as time and velocity. It is a cost-effective, accurate and globally available tool for collecting spatial data where data are not already available in analogue map form or where field surveys are still required to collect data in the first place. GNSS data can be used not only for mapping, but also for georeferencing and ground truthing of environmental data. More information about GNSS, particularly GPS, is provided in Chapter 3.

With a dramatic increase in spatial data from diverse sources, effective management of large and complex spatial databases in GIS became critical. Because of this, GIS was moving to full integration with the relational DBMS using spatial extensions, which allow spatial data to be integrated within a relational DBMS. Examples of such DBMS systems include IBM's DB2 Spatial Extender, and Oracle 7 with a spatial data option (now called Oracle Spatial and Graph). Multimedia mapping was attempted before the Internet mapping started in 1993 when the first interactive map appeared on the World Wide Web (or Web). Multimedia mapping involves interactive integration of sound, animations, text, images and video clips with the map. It links multimedia information to the geographical features at different locations represented on the map. A multimedia map provides a digital multimedia environment which allows users to explore interactively the mapped environmental phenomena and show dynamically the environmental processes. Internet mapping can easily integrate multimedia in and with Web maps, and deliver up-to-date or real-time environmental information accessible to anyone around the world via the Internet.

By the late 1990s, GNSS and remote sensing imagery had been fully integrated with GIS. The launch of the IKONOS satellite in 1999 started a new generation of very high-resolution satellite remote sensing, which provides remote sensing data with a spatial resolution of 1m or below. Also in the late 1990s, airborne LiDAR (light detecting and ranging) technology revolutionised the collection of high-resolution and very accurate digital surface and digital terrain models. The availability of very high-resolution remote sensing imagery and LiDAR data (discussed in Chapters 3 and 6) opened up a new level of GIS analytical capability allowing more accurate and more detailed environmental analysis and modelling in the 2000s.

Since 2006, a number of online mapping platforms have emerged, including OpenStreetMap, Google Earth, Google Maps, Microsoft's Virtual Earth and Wikimapia. They allow users to add their own spatial information to maps served on the Web – for example, selecting an area on the map displayed on the Web and providing it with a description or adding links to photographs, narrative text and video clips about the area. Such user-created information is termed volunteered geographical information (Goodchild 2007). The method for data collection through these Internet-based platforms is called crowdsourcing mapping (Crooks et al. 2009). It provides an innovative approach to the collection of environmental data by engaging with the broad community (discussed in more detail in Chapter 3). Box 1.2 provides an example of a crowdsourcing mapping system for collecting data on specified indicator plant and animal species in Australia.

Box 1.2   ClimateWatch   CASE STUDY

In Australia, unique flora and fauna inhabit a low-lying, fragmented landscape, and they are particularly vulnerable to climate change. The observation of phenological phases of plants and animals has been recognised as an effective way to monitor climate change and ecological responses. This requires large-scale data collection. The scale of the data gathering suggests that success is only possible through a multi-sector approach that engages the community. In response to this challenge, Earthwatch Australia, in partnership with the Bureau of Meteorology and the University of Melbourne, developed ClimateWatch – an online crowdsourcing system for collecting, storing and reporting indicator plant and animal species in Australia. Amateur naturalists, bushwalkers, schools and businesses form part of an extensive network of data collectors who gather data on specified indicator species and report their observations on the ClimateWatch website, where scientists, policy makers and land managers can download the data for analysis and policy making.

Figure 1.10 ClimateWatch Web interfaces (http://www.climatewatch.org.au/).

ClimateWatch was developed as a Web-based crowdsourcing system with the content management capability. It allows for observations on the nominated indicator species to be entered by registered users, reviewed, and made available to a wider audience. Figure 1.10 shows its Web interfaces, including the data entry form and map display of observation records.

The system has established which indicator species to use, and provides species images and sample bird, frog and mammal calls to help data collectors identify the species. In addition to the Web-based data entry, it offers a free iPhone/Android app for data collectors to record observations, take photos and submit the records with the photos using smartphones in real time and anywhere. Data in ClimateWatch can be reviewed by location or by species through a map interface. When a user chooses a region, the map zooms in to that region and displays the point locations where observations are recorded. Each location can be clicked on to display a species listings for that location, as shown in Figure 1.10. If a user chooses an indicator species, the map will display the locations where the species are recorded. Registered users can also edit their own records.

ClimateWatch has enabled the tracking of animal, bird, plant and insect species to determine the impact of climate change on their distribution and behaviour in Australia since 2010. By October 2013, it had over 25,000 data points recorded on the website and over 5,000 registered users.

Source: Donnelly et al. (2011).

With the continued proliferation of GNSS-enabled devices and advanced modern data acquisition techniques, including very high-resolution remote sensing and Internet-based volunteered geographical information, voluminous spatial data have been and continue to be collected. This has resulted in large spatial databases. To meet the need to manage the ever-growing amount of spatial data effectively, in recent years almost all major known DBMS systems have provided support for spatial data types and offered the capability of storing and managing spatial data with spatial data query and manipulation operators. Microsoft is the most recent entrant into the spatial database game, with the 2008 release of SQL (Structured Query Language) Server including spatial data types, functions and indexes. In the meantime, there has been an increasing interest in developing automatic or semiautomatic tools, by utilising methods at the intersection of artificial intelligence, machine learning, spatial statistics and spatial database systems, to enhance the exploratory and analytical capabilities of GIS. These tools aim to provide GIS with the ability to suggest plausible hypotheses, identify spatial patterns or to discover knowledge from large spatial databases (Miller and Han 2009). They are included in a GIS system as standard built-in functions, optional toolsets or add-ins. Many GIS software systems offer software development kits (SDKs), programming languages and programming interfaces to develop one's own analytical tools.

In addition, spatial visualisation techniques continue to advance. Virtual reality and virtual globes emerged as part of 3D extension of GIS mapping, which allow virtual displays of a real landscape in great detail to provide contextual information for better understanding of environmental phenomena and environmental processes. Unlike a static 3D map, virtual reality can create a 3D landscape model in which users can be immersed and navigate among movable landscape elements and objects. By flying through the virtual model of a real landscape, users can not only explore the properties of landscape and its overlying environmental phenomena in visual terms, but also discover their relationships and patterns. Google Earth was the first virtual globe, launched in 2006. It was immediately seen as a new medium for visualising environmental and Earth features. Virtual globes, including Google Earth, Microsoft Virtual Earth, NASA World Wind and ArcGIS Explorer, are freely available. They allow multiple viewing angles, panning and zooming, and adopt an efficient data exploration strategy by providing an overview first, then zooming in to provide details on demand, to avoid visual clutter. They provide a virtual environment for integrating digital thematic maps with very high-spatial resolution contextual data and remote sensing imagery, thus enabling intuitive display and effective delivery of environmental information to a broad audience. Both virtual reality and virtual globes have been used to interactively explore environmental processes, monitor natural hazards, produce virtual realities, demonstrate various scenarios of environmental management and its impacts, and to reconstruct landscapes before human activity.

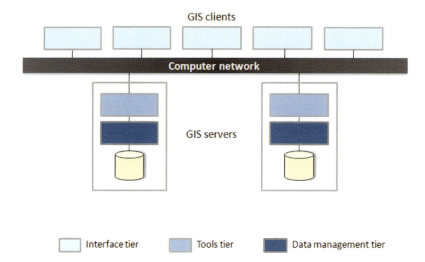

GIS clients

Computer network

GIS servers

☐ Interface tier   ▨ Tools tier   ■ Data management tier

Figure 1.11 Client–server architecture of distributed GIS.

With the maturity of the Web and distributed computing technologies as well as the widespread use of wireless networks in homes, education institutes and urban public spaces, today's GIS can run not only on desktop computers, but also on mobile devices or over the Web from remote locations. Distributed GIS is a system where components of a GIS are distributed among computers at multiple geographical locations linked together via a computer network (such as the Internet). Figure 1.11 shows a typical client–server architecture of distributed GIS. With this architecture, the tools and data management components of a GIS system reside on a GIS server hosted at one location, while the interface component of the GIS is installed on multiple desktop computers called clients distributed at other locations, through which users access the GIS tools, spatial databases and database management functions on the server. Clients and servers communicate over the computer network. For example, a distributed GIS can provide Web maps, data services and spatial data processing capabilities through a dedicated GIS server or servers. Today, almost everyone has used a form of distributed GIS, from generating driving directions using Google Maps to viewing aerial photography via a remote GIS server using a Web browser as a client. Distributed GIS allows different organisations or different departments within an organisation to share GIS software and data resources, and enables geographically distributed GIS functions and spatial databases to be linked together to produce complete environmental applications.

The concept of distributed GIS has gradually evolved into cloud GIS since the beginning of the 2010s (Bhat et al. 2011). Cloud GIS uses cloud computing for distributed processing of large spatial datasets. Cloud computing is a type of distributed system consisting of a collection of virtualised computers that are interconnected and provided as one or more unified computing resources for data-intensive and computationally demanding applications (Hayes 2008). It basically moves software, data and applications from local computers to a dispersed set of remote server computers and data centres, and delivers them as a service over the Internet. There are three major types of cloud computing: software as a service (SaaS), platform as a service (PaaS) and infrastructure as a service (IaaS). SaaS, also referred to as on-demand software, delivers software and associated data centrally hosted on the cloud. Users are provided access to the required software via the Web browser. The cloud providers manage the infrastructure and platforms on which the software runs. Therefore, SaaS removes the need to install and run the software on the user's own computers. PaaS provides a computing platform as a service on which developers can build and deploy applications using tools provided by the cloud providers. It facilitates the development and deployment of applications without the need for users to buy and manage the underlying hardware and software tools. IaaS offers computers, storage, software bundles, networks or other IT infrastructure for users to deploy and maintain the operating systems and application software. Cloud services are provided based on service-level agreements, usually priced on a pay-per-use basis. Cloud GIS systems allow users to use GIS software, hardware and spatial data delivered on demand as services via the Internet that can be accessed from a variety of networked devices, such as desktop computers, laptops, tablets, mobile phones and other remote servers (Figure 1.12). ESRI Business Analyst Online is an example of cloud GIS offering SaaS (http://www.esri.com/software/bao). It enables on-demand analysis of trade areas, competition, supply and demand using GIS, and demographic, consumer spending and business data for the United States and Canada hosted by ESRI (Environmental Systems Research Institute). Users do not need to manage data and GIS software, and reports and maps resulting from the on-demand analysis are delivered to the users over the Web. The EyeOnEarth network (http://network.eyeonearth.org/en-us/Pages/Home.aspx) is another example of cloud GIS, hosted by the European Environment Agency (EEA). This cloud GIS also offers SaaS, which provides GIS tools for producing maps and accessing environmental datasets (such as land use/land cover, air quality, water quality, nitrogen from different sources, sea surface temperature and heat wave risk of cities in Europe). The network allows people to share environmental data with the public and among groups. Relevant stakeholder groups can use the network as an environmental platform to work together to understand problems, design solutions and take actions. Such types of cloud GIS can be used to support citizen engagement and collaborative environmental problem solving. They allow environmental data, information and knowledge to be accessed, shared, analysed and used anywhere and at any time.

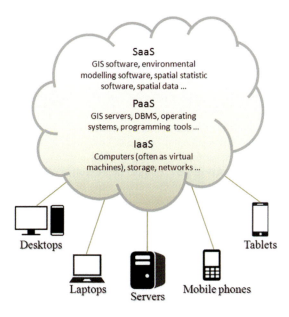

Figure 1.12 Cloud GIS.

1.3 ROLES OF GIS IN ENVIRONMENTAL RESEARCH, MANAGEMENT AND PLANNING

The value of GIS for environmental problem solving and decision making increases as the scale and complexity of the environmental problem increase. Challenges currently faced by society, such as global warming, resource shortages and loss of biodiversity, require the integration of a wide range of data and information, addressing problems from multiple perspectives and evaluating alternative solutions in a holistic, comprehensive, systematic, analytic and visual manner. GIS provides computerised tools for organising and integrating data, recognising spatial patterns, and visualising spatial patterns and processes that enable decision makers to address these challenges and make effective judgements. It allows various types of data relating to the environmental problems and their geographical contexts to be considered together, thus enabling the analysis of the problems to have a greater depth and wider scope.

Essentially, GIS deals with environmental processes and patterns in a spatially distributed manner. In other words, GIS uses spatial data to characterise, explore, analyse and model spatial variations, processes and spatial relationships of environmental phenomena at various times and at various spatial scales, ranging from local (such as a field or a forested area) to regional (such as a catchment) to global scales. There is a very diverse range of environmental application areas of GIS, depending on the particular environmental sphere (atmosphere, biosphere, hydrosphere or lithosphere) and the specific environmental issue being investigated. However, the role of GIS in environmental research, management and planning can be generalised into three categories:

1. inventory and monitoring;
2. analysis and modelling;
3. visualisation and communication.

Inventory and monitoring

It is a well-accepted principle that environmental data and information at an appropriate spatial scale are required to underpin complex decision making in environmental management and planning. A compilation of environmental data pertinent to a specific geographical area is called an environmental inventory. Examples include land resource, land account, wetland, wildlife, pollutant and landslide inventories. An environmental inventory typically comprises data that represent both the thematic and locational attributes of environmental phenomena or resources. For example, an air emissions inventory is a database containing detailed information about pollutants discharged into the atmosphere by each source type during a given period and at a specific location. A forest inventory contains forest attribute data for each forest stand in a given area, including stand attributes of a unique identifier, stand age, stand species, stand size, stocking, soil type, the level of reconnaissance (the method used to inventory the stand) and other forest attributes or variables depending on the purpose of the inventory. Maps, imagery and related tabular data are the centrepiece of most environmental inventories. Traditionally, environmental inventories were stored in paper maps, orthophotographs and tally sheets, and updated by aerial photography, field sampling methods and manual drafting. GIS has been used to store, organise, document, retrieve, integrate, summarise, update and map environmental inventory data since the 1960s, when early GIS systems were mainly developed for land and natural resource inventory. Modern GIS not only provides spatial database management tools for compiling, managing and maintaining environmental inventories, but also allows the use of alternative source data, such as high-resolution digital satellite imagery and LiDAR data, to automatically capture inventory data over a large and/or inaccessible area, which would be infeasible to collect using traditional field inventory and mapping methods.

GIS-based environmental inventory ensures accurate environmental reporting with improved data collection, and supports the monitoring of key environmental variables that is essential for resource and environmental management. Environmental inventory is often a foundation for environmental analysis and modelling.

Analysis and modelling

Environmental analysis and modelling with GIS are associated with a process of applying particular GIS analytical tools to environmental data. GIS provides a large range of spatial data analysis capabilities, as discussed in Section 1.1. They can be used for descriptive analysis, predictive modelling and decision analysis.

Descriptive analysis attempts to examine environmental data for patterns, trends, correlations or other relationships. GIS is a flexible tool for descriptive analysis. For example, we may retrieve soil data in a given area from a GIS database and display them on a map, from which the spatial pattern of soils in the area can be visually examined. We can use a GIS to quickly generate a map of rainfall based on data measured at a set of rain

gauge stations at a specific time. If a series of rainfall maps for different times is produced, they can be overlaid on each other, which allows us to analyse the temporal trend of rainfall in the area over a certain period of time. We may also use a GIS to overlay a soil map, an elevation map and a rainfall map of the same area to visually analyse the relationships between soil, elevation and rainfall distributions. Descriptive analysis with GIS seeks to describe, measure and explain what is where or where is what, rather than what should be where or where might be what. Water quality assessment, land evaluation and demographic movement are examples of descriptive analysis. Descriptive analysis can give a researcher intimate knowledge of the environmental data and a good understanding of environmental conditions before embarking upon further investigation.

Predictive modelling with GIS is used to predict where particular types of environmental phenomena or events could or should occur under certain environmental conditions, or what will happen at different locations given projected changes in environmental conditions. The former type of predictive modelling involves establishing statistically valid, causal or covariable relationships between environmental conditions such as land cover, elevation, slope, aspect, soils, vegetation, water availability and geomorphology, along with the observed presence of the environmental features, then using that information to predict where the environmental features might be or are likely to occur. The species distribution modelling described in Box 1.1 is an example of this type of predictive modelling. Other examples include forest fire modelling, landslide risk assessment and wildlife habitat modelling.

The second type of predictive modelling is mainly used for scenario analysis or 'what if' analysis. It involves the use of models to produce the outcomes of potential future alternatives (scenarios) or to analyse possible future environmental events by considering alternative possible environmental outcomes or states. It is particularly useful for generating predictive scenarios for environmental impact studies. For example, GIS has been used to estimate the surface runoff and sediment yield of a river catchment under different scenarios of future urban development in order to evaluate the impacts of urban growth patterns on surface water hydrology (Kepner et al. 2004), to assess the ecological effects of alternative water management strategies by integrating GIS with water allocation, landscape dynamics and habitat suitability models (Schluter and Ruger 2007), to simulate crop responses to climate change scenarios to assess the impacts of climate change in agricultural systems (Hodson and White 2010) and to generate socio-economic scenarios of land cover change (Swetnam et al. 2011). Scenario analysis is typically used in the context of environmental and resource planning over a long period, or environmental decision making involving factors and trends of interactions and human consequences that may impact the future. It does not try to present one exact picture of the future. Instead, it shows several alternative future developments and possible outcomes.

Decision analysis is a process that allows decision makers to select a desired option from a set of possible decision alternatives. However, decision making in environmental management is a complex process, and often requires consideration of trade-offs between environmental, socio-political and economic impacts. It invariably involves multiple issues, multiple criteria, multiple objectives and multiple stakeholders. GIS, when combined with decision analysis techniques such as multi-criteria decision analysis and multi-objective optimisation (discussed in Chapter 9), can be used to support the design and evaluation of environmental decision alternatives by integrating spatial information derived from descriptive analysis and/or predictive modelling with stakeholders' preferences. Multi-criteria decision analysis is to evaluate a set of decision alternatives based on conflicting and incommensurate criteria. Multi-objective optimisation aims to maximise certain outcomes while minimising others. For example, coupled with a particular multi-criteria decision analysis or multi-objective optimisation technique, GIS has been used for landfill siting (Chang et al. 2008), demarcation of greenway (Giordano and Riedel 2008), prioritisation of forest areas for conservation (Phua and Minowa 2005) and land use optimisation (Seppelt and Voinov 2002; Cao et al. 2012).

Visualisation and communication

GIS is also an effective visualisation and communication tool offering unique spatial visualisation functions, particularly mapping functions. Both interactive and static maps can be made quickly and easily using GIS, with users having full control over the map symbols and map styling. Maps can enable users to get their points across quickly and efficiently, and help them examine the data in different ways to gain additional insight. Therefore, spatial visualisation in GIS can facilitate analysis, envisioning, reasoning and deliberation using spatial data and information. It can produce traditional maps with high levels of conceptual abstraction, or separate spatial data into themes representing different aspects of the environment, or create a visual simulation that is less abstract and highly reminiscent of real views of real-life situations. Spatial visualisation in GIS provides a powerful tool to communicate complex environmental data in a form that makes the data intelligible not only to scientists and decision makers, but also to the public. As environmental decision making becomes increasingly an exercise in public consultation and requires that all aspects of an environmental project be clearly understood by the public, spatial visualisation with GIS provides an effective means for improving awareness and facilitating conflict resolution, public participation and capability building.

1.4 A GENERAL FRAMEWORK FOR ENVIRONMENTAL PROBLEM SOLVING WITH GIS

Environmental problems are complex and diverse. But they share some basic features, including human actions leading to environmental changes, community concerns about those changes, and needs for tools to acquire and analyse relevant environmental information and methods for predicting environmental responses to human actions (National Academy of Sciences 1986). As discussed above, GIS has evolved into a transforming technology that enables integration, management, access, sharing and visualisation of environmental data from multiple sources, empowers spatial analysis and exploration of environmental patterns, processes and trends, allows prediction of environmental changes or environmental responses due to human actions, and facilitates communication and decision making. As one of the scientific methods for environmental problem solving and decision making, GIS supports systematic observation, measurement and experiment, and the formulation, testing and modification of hypotheses with spatial data and spatial analysis and modelling tools. However, the usefulness of GIS is less a function of the technology itself than of the way in which it is applied. The requirements and use of a GIS depend on the specific nature of the environmental problem and the objectives of the environmental problem solving or decision making. Generally, environmental problem solving with GIS involves six (often iterative) steps:

1. defining the problem;
2. structuring the problem;
3. acquiring and preparing data;
4. designing a solution process;
5. implementing the solution process;
6. communicating the results (Figure 1.13).

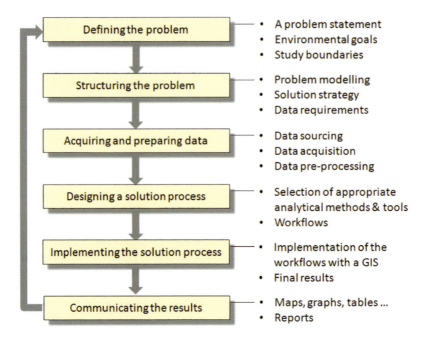

Defining the problem	• A problem statement • Environmental goals • Study boundaries
Structuring the problem	• Problem modelling • Solution strategy • Data requirements
Acquiring and preparing data	• Data sourcing • Data acquisition • Data pre-processing
Designing a solution process	• Selection of appropriate analytical methods & tools • Workflows
Implementing the solution process	• Implementation of the workflows with a GIS • Final results
Communicating the results	• Maps, graphs, tables … • Reports

Figure 1.13 Process of problem solving with GIS.

Defining the problem

Defining and understanding the problem to be studied using GIS is a crucial step. A clear understanding of the problem can lead to success in finding an acceptable solution or reaching a conclusion. When a group of people are involved in the problem solving, consensus needs to be reached on the problem definition. Defining the problem involves developing a problem statement, stating the environmental goals and establishing study boundaries. A problem statement summarises the problem in a clear and concise manner. An environmental goal focuses on what the successful end of the problem solving process should be.

Environmental analysis with GIS also needs to establish a set of spatial and temporal boundaries. The spatial boundary defines the area within which the study is to be carried out and what areal unit is to be used for analysis (for example, at farm level, at shire or county level or at catchment level). Different areal units may result in different measured characteristics of the geographical features of interest. For example, when timber is managed on a forest-wide basis, rather than by stands, yields are higher and more consistent. The temporal boundary defines the period over which the analysis is to be performed, and whether it is to reconstruct the past, to look at the present, to predict the future or to examine temporal changes and trends.

For example, the Woori Yallock Catchment is one of the catchments in Yarra Valley, located to the north east of Melbourne, Australia. It covers an area of 362km^2. The key management issues for this catchment include the loss of native vegetation, bed and bank erosion, water quality deterioration, weeds, fish barriers and maintaining stream flows. In particular, agricultural activities, settlements and septic tanks on rural properties are major point and non-point sources of pollution of the aquatic environments and have significant impacts on the water quality of the rivers and streams in the catchment. Riparian vegetation can act as a filter to trap sediment, nutrients and other contaminants, reducing their movement into streams. It also protects river banks from surface erosion by rain, water flow or stock, helps dry and reinforce bank soils to prevent cracking and slumping, and maintains natural levels of light intensity and water temperature for healthy in-stream ecosystems. However, many parts of the river system in the catchment are not protected by natural vegetation buffers, and the water quality standards mean the river ranks rather low on the river health scale. Identifying the critical areas of riparian zones that are susceptible to the threat of pollution is the first step in addressing those environmental issues and developing water resource management and biodiversity conservation strategies for the catchment. A problem statement could be drawn up as follows:

Many parts of the river system in the Woori Yallock Catchment are susceptible to the threat of pollution from adjacent land uses and septic tanks, and they need to be identified for riparian vegetation plantation or restoration. The environmental goal is to protect and maintain water quality and biodiversity. The spatial boundary is the whole catchment and the temporal boundary is the present.

In environment applications, a problem statement reflects a particular point of view or a specific focus of research, and it is relevant to taking purposeful actions in a problem situation. A purposeful action means a deliberate, decided or willed action by an individual or by a group. Therefore, a problem statement accounts for some ways of perceiving a problem. In many cases, information needs to be gathered to define the problem by various means, including literature reviews, interviews, focus groups and statistics.

Structuring the problem

Solution of an environmental problem with GIS is possible only after it is properly and explicitly structured. It is critical to begin with an understanding of the problem structure rather than a determination to apply readily available GIS tools and spatial data. Problem structuring is a process of arriving at a sufficient understanding of the components of a particular problem to allow specific research actions. There are two main problem structuring techniques which have been widely used in problem solving: decomposition and modelling.

The decompositional approach to problem structuring (Smith 1989) suggests that a complex problem can be decomposed into several subproblems, which are less complicated and can be solved separately ('divide and conquer'). After the results of each subproblem are obtained, they are integrated into a solution. Given a way of decomposing problems into finite sets of simple subproblems, decompositional structuring facilitates identification of a successful solution.

The modelling approach sees problem structuring as a part of or a prelude to modelling (Woolley and Pidd 1981). Modelling entails development of a formal representation of the problem, followed by manipulation and solving of this representation and application of the results to the original problem. Problem structuring is essentially a process of identifying a collection of elements with which to build a problem model. The modelling then consists of defining the relationships between the elements. Defining the problem elements and their interrelations yields the components of a model. The

modelling approach also implies a solution strategy. Graphs are usually employed in problem model-based representation and structuring, as the graphical representation of a problem model is intuitive and can be easily understood by decision makers. Here we introduce the concept of spatial influence diagrams for problem structuring proposed by Zhu (1997).

A spatial influence diagram is a graph consisting of nodes and directed arcs with no cycles. The nodes in the diagram represent the environmental variables or factors relevant to a particular environmental problem, and the directed arcs between nodes represent influences, dependencies or relevance among the environmental variables. There are three types of nodes: chance, value and border nodes. The value node represents the expected outcome in solving the environmental problem. Chance nodes represent environmental variables, which influence (directly or indirectly) the expected outcome and enable the computation of an outcome of the value node. Border nodes represent environmental variables that correspond to available or acquirable data.

Figure 1.14 is a spatial influence diagram constructed to structure the problem of identifying the critical areas of riparian zones in the Woori Yallock Catchment. To differentiate the three types of nodes visually, the value node is drawn as a rounded rectangle, chance nodes are drawn as rectangles and border nodes are drawn as ellipses. In this example, the desired outcome is the identified critical riparian areas in the catchment, which is the value node. It will be derived by comparing the riparian vegetation buffer delineated with desirable widths with the existing riparian vegetation buffer. Those riparian areas without existing native vegetation or with existing native vegetation whose width is less than the desirable width are critical areas. The desirable width of a riparian vegetation buffer is dependent on site characteristics, including slope, soil texture and erodibility, drainage area, adjacent land use and existing vegetation. Therefore, the riparian vegetation buffer with desirable widths can be delineated based on these site characteristics and the drainage network. The existing riparian vegetation buffer is derived based on the existing vegetation and drainage network. Slope will be derived from the elevation data, soil texture and erodibility from the soils data, adjacent land use from the land use/cover data, existing vegetation from the land use/cover data, and drainage network and drainage area from the hydrology and elevation data.

The spatial influence diagram brings problem considerations together and shows how they are related. It can be seen as an environmental problem model, depicting the overall structure and nature of the dependencies among variables. A spatial influence diagram not only

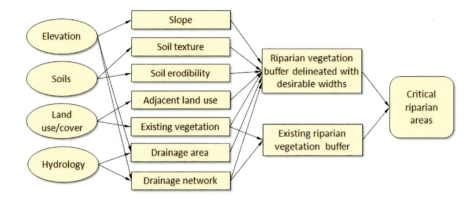

Figure 1.14 A spatial influence diagram for identifying the critical areas of riparian zones in the Woori Yallock Catchment.

provides an intuitive means of presenting the problem components and their relationships in a comprehensible way, but also offers an effective way to identify the flow of information and 'chains' of analysis (that is, links among the environmental variables).

Problem structuring with a spatial influence diagram may start from the expected or desired outcome – the value node. That is, we first assume what is sought has been found. Then we ask from what antecedent the desired result could be derived. If the antecedent is found, then we ask from what antecedent *that* antecedent could be obtained. This process is repeated until the data or border nodes are reached. The solution to the original problem is obtained by traversing this process backwards to the desired outcome. This approach allows us to avoid the tendency to let the available data shape the final outcome. Data availability should be treated as a modifier of the problem formulation rather than an initial driver. The whole process should be iterative and viewed as a voyage of discovery. In many environmental applications, the knowledge domain under study is in fact a system characterised by rich interaction and feedback among all the variables of interest. The choice of the value node results from a clear specification of the problem. Intervening environmental variables are also chosen based on the purpose of the analysis.

Acquiring and preparing data

A spatial influence diagram resulting from the problem structuring process captures the definitional structure of an environmental problem. It also identifies the data required for problem solving. The next step is to source and capture the data. There are a number of issues related to data sourcing and capturing, including spatial data

acquisition methods, spatial data quality, spatial resolution or map scale, spatial data models, georeferencing systems and the age of spatial data.

Some of the data may be available from different sources, and some may not exist. If the required data do not exist, they need to be captured by using one or several of the spatial data collection techniques discussed in Chapter 3. If the required data are available from third parties, the quality of the data must be assessed. Existing spatial data may contain errors, distortions or uncertainties. Chapter 3 will discuss the issues of spatial data quality.

Spatial resolution or map scale determines the accuracy of the data, and the amount of details about geographical features represented in the data. As a general rule, the smaller the study area, the more detail is required, and the higher the resolution or the larger the map scale the data will need to have. Conversely, the larger the study area, the lower the resolution or the smaller the map scale required. A spatial data model determines how spatial data are represented, structured and organised. Spatial data can be represented and organised in terms of discrete geographical entities, or groups of grid cells. A spatial data model also determines how spatial data are stored and manipulated in GIS. A georeferencing system is used to define locations of geographical features in the spatial data with reference to a map projection or a projected coordinate system. All these concepts are discussed in more detail in Chapter 2.

When acquiring spatial data, you also need to consider the age or currency of the data. The temporary boundary of the analysis defined in the first stage of problem solving determines what period(s) of time your analysis covers, and whether you need data from thirty years ago, every five years since 1980 or the most currently available. All the above issues related to data need to be addressed according to the purpose of the analysis. Spatial data from different sources need to be compatible in temporal and geographical coverage and data quality. All acquired data need to be processed to have the same spatial and temporal boundaries, the same spatial resolution, the same georeferencing system and data formats compatible with the GIS software in use before they are used for analysis. This can be done by using spatial data editing and transformation functions in GIS, as discussed in Chapter 3.

Designing a solution process

After the required data are obtained and preprocessed, appropriate analytical methods, tools, procedures and models available in a GIS software system (or with add-ons and external special-purpose modelling software) need to be selected. Selection is based on the problem model formulated in the problem structuring stage. When a spatial influence diagram is used, it commences with the data or border nodes, and follows the links to select an appropriate GIS function, tool or model to derive the value of each chance or value node with its predecessors as the inputs. During this process, new

chance nodes may be added to represent immediate results, or a chance node may be decomposed into two or more new chance nodes so that a single GIS function or tool or model can be used for each link. If the links are labelled with the names of selected GIS functions, tools or models, the re-worked spatial influence diagram shows a workflow from the data to the outcome. It represents a solution process for the problem.

Using Figure 1.14 as an example, we may use the slope function to calculate slope based on the elevation data, the spatial query function to derive soil texture and soil erodibility from the soils data, the neighbourhood function to identify adjacent land uses based on the land use/cover data, the hydrologic tools to delineate drainage areas and drainage networks based on the elevation and hydrology data, the buffer function to create the riparian buffer, the spatial query function to extract existing vegetated areas from the land use/cover data, the overlay operation to produce the existing riparian vegetation buffer based on the riparian buffer and the existing vegetation, the overlay tool to combine slope, soil texture, soil erodibility, adjacent land use, existing vegetation and drainage area to calculate the desirable widths of the riparian vegetation buffer at every location according to certain rules or a pre-defined model, the buffer

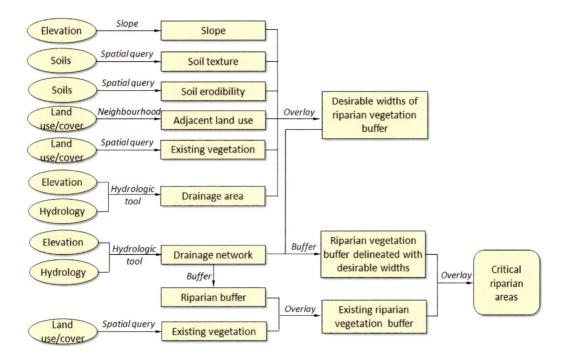

Figure 1.15 A solution process for identifying the critical areas of riparian zones in the Woori Yallock Catchment.

function to generate the riparian vegetation buffer with variable desirable widths around the drainage network, and finally the overlay operation to intersect the riparian vegetation buffer with variable desirable widths and the existing vegetation buffer. Figure 1.15 shows the result of the process. In this process, a new chance node, 'Riparian buffer', is added. For the purpose of clarity and simplicity, some nodes are drawn several times as inputs for deriving the values of different variables.

It should be noted that different GIS software packages may have different sets of analytical functions and tools, may use different names and different input parameters for the same type of functions, and may impose different constraints on the values of inputs and outputs. In addition, a problem can be solved by different combinations of GIS functions and tools. Therefore, a solution process designed based on a problem model with the approach discussed above is not unique. Different people may come up with different solution processes for the same problem by selecting different sets of methods and tools. However, the selected methods and tools should be simple, effective, valid, and fit for the problem and objectives.

Implementing the solution process and communicating the results

The solution process designed in the previous stage is implemented with GIS software to produce the expected outcome. The results can be presented as maps and other forms of spatial visualisation, summarised in graphs and tables, and interpreted in a report in a clear, concise, accurate and comprehensible way. They will then be used by decision makers to develop an action plan to tackle the problem, or used by researchers for further studies.

For many environmental problems, the final stage is not the end of the process, but the start of a new round of the process, whereby the problem is revisited and the entire process or parts of it are iterated. Not only is the process iterative, but at each stage one often looks back to the previous step and re-evaluates the validity of the decision made. Therefore, in reality, the problem solving process is more complex and iterative than this summary of steps suggests.

1.5 GIS SOFTWARE

Categories of GIS software

GIS software is a vital component of a GIS project in environmental studies. There are many different types of GIS software, ranging from desktop GIS to server GIS,

mobile GIS, spatial DBMS packages and GIS software libraries.

Desktop GIS software runs on a personal computer and is the most widely used category of GIS software. It offers complete GIS functionality in a single-user environment, adopts standard desktop user interface styles (such as Microsoft Windows) and interoperates with other desktop applications (such as Microsoft Excel). Users can interact directly with the software via a user interface to perform ad hoc GIS tasks. With desktop GIS software, all the data and software components reside in one personal computer and all the tasks are carried out on the same computer.

As its name suggests, server GIS software runs on a computer server, whose functions can be accessed and used by remote networked users. It is mainly used by software developers and Web developers to create Web and mobile applications with well-defined tasks, such as routing, finding addresses/places, mapping, map publishing and data editing. It has also been used to build geoportals as gateways to spatial data resources. The simplest types of server GIS applications are Web or Internet mapping systems, which are used to serve or publish maps and/or images over the Internet that can be browsed and downloaded using standard Web browsers. More advanced server GIS systems allow maps and images to be integrated or mashed up with other datasets and Web services. The new generation of server GIS software can provide complete GIS functionality in a multi-user environment for distributed spatial data editing, management, analysis and visualisation. It enables sharing of GIS resources among an organisation or a group of stakeholders, which can be deployed in a client–server distributed computing or cloud computing environment.

Mobile GIS software runs on a mobile platform with an internal or external GPS, such as a tablet or a smartphone. It is mainly designed for spatial data acquisition and updating in the field, with positioning, spatial data entry, spatial data transfer, map display, feature editing, map query and simple analytical functions. Some mobile GIS software also provides links to a GIS server, enabling field-collected data to be uploaded directly to the server and the data stored on the server to be accessed instantly and updated via the wireless network.

Spatial DBMS software is used to develop and manage large spatial databases for enterprise applications in organisations, such as in a business or government. But they generally lack spatial data editing, analysis and mapping functions. GIS software libraries are collections of GIS software components, which implement different GIS functions, such as reading and writing spatial datasets in different data formats, performing carto-graphic projections (see Section 2.2), generating buffers, overlaying map data layers and producing a contour map. They are intended to be used by software developers to design and build specific GIS applications. GIS software libraries do not allow direct use of functions. They can only be embedded in application software.

Table 1.1 lists typical tasks that can be carried out with different types of GIS software and their potential uses. Here we classify the basic GIS tasks into data capturing, editing, storage, query, exploration (viewing and exploring spatial patterns), manipulation, analysis and mapping. Data editing, storage and query are data management functions. We also list Internet mapping as a category separate from server GIS, as its functionality is limited to serving maps over the Internet. In some environmental applications, different categories of GIS software are used in combination or together with other software for database management, statistical analysis and environmental modelling.

Proprietary GIS software

A large number of GIS software packages are commercially or freely available. A proprietary GIS software package is a commercial product licensed under the exclusive legal right of the copyright holder. The licensee is given the right to use the software under certain conditions, while restricted from other uses, such as modification and further distribution. Its source code is almost always kept secret. The price of proprietary GIS software typically ranges from $500 to $1,500 for desktop software and from $5,000 to $25,000 for server products. Although it is costly, proprietary GIS software is generally packaged, robust, comprehensive, well-documented, regularly and easily updated, and has reliable professional service and support. Below is a brief introduction to some popular proprietary GIS software packages.

ArcGIS

ArcGIS is a comprehensive suite of GIS software products developed by ESRI (http://www.esri.com/). It includes ArcGIS for Desktop (desktop GIS software), ArcGIS for Server (server GIS software), ArcGIS Online, ArcGIS for Developers and ArcGIS Solutions.

ArcGIS for Desktop is developed exclusively for the Microsoft Windows (hereafter simply Windows) operating system. However, the ArcGIS desktop software has been widely adopted by a large user base since the launch of ARC/INFO in the 1980s. It is a popular choice for environmental analysis and modelling as it includes a wide range of analytical functions and provides scripting tools and application programming interfaces (APIs) for adding user-developed functions. Many extensions and add-ons to ArcGIS desktop software have been developed to incorporate specific environmental analysis and modelling capabilities, some of which will be introduced later. Due to its popularity and wide use in a variety of organisations worldwide, ArcGIS for Desktop has been chosen as the basis for this book, and Section 1.6 will provide a guided tour through it.

ArcGIS for Server provides complete GIS functionality similar to ArcGIS for Desktop, but in a server environment which allows centralised management and Web-based delivery of GIS services for distributed mapping, spatial data management, spatial analysis and modelling. ArcGIS Online provides a cloud GIS computing environment for mapping. With ArcGIS Online, no GIS software installation and setup are required. The software is delivered as a service, providing users with mapping tools to create and publish maps on demand. ArcGIS for Developers provides tools for building Web, mobile and desktop applications, developer APIs, ready-to-use content and self-hosted solutions. ArcGIS Solutions provides a series of ready-to-use maps and applications tailored to many government services and industries, such as water utilities, public works, fire services, emergency management, planning and development.

IDRISI

IDRISI is an integrated desktop GIS and remote sensing software system developed by Clark Labs at Clark University, which runs on the Windows platform (http://www.idrisi.com/). Since 1987, it has been widely used in over 180 countries worldwide for environmental management, sustainable resource management and equitable resource allocation. It contains a comprehensive set of GIS analysis tools, including those for surface analysis, statistical analysis, decision analysis, change and time series analysis, a digital image processing toolkit and a variety of environmental modelling tools, including the Earth Trends Modeler for modelling time series of environmental trends, the Land Change Modeler for land change analysis and prediction, the Image Calculator for constructing environmental models using maps as variables, the Macro Modeler for graphically building models using all of its functions or GIS modules as objects, and a set of prediction tools such as cellular automata and neural networks (discussed in Chapter 9) as well as modelling tools for 3D visualisation, data assessment and quality control.

Table 1.1 Typical tasks that can be carried out with different types of GIS software

	Capturing	Editing	Storage	Query	Exploration	Manipulation	Analysis	Mapping	Used by
Desktop GIS	×	×	×	×	×	×	×	×	Individuals for GIS analysis and mapping
Internet mapping				×	×			×	Networked users to access and explore data
Server GIS	×	×	×	×	×	×	×	×	Networked users to access and use GIS functions as remote services, or software developers and Web developers to create Web and mobile GIS applications
Mobile GIS	×	×		×	×			×	Individuals for field data collection
Spatial DBMS	×	×	×	×		×			Enterprise users to manage large spatial databases
GIS software libraries			×			×	×	×	Software developers to develop specific GIS applications

MapInfo

The MapInfo GIS software product line includes MapInfo Professional, MapBasic, MapInfo Manager, MapInfo Stratus, MapInfo Spatial Server and MapXtreme, now owned by Pitney Bowes Business Insight (http://www.pbinsight.com/). They all run on the Windows platform. MapInfo Professional was the first desktop GIS software product on the market. It was first released in 1986 as the Mapping Display and Analysis System (MIDAS), replaced by MapInfo for Windows in 1990 and renamed to MapInfo Professional in 1995. It is well-distributed and has been widely used in demographic analysis, business and service planning. MapBasic is a BASIC-like programming language used to create additional tools, new functionality and custom applications for use with MapInfo Professional. MapInfo Manager is a spatial DBMS package for building, managing and maintaining a centralised spatial database. MapInfo Stratus is Web mapping software which leverages the power of MapInfo Professional to deliver Web-based interactive maps. MapInfo Spatial Server is server GIS software used for developing Web-based geoportals. MapXtreme is a GIS software library which provides GIS software components to build custom GIS applications.

GeoMedia

The GeoMedia product suite from Intergraph Corporation consists of GeoMedia, GeoMedia Professional and a set of add-on modules (http://www.intergraph.com/). GeoMedia is a Windows-based desktop GIS software package. GeoMedia Professional provides all the functionality of GeoMedia, and adds spatial data capturing and editing functions. The add-on modules include GeoMedia 3D for 3D visualisation and analysis, GeoMedia Image for image processing, GeoMedia Terrain for terrain analysis and visualisation, GeoMedia WebMap and GeoMedia WebMap Professional for building Web-based mapping and GIS applications, GeoMedia OnDemand for mobile applications, GeoMedia SDI Pro for extending GeoMedia WebMap to allow the implementation of spatial data infrastructure (SDI), GeoMedia Fusion for integrating and maintaining spatial data, GeoMedia Grid for raster data analysis, GeoMedia Map Publisher for map publication, GeoMedia Parcel Manager for parcel management, GeoMedia PublicWorks for network data analysis, GeoMedia Transaction Manager for transaction management, and GeoMedia Transportation Analyst and GeoMedia Transportation Manager for analysis and management of transportation infrastructure.

Maptitude

Maptitude is a desktop mapping software package developed by Caliper Corporation (http://www.caliper.com/). It can run under the Windows, OSX, Linux and Unix operating systems. The package has many common features of desktop GIS software, and also provides MapWizard for thematic mapping and the GIS Developer's Kit (GISDK) to enable developers to extend its functionality and develop new applications.

Open-source GIS software

Over the last decade an increasing number of free and open source GIS software packages have been made available. These software packages have their source code freely available for modification and redistribution by the general public. They do not charge licence fees, they use open standards that facilitate integration with other systems and are easily customisable. They are often developed and maintained by a community of shared interest. However, they do not offer professional support and lack release co-ordination. In spite of these disadvantages, open source GIS provides unique opportunities to serve the low-cost, high-control niches that are difficult, if not impossible, to cater for through commercial, proprietary avenues. The following is a brief summary of some open source GIS software packages that are useful for environmental applications.

GRASS

GRASS GIS was originally developed by the US Army Corp of Engineers as a tool for land management and environmental planning by the military, and it is now an official project of the Open Source Geospatial Foundation (OSGeo; http://grass.osgeo.org/). Since 1982, the GRASS Development Team has gradually grown into a multinational team consisting of developers at numerous locations across the world. The system has evolved into a software suite with a wide range of applications in resource management, environmental planning and scientific research. GRASS GIS is currently released under the GNU General Public License (http://www.gnu.org/licenses/gpl.html). It can be used on Windows, OSX, Linux and Unix platforms. The system contains over 350 modules and more than 100 add-on modules for spatial data manipulation, geocoding, remote sensing data processing and analysis, spatial analysis and modelling, mapping and 3D visualisation. It also provides interfaces to statistical packages (such as R, an open source statistical software package with extensions for spatial statistics), database management systems (such as PostgreSQL and mySQL) and environmental modelling tools (such as SWAT, Soil & Water Assessment Tool). GRASS GIS offers both a graphical user interface and command-line syntax.

GDAL/OGR

GDAL/OGR is a software library for reading and writing spatial data in different data formats, released under the open source licence by OSGeo (http://www.gdal.org/). The library comes with a variety of command-line utilities for spatial data translation and processing. It supports over 200 data formats, which provides wide-ranging capabilities for spatial data exchange. The GDAL/OGR software library has been extensively used in both commercial and non-commercial GIS communities.

Quantum GIS

Quantum GIS is another OSGeo project that is also open source under the GNU General Public License (http://qgis.org/). It offers basic desktop GIS functions and supports many spatial data formats, including the ArcGIS and MapInfo non-proprietary data formats. Quantum GIS provides integration with other open source GIS packages, including GRASS and GDAL/OGR, along with plugins developed by users in the C++ and Python computer programming languages. It also offers interfaces with PostgreSQL and mySQL DBMS. The package runs on multiple platforms, including Linux, Unix, OSX, Windows and Android. It is maintained by an active group of volunteer developers. In addition, it has been released in forty-eight languages.

MapGuide Open Source

MapGuide Open Source is a server GIS software package for Web-based mapping and spatial data delivery (http://mapguide.osgeo.org/). It was introduced as open source by Autodesk (a company best-known for the AutoCAD software product family) in 2005, and the code was contributed to OSGeo under the GNU Lesser General Public License (http://www.gnu.org/licenses/lgpl.html). MapGuide features an interactive map viewer, which supports map browsing, feature selection, spatial query, buffer generation and measurement. It includes an XML (Extensible Markup Language, a markup language that defines a set of rules for encoding documents) database for managing content, and supports most popular spatial data formats, databases and standards. The software can be deployed on Linux or Windows, supports popular Web servers and provides extensive APIs for application development.

GeoTools

GeoTools is a GIS software library written in the Java programming language (http://www.geotools.org/). It is another project of OSGeo under the GNU Lesser General Public License. It contains Java packages or software modules for reading and writing spatial datasets in a number of data formats, enabling georeferencing and transforming spatial data with an extensive range of map projections or projected coordinate systems, filtering and analysing data in terms of spatial and non-spatial attributes, displaying maps and making map-based queries. GeoTools has been used for desktop and Web-based mapping applications.

1.6 A TOUR OF ARCGIS FOR DESKTOP

ArcGIS for Desktop comprises a set of integrated software modules, the main ones being ArcMap, ArcCatalog and ArcToolBox. ArcMap provides basic mapping and analysis tools, and is usually used to launch other modules and access the ArcGIS for Desktop extensions. ArcCatalog resembles Microsoft Windows Explorer, but it is designed for managing files and folders containing spatial data, and for browsing, previewing and documenting spatial data and their metadata (data about data). This module makes it simple to organise, access and modify spatial data files. Both ArcMap and ArcCatalog are accessible from the Windows **Start** menu. ArcToolbox contains an extensive set of tools for spatial data manipulation and analysis, including data conversion, map projection and transformation, overlay analysis, proximity analysis, network analysis, spatial statistics and many more. ArcToolbox can be launched from either ArcMap or ArcCatalog.

This section will take you on a guided tour of ArcGIS for Desktop using the sample database of a hypothesised virtual catchment included on the website that accompanies this book (www.routledge.com/cw/zhu). Download the data from the website and install them to the local drive C:\.

Starting ArcMap

ArcMap can be started by either double-clicking with the left mouse button on the ArcMap icon on the desktop or by clicking on the windows **Start** menu and then selecting **All Programs > ArcGIS > ArcMap**. The symbol > means that the right-side menu item or command is one of the menu items or commands under the left-side

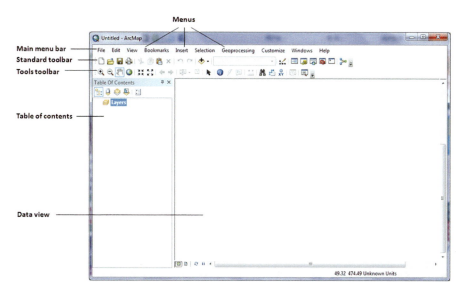

Figure 1.16 The ArcMap opening screen and screen elements.

menu item or command. By default, a getting-started dialog box appears once ArcMap is loaded. This dialog box allows the user to select an existing map to open or a template to make a new map. Clicking the Cancel button opens and loads a blank map, as shown in Figure 1.16.

ArcMap screen elements

ArcMap makes extensive use of the graphical user interface, with menus, buttons, icons and different mouse pointers. As displayed in Figure 1.16, the main elements of the ArcMap screen include the main menu bar, standard toolbar, tools toolbar, the table of contents and data view. The main menu bar consists of an array of menu choices, each having a number of menu items corresponding to particular commands or functions or tools. For example, clicking on the File menu will show commands that deal with map document files.

Toolbars are rows of buttons with small icons. Each icon represents a specific function. The standard toolbar consists of functions for creating, opening and saving map documents, and for adding data and launching other software modules. The tools toolbar provides map browsing and query functions, such as zoom and pan. Simply clicking a button will implement the function it represents. You may add additional toolbars by clicking Customize > Toolbars on the main menu and selecting those that perform the tasks you wish. Toolbars can be docked at the top or bottom or to the left or right side of the ArcMap window. They can also float on the Windows

desktop. Mastering the use of the toolbars can dramatically increase the speed with which you can work with ArcMap.

The data view is a map window for displaying map contents. A map is made up of a number of data layers. A data layer is a spatial dataset of a particular theme, such as rivers, roads, land parcels or terrain (discussed further in Section 2.3). In ArcMap, a map is a data frame which displays a collection of symbolised layers drawn in a certain order for a given map extent and map projection or coordinate system, as shown in Figure 1.17. In an ArcMap session, several data frames can be included, but the data view can only show the active data frame. The data view presents spatial data without all the map elements such as titles, scale bars and north arrows. We work with the data view for spatial data manipulation and analysis.

ArcMap also provides a layout view for composing maps. A layout is a set of map or cartographic elements, such as data frames (map bodies), map title, scale bar, north arrow and legend, arranged on a page (Figure 1.18). You can use the toggle buttons to switch between the layout view and data view, as shown in Figure 1.18.

The table of contents lists all the data frames and data layers in each data frame in the current project. As shown in Figure 1.17, five data layers in the data frame 'Layers' and two data layers in the data frame 'World' are listed in the table of contents. The symbols representing the geographical features on each layer are also displayed. The checkbox next to each layer indicates whether its display

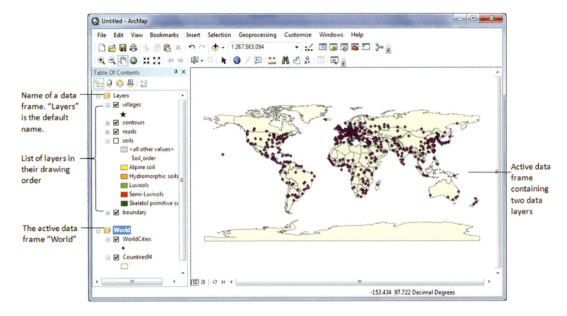

Name of a data frame. "Layers" is the default name.

List of layers in their drawing order

The active data frame "World"

Active data frame containing two data layers

Figure 1.17 Data view of ArcMap.

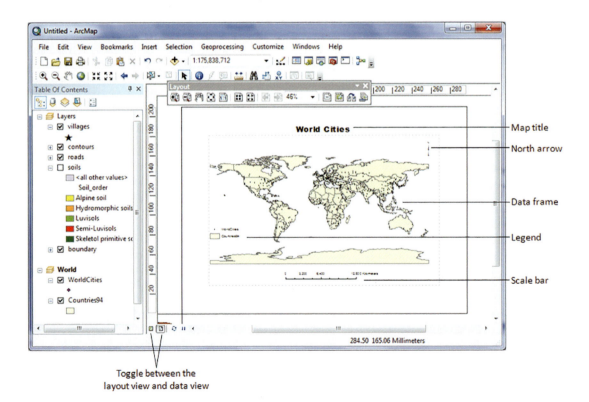

Map title

North arrow

Data frame

Legend

Scale bar

Toggle between the layout view and data view

Figure 1.18 Layout view of ArcMap.

is currently turned on or off. The order of layers specifies their drawing order in the data frame. As a general rule, point and line feature layers should be placed above the area or surface feature layers.

Adding and removing spatial datasets

There are two main ways to add spatial datasets. One is to add spatial datasets from ArcCatalog. ArcCatalog can be launched from within ArcMap by clicking the **Catalog** window button 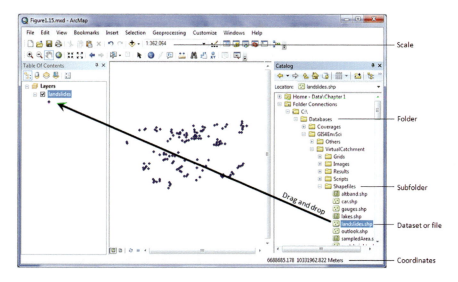 on the standard toolbar. Figure 1.19 shows an example of an ArcCatalog window opened in ArcMap. The ArcCatalog window displays the contents of folders and subfolders where spatial data are stored in the form of a catalogue tree. You can navigate through the catalogue tree to the desired dataset, and then drag the dataset and drop it into the data frame listed in the table of contents. For example, navigate to **C:\Databases\GIS4EnvSci\VirtualCatchment\Shapefiles**, then drag and drop **landslides.shp** to the table of contents. The dataset is added and displayed in the data view using the default symbology, as shown in Figure 1.19. ArcCatalog also allows you to copy, move, rename and otherwise manage your spatial data files and folders.

Another way to add spatial data is to use the **Add Data** tool on the standard toolbar. Once you click the tool, the **Add Data** dialog box (or simply dialog) pops up. Navigate to the desired dataset in the dialog and click the **Add** button. The dataset is added to the table of contents and displayed in the data view. You may try to use this tool to add any dataset in **C:\Databases**. Once a dataset is added, it is placed at the top of the layer list under the data frame in the table of contents. It also has the box to its left checked. If you uncheck the box, the layer will become invisible in the data view. In addition, the scale of the current display in the data view is shown in the combo box next to the **Add Data** button, as indicated in Figure 1.19. Combo boxes in ArcMap allow you to select an option from a drop-down list. Therefore, you can use this combo box to specify a scale for displaying the data.

As you move the mouse over the map in the data view, the coordinates of the current position of the cursor are updated and displayed at the bottom right of the ArcMap window. In the example shown in Figure 1.19, the coordinates of the cursor's current position are (6688685.178, 10331962.822) in units of metres (Section 2.2 will provide detailed discussion on map coordinate systems). ArcGIS can read spatial data in ESRI Shapefiles, ArcInfo Coverages, ESRI Grid and more than seventy other spatial data formats (see Section 2.3).

To remove a data layer, you may right-click on the name of the layer in the table of contents. A context menu (also called a contextual, shortcut or pop-up menu) pops up, as shown in Figure 1.20. It is a floating menu that appears at the location of the pointer when you right-click. From the pop-up menu, select **Remove**. The layer is then removed from the table of contents and the data view.

Figure 1.19 Catalog tree in ArcCatalog.

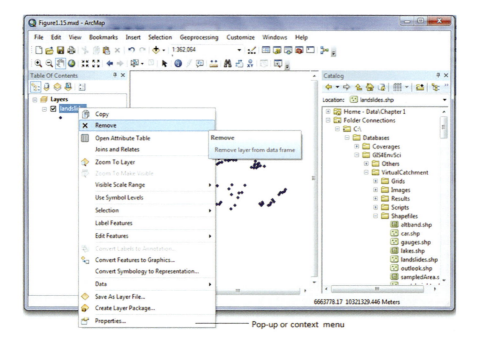

Figure 1.20 Pop-up menu in ArcMap.

Using ArcToolbox

The majority of spatial data manipulation and analysis tools are accessible from ArcToolbox. In ArcMap, you launch ArcToolbox by clicking the ArcToolbox window button found on the standard toolbar or by clicking **Geoprocessing > ArcToolbox** on the main menu. As shown in Figure 1.21, the ArcToolbox window displays all available toolboxes and toolsets packaged in your copy of the software. The tools, toolsets and toolboxes are organised in a tree structure. Just like navigating through the catalogue tree in the **Catalog** window, you expand toolboxes and toolsets until you find your desired tool, then double-click the tool to open its dialog box.

For example, click the plus sign beside **Analysis Tools** to expand this toolbox, next similarly expand the **Proximity** toolset, then double-click on the **Buffer** tool. This will open the **Buffer** dialog box, as shown in Figure 1.22. To run the tool, you need to add the input features by selecting a data layer from a drop-down list (by clicking the down arrow of the combo box under the input features), or from disk (by clicking the browse button beside the combo box to navigate through the directory tree to get the desired dataset). Then you need to type in the name of the output feature class (output dataset) with the full directory path, enter a number to

specify the distance and select the unit of distance. Other parameters are optional, with default values calculated or set by the tool. You can either accept the default value by doing nothing or enter a new value. Once you click **OK**, the tool will run and generate the results.

In most dialog boxes, there are several standard buttons, including **OK**, **Cancel**, **Environments** and **Tool Help**, as shown in Figure 1.22. The **OK** button is used to run the tool. Clicking the **Cancel** button closes the dialog box and cancels the task. Pressing the **Tool Help** button shows information about the tool. The **Environments** button opens the **Environment Settings** dialog box, which allows you to set additional parameters that affect the tool's results. For instance, you may want to use the **Environment Settings** dialog to set workspaces for inputs and outputs, define a specific geographical area for the analysis, and specify the coordinate system for the output data.

Tools can be run as either foreground or background processes. When a tool is running in foreground mode, you must wait until the tool stops running before you can continue with other work in ArcMap. When a tool is executing in background mode, you can continue working with ArcMap while the tool is running in the background. You can decide whether tools run in foreground or background mode by clicking **Geoprocessing > Geoprocessing Options** on the **Standard** toolbar.

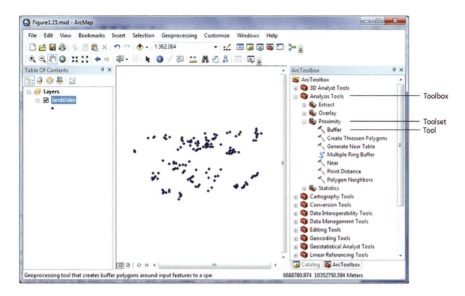

Figure 1.21 ArcToolbox.

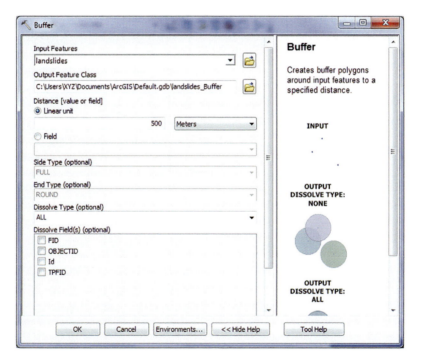

Figure 1.22 Example of a dialog box.

Saving and opening a map document

A map document is a data file storing the specifications for the data layers, data frames and layouts, including information on what data layers are used and created, and how they are displayed in the current ArcMap session. It is not automatically saved at the end of an ArcMap session. Before saving your current ArcMap session as a map document, you should first decide how the document references data. To do so, you open the **Map Document Properties** dialog box by clicking **File > Document Properties** on the main menu bar. In the dialog, you can check **Store relative pathnames to data source** if you want the map document to reference data relative to the document's current location in the file system rather than by full paths that include a drive letter or machine name, or otherwise uncheck **Store relative pathnames to data**. Using relative paths makes the map document more portable across different computers. After setting the document property, you can go to the **File** menu on the main menu bar and select **Save**, or click on the **Save** button 🖫 on the standard toolbar to save the current ArcMap session. The file name extension (.mxd) is automatically appended to your map document name. Note that the data layers are not stored in the.mxd file. If you copy the map document to a disk, you must copy all the dataset files used in the ArcMap project to that disk as well.

When you have completed your ArcMap session and wish to close the program, click **File > Exit** on the main menu. ArcMap will prompt you to save the session or changes to a map document. An existing map document can be opened by double-clicking it. This will start a new ArcMap session with the data layers and display properties specified in the .mxd file.

1.7 SUMMARY

- GIS is a computerised system for collecting, managing, manipulating, analysing, presenting and communicating spatial data – an important tool for environmental research, management and planning. GIScience is the science behind GIS technology.
- GIS consists of hardware, software, data, people and procedures. It is typically composed of the following functional components: spatial data input, data storage and management, data manipulation and analysis, data output and interfaces.
- The roots of GIS lie in computer mapping. Since the 1960s, GIS has evolved from a simple computer mapping system to a sophisticated computerised system with spatial database management, advanced spatial analysis and modelling and 3D visualisation, and GIS computing platforms have moved from mainframe computers through workstations and desktops to distributed and cloud computing.
- GIS deals with environmental processes and patterns in a spatially distributed manner. Environmental applications of GIS can be generalised into three

categories: inventory and monitoring, analysis and modelling, and visualisation and communication.
- Environmental problem solving with GIS involves (1) defining the problem, (2) structuring the problem, (3) acquiring and preparing data, (4) designing a solution process, (5) implementing the solution process and (6) communicating the results.
- There is a variety of GIS software, including desktop GIS, server GIS, mobile GIS, spatial DBMS software and GIS software libraries. Proprietary GIS software is commercial and licensed under the exclusive legal right of the copyright holder. Open source GIS software has its source code freely available and can be used with no licence fees.

REVIEW QUESTIONS

1. What is GIS? How is GIS different from other types of computerised information systems?
2. Describe each of the GIS functional components.
3. Describe the three-tier architecture of GIS.
4. Describe how GIS technology has evolved.
5. What is volunteered geographical information?
6. What are distributed GIS and cloud GIS?
7. Discuss the roles of GIS in environmental research, management and planning.
8. Describe the process of environmental problem solving with GIS.
9. Examine the case study presented in Box 1.1. In what ways would GIS help? What are the advantages of using GIS in this case study?
10. What are the differences between desktop GIS, Internet mapping, spatial DBMS and GIS software libraries in terms of their functionality?

REFERENCES

Berry, J.K. (1987) 'Fundamental operations in computer-assisted map analysis', *International Journal of Geographical Information Systems*, 1: 119–136.

Bhat, M.A., Shah, R.M. and Ahmad, B. (2011) 'Cloud computing: a solution to geographical information systems (GIS)', *International Journal on Computer Science and Engineering*, 3(2): 594–600.

Cao, K., Huang, B., Wang, S. and Lin, H. (2012) 'Sustainable land use optimization using Boundary-based Fast Genetic Algorithm', *Computers, Environment and Urban Systems*, 36(3): 257–269.

Chang, N.B., Parvathinathan, G. and Breeden, J.B. (2008) 'Combining GIS with fuzzy multicriteria decision-making for landfill siting in a fast-growing urban region', *Journal of Environmental Management*, 87: 139–153.

Chrisman, N.R. (1999) 'What does "GIS" mean?', *Transactions in GIS*, 3(2): 175–186.

Coppock, J.T. and Rhind, D.W. (1991) 'The history of GIS', in D.J. Maguire, M.F. Goodchild and D.W. Rhind (eds) *Geographical Information Systems: Principles and Applications*, Vol. 1, London: Longmans, 21–43.

Cowen, D.J. (1988) 'GIS versus CAD versus DBMS: what are the differences?', *Photogrammetric Engineering and Remote Sensing*, 54(2): 1,551–1,555.

Crooks, A.T., Hudson-Smith, A.M., Milton, R. and Batty, M. (2009) 'Crowdsourcing spatial surveys and mapping', in D. Fairbairn (ed.) *Proceedings of the 17th Geographical Information Systems Research UK Conference*, Durham University, England.

Dangermond, J. (1983) 'A classification of software components commonly used in geographic information systems', in D. Peuquet and J. O'Callaghan (eds) *Design and Implementation of Computer-Based Geographic Information Systems*, Amherst, NY: IGU Commission on Geographical Data Sensing and Processing.

Densham, P.J. (1991) 'Spatial decision support systems', in D.J. Maguire, M.F. Goodchild and D.W. Rhind (eds) *Geographical Information Systems: Principles and Applications*, Vol. 1, London: Longmans, 403–412.

Donnelly, A., Chambers, L., Keatley, M., Maitland, R. and Weatheril, R. (2011) *Climate Witness: A Dispersed National Observer Network for NRM Phenology (Climatewatch)*, Publication no. 11/048, Canberra: Rural Industries Research and Development Corporation.

Dueker, K.J. (1979) 'Land resource information systems: a review of fifteen years of experience', *Geo-Processing*, 1(2): 105–128.

Dueker, K.J. and Kjerne, D. (1989) *Multipurpose Cadastre: Terms and Definitions*, Falls Church, VA: American Society for Photogrammetry and Remote Sensing and American Congress on Surveying and Mapping.

Forsyth, D.M., Barker, R.J., Morriss, G. and Scroggie, M.P. (2007) 'Modelling the relationship between fecal pellet indices and deer density', *Journal of Wildlife Management*, 71(3): 964–970.

Gatrell, A.C. (1991) 'Concepts of space and geographical data', in D.J. Maguire, M.F. Goodchild and D.W. Rhind (eds) *Geographical Information Systems: Principles and Applications*, Vol. 1, London: Longmans, 119–134.

Giordano, L.D. and Riedel, P.S. (2008) 'Multi-criteria spatial decision analysis for demarcation of greenway: a case study of the city of Rio Claro, Sao Paulo, Brazil', *Landscape and Urban Planning*, 84: 301–311.

Goodchild, M.F. (1990) 'Geographic information systems and cartography', *Cartography*, 19(1): 1–13.

Goodchild, M.F. (1992) 'Geographical information science', *International Journal of Geographical Information Systems*, 6: 31–45.

Goodchild, M.F. (2007) 'Citizens as sensors: the world of volunteered geography', *GeoJournal*, 69: 211–221.

Harmon, J.E. and Anderson, S.J. (2003) *The Design and Implementation of Geographic Information Systems*, Hoboken, NJ: Wiley.

Hayes, B. (2008) 'Cloud computing: as software migrates from local PCs to distant Internet servers, users and developers alike go along with the ride', *Communications of the ACM*, 51: 9–11.

Hodson, D. and White, J. (2010) 'GIS and crop simulation modelling applications in climate change research', in M.P. Reynolds (ed.) *Climate Change and Crop Production*, Wallingford, England: CAB International, 245–262.

Hunsaker, C.T., Nisbet, R.A., Lam, D.C.L., Browder, J.A., Baker, W.L., Turner, M.G. and Botkin, D.B. (1993) 'Spatial models of ecological systems and processes: the role of GIS', in M.F. Goodchild, B.O. Parks and L.T. Steyart (eds) *Environmental Modelling with GIS*, New York: Oxford University Press, 248–264.

Kepner, W.G., Semmens, D.J., Bassett, S.D., Mouat, D.A. and Goodrich, D.C. (2004) 'Scenario analysis for the San Pedro River, analyzing hydrological consequences of a future environment', *Environmental Monitoring and Assessment*, 94(1): 115–127.

Longley, P.A., Goodchild, M.F., Maguire, D.J. and Rhind, D.W. (2011) *Geographic Information Systems and Science*, 3rd edn, Chichester, England: John Wiley.

Marble, D.F., Peuquet, D.J. and Calkins, H.W. (1984) *Basic Readings in Geographic Information Systems*, Williamsville, NY: SPAD Systems.

Miller, H.J. and Han, J. (eds) (2009) *Geographic Data Mining and Knowledge Discovery*, 2nd edn, London: Taylor & Francis.

National Academy of Sciences (1986) *Ecological Knowledge and Environmental Problem-Solving: Concepts and Case Studies*, Washington, DC: National Academy Press.

Openshaw, S. (1998) 'Building automated geographical analysis and exploration machines', in P.A. Longley, S.M. Brooks, R. McDonnell and B. Macmillan (eds) *Geocomputation: A Primer*, Chichester, England: Wiley, 95–115.

Peucker, T.K. and Chrisman, N. (1975) 'Cartographic data structures', *The American Cartographer*, 2(1): 55–69.

Peuquet, D.J. (1984) 'A conceptual framework and comparison of spatial data models', *Cartographica*, 21(4): 66–113.

Phua, M.H. and Minowa, M. (2005) 'A GIS-based multi-criteria decision making approach to forest conservation planning at a landscape scale: a case study in the Kinabalu Area, Sabah, Malaysia', *Landscape and Urban Planning*, 71: 207–222.

Pielke, R.A., Baron, J., Chase, T., Copeland, J., Kittel, T.G.F., Lee, T.J., Walko, R. and Zeng, X. (1996) 'Use of mesoscale models for simulation of seasonal weather and climate change for the Rocky Mountain states', in M.F. Goodchild, L.T. Steyaert, B.O. Parks, C. Johnston, D. Maidment, M. Crane and S. Glendinning (eds) *GIS and Environmental Modeling: Progress and Research Issues*, Boulder, CO: GIS World Books, 99–104.

Robinson, A.H., Morrison, J.L., Muehrcke, P.C., Kimerling, A.J. and Guptill, S.C. (1995) *Elements of Cartography*, 6th edn, New York: Wiley.

Schluter, M. and Ruger, N. (2007) 'Application of a GIS-based simulation tool to illustrate implications of uncertainties for water management in the Amudarya River Delta', *Environmental Modelling Software*, 22(2): 158–166.

Seppelt, R. and Voinov, A. (2002) 'Optimization methodology for land use patterns using spatially explicit landscape models', *Ecological Modelling*, 151(2): 125–142.

Smith, G.F. (1989) 'Defining managerial problems: a framework for perspective theorizing', *Management Science*, 35(8): 963–981.

Sugarbaker, L.J. (2005) 'Managing an operational GIS', in P.A. Longley, M.F. Goodchild, D.J. Maguire and D.W. Rhind (eds) *Geographical Information Systems: Principles, Techniques, Management and Applications*, abridged edn, Hoboken, NJ: Wiley, 611–620.

Sui, D.Z. and Maggio, R.C. (1999) 'Integrating GIS with hydrological modeling: practices, problems, and prospects', *Computers, Environment and Urban Systems*, 23: 33–51.

Swetnam, R.D., Fisher, B., Mbilinyi, B.P., Munishi, P.K., Willcock, S., Ricketts, T., Mwakalila, S., Balmford, A., Burgess, N.D., Marshall, A.R. and Lewis, S.L. (2011) 'Mapping socio-economic scenarios of land cover change: a GIS method to enable ecosystem service modelling', *Journal of Environmental Management*, 92(3): 563–574.

Tomlinson, R.F. (1967) *An Introduction to the Geographic Information System of the Canada Land Inventory*, Ottawa, Canada: Department of Forestry and Rural Development.

Wilson, J.P. (1996) 'GIS-based land surface/subsurface models: new potential for new models?', paper presented at the Third International Conference/Workshop on Integrating GIS and Environmental Modeling, Santa Fe, New Mexico, 21–26 January 1996.

Woolley, R.N. and Pidd, M. (1981) 'Problem structuring: a literature review', *Journal of the Operational Research Society*, 32: 197–206.

Zhu, X. (1997) 'An integrated environment for developing knowledge-based spatial decision support systems', *Transactions in GIS*, 1(4): 285–300.

Zhu, X. and Dale, A.P. (2000) 'Identifying opportunities for decision support systems in support of regional resource use planning: an approach through soft systems methodology', *Environmental Management*, 26(4): 371–384.

CHAPTER TWO

Spatial data and data management

2

This chapter discusses the nature of spatial data, introduces the concepts of georeferencing, geographical coordinate systems, map projections and projected coordinate systems, discusses the data models and data structures for representing and managing spatial data in computers, and provides a brief introduction to spatial databases and database management.

LEARNING OBJECTIVES

After studying this chapter, you should be able to:

- explain the characteristics of spatial data;
- conceptualise geographical space with the object and field views;
- understand the concepts of map scale, map generalisation and their impacts on the representation of geographical space;
- understand how spatial locations are defined and measured in geographical and projected coordinate systems;
- select appropriate map projections and projected coordinate systems for specific GIS applications;
- identify and transform georeferencing systems associated with spatial data in GIS;
- know how spatial data are represented, structured and organised using computers;
- understand differences between vector and raster representations and compare their advantages and disadvantages;
- apprehend how spatial data are managed in spatial databases.

2.1 THE NATURE OF SPATIAL DATA

The unique feature of GIS is its ability to handle spatial data. Spatial data describe the spatial distributions of geographical features and their attributes observed or measured at a particular time or during a certain period. For example, the spatial data shown in Figure 1.1 list some of the records of the landslides that have occurred in Australia since 1900. Each record comprises the date, type and location of a landslide event. The location is specified by the place name and a pair of latitude and longitude. With the geographical coordinates of latitude and longitude (discussed in Section 2.2), each landslide can be accurately located and plotted on the map. The type and place name of a landslide can be considered as its two attributes. The date of a landslide is its temporal dimension. Therefore, spatial data consist of three components: location, attribute and time. The location component defines the positions and shapes of geographical features on the Earth's surface, the attribute component describes the properties or thematic values of the geographical features, and the time component specifies the time when the geographical features occurred, or were measured or observed, or when the spatial data were collected.

Spatial data are traditionally recorded, represented and stored in maps. The spatial data describing the political, social, economic, cultural, environmental and ecological features in a region are often represented in a series of thematic maps, such as administration maps, population maps, land use maps, transportation maps, cadastral maps, soil maps, vegetation maps, geological maps and ecosystems maps. GIS uses maps as one of the major sources of spatial data, as a means of representing and visualising spatial data, and as a conceptual framework for structuring, organising, storing, managing and manipulating spatial data. The fundamental spatial analysis functions in GIS are also largely based on maps. Although spatial data may come from other sources, such as remote sensing, GPS, surveying and sampling, and may represent various types of geographical entities, they have three common features.

First, spatial data are usually 'observational' rather than 'experimental' (Haining 2009). This means that spatial data are generally not collected under controlled situations. They characterise the occurrence or attributes of a geographical feature or set of geographical features through a data collection event at a location in natural settings or naturally occurring situations. Observations are not necessarily independent entities, and could be linked via characteristics such as time, place, proximity and co-occurring geographical features. All this has implications for the quality of spatial data (discussed in Section 3.5).

Second, spatial data capture the complexity of the real world in finite form through a particular process of conceptualisation and representation. It is impossible and impractical for spatial data to capture all of the infinite detail of the real world and to provide a perfect and complete representation of geographical features, their attributes and spatial variations. Geographical features characterised in spatial data are selected and generalised, and represented based on a particular way in which the geographical space is conceptualised (discussed below in this section).

Third, all spatial data record locations of geographical features according to a particular coordinate system that has a fixed relationship to the Earth's surface and takes the Earth's complex shape into account. Such a coordinate system is called a spatial measurement framework, or georeferencing system. Different georeferencing systems may be established based on different definitions or specifications of the Earth's size and shape (discussed in Section 2.2). All spatial data in an environmental project must be georeferenced and aligned to a known georeferencing system so that they can be viewed, mapped, queried, analysed and integrated. These special properties of spatial data need to be thoroughly understood before they can be appropriately used for problem solving with GIS.

Conceptualisation of geographical space

There are two views of geographical space: the object and the field view. The object view conceptualises geographical space as populated by individual objects. For example,

we may consider the geographical space of a city consisting of many discrete objects, such as buildings, streets, highways, railways, rivers and streams, sport grounds and parks. According to the geometric shape or dimensionality of an object, we may classify the objects into point objects (such as individual buildings), line objects (such as streets and streams) and area objects (such as parks and sport grounds). They are also called point, line and area features. Point features occur at single locations with no measurable length, line features have a measurable length and area features have a measurable area. The map in Figure 2.1 presents an object view of a city area in the downtown of Austin, Texas, in which small buildings are represented as point features using point symbols; streets, highways, railways and streams are depicted as line features using line symbols; and built-up areas, the river and forested areas are shown as area features using area symbols. All objects in the object view can be located precisely, have well-defined boundaries and are countable. Most of the tangible natural and human phenomena can be represented with the object view. Spatial data based on the object view record the locations and attributes of each identifiable object.

However, many environmental phenomena in geographical space, such as terrain, temperature, population density and soil pH, cannot be conceptualised with the object view. For example, can we identify individual discrete objects with well-defined boundaries from the terrain surface depicted in Figure 2.2a? There are low and high grounds, and smooth and steep slopes. But how

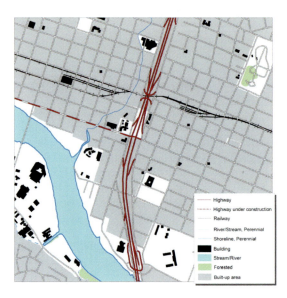

Figure 2.1 The object view of geographical space.

can we delineate their boundaries? It is not effective to conceptualise the terrain surface as a set of discrete and countable objects. Instead, we view the terrain surface as a continuous surface. This view represents the field view. In other words, the field view conceptualises geographical space as covered by continuous surfaces in which the attribute values can be defined at every location. The field view of geographical space can be presented using a contour map as shown in Figure 2.2b (a contour line links the points of equal value), or a raster map (a grid of cells, simply called a raster) as shown in Figure 2.2c. Spatial data based on the field view reduce a field to a finite number of bits of data, which represent the surface using a finite number of sample points at which the attribute is recorded and from which contours can be interpolated, or using a grid of cells in which cell values represent the attribute. The cells constituting a raster are often squares of equal size. (We will discuss rasters and interpolation in Sections 2.3 and 4.5 respectively.) A remote sensing image actually provides a field view of the area it covers. The conceptualisation of a geographical space as a field or as an object is largely determined by the attribute.

Map scale and spatial resolution

Any geographical representation, based on either the object view or field view, must be a reduction of reality.

Geographical features, such as rivers, lakes, roads and forests, must be shown on a map or image proportionately smaller than they really are, as the size of the sheet of paper or computer screen is very limited. The amount of reduction between the real world and its map or image representation is called map scale. Map scale is also referred as the representative fraction. It is defined as the ratio of a distance on the map or image to the corresponding distance on the ground. For example, on a 1:100,000 scale map, 1cm on the map equals 1km on the ground. It also means that the size of objects on the map is 1/100,000 of their size on the ground. Map scale is often confused or interpreted incorrectly, perhaps because the smaller the map scale, the larger the reference number, and vice versa. For example, 1:100,000 is considered a larger scale than 1:250,000.

Spatial data are all captured or collected at a certain map scale. The smaller the map scale, the fewer geographical features and the less detail about them can be captured and represented. For example, a river with a width of less than 100m can only be represented using a single line on a map of 1:100,000, and all details of the river bank will be lost. Map scale also determines the accuracy of measurement on a map. The larger the map scale, the more accurate the measurements of position, distance, area, perimeter, shape and other characteristics of geographical features; the smaller the map scale, the less accurate the

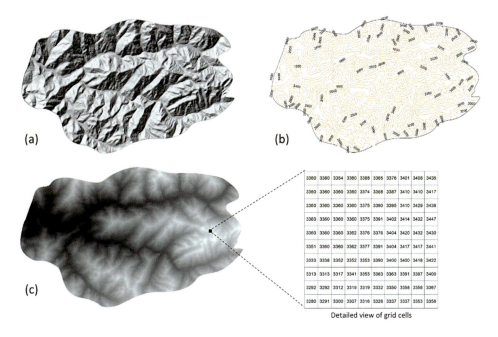

Detailed view of grid cells

Figure 2.2 The field view of geographical space.

measurements. Maps are more abstract as scales decrease. Figure 2.3 shows Pulau Ubin, a small island situated in the northeast of Singapore, mapped at two different scales. The coastline of the island captured at 1:50,000 shows a lot of small features that are smoothed out on the map at 1:1,000,000. The map at the smaller scale shows only a general outline of the island. Obviously, the length of the coastline and the area of the island can be measured much more accurately on the larger-scale map than on the smaller-scale one. As the map scale becomes larger, more details of the coastline can be revealed and more accurate measurements can be made.

When a raster is used to represent geographical features based on the field view, grid cells are laid down independently of the underlying field and its surface variation. All spatial variations in a cell are lost. How well a surface feature is captured depends on the size of the cell in relation to the surface variability. Therefore, the process of discretising a continuous surface with a raster involves a loss of information on surface variability. Larger grid cells incur more losses of detail on variability. The size of a cell is called spatial resolution. If the cell of a raster has a size of 10m × 10m, its spatial resolution is said to be 10m. In such a raster, a geographical feature that is smaller than $10 \times 10m^2$ may not be shown. In a satellite image with a spatial resolution of 100m, any geographical features of less than 100m × 100m are indiscernible. The smaller the size of a cell, the higher the spatial resolution. Both map scale and spatial resolution determine the accuracy of spatial data and the amount of information contained in spatial data.

Generalisation of geographical features

Representing real-world phenomena on a map at a certain map scale or a raster grid with a particular spatial resolution means that the geographical features have to be generalised or simplified. As discussed above, it is impossible to completely represent all geographical features and their intricate details in the limited space of the display medium of a map. Instead, the most important geographical features and their attributes relevant to the theme and purpose of the map are selected and represented in a way that adapts to the map scale and preserves the distinguishing characteristics and spatial pattern of the environmental phenomena being mapped. This process is called cartographic generalisation or abstraction. Cartographical generalisation involves selection, simplification, exaggeration, combination, displacement and classification of geographical features.

Selection is the process of choosing and retaining geographical features that are significant, relevant

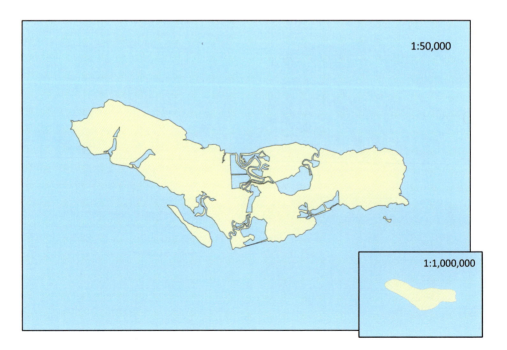

Figure 2.3 Pulau Ubin mapped at different scales.

and necessary given the theme and purpose of the map and are recognisable according to the map scale, while eliminating those that are insignificant, unnecessary and unrecognisable. For example, as shown in Figure 2.3, the 1:50,000 scale map includes the small lakes and offshore islands on Pulau Ubin, while the 1:1,000,000 scale map removes them as they are too small to be represented at this small scale. Generally, larger-scale maps retain more geographical features than smaller-scale maps. For another example, a water pipeline network map designed for the purpose of planning may only show water mains and hydrants, and omit other water pipes, meters, fittings, valves, storage tanks and other facilities in the network. Selection may also be influenced by geographical conditions. For example, wells are an important feature shown on topographical maps of desert and semi-desert

areas. However, they are insignificant in areas with good water supplies, and therefore generally not represented on topographical maps of these areas.

Simplification aims to streamline the shapes of the selected geographical features in order to enhance interpretability and reduce the complexity. It involves removing or smoothing out insignificant and unnecessary details along the boundary of a geographical feature, such as small angular turns along a river, a road, a coastline and the boundary of a forest stand, as shown in Figure 2.4a. Smaller-scale maps have more simplified and smoother features than larger-scale ones, which is clearly illustrated in the case of the coastline of Pulau Ubin in Figure 2.3. Simplification also involves decreasing the dimensionality of geographical features. A large river whose width can be truly represented on a large-scale

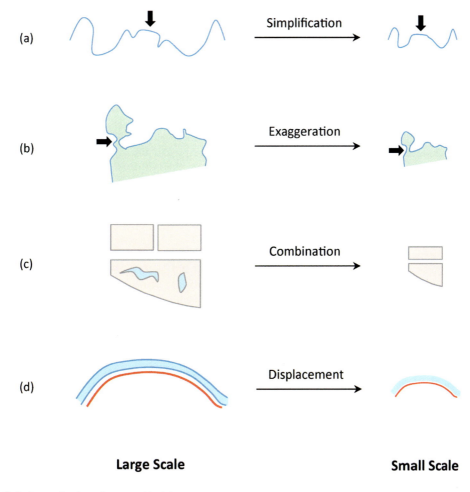

Figure 2.4 Generalisation of geographical features.

map can be depicted as an area feature, but has to be represented as a line feature on a small-scale map on which its width cannot be mapped and measured. A city can be represented as an area feature on a large-scale map where the city boundary can be drawn, while it is reduced to a point feature on a small-scale map.

Exaggeration is used to retain and enlarge some small but significant features that cannot be scaled to be drawn on the map. Only very large-scale maps can show such features as roads, buildings and small streams without greatly enlarging them. The purpose of exaggeration is to enhance or emphasise important characteristics of geographical features. For example, a section of coastline may contain many small headlands and curved inlets which cannot be represented on a small-scale map. If we remove all of them, this section of coastline may become a straight line. In order to preserve the true character of the coastline, some headlands and inlets which can characterise its shape are selected to be represented in an enlarged form. In Figure 2.4b, a tombolo is attached to the mainland by a narrow spit. The spit is too narrow to be represented on the small-scale map. However, removing the spit will make the tombolo an unattached island. In order to show it as a true tombolo, the spit is enlarged on the small-scale map.

Combination involves merging geographical features of the same class when the distance separating them is too narrow to be shown on the map, or when their separation is irrelevant or unnecessary to serve the map's purpose. For example, small city blocks are merged into large city blocks, as shown in Figure 2.4c. Combination reduces the number of geographical entities being mapped. Therefore, due to scale limitations, combined entities are not precise depictions of geographical features.

Displacement is required when different classes of geographical features, such as highways and railroads, are too close to separate on a small-scale map, but they have to be distinguishable. It involves small positional shifting of these features while minimising loss of positional accuracy, avoiding overlap and intersection, and maintaining their general defining characteristics, including shape, orientation, smoothness and pattern of distribution. Figure 2.4d gives an example.

Classification seeks to group geographical features by their attributes and attribute values – for example, categorising farm fields according to soil types, or grouping world countries in terms of CO_2 emissions per capita. Categorical attributes (such as land use types and soil types) are often grouped into classes and subclasses. For example, agriculture and settlement are two classes of land use. Agriculture can be further classified into subclasses, such as cropping, plantation forestry, pastures and horticulture. Settlement may have two subclasses:

urban settlement and rural settlement. Cropping can be further divided into subclasses, such as cereals, oil seeds, cotton and tobacco. Wetland may have four subclasses: bog, marsh, swamp and lake. Mapping involves selecting a certain classification scheme to classify geographical features according to their attributes and then using different colours or symbols to represent different classes.

For numerical attributes, groups of geographical features are defined in terms of ranges of their attribute values. For example, the values of CO_2 emissions per capita per year can be grouped into six ranges, also known as classes: < 2, $2–5$, $5–10$, $10–15$, $15–20$ and ≥ 20 tonnes. Instead of mapping the CO_2 emissions of 150 of the countries in the world using 150 different colours or 150 shades of a single colour, we start with grouping the 150 countries according to their CO_2 emissions per capita per year into the six classes listed above. The countries in the same class are then mapped using the same shade of a colour. The resultant map shows the global pattern of CO_2 emissions in six different shades of the colour, which is easy to interpret and may serve the purpose of the map well. There are different methods for classifying geographical features according to their numerical attribute values, which will be discussed in detail in Section 8.2.

The process of classification is a natural consequence of the need to handle detail. However, different ways of classifying data will lead to different patterns on the map. Also, maps with different scales have classifications with different levels of detail. Larger-scale maps may show a larger number of detailed classes, while smaller-scale maps involve reclassifying the attributes of geographical features into a smaller number of broad classes. For example, cereals, oil seeds, cotton and tobacco can be mapped on a large-scale map, but have to be combined into a single, broader class of cropping to be represented on a small-scale map. On a large-scale map, rural settlements may be mapped in four classes categorised in terms of population: < 100, $100–500$, $500–1,000$ and $\geq 1,000$ people. However, they may be simplified into two classes on a small-scale map: < 500 and ≥ 500 people.

It is not difficult to understand that some cartographic generalisation methods produce drastic changes in the form and composition of the map (such as selection and classification), while others create more subtle changes (such as simplification and displacement). Therefore, it is important to note that all spatial data captured from maps or images are not perfect, but generalised and simplified representations of geographical features due to scale limitations. Every spatial dataset contains a certain amount of detail about geographical features corresponding to the map scale or spatial resolution at which it is captured. The accuracy of geographical representation in spatial data will not be better than the source maps or images from which they are derived, and is determined by the original map scales and spatial resolutions. It is also necessary to understand that in addition to map scale, cartographic generalisation is affected by reality, map purpose, quality and quantity of available data, audience, conditions of use and technical limits (Robinson et al. 1995, p. 458).

2.2 GEOREFERENCING SYSTEMS

Spatial data record locations of geographical features using georeferences defined with a particular georeferencing system. A georeference, also called a geocode, is a reference to a location on the Earth's surface. The process of assigning georeferences to geographical features is called georeferencing, or geocoding. Although street addresses, place names, postal codes and jurisdictional boundaries can all be used as georeferences for specifying locations, metric georeferencing systems, which define location using measures of distance from fixed places, are most fundamental in GIS. Only metric georeferencing systems allow maps to be made and distances and other spatial relationships to be calculated. A metric georeferencing system is a coordinate system. There are two main categories of metric georeferencing systems commonly used in GIS: geographical coordinate systems and projected coordinate systems. Both are established based on the shape of the Earth.

The shape of the Earth

The Earth is not flat, but an approximately spherical object with some deviations from that shape. For many mapping applications, the size and shape of the Earth are approximated using an ellipsoid or spheroid, which is

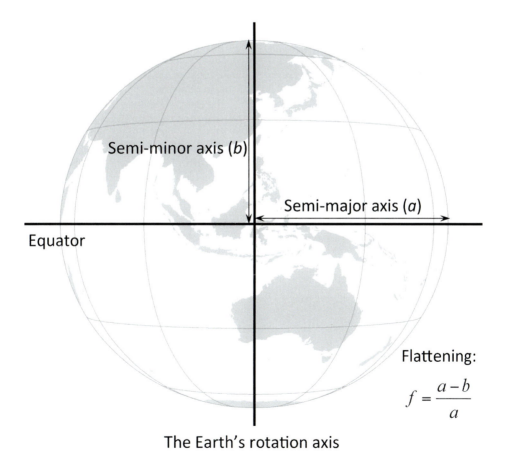

Semi-minor axis (b)

Semi-major axis (a)

Equator

Flattening:

$$f = \frac{a-b}{a}$$

The Earth's rotation axis

Figure 2.5 Definition of the spheroid.

formed by rotating an ellipse about its shorter axis that corresponds to the Earth's rotation axis. The spheroid, also called the oblate spheroid, is usually defined in terms of three basic parameters: semi-major axis, semi-minor axis and flattening, as shown in Figure 2.5. Flattening represents the reduction in the semi-minor axis relative to the semi-major axis.

Various spheroids are used to approximate the size and shape of the Earth. Table 2.1 lists some major spheroids used in different countries. The World Geodetic System of 1984 (WGS 1984) spheroid is most recent and widely adopted, but many others remain in use in some countries, and are still used with older spatial data.

Once a spheroid is defined, its position and orientation with reference to the Earth is determined so that it best fits the Earth's surface. The specifications of an Earth reference spheroid and its position and orientation in 3D space constitute a datum. A datum provides a frame of reference for measuring locations on the surface of the Earth, which serves as a basis for a geographical coordinate system.

Geographical coordinate systems

A geographical coordinate system is established by placing a graticule around the Earth spheroid, which is formed by meridians and parallels. Meridians are halves of imaginary great ellipses (the intersections of the spheroid and planes which pass through the centre point of the spheroid) on the Earth's surface, which converge at the North Pole and the South Pole. In 1884, the International Meridian Conference held in Washington,

DC decided to set the meridian that runs through the line marked on the ground at the Royal Observatory in Greenwich, England as the prime or zero meridian. Parallels are imaginary lines, perpendicular to meridians, extending around the Earth's surface parallel to the equator. The graticule determines the latitude and longitude of location on the Earth's surface (Figure 2.6).

For a particular point P on the Earth's surface, its longitude is defined by the angle, in the equator plane, between the prime meridian and the meridian where it is located. Its latitude is the angle, in a meridian plane, between the equator and the line drawn through the point perpendicular to the surface of the spheroid at this point. This line passes a few kilometres away from the centre of the spheroid except at the poles and the equator, where it passes through the spheroid's centre. Latitude and longitude are measured in degrees, minutes and seconds. The longitude of the prime meridian is 0°, and the longitude of its antipodal meridian (on the opposite side of the Earth) is both 180°W (west) and 180°E (east). The great ellipse formed by the two meridians divides the Earth into Western and Eastern Hemispheres. The meridians east of the prime meridian (in the Eastern Hemisphere) have a longitude ranging from 0° to 180°E, and those to the west of the prime meridian (in the Western Hemisphere) have a longitude of 0° to 180°W.

The equator's latitude is 0°. It divides the Earth into Northern and Southern Hemispheres. The latitude of the North Pole is 90°N (north), and that of the South Pole is 90°S (south). Therefore, the latitude of a location in the Northern Hemisphere ranges from 0° to 90°N, and that of a location in the Southern Hemisphere is within

Table 2.1 Major Earth spheroids

Name	Semi-major axis (m)	Semi-minor axis (m)	Flattening
Airy 1830	6,377,563.396	6,356,256.90923729	1/299.3249646
Bessel 1841	6,377,397.155	6,356,078.963	1/299.1528128
Clarke 1866	6,378,206.4	6,356,583.799999	1/294.9786982
Helmert 1906	6,378,200.0	6,356,818.16962789	1/298.3
International 1924	6,378,388.0	6,356,911.94612795	1/297.0
Krassowski 1940	6,378,245.0	6,356,863.01877305	1/298.3
Australian National	6,378,160.0	6,356,774.719	1/298.25
GRS80	6,378,137.0	6,356,752.31414036	1/298.257222101
WGS 1984	6,378,137.0	6,356,752.3142	1/298.257223563

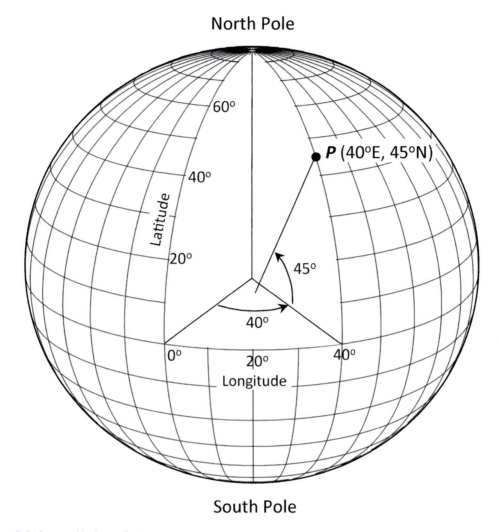

Figure 2.6 Geographical coordinate system.

the range of 0° to 90°S. In GIS, in order to simplify the computation and storage of geographical coordinates, degrees of latitude in the Southern Hemisphere and degrees of longitude in the Western Hemisphere are assigned negative values. Also, degrees of latitude and longitude are stored as decimal degrees – that is, minutes and seconds are expressed as decimal values.

Longitude and latitude are geographical coordinates. They can uniquely define any location on the surface of the Earth. However, they are dependent on the datum, which specifies the reference spheroid and its position and orientation with reference to the Earth. There are hundreds of locally or regionally developed datums around the world, such as the Geocentric Datum of

Australia 1994 (GDA94), the North American Datum 1983 (NAD83), European Terrestrial Reference System 1989 (ETRS89), China Geodetic Coordinate System 2000 (CGCS2000) and Singapore Survey 21 (SVY21). Each of them provides the best fit of the spheroid to a local area of the geoid, which is a 3D surface approximated by mean sea level. The shape of a geoid is affected by the Earth's mass, rotation and gravity anomalies, resulting in a highly irregular surface. Therefore, a local or regional datum may fit the local area extremely well, but may be a poor approximation of the geoid in other parts of the world. As a consequence, a specific point on the Earth can have substantially different geographical coordinates, depending on the datum used to make the measurement. Contemporary datums, based on increasingly accurate measurements of the shape of the Earth, are intended to best fit the entire Earth, and do not favour any particular area on the Earth, such as the WGS84 datum based on the WGS 1984 reference spheroid and the GDA based on the GRS80 reference spheroid. The study of the Earth's size, shape and geoid is the science of geodesy. A comprehensive introduction to geodesy can be found in Torge and Müller (2012). As users, we do not have to know much about geodesy in order to use GIS effectively. However, we must make sure that all our data are based on the same spheroid and datum so that geographical coordinates are measured in the same frame of reference. Thus, GIS users may need to convert between datums, and functions to do that are generally available in GIS software packages. Box 2.1 shows how to obtain information about the geographical coordinate system and datum used with a particular spatial data layer and how to convert between datums using ArcGIS.

Box 2.1   Geographical coordinate systems, datums and their transformation with ArcGIS   **PRACTICAL**

Geographical coordinate systems (GCSs) in ArcGIS are defined in terms of a datum, spheroid, units of measure and a prime meridian. The unit of measure in a GCS is decimal degrees, which is an angular unit. Spatial data with coordinates in decimal degrees exist in a GCS. They can be created on different datums. The GCS of a data layer is often defined in a projection file associated with the dataset. For shapefiles, the specification of a coordinate system is recorded in a projection file with the .prj extension (see Section 2.3). ArcGIS supports more than 600 GCSs and datums used for all parts of the world (Maher 2010).

Identifying the geographical coordinate system associated with a data layer

1. Start ArcMap with a new, empty map.
2. Click on the **Add Data** button ✛ , and add a data layer from **C:\Databases\GIS4EnvSci\ VirtualCatchment\Geodata.gdb** (every layer in the geodatabase has a GCS defined in its projection file).
3. Right-click the name of the layer in the table of contents. In the pop-up menu, select **Properties. . . .**
4. In the **Layer Properties** dialog box, click the **Source** tab. The information about the GCS associated with the layer appears in the panel. Here is an example:

```
Geographic Coordinate System: GCS_WGS_1984
Datum: D_WGS_1984
Prime Meridian: Greenwich
Angular Unit: Degree
```

In this example, the name of the GCS used is GCS_WGS_1984. It is based on the WGS84 datum. Its prime meridian is at Greenwich, and the coordinates are in decimal degrees.

Defining a geographical coordinate system

Every spatial data layer is in a coordinate system. ArcGIS, like any other GIS software system, cannot determine on its own what coordinate system a data layer uses unless it is specified in a projection file. Generally, if you have a data layer which does not have a coordinate system defined and you know which coordinate system it uses or you can obtain the information about its coordinate system from the data source, you need to define the coordinate system as in the following example.

1. Click on the **Add Data** button ✛ , and add the data layer **singapore.shp** from **C:\Databases\ GIS4EnvSci\Others**. This layer contains the coastline of Singapore, which is encoded in the

GCS based on the SVY21 datum, but not defined in the dataset. When the **Unknown Spatial Reference** dialog pops up, dismiss it by clicking **OK**.

2. Right-click **singapore** in the table of contents. In the pop-up menu, select **Zoom To Layer**.
3. Click on the **ArcToolbox** button 🔴 to open the ArcToolbox window.
4. In the ArcToolbox window, expand **Data Management Tools** > **Projections and Transformations**, then double-click **Define Projection**.
5. In the **Define Projection** dialog box:

   1) Enter **singapore** as the name of the input dataset. The coordinate system is shown as **Unknown**.
   2) Click the button for the coordinate system 🗺️ . The **Spatial Reference Properties** dialog box pops up, as shown in Figure 2.7.

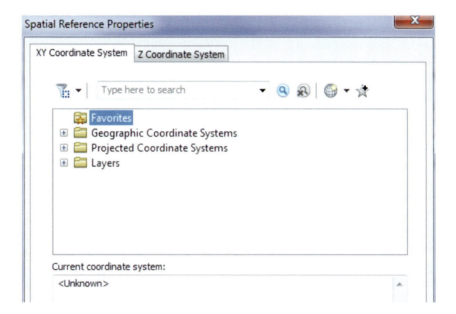

Figure 2.7 Example of a Spatial Reference Properties dialog box.

   3) In the **Spatial Reference Properties** dialog box, open the **Geographic Coordinate Systems** folder and then select the GCS that is used for the data layer. In this case, expand the **Asia** folder and select **SVY21**. Click **OK**.
   4) Click **OK**. Now, the GCS for the data layer is defined and the information on the GCS is stored with the data layer.

Transforming the geographical coordinate system

1. Click the **New** button 🗋 on the standard toolbar to create a new map document. In the **New Document** dialog, click **Blank Map** under **My Templates,** then click **OK**. When prompted to save changes to untitled, click **No**.

(continued)

(continued)

2. Add the data layer **singapore.shp** again.
3. Open the **ArcToolbox** window.
4. In the ArcToolbox window, expand **Data Management Tools** > **Projections and Transformations** > **Feature**, then double-click **Project**.
5. In the **Project** dialog box:

    1) Enter **singapore** as the name of the input data layer. Its associated GCS appears in the **Input Coordinate System** text box. Then enter a name of the output layer.
    2) Click the button for the output coordinate system. The **Spatial Reference Properties** dialog box pops up.
    3) In the **Spatial Reference Properties** dialog box, open the **Geographic Coordinate Systems** folder, and then expand the **Asia** folder and select **Kertau** (the datum used in Singapore for cadastral purposes before August 2004). Click **OK**. The selected GCS appears in the **Output Coordinate System** text box (Figure 2.8).
    4) Click **OK**. The output layer is produced with the transformed GCS.

Projected coordinate systems

Geographical coordinates are spherical coordinates. They provide a convenient reference system for describing locations on the approximately spherical surface of the Earth. While the Earth has a spherical surface, map sheets and computer screens are 2D flat surfaces. How can a flat map be used to describe locations on the Earth's surface? A common approach is to systematically transform the Earth's spherical surface to a 2D flat surface. Such a transformation is called a map projection. After the transformation, a plane coordinate system is established, which translates latitude and longitude (non-linear measures) to orthogonal Cartesian (x, y) coordinates (linear measures). A plane coordinate system established based on a map projection is called a projected coordinate system, also called a plane coordinate grid.

Map projections

Map projection involves mathematically projecting the surface of the Earth onto a flat surface or a surface that

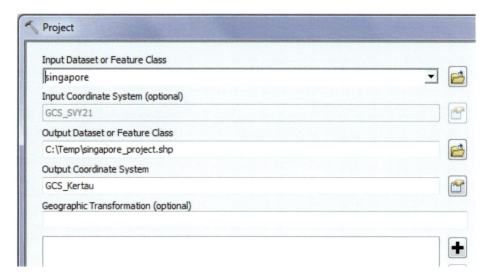

Figure 2.8 Example of a Project dialog box.

can be made flat by cutting. Such a surface is called a developable surface. The commonly used developable surfaces for map projection include cylinder, cone and plane, which lead to three families of map projection: cylindrical, conic and azimuthal (planar). Conceptually, a map projection is created by placing a light source at the centre of the Earth, then projecting an image of the surface of the Earth onto a developable surface, and then cutting the developable surface and laying it out flat (a plane is already flat) (Figure 2.9).

Figure 2.9 illustrates the simplest map projections, where the developable surfaces are tangent to the surface of the Earth along a line of latitude for cylindrical and conic projections, or at one pole point (here the North Pole) for an azimuthal projection. These projections are called tangent projections. The lines of contact are called standard parallels or standard lines. The point of contact is called the standard point. Scale is best preserved along standard parallels and at standard points. After the surface of the Earth is projected onto a developable surface, the developable surface (except the plane used in an azimuthal projection) is cut along a particular meridian to make the final projection. The meridian opposite the cutting line is called the central meridian. Map projections can also be made by intersecting a developable surface with the surface of the Earth, and the resulting projections

are called secant projections (Figure 2.10). Secant cylindrical and conic projections have two standard parallels, and secant azimuthal projections have one standard parallel. A cylindrical projection projects meridians and parallels as straight lines, a conic projection maps meridians as straight lines meeting at the apex of the cone and parallels as circular arcs, and an azimuthal projection projects meridians as straight lines converging at the point of contact and parallels as concentric circles.

The above projections align the axis of the developable surface with the Earth's polar or rotation axis, often called normal projections. A map projection may be transverse when the axis of the developable surface is perpendicular to the Earth's polar axis, or oblique when the axis of the developable surface is at any other angle (see Figure 2.10). Meridians and parallels are projected as complex curves in transverse and oblique projections.

Distortions in map projections

As you might expect, all map projections involve spatial distortions, which misrepresent the 3D surface of the Earth. Distortions may occur in terms of direction, distance, area or shape. In general, there is no distortion of any kind on the standard lines or at the point of contact. Distortion increases with the distance from the point or

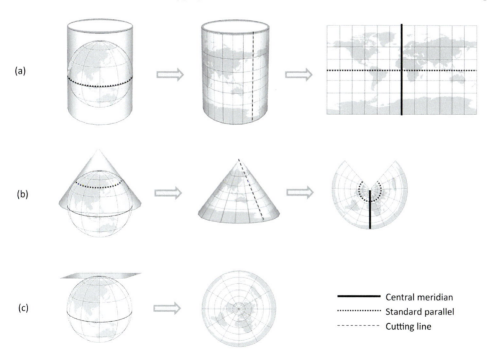

Figure 2.9 Map projections: (a) cylindrical projection, (b) conic projection and (c) azimuthal projection.

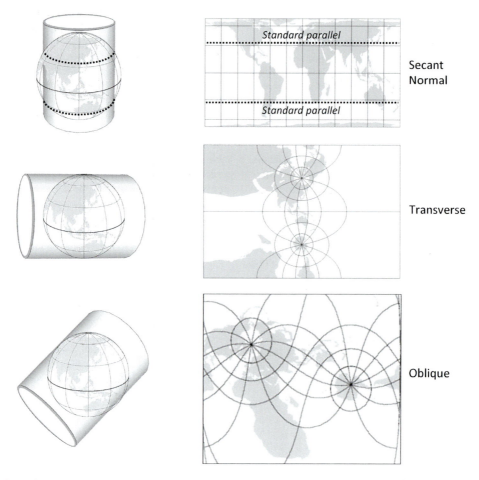

Figure 2.10 Secant, normal, transverse and oblique map projections.

lines of contact. For example, in a normal tangent cylindrical projection, all the locations on the equator have no distortion. The further away a point is from the equator, the greater the distortion it has.

Although distortions are inherent in all projections, a map projection can be made to accurately portray one spatial property at the expense of another – that is, to preserve one of the properties: direction, distance, area or shape. A map projection that preserves local shape is said to be conformal. In a conformal projection, the angles between lines on the projected flat surface or map are the same as the angles between the original lines on the Earth's curved surface. In other words, correct directions around any given point are maintained. Therefore, projected meridians and parallels intersect at the right angle. A conformal projection retains correct shapes of small areas such as a lake, part of a meandering river or

a greenbelt area, but not shapes of larger regions. This type of map projection is suitable for weather maps, navigation charts, topographic mapping and large-scale surveying.

A map projection preserving area is called equal area, or equivalent. In an equivalent projection, the area sizes of geographical features on the map are equal to the area sizes (at the map scale) of the corresponding features on the curved surface of the Earth. Equivalent projections allow accurate measurements of the areas on any part of the Earth's surface, and are useful for mapping geographical features whose relative size and area accuracy are important, and for general thematic mapping, such as population density, soil, vegetation, land use and geological maps.

A map projection is equidistant if it maintains correct distances from the centre of the projection or between points along one or a few lines (which may be straight or curved). For example, in an equidistant azimuthal projection, all points on the map are at correct distances from the centre point. In an equidistant cylindrical projection (also known as equirectangular), the true distances are represented along the meridians. No map projection is equidistant to and from all points on a map. Equidistant map projections are often used for air and sea navigation charts, and radio and seismic mapping.

A particular map projection can be conformal or equivalent or equidistant. However, no projection can be both conformal and equivalent or both equidistant and equivalent at the same time. Each projection is a compromise best suited for a specific purpose. In addition, no map projection can maintain correct scale all over the map, as there is simply no way to flatten out the Earth's spheroid surface without stretching, tearing or shearing. The scale at the standard point or the locations on the standard lines is often referred to as the principal scale or nominal scale, which is the stated representative fraction of a map. At all other locations, the scale varies and is referred to as the local scale. In other words, all projections contain scale distortions, and it is impossible to have a constant scale throughout the map. Scale distortions can be measured using the scale factor. It is calculated as the ratio of the local scale at a location to the principal scale of the map. A scale factor of 1 indicates no scale distortion at this point. Therefore, at the standard

point or along the standard lines, the scale factor is equal to 1.0. At all other locations on the map, the scale factors are greater or less than 1.0. A scale factor of greater than 1 means that this location has been exaggerated in scale, while a scale factor of less than 1 indicates the location has been reduced in scale. The values of the scale factors and the rate at which they change across the mapping area depend on the type of map projection and the placement of the standard point or lines. Generally, a secant map projection has less overall distortion than a tangent map projection.

The distortion pattern of a projection is often described using lines of equal distortion and visualised by distortion ellipses, also known as Tissot's indicatrices. Each ellipse represents the distortion at the point it is centred on. Figure 2.11 shows distortion ellipses on a secant Mercator projection, a conformal cylindrical projection named after Flemish cartographer Gerardus Mercator (1512–1594). All the distortion ellipses are shown as circles, indicating that the local angles, and consequently shapes of small areas, are correctly represented. The projection has two standard parallels at 30° latitude, where the scale factor is equal to 1 and the circles have no distortion. Between the two standard parallels, the scale factor is less than 1 and circles appear smaller. Moving from the standard parallels to the poles, the scale factor becomes greater than 1 and circles appear larger. All circles on the same parallel have the same size.

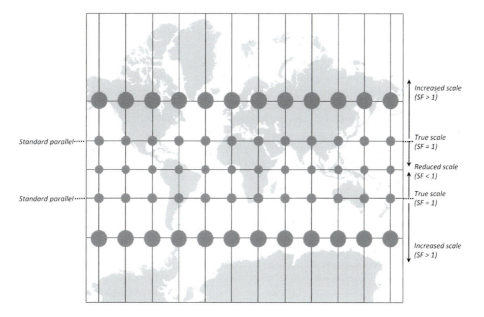

Figure 2.11 Mercator (conformal cylindrical) projection (SF – scale factor).

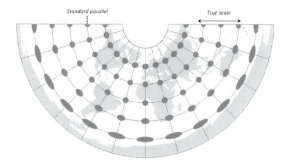

Figure 2.12 Equidistant conic projection.

Therefore, lines of equal distortion in a cylindrical projection are straight lines parallel to the standard lines. As a result, normal cylindrical projections are most suitable for low-latitude areas having a small extent in longitude and unusable for the North and South Polar regions.

Figure 2.12 is an equidistant conic projection with a single standard parallel at 30° latitude. The distortion ellipses plotted on the projection are distorted in size and shape except those on the standard line. However the north–south axis of the ellipses along meridians has an equal length. This indicates that meridians are true to scale and distances along these lines are accurate. Apparently, lines of equal distortion are circular rings parallel to the standard line. Normal conic projections are suitable for regional or national maps of middle-latitude zones such as maps of Europe, Australia, China and the USA. Conic projections are rarely used for world maps.

Azimuthal projections also have a circular pattern of lines of equal distortion, as shown in Figure 2.13. In this normal equivalent azimuthal projection, the distorted ellipses on the same parallel have the same shape and size. Although all the ellipses are distorted, each of them occupies the same amount of area. Because lines of equal distortion in azimuthal projections are concentric around the point of contact or the centre of the standard line (which is actually a circle), they are particularly useful for areas with an approximately circular shape and centred at the centre of the standard line or the contact point to minimise distortion for those areas. A normal azimuthal projection is most suitable for polar regions. As a general principle, the shape of a mapping area should be matched with the distortion pattern of a specific projection. Table 2.2 lists some map projections recommended for different regions.

Establishment of a projected coordinate system

After a map is made by transforming the Earth's spherical surface to a flat plane by map projection, a plane coordinate system is established over the map. The origin of

Table 2.2 Recommended map projections for small- and medium-scale mapping

Regions	Projections
The world	**Robinson Pseudocylindrical Projection:** parallels are projected as straight lines, while meridians resemble elliptical arcs, concave towards the central median. It does not have equal area, conformal, or equidistant properties. It is a compromise projection that attempts to achieve balance between these properties. This projection is only used for world thematic maps, not intended to be used for distance or area measurement.
	Miller Cylindrical Projection: a compromise between the Mercator and other cylindrical projections, neither conformal nor equivalent. The projection attempts to eliminate some of the scale exaggeration of the Mercator projection, particularly in the polar regions. The two poles are represented as straight lines. It is mainly used for general-purpose world maps.
	Behrmann Equal Area Cylindrical Projection: standard parallels are at 30°N and S. Shapes are distorted north–south between the standard parallels and distorted east–west above 30°N and below 30°S. Directions are generally distorted, except along the equator. It is only useful for world maps.
	Mollweide Projection: a pseudocylindrical equal area projection. All parallels are projected as straight lines, while all meridians are equally spaced elliptical arcs, except the central meridian, which is a straight line. Shape and local angles are true at the intersection of the central meridian and latitudes of 40°44′N and S, and distorted elsewhere. Scale is true along latitudes of 40°44′N and S.

Polar regions, oceans and hemispheres	**Orthographic Projection:** an azimuthal projection assuming the light source is located at an infinite distance from the point of tangency, providing a 'view from space'. It is neither conformal nor equal area, but direction from the central point is true. **Stereographic Projection:** a conformal azimuthal projection assuming the light source is located at the point on the Earth's surface opposite the point of tangency. **Lambert Azimuthal Equal Area Projection:** an azimuthal projection that preserves area and maintains true direction from the central point. This projection is often used for oceanic mapping of energy, minerals, geology and tectonics.
Continents	**Transverse (equatorial) Lambert Azimuthal Equal Area Projection** for Central America, the Caribbean, Africa and Southeast Asia **Oblique Lambert Azimuthal Equal Area Projection** for Asia, Europe and North America **Lambert Conformal Conic Projection:** a secant normal conformal conic projection for Europe, North America and Eurasia.
Countries or regions	For regions that are mainly east–west in extent in the middle north or south latitudes (for example, the USA, Canada, Russia, Australia and China): **Albers Equal Area Conic Projection:** a secant normal equivalent conic projection. **Lambert Conformal Conic Projection** For regions that are long, thin and mainly north–south in extent: **Transverse Mercator (also known as the Gauss-Krüger projection):** a transverse conformal cylindrical projection.

the coordinate system is usually set somewhere near the centre of interest on the map. The horizontal and vertical axes intersected at the origin are called the X-axis and Y-axis respectively (Figure 2.14a). The orthogonal (x, y)

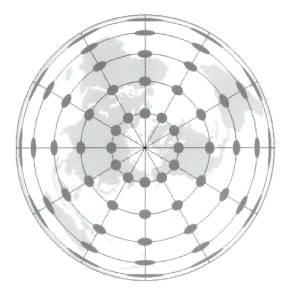

Figure 2.13 Equal area azimuthal projection (tangent at the North Pole).

coordinates of a location are defined as the positions of the perpendicular projections of the location onto the two axes, expressed as x metres east or west and y metres north or south of the origin. The coordinates at the origin are $x = 0$ and $y = 0$; x values are called eastings and y values are referred to as northings. Figure 2.14a provides an example of a projected coordinate system for Australia established based on the Lambert Conformal Conic Projection with Geocentric Datum of Australia 1994 (GDA94). The origin is located at longitude 134°E and latitude 27°S. The easting of Location 1 is 769,886m, and its northing is 630,548m. The easting and northing coordinates of Location 2 are −1,040,911m and −446,357m respectively.

In order to simplify calculation, the true origin is generally shifted to somewhere off the southwest corner of the map area. This move effectively adds some amount to each coordinate so that all eastings and northings in the map area are positive. The new origin is called the false origin, where the easting and northing are zero. The x-shift and y-shift values of the true origin are called the false easting and the false northing. Figure 2.14b shows the new projected coordinate system transformed by shifting the true origin at (134°, −27°) of the coordinate system depicted in Figure 2.14a to the false origin at (106.18°, −43.59°). Its false easting is 2,300,000m and the false northing is 2,100,000m.

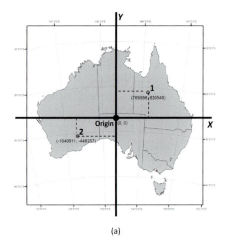

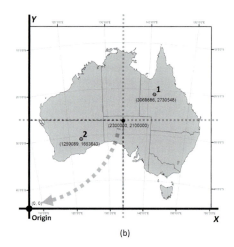

(a)   (b)

Figure 2.14 Projected coordinate systems based on the Lambert Conformal Conic Projection and GDA94 (the central meridian: 134°, the first standard parallel: −18°, the second standard parallel: −36°, the linear unit: metre): (a) the true origin (134°, −27°), (b) the false origin (106.18°, −43.59°), false easting 2,300,000m, false northing 2,100,000m.

In the new coordinate system, the easting of Location 1 is (769,886 + 2,300,000) = 3,069,886m, and its northing is (630,548 + 2,100,000) = 2,730,548m; the easting of Location 2 is (−1,040,911 + 2,300,000) = 1,259,089m, and its northing is (−446,357 + 2,100,000) = 1,653,643m.

Every projected coordinate system is associated with a particular map projection, which is based on a specific datum. Like geodetic datums, there are many local or national projected coordinate systems for mapping and spatial data collection, each of which is established to minimise distortion errors and increase measurement accuracy locally or nationally. Here we introduce a worldwide projected coordinate system used by many countries for large-scale mapping.

Universal Transverse Mercator (UTM) system

The UTM system is a global coordinate system based on transverse Mercator projections. It was recommended for topographic mapping by the United Nations Cartography Committee in 1952, and has been widely used for large-scale topographic maps, remote sensing images, and environmental and natural resource databases in GIS. Almost all GPS devices allow locations to be recorded in UTM coordinates. The UTM system divides the Earth's surface into sixty north–south zones extending from 80°S to 84°N latitude, each of them 6° of longitude wide (Figure 2.15). They are numbered from 1 to 60 eastward, starting at 180° longitude. For example, Zone 1 spans from 180° to 174°W longitude and has a central

meridian at 177°W. Zone 50 extends from 114° to 120°E longitude, and its central meridian is located at 117°E.

Each of the sixty UTM zones is mapped onto a transverse Mercator projection, which is a secant with two standard lines parallel to and approximately 180km to each side of the central meridian. The central meridian and equator in each zone are projected to two perpendicular straight lines, whose intersection is the true origin of the zone. However, the coordinate systems are established separately for the north (designated as N) and south zones (designated as S) divided by the equator. In order to avoid negative coordinates, a false easting of 500,000m is applied to both north and south zones. A north zone has a false northing of 0, while a south zone has a false northing of 10,000,000m. In other words, in a north zone, easting and northing coordinates are measured from a false origin placed at the equator and 500,000m west of the zone's central meridian. In a south zone, easting and northing coordinates are determined from a false origin placed at 10,000,000m south of the equator and 500,000m west of its central meridian.

UTM is conformal with minimal area distortion within each zone. The scale factor of the central meridian of a UTM zone is 0.9996. Two standard lines parallel to the central meridian have a scale factor of 1. Distances and directions between two points in a UTM zone can be measured to an accuracy of one part in 2,500. Error and distortion increase for regions that span more than one UTM zones. Therefore, the UTM is not designed for areas that cover two or more zones, and not suitable for mapping a continent or the entire area of a country like

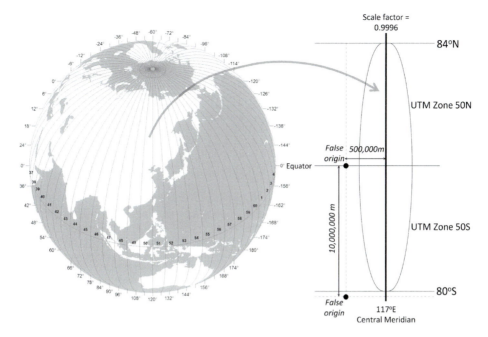

Figure 2.15 Universal Transverse Mercator system.

the USA, Canada, Australia, China or Russia (Table 2.2 provides the recommended map projections for these countries).

The UTM is based on a spheroidal model of the Earth. Many countries use local UTM zones based on the official datums to best suit their needs. For example, in Australia, the Map Grid of Australia 1994 (MGA94) provides the UTM coordinate system for local UTM zones (UTM Zone 49S to UTM Zone 56S) using the GRS80 ellipsoid based on the GDA94 datum. The New Zealand UTM coordinate system is currently based on the New Zealand Geodetic Datum 2000 (NZGD2000). The local UTM zones (UTM Zone 7N to UTM Zone 22N) in Canada use the coordinate system based on North American Datum of 1983 (NAD83).

Spatial data from different sources may adopt different map projections or projected coordinate systems. They must be converted to a common projected coordinate system before they can be integrated. The ability to convert spatial data to a common georeferencing system is an important consideration in GIS. GIS software packages generally provide the ability for conversion between a geographical coordinate system and a projected coordinate system, or among projected coordinate systems. Box 2.2 shows how to obtain information about a projected coordinate system used with a data layer in ArcGIS, and introduces the basic utilities available in ArcGIS for defining, transforming and customising coordinate systems.

Box 2.2 Projected coordinate systems and projection utilities in ArcGIS

PRACTICAL

Projected coordinate systems in ArcGIS are defined in terms of a number of projection parameters, such as the central meridian, standard lines, latitude of the projection's origin, scale factor, false easting, false northing, units of measure (usually in metres or feet) and associated GCS. ArcGIS is installed with a wide variety of predefined map projections or projected coordinate systems that can

(continued)

(continued)

be used to define a projected coordinate system for data layers covering different geographical areas or the entire world (Maher 2010). The parameter values of a projected coordinate system for a data layer are often stored in a projection file associated with it.

Identifying the projected coordinate system associated with a data layer

1. Start ArcMap with a new, empty map.
2. Click on the **Add Data** button ✛ , and add a data layer from **C:\Databases\GIS4EnvSci\ VirtualCatchment\Geodata.gdb** (every layer in the geodatabase has the same projected coordinate system defined in its projection file).
3. Right-click the name of the layer in the table of contents. In the pop-up menu, select **Properties. . . .**
4. In the **Layer Properties** dialog box, click the **Source** tab. The information about the projected coordinate system associated with the layer appears in the panel. The information is displayed in three parts: the name of the projected coordinate system, the parameters of the map projection it uses and the specifications of the GCS it is based upon. Here is an example:

```
Projected Coordinate System: World_Equidistant_Conic
Projection: Equidistant_Conic
False_Easting: 0.00000000
False_Northing: 0.00000000
Central_Meridian: 0.00000000
Standard_Parallel_1: 60.00000000
Standard_Parallel_2: 60.00000000
Latitude_Of_Origin: 0.00000000
Linear Unit: Meter
Geographic Coordinate System: GCS_WGS_1984
Datum: D_WGS_1984
Prime Meridian: Greenwich
Angular Unit: Degree
```

In this example, the name of the coordinate system is World_Equidistant_Conic. It is an equidistant conic projection. The central meridian of the map projection is at 0°. The standard parallel is set at 60°N. The true origin of the projection is located on the equator as the latitude of the origin is 0°. Its false easting and false northing are 0. Eastings and northings are in metres (linear units). The GCS uses the WGS84 datum.

Using the project-on-the-fly utility

Many GIS software packages provide a project-on-the-fly utility to project data layers in different coordinate systems on the fly and display them with a common coordinate system. In ArcGIS, each data frame uses a coordinate system in order to display data correctly. When ArcMap is started with a new, empty map, the coordinate system for the default data frame is not set. The first data layer added to an empty data frame defines its coordinate system if the coordinate system of the data layer is correctly defined in its associated projection file. When subsequent data layers, which also have their coordinate systems correctly defined, are added to the data frame, they are projected automatically on the fly to the current coordinate system of the data frame and displayed using the data frame's coordinate system.

    The project-on-the-fly utility in GIS facilitates quick map browsing and map making with data in different coordinate systems. However, it does not change the coordinate system specification of a data layer. In other words, the coordinates encoded in the data layer have not actually changed. If the data layers to be used in GIS analysis are in different coordinate systems, all of them must be converted to the same coordinate system. Conversion between a GCS and a projected coordinate system or between projected coordinate systems can be done using the **Project** tool in ArcToolbox as described for transforming the geographical coordinate system in Box 2.1. ArcGIS provides options of projected coordinate systems grouped into world, continental, polar, national grids, Gauss-Krüger, UTM, and some specific state and county systems. Predefined projected coordinate systems are listed in the **Projected Coordinate Systems** folder in the **Spatial Reference Properties** dialog box, as shown in Figure 2.7. If a data layer has a projected coordinate system, but not defined, it needs to be defined using the **Define Projection** tool in ArcToolbox in the same way as a GCS is defined (see Box 2.1). This is required for working with that data layer in ArcGIS.

Changing the data frame coordinate system

When the first data layer with a valid coordinate system is added to an empty data frame in ArcGIS, the coordinate system is set as the one for the data frame. But it can be changed by following these steps:

1. Right-click the data frame name in the table of contents and select **Properties** from the pop-up menu. The **Data Frame Properties** dialog box appears (Figure 2.16).
2. Click the **Coordinate System** tab and navigate to the desired coordinate system for map display in the data frame.

    1) To set the data frame coordinate system to be the same as a particular data layer in the data frame, expand the **Layers** folder, click the desired coordinate system, and select the data layer that references it.
    2) To set the data frame coordinate system to be the same as a data layer that is not in the data frame, click the **Add Coordinate System** 🌐 drop-down menu, select **Import**, and browse to a data source that is defined with the desired coordinate system. **Import** can also be used to read a previously saved .prj file for the coordinate system specification.
    3) To create a new coordinate system for the data frame, click the **Add Coordinate System** 🌐 drop-down menu, and select **New**. You may select one from the existing predefined geographical or projected coordinate systems and modify its parameters to create a customised coordinate system as shown below.

Customising a coordinate system

Both geographical coordinate systems and projected coordinate systems can be customised in ArcGIS by modifying their parameters.
   To customise the data frame coordinate system:

1. In the **Data Frame Properties** dialog box, click the **Coordinate System** tab.
2. Click the **Add Coordinate System** 🌐 drop-down menu, and select **New**.
3. Select **Projected Coordinate Systems. . .** to bring up the **New Projected Coordinate System** dialog box (or you may select **Geographical Coordinate Systems. . .** to customise a GCS).
4. In the **New Projected Coordinate System** dialog box:

    1) From the **Projection Name** drop-down menu, select an existing projected coordinate system. The default parameter values for the selected coordinate system are shown in the table beneath the projection name.

(continued)

(continued)

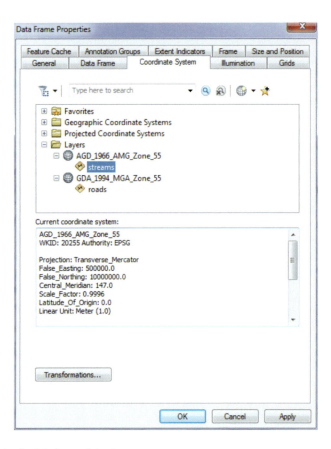

Figure 2.16 Example of a data frame dialog box.

2) Change the parameter values to modify the coordinate system.
3) Change the properties of the geographical coordinate system if necessary by clicking the **Change. . .** button.
4) Click **OK**.

5. Click **OK** to finish. Figure 2.17 shows an example.

In this example, the Lambert Conformal Conic Projection is modified by changing the central meridian from 0° to 134°E (from the prime meridian to the centre of Australia), and the two standard parallels from 60°N to 18°S and 36°S (from a tangent projection to a secant one). The projection is renamed as Australia_Conformal_Conic.

To customise the coordinate system of a data layer:

1. Click on the **ArcToolbox** button to open the **ArcToolbox** window.
2. In the **ArcToolbox** window, expand **Data Management Tools** > **Projections and Transformations** > **Feature**, then double-click **Project**.

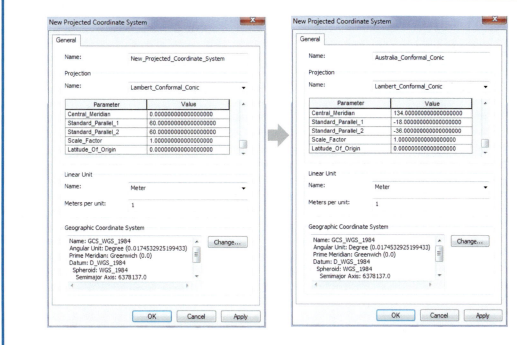

Figure 2.17 Example of customisation of a coordinate system.

3. In the **Project** window:
    1) Enter the name of the input data layer.
    2) Click the button for the output coordinate system. The **Spatial Reference Properties** dialog box pops up (Figure 2.7).
    3) In the **Spatial Reference Properties** dialog box, click the **Add Coordinate System** drop-down menu, and select **New**.
    4) Follow initial steps 3 and 4 in 'Customising a data frame coordinate system' as described above.

2.3 SPATIAL DATA MODELS

Spatial data are stored and maintained in GIS according to specific spatial data models. A spatial data model is a collection of concepts for describing how spatial data represent geographical features, and how spatial data are structured, related and organised. There are three types of spatial data models used in GIS software systems: vector, raster and object-oriented data models.

Vector data model

The vector data model is based on the object view of geographical space. The basic concepts of the data model include feature classes, point coordinates, attributes, topological relationships and feature–attribute relationships. A feature class is a collection of geographical features of the same geometric type (point, line and polygon) with the same set of attributes. The attributes of a geographical feature hold values that describe the feature. The vector data model typically stores spatial data as feature classes. For example, a point feature class may represent meteorological stations as points with observed values of a set of meteorological variables over a certain period of time, a line feature class may represent streams with a set of stream attributes (such as reach length, elevation, gradient, sinuosity and undercut depth), and a polygon feature class may represent land parcels with land use types.

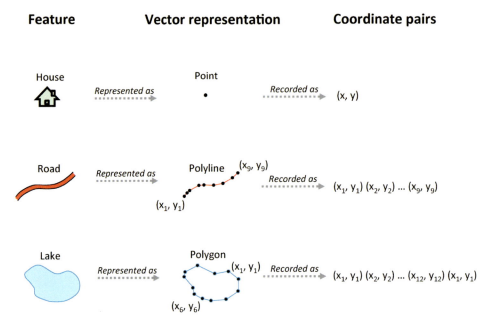

Figure 2.18 Spatial data describing locations of geographical features with the vector data model.

Point coordinates are used to define locations of geographical features in each feature class. The location of a point feature is defined using a pair of coordinates (for example, latitude and longitude, or easting and northing). The location of a line feature is recorded using a series of ordered points, each being defined using a coordinate pair. The shape of the line feature is approximated by a polyline, which links the ordered points with straight lines. The location of an area feature is specified using an enclosed series of ordered points with the starting and ending points being the same point, and its boundary or shape is approximated by a polygon which is formed by linking these points. Figure 2.18 shows how locations of point, line and area features are defined with point coordinates in the vector data model.

Topological relationships refer to the spatial relationships between adjacent or neighbouring geographical features, which remain unchanged when they are projected onto a map from the Earth's surface, or when the map is stretched, bended or distorted (Corbett 1979). Three topological relationships often encoded in spatial data include: adjacency, connectivity and containment. If two features share a common boundary, they are said to be adjacent, such as the forest and grassland in Figure 2.19a. If one feature is completely contained within another feature, they have a containment

relationship. For example, the lake is contained in the grassland in Figure 2.19a. Figure 2.19b shows an example of connectivity, where links of a pipeline network are connected with each other at nodes. Connectivity can be used to follow a path along a network (such as a stream network, a sewer network or road network). Topological information can be explicitly or implicitly represented in spatial data.

In the early GIS systems, spatial data in the vector data model (simply called vector data) contained no explicit topological information. They used non-topological data structures to store and organise vector data in a computer. The simplest non-topological data structure is cartographic spaghetti (Peuquet 1984). It represents each geographical feature as a string of coordinate pairs, as shown in Figure 2.18, and stores the coordinates of all the geographical features sequentially in a single file, using a flag or code to signify the end of a feature. It has no structure, no attributes and no spatial relationships. The shapefile is now a widely used non-topological data structure (ESRI 1998). In a shapefile, a point feature is represented by a vertex, defined by a pair of coordinates. A line feature or polyline is represented by an ordered sequence of vertices, which are connected by non-branching line segments. An area or polygon feature is represented by one or more geometric rings. Each geometric ring

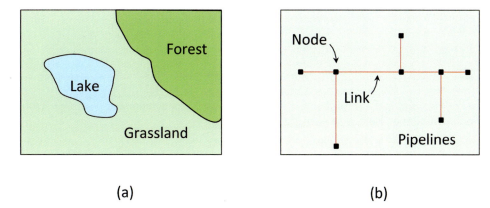

(a)   (b)

Figure 2.19 Topological relationships: (a) adjacency and containment, and (b) connectivity.

is a connected sequence of vertices that form a closed, non-self-intersecting loop. A geometric ring representing a single polygon maintains a consistent, clockwise ordering of vertices so that the area to the right (as one 'walks' along the boundary in vertex order) is inside the polygon and the left is outside. Holes in a polygon are represented by geometric rings whose vertices are in a counter-clockwise direction. A shapefile stores both non-topological geometry and attribute information for the geographical features in a data layer (see Box 2.3). Attribute data are structured in the form of a table or spreadsheet, in which a row or record corresponds to a geographical feature, and a column stores values of an attribute. Such tables are also called feature attribute tables. Attribute data and geometric data are linked by GIS via feature IDs.

Box 2.3   ESRI shapefile   **TECHNICAL**

A shapefile stores spatial data in three core files and several optional files. All filenames adhere to the 8.3 naming convention: an eight character filename prefix, a period and a three-character filename suffix such as 'shp'. All files have the same prefix and must be located in the same folder.

Core files

The three core files include:

- The *.shp* file – this is the main file, storing the primary geometric data, including shape types or feature classes and point coordinates (in terms of x and y). Each feature has a record of coordinates.
- The *.shx* file – this is the index file, storing the positional index (the offset and content length) of each record of coordinates in the *.shp* file so that the coordinates for a feature can be quickly retrieved.
- The *.dbf* file – this is the dBASE table, storing attribute data for each feature in dBase IV format, which can be edited in Microsoft Excel. Each attribute record has a one-to-one relationship with the associated feature record in the *.shp* and *.shx* files.

(continued)

(continued)

Optional files

The following three optional files are often included in a shapefile:

- The *.prj* file – this is the projection file, storing the coordinate system and projection information in plain text format. Here is an example:

```
PROJCS["Africa_Albers_Equal_Area_Conic",
GEOGCS["GCS_WGS_1984",
DATUM["D_WGS_1984",SPHEROID["WGS_1984",6378137.0,298.257223563]],
    PRIMEM["Greenwich",0.0],
    UNIT["Degree",0.0174532925199433]],
PROJECTION["Albers"],
    PARAMETER["False_Easting",0.0],
    PARAMETER["False_Northing",0.0],
    PARAMETER["Central_Meridian",24.0],
    PARAMETER["Standard_Parallel_1",-18.0],
    PARAMETER["Standard_Parallel_2",-32.0],
    PARAMETER["Latitude_Of_Origin",0.0],
    UNIT["Meter",1.0]]
```

- The *.sbn* and *.sbx* files – these are two spatial index files, containing the spatial index of the geographical features. The *.sbn* (spatial bin) file divides the area the data covers into rectangular regions (bins). Each bin contains the record numbers of the features in the *.shp* file that fall in its area. The *.sbx* (spatial bin index) file contains rows, each row storing the record number and the length in bytes of the corresponding bin record in the *.sbn* file.
- The *.shp.xml* file – this is the metadata file, containing information about the basic characteristics of the data layer in XML format, which may include the identification, the extent, the quality, the georeference and custodian of the dataset.

Non-topological data structures store geographical features as isolated, spatially unrelated objects. Therefore, adjacent polygons that share common boundaries duplicate common vertices, which may create gaps and overlaps between the boundaries when simplifying a boundary. Topological relationships between geographical features are established by computing their spatial intersections using computer programs (Theobald 2001). Many contemporary desktop GIS packages support non-topological data structures due to their simplicity (simple file structures), interoperability (easy data transfer between different GIS software packages) and the availability of efficient computer algorithms that can be used to build topology on the fly.

Topological data structures explicitly store topological relationships between adjacent features. They represent a point feature by a pair of coordinates, a line feature as an arc or chain and a polygon as a closed sequence of connected arcs. An arc is a polyline consisting of a sequence of ordered vertices (Figure 2.20). The endpoints of an arc are called nodes. The starting point of an arc is the from-node, and the end point is the to-node. The direction of an arc is the order of the vertices from the from-node to the to-node. Topological data structures describe all three topological relationships, as depicted in Figure 2.19. A typical topological data structure uses the arc as the basis for data storage. Figure 2.20 shows a topological data structure for line features. Arc coordinates define geometry of line features, and arc-node topology describes connectivity. Connected arcs intersect at common nodes. For example, in Figure 2.20 arcs 1, 2 and 3 are all connected at node 2, arcs 3, 4 and 5 are connected at node 4, but arcs 4 and 5 are not directly connected with arcs 1 and 2 because they do not share a common node.

A topological data structure for area features is shown in Figure 2.21, where an area or polygon is represented as

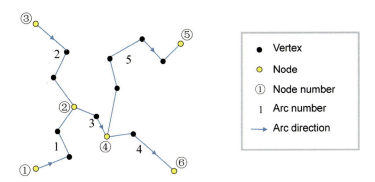

Arc-Node Topology

Arc	From-node	To-node
1	1	2
2	3	2
3	2	4
4	4	6
5	4	5

Arc Coordinates

Arc	(x, y) coordinate pairs
1	(20, 50) (175, 110) (120, 226) (205, 330)
2	(21, 725) (170, 590) (105, 475) (205, 330)
3	(205, 330) (302, 290) (350, 200)
4	(350, 200) (480, 220) (675, 89)
5	(350, 200) (400, 430) (370, 560) (523, 651) (624, 550) (700, 630)

Figure 2.20 A topological data structure for line features.

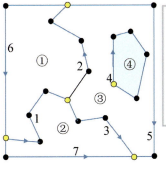

Polygon-Arc Topology

Polygon	Arcs
1	1, 2, 6
2	1, 3, 7
3	2, 3, 5, 0, 4
4	4

Left-Right Topology

Arc	Left-polygon	Right-polygon
1	1	2
2	1	3
3	3	2
4	3	4
5	0	3
6	1	0
7	2	0

Arc Coordinates

Arc	(x, y) coordinate pairs
1	(20, 50) (160, 40) (125, 150) (200, 251) (302,200)
2	(302,200) (401, 348) (375, 490) (226, 575) (310, 650)
3	(302,200) (351, 125) (480, 130) (620, 5)
4	(525, 300) (524, 500) (601, 530) (675, 425) (626, 225) (525, 300)
5	(300, 650) (710, 650) (710, 5) (620, 5)
6	(300, 650) (20, 650) (20, 50)
7	(20, 50) (20, 5) (620, 5)

Figure 2.21 A topological data structure for area features.

an ordered sequence of arcs. Similarly, arc coordinates define geometry of area features. Polygon-arc topology tells which polygon is made up of which arcs and provides indications of containment, while left-right topology describes adjacency. For example, in Figure 2.21, left-right topology indicates that polygon 1 is on the left of arc 1 and polygon 2 is on its right. Therefore, polygon 1 is adjacent to polygon 2. Each arc appears in two polygons as their common boundary. It ensures that the boundaries of adjacent polygons are stored only once and do not overlap, thus avoiding duplication in digitising common boundaries of two polygons and solving problems when the two versions of the common boundary do not coincide. Note that the area outside the boundary of the study area is labelled as polygon 0, often called the external or universe polygon.

The polygon-arc topology in Figure 2.21 shows that polygon 4 is an isolated polygon as it is formed by a single arc (arc 4). On the other hand, polygon 3 is made up of arcs 2, 3, 5 and 4 (0 before 4 indicates that arc 4 forms an island in the polygon). Therefore, it can be inferred that polygon 4 is contained in polygon 3.

Topological data structures reduce data storage for area features because boundaries or arcs between adjacent polygons are not stored twice, maintain explicit topological relations, and improve performance of adjacency analysis. Like non-topological data structures, they store attribute data in feature attribute tables, and link attribute and geometry data via feature IDs. The topological data structures described above are used in the ArcInfo coverage, or simply coverage, one of several spatial data formats supported by ArcGIS (Box 2.4). Other well-known topological data structures include POLYVRT (Peucker and Chrisman 1975), DIME (Peucker and Chrisman 1975) and TIGER (Marx 1990).

Box 2.4   ArcInfo coverage   **TECHNICAL**

A coverage is composed of a set of files storing geometric, topological and attribute information about a set of feature classes. All the files are stored in a directory named as the coverage name, which must be shorter than fourteen characters, contain no spaces, have no extension and be in all lowercase letters.

A collection of coverages is called a workspace. Figure 2.22 shows an example of a workspace containing several coverages. Each workspace has an INFO database (a relational database that can be perceived as a collection of tables; see Section 2.4) stored under the INFO directory. Feature

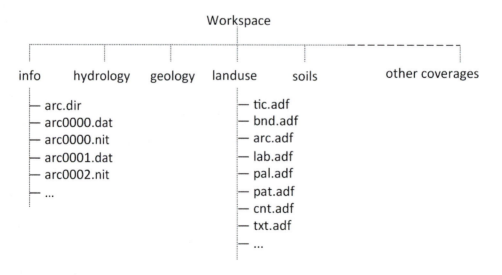

Figure 2.22 Example of coverage workspace.

Table 2.3 Common files used to store a coverage

Feature class	Spatial data in coverage directory	Feature attribute table in coverage directory	Feature attribute table in INFO database
Point	lab.adf	pat.adf	.nit and .dat
Arc	arc.adf	aat.adf	.nit and .dat
Polygon	pal.adf, cnt.adf, lab. adf, arc.adf	pat.adf	.nit and .dat
Tic	tic.adf	tic.adf	.nit and .dat
Coverage extent	bnd.adf	bnd.adf	.nit and .dat

attribute tables are stored in the .adf files in coverage directories. Other attributes are stored in tables in the INFO database. The features in a coverage exist in a one-to-one relationship with the corresponding records in the feature attribute table. Each .adf file in a coverage directory is associated with a pair of the .dat and .nit files in the INFO directory. The arc.dir file in the INFO directory records which pair of .dat and .nit files is related to which .adf file.

Table 2.3 lists the common files used to store a coverage for point, arc (line) and polygon (area) feature classes. It also lists the files for the tics and coverage extent. Tics are control points used to georeference and transform the coordinates in a coverage. They allow coverage coordinates to be registered to a common coordinate system. Tics are important for registering map sheets during digitising and editing and transforming coordinates from units such as digitiser centimetres into real-world coordinates such as UTM metres. Coverage extent defines the outer boundary of a coverage. It is the minimum bounding rectangle that defines the coordinate limits (that is, the minimum and maximum x, y coordinates) of the coverage.

The coverage is a legacy proprietary spatial data format supported by ArcGIS. Only the data in feature attribute tables are accessible to users. Other files are in binary format and maintained automatically by ArcGIS. However, users do not need to know which files are used. They only need to understand which feature classes are present in a coverage, and what geographical features they represent.

Triangulated irregular network (TIN)

The triangulated irregular network is a vector data model developed for representing surface morphology. It represents a surface using a set of contiguous, non-overlapping triangles (Figure 2.23). Within each triangle, the surface is represented by a plane with a constant gradient. A TIN is constructed from a set of points, called mass or data points. Each point is defined with 3D coordinates (x, y, z); (x, y) represents the location of a point, and z is a value of a particular variable represented by the surface (for example, elevation) at that location. These points are used as vertices (or nodes) and connected into triangles, typically through Delaunay triangulation.

Delaunay triangulation connects three data points into a triangle if and only if the circle which passes through them contains no other data point. In this way, the minimum interior angle of all triangles is maximised, and long and thin triangles are avoided as much as possible. The centre of the circle passing through three data points is the common vertex of three Voronoi (or Thiessen) polygons corresponding to the three points (Figure 2.24). A Voronoi polygon contains a single data point. All the locations in a Voronoi polygon are closer to the data point within the polygon than to any other data point. Therefore, Delaunay triangles can be created from Voronoi triangles, and vice versa. (For the mathematics and algorithms for Delaunay triangulation, see de Berg et al. 2008.)

The TIN is also a topological data structure. It stores not only (x, y, z) coordinates of the nodes, but also triangle-node topology (Figure 2.25). Triangle-node topology records the

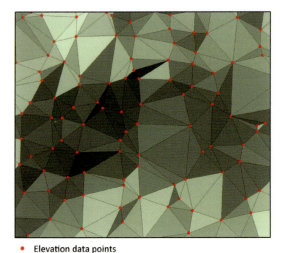

• Elevation data points

Figure 2.23 Part of a TIN constructed from a set of elevation points.

nodes forming each triangle and their neighbouring triangles. Each triangle possesses three attributes – area, slope

and aspect – which may be stored in a separate file or calculated on the fly when required.

The edges of a TIN can be used to capture the position of line and area features, such as roads, ridgelines, river courses, lake boundaries and coastlines. Therefore, accurately located features on a surface, whether they are point, line or area features, can be used as input features to the TIN. The input features used to create a TIN remain in the same position as the nodes or edges in the TIN. This allows a TIN to preserve all the accuracy of the input data while simultaneously modelling the values between the known data points. Input data to a TIN should be in units of metres or feet, not decimal degrees. Delaunay triangulations are invalid when constructed using geographical coordinates. Boxes 7.2 and 8.8 illustrate how to construct TINs in ArcGIS.

Because a TIN is constructed using a set of irregularly placed data points over a surface, it can effectively represent spatial variations of a geographical phenomenon by placing more and denser data points in areas where a surface is highly variable (for example, mountainous areas) and fewer and sparser data points in areas that are less variable (for example, flood plains).

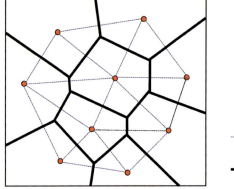

• Data point

········· Delaunay triangle edge

—— Voronoi polygon edge

Figure 2.24 Delaunay triangulation and Voronoi polygons.

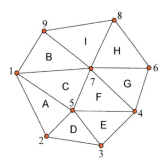

Triangle-node topology

Triangle	Nodes	Neighbours
A	1,2,5	-, C, D
B	1,7,9	-, I, C
C	1,5,7	A, B, F
D	2,3,5	-, A, E
...	...	...

Coordinate list

Node	x, y, z
1	1500, 2990, 444
2	2100, 1750, 420
3	3280, 1580, 382
4	4002, 2200, 358
...	...

Figure 2.25 A TIN data structure.

Raster data model

The raster data model is based on the field view of geographical space. It covers the geographical space using a grid of cells, each cell having a value. Therefore, the raster data model records the geographical features and their properties at every location. It involves the concepts of cells, cell values and cell sizes. A grid is often called a raster.

The cells of a raster are typically squares of equal size. As shown in Figure 2.26, a point feature is represented by a single cell, a line feature by a sequence of connected cells, and an area feature by a group of contiguous cells. Each cell has a value, representing a property of a geographical feature in the cell. A cell value can be a numerical value (for example, representing slope), or a code (for example, 1 representing agricultural land, 2 urban land and so on, as in Figure 2.26) or a label (for example, the name of a soil type). Cell values can also be used to represent the presence or absence of a geographical feature. For example, in the rasters representing houses and roads in Figure 2.26, if a cell is occupied by a house or road, it is assigned a cell value of 1; otherwise it gets a cell value of 0.

The location of a cell is defined by its row and column numbers (Figure 2.27). Generally, a raster is oriented with north at the top. Rows are parallel with the west–east direction and columns parallel with the north–south direction. The cell size is defined as the side length of a cell, often in metres. When the cell size, orientation and real-world coordinates of the upper-left corner of the raster are known, the real-world coordinates of a cell can be calculated; hence the location of a geographical feature on the Earth's surface can be derived. Topological relationships are not explicitly represented in the raster data model, but they can be inferred based on the cell locations. For example, if we know the row and column numbers of a cell, its neighbouring cells can be easily identified.

The cell size is the spatial resolution of a raster. It determines how accurate a raster representation is. As discussed in Section 2.1, the larger the cell size, the less accurate the raster representation. Figure 2.28 illustrates three raster representations of the same lake (containing a small island) with different cell sizes. It is obvious that a larger cell size or smaller spatial resolution will lead to loss of more details about the lake. In the raster representation with the smallest cell size, the island is captured, while it is lost in the other two raster representations with larger cell sizes. The shape of the lake is also severely distorted in the two raster representations with larger cell sizes. However, no matter how small the cell size is, each cell can only have one value. Therefore, spatial variations of the geographical feature in a cell are all lost. Each cell is considered homogeneous. When a cell contains two or more categories of features, two rules are often applied to assign cell values:

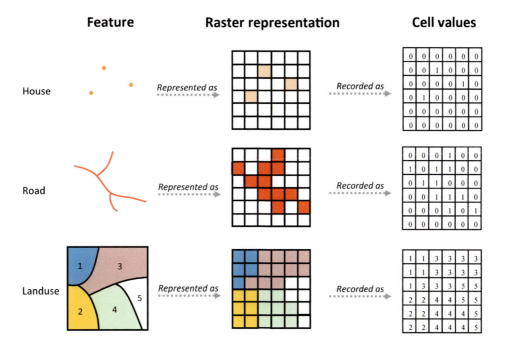

Figure 2.26 Spatial data describing geographical features with the raster data model.

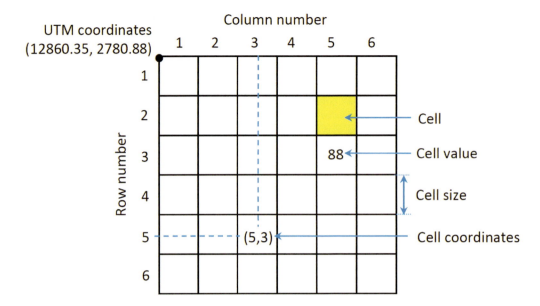

Figure 2.27 Elements of a raster.

the maximum area rule and the centre location rule. The maximum area rule assigns the cell value to the category which occupies the largest percentage of the cell's area. The centre location rule assigns the cell value to the category which is located at the centre of the cell. In Figure 2.28, the maximum area rule has been applied.

Spatial data represented in the raster data model are referred to as raster data. There are mainly three methods for structuring and storing raster data which have been used in GIS: cell-by-cell encoding, run-length encoding and quadtree.

Cell-by-cell encoding

The simplest raster data structure is a data array storing the matrix of cell values in a single file. In other words, all cell values are recorded and stored by row and column. This method is called cell-by-cell encoding. The data file usually has a header describing the number of rows

(nrows) and columns (ncols), cell size, real-world coordinates of the lower-left corner of the raster (xllcorner and yllcorner) and a NoData value (indicating that no measurements have been taken for that cell), as shown in Figure 2.29. Digital elevation models (DEMs – rasters whose cell values are elevation values) are generally stored using this simple data structure. Most GIS software systems supporting the raster data model, such as ArcGIS, GRASS and IDRISI, can use the structure to store raster data. However, this raster data structure requires every individual cell value to be stored. If a raster representing the distribution of forty types of soil in a region has m rows and n columns, it has to store $m \times n$ cell values. Using a raster of the same size to represent a single lake in the same region will entail storing the same amount of data. The data structure in many cases contains a lot of redundant data, which may lead to a large file size. The following two raster data structures may overcome the data redundancy issue.

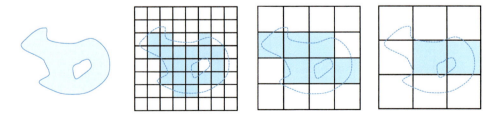

Figure 2.28 Raster representations with different cell sizes.

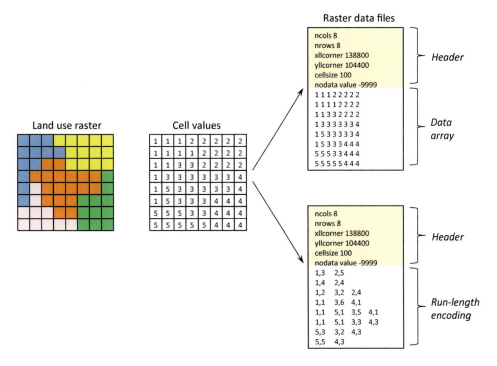

Figure 2.29 Cell-by-cell and run-length encoding.

Run-length encoding

Run-length encoding groups sequences of consecutive cells with the same value from left to right, and stores every sequence as a single data value and count on a row-by-row basis. For example, in the file containing the run-length encoding data in Figure 2.29, the first line in the data part contains two pairs of data, which indicate that the first row of the raster starts with three cells with the value of 1, followed by five cells with the value of 2. The second line of the data shows that the second row of the raster starts with four cells with the value of 1, followed by four cells with the value of 2, and so on. Run-length encoding does not require every individual cell value to be stored, thus compressing the raster data. However, its effectiveness is reduced when there is a high degree of variation in adjacent cell values, such as in a DEM. In such cases, the cell-by-cell encoding method that preserves all cell values is more suitable. ESRI Grid, supported by ArcGIS, is a raster data format using run-length encoding (see Box 2.5). GRASS and IDRISI also support this data structure.

Box 2.5 ESRI grid   **TECHNICAL**

The grid is the native raster data format of ArcGIS. It uses the cell-by-cell or run-length encoding method to store raster data depending on which is more efficient. A grid whose cell values are integer numbers (without a fractional or decimal component, for example 24 and 80) is called an integer grid. Integer grids are mainly used to represent categorical or discrete data. A grid is called a floating-point grid if its cell values are floating-point numbers (with a fractional or decimal component, for example 24.4 and 80.588). Floating-point grids are primarily used to represent continuous data. Integer grids usually have associated with them a value attribute table (VAT), while floating-point grids do not.

(continued)

(continued)

A VAT typically contains cell values that define a class, group or category. It contains a minimum of three fields: OID, VALUE and COUNT. Figure 2.30 illustrates a soil grid with VAT. OID is a unique, system-defined object ID for each row in the table. VALUE stores each unique cell value in the raster. COUNT represents the number of cells in the raster with the cell value in the VALUE field. For example, the first row in the VAT shows VALUE = 1 and COUNT = 7294, indicating that 7,294 cells have a value of 1 representing bedrock in the grid. Cells without measurements or actual values are assigned NoData. NoData is a special, system-defined value, which is not 0 (zero). 0 is a valid cell value. NoData cells are not represented in the VAT.

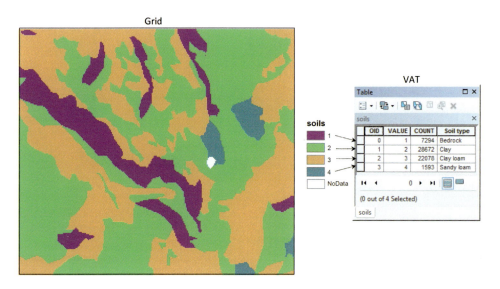

Figure 2.30 Grid with VAT.

The coordinate system of a grid is defined by the cell size, the number of rows and columns and the *x*, *y* coordinate of the upper-left corner. The rows and columns are parallel to the *x* and *y* axes of the coordinate system.

A grid, like a coverage, is stored as a directory named as the grid name, with associated tables and files that contain specific information about the grid. A grid directory basically contains the BND table (dblbnd.adf), which stores its boundary; the HDR file (hdr.adf), which contains the cell size and other specific information about the grid; the STA table (sta.adf), which stores statistics for the grid; the VAT table (vat.adf) if it is an integer grid; the LOG file, which records the operations that has been performed on the grid; and the data file (w001001.adf), which stores the cell data and the accompanying index file (w001001x.adf). When a grid is modified, the data stored in the files and tables are updated immediately.

Quadtree

Quadtree is a tree data structure which partitions a raster by recursively subdividing it into four quadrants, subquadrants and so on until each quadrant contains cells with the same value. Figure 2.31 illustrates a quadtree constructed for storing a raster representing a water body (in blue) surrounding by land (in white). The root (top) node of the tree represents the entire raster. A non-leaf node represents a quadrant with multiple cell values and is subdivided into four subquadrants. Leaf nodes correspond to those quadrants which have a single cell value and for which no further subdivision is required.

Many indexing schemes have been developed to identify quadrants. The example in Figure 2.31 is only one of them. With this scheme, the northwest quadrant is numbered as 0, the northeast quadrant as 1, the southwest quadrant as 2 and the southeast quadrant as 3. The quadrant numbers are used as the codes or indices for identifying their positions. Each quadrant is partitioned into four subquadrants, which are coded using their parent quadrant number appended with their own quadrant number. For example, the code 31 represents the second-level northeast quadrant (code 1) as a subquadrant of the first-level southeast quadrant (code 3). This hierarchical indexing system allows the location of each quadrant to be calculated relative to the origin of the raster and facilitates the determination of neighbourhood information (Lo and Yeung 2007).

Quadtree is best suited for areas where the data are relatively homogeneous and for applications that require frequent search of the database. Data storage, indexing and search techniques by quadtree have been well researched and documented (Samet 1990a, 1990b). The quadtree data structure was used in SPANS GIS (Ebdon 1992), which is now incorporated into PCI's Geomatica.

Object-oriented data model

Similar to the vector data model, the object-oriented data model conforms to the object view of geographical space. It consists of several object-oriented concepts, including object and object identifier, attributes and methods, class, class hierarchy and inheritance.

With an object-oriented data model, any geographical entity is uniformly modelled as an object associated with a unique identifier. The object identifier (OID) distinguishes the object from every other object in the system. It is independent of the values of its attributes and is invisible to the user. Every object has a state (the set of values for the attributes of the object) and a behaviour (the set of methods or functions which operate on the state of the object). For example, a road is an object which has a set of attributes including geometry (describing its location and shape), topology (describing its connectivity with other roads), surface materials and access status. For a particular road, the values of the geometry, topology,

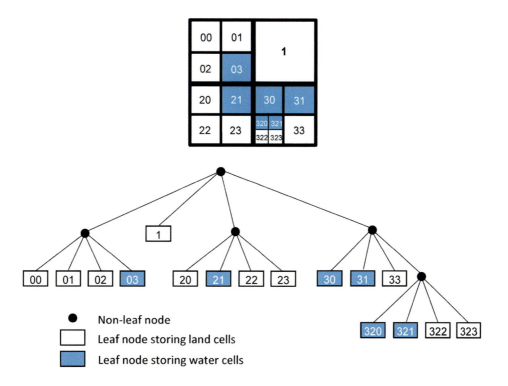

Figure 2.31 Quadtree.

surface materials and access status represent its state. Therefore, unlike the vector data model discussed above, geometry and topology are treated as attributes in an object-oriented data model, and in some systems they are stored as a binary large object (BLOB – a data type used to store a large collection of binary data as a single entity in a database). Methods define the behaviour of the object. They can be used to change the object's state by modifying its attribute values, or to query the value of selected attributes. For example, we may have methods to draw the road on a computer screen using a line symbol, determine its centre line, calculate its length, and find other roads it intersects based on the state of the road. The state and behaviour are encapsulated in an object, and can be accessed or invoked from outside the object only through explicit message passing. A message is simply a request from one object to another object asking the second object to execute one of its methods. It is the means by which objects communicate.

Every object belongs to a particular class. All the objects in a class share the same set of attributes and methods and respond to the same messages. An object must belong to only one class. For example, all road objects would be described by a single Road class. The objects in a class are called instances of the class. Each

instance has its own value for each attribute, but shares the same attribute names and methods with other instances of the class, as shown in Figure 2.32.

In addition, a new class (subclass) can be derived from an existing class (superclass). The subclass inherits all the attributes and methods of the superclass, and may have additional attributes and methods. Classes can be organised into a class hierarchy where superclasses are at the top and subclasses branch off to the bottom. In this way, inheritance can take place, where superclass methods can be defined and then used in all subclass objects.

Figure 2.33 gives an example in which GeoEntity is the superclass and Point, Line and Area are its subclasses. The class GeoEntity has three attributes – name, coordinates of the centroid and georeferencing system – and seven methods, including draw, delete, move, getCentroid, getColour, setColour and getSize. The Point, Line and Area classes inherit the attributes and methods from GeoEntity, but define their own attributes and methods. The Line and Area classes add three attributes: the number of point coordinates, the array of x coordinates and the array of y coordinates. The two classes have a new method, getBound, for getting the maximum and minimum values of x and y of a line or area object. The Area class has another new method, contains, which is

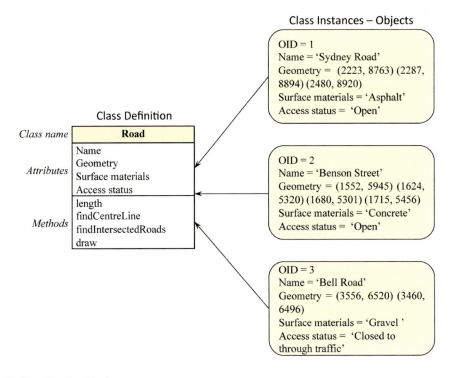

Figure 2.32 Class Road and its instances.

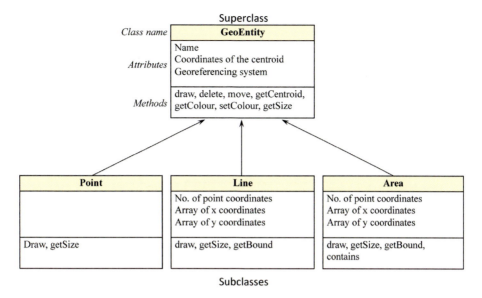

Figure 2.33 Inheritance and method overloading.

used to determine whether a particular point is located within an area object. The three subclasses redefine the draw and getSize methods. The Point class's draw method may draw a point object with a point symbol, and its getSize method may return a value of 0. The draw method of the Line class may show a line object with a line symbol of a certain width, and its getSize method measures the length of the line object. For an area object, the Area class's draw method depicts it using an area symbol, and its getSize method calculates the area size of the object. This ability of one method to perform different tasks with different input and output is called method overloading.

The object-oriented data model stores spatial data (including geometry, topology and attributes) as objects. An object encapsulates both state and behaviour, and can store all the relationships it has with other objects. Each object is defined as a separate component or building block, which can be used with other objects to form complex objects. These features allow users to create their own object-oriented models that extend a base model according to their own view and classification of the geographical features they study. In addition, they allow spatial data to be manipulated directly using an object-oriented programming language, which has become the preferred programming technique used by GIS developers, largely due to its superior efficiency, flexibility and its ability to unify the application and database development into a seamless data model and programming language environment. The geodatabase data model developed by ESRI is an example of the object-oriented data model (see Box 2.6). The object-oriented data model has also been used in SmallWorld GIS, Oracle Spatial and IBM DB2 Spatial Extender.

Box 2.6   The geodatabase data model   **TECHNICAL**

Geodatabase is an object-oriented data model created by ESRI for representing real-world geographical features and storing spatial data, as part of ArcObjects, the underlying component technology for ArcGIS products (Zeiler 2010). As shown in Figure 2.34, the base model of the geodatabase contains the following key classes:

(continued)

(continued)

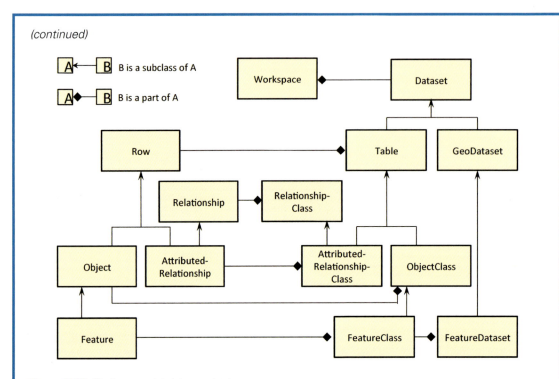

Figure 2.34 The base model of the geodatabase.

- Workspace – a workspace corresponds to a geodatabase, an ArcInfo coverage workspace or a folder with shapefiles. This class is a container of Dataset objects representing spatial and non-spatial datasets, and provides methods to instantiate existing datasets and to create new datasets. A workspace may contain zero or several datasets.
- Dataset – this is a collection of data in a workspace. This class is the highest-level data container, whose instances are collections of different types of datasets.
- GeoDataset – this is a subclass of the Dataset class, representing a type of dataset that contains spatial data.
- FeatureDataset – this feature dataset stores vector data in feature classes. It is the subclass of the GeoDataset class. Its instances are collections of feature classes with the common georeferencing systems.
- FeatureClass – generally, feature classes are collections of points, lines or polygons. Feature classes which store topological features, for example those representing networks consisting of a set of connected line and point features along with connectivity rules, must be contained within a feature dataset to ensure a common georeferencing system. Feature classes that store simple features can be organised either inside or outside a feature dataset. Those outside a feature dataset are called stand-alone feature classes, which are subclasses of ObjectClass.
- Table – a table is a dataset. It is a collection of rows that have attributes stored in columns, referred to fields. This class is a Dataset object. A Table object represents an ObjectClass or an AttributedRelationshipClass in a geodatabase.
- Row – this is a record in a table. A Row object is an instantiated object that represents a persistent row in a Table object. A Row has a set of Fields. All rows in a table share the same set of fields. The class Row is a member of the class Table.

- ObjectClass – this is a table whose rows represent entities, modelled as objects with states and behaviours. It may contain a subtype field which can be used to classify its instances into a number of subtypes. All subtypes share the same field definition and are stored in the same table. Subtyping is an alternative to creating multiple subclasses, each represented by its own object class. It can also be used in defining attribute and connectivity rules that apply to the instances of ObjectClass.

A FeatureClass is an ObjectClass whose objects are features. A feature class has one distinguished field of type Geometry, referred to as the Shape field. The Shape field stores the geometry for the features in the FeatureClass. Figure 2.35 is an example of a polygon feature class stored as a table. Each row represents one polygon feature. In this polygon feature class table, the Shape field holds the polygon geometry for each feature. The value Polygon indicates that the field contains the coordinates and geometry that defines one polygon in each row.

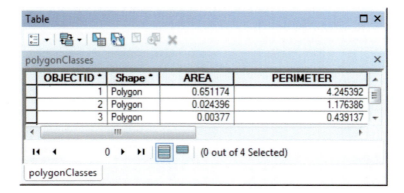

Figure 2.35 Example of a polygon feature class table.

- Object – this is a row with a non-negative object class ID (OBJECTID) which is unique within the geodatabase as shown in Figure 2.35. It is a subclass of the Row class and a member of ObjectClass.
- Feature – this is a spatial object with a geometric shape. This class is a subclass of the Object class. It is also a member of FeatureClass. The type of shape of a feature is defined by the feature class, which could be Point, Multipoints (features that are composed of more than one point), Polyline or Polygon. A feature is stored as a row in the feature class table, as shown in Figure 2.35.
- RelationshipClass – this represents an association between two object classes/tables.
- Relationship – this class represents a set of relationships between the objects belonging to two classes/tables (one is the origin class/table and the other the destination class/table). It is a member of RelationshipClass. There are two types of relationship: simple relationship and attributed relationship. Simple relationship represents one-to-one or one-to-many relationships between objects via a field in the destination table that uniquely identifies a row of the origin table (Figure 2.36a). Attributed relationship represents one-to-one or one-to-many

(continued)

(continued)

relationships between objects using a table known as a relationship table, in which one field uniquely identifies a row of the origin table, and the other uniquely identifies a row of the destination table (Figure 2.36b).

- AttributedRelationshipClass – this is a subclass of RelationshipClass and also a relationship table, as shown in Figure 2.36.
- AttributedRelationship – this is a member of AttributedRelationshipClass, and also a subclass of Row and Relationship classes. It is a kind of row that represents a pair of related objects or features (see Figure 2.36).

Geodatabase provides hundreds of other objects for building, modelling and structuring spatial data. By storing feature classes within a feature dataset, geodatabase allows users to build spatial datasets with topological structures and add behaviour to them with no programming required. Such datasets include:

- Geometric networks – these consist of a set of connected edges (lines) and junctions (points) along with connectivity rules; representing and modelling the behaviour of utility networks such as sewage, water and gas pipelines.
- Network datasets – these consist of a set of connected edges and junctions as well as turn features along with connectivity rules; representing and modelling the behaviour of transportation networks.
- Terrain – this is represented and structured in TIN.
- Cadastral Fabric – this consists of connected parcel features that represent the record of survey for an area of land.

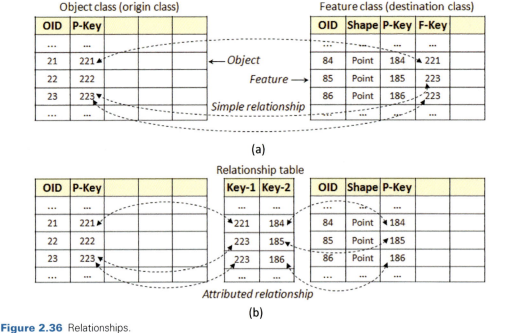

Figure 2.36 Relationships.

Topological rules can also be defined and applied within a feature class or between two or more feature classes: for example, buildings (one feature class) must not overlap with easements (another feature class) (ESRI 2003).

With the geodatabase data model, real-world objects can be defined with different attributes and behaviours. A number of application-specific data models have been developed by expanding the base model of the geodatabase. They currently cover the areas of agriculture, atmospheric science, biodiversity, energy utilities, forestry, geology, groundwater, hydrology, petroleum and so on. Their specifications are available at http://support.esri.com/en/knowledgebase/techarticles/detail/40585.

2.4 SPATIAL DATABASE MANAGEMENT

A collection of spatial data represented and structured in one or more spatial data models described above constitutes a spatial database. According to how it is organised and managed, there are two types of spatial databases that are predominantly used in GIS: geo-relational databases and object-relational databases. They are based on the concepts and principles of relational databases.

Relational databases

A relational database organises data as a collection of interrelated tables. Each table is called a relation, which contains data about a set of objects (for example, forest stands) with the same attributes or properties. Rows (also called tuples or records) in a relation correspond to individual objects (for example, one forest stand per row), and columns (also called fields) correspond to attributes (for example, the age, dominant species, average height, average diameter at breast height, soil type and site quality of a forest stand). The set of values the attributes are allowed to take is called a domain. Domains may be distinct for each attribute, or two or more attributes may have the same domain. Rows and columns can appear in any order and the relation will still be the same. However, an attribute must be unique in a relation, and the rows cannot be identical. Also, each row can only contain a single value for each of its attributes.

A relational database contains multiple relations. A relation can have a designated single attribute or a set of attributes which can be used to uniquely identify each row in the relation. This attribute or set of attributes is called a primary key. If one or more attributes in a relation represents a primary key in another relation, it is called a foreign key. Any attribute can be a key, or multiple attributes can be grouped together into a compound key. It is not necessary to define all the keys in advance. An attribute can be used as a key even if it was not originally designated. Figure 2.37 shows two tables representing the stand and drainage relations. StandID is the primary key of the stand relation and DCode is the primary key of the drainage relation. DCode is also a foreign key in the stand relation. It links information on soil drainage conditions to the details of forest stands, which

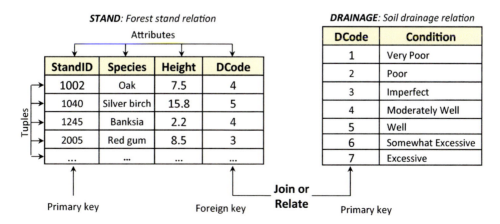

Figure 2.37 Key concepts relational databases.

allows us to match each stand to its drainage condition. When an attribute appears in more than one relation, its appearance usually represents a relationship between tuples of the two relations, just like DCode in the stand and drainage relations.

Keys are commonly used to join or relate data from two or more relations. A join operation in the relational database system combines two or more relations to form a new relation, while a relate operation links relations without forming a new relation. The ability to join or relate tables through the use of keys is the essence of a relational database. Such relational joins form the basis for database querying and fast retrieval of data from large tables.

Relational databases use SQL (Structured Query Language) to retrieve, combine, update and manage data (Gillenson 2012). The general form of SQL for simple database query follows a `SELECT * FROM <table_name> WHERE` clause. For example, the following SQL statement performs a join operation on the two relations in Figure 2.37 to combine the data from both the stand and drainage relations:

```
SELECT StandID, Species, Height,
Condition
FROM stand, drainage
WHERE stand.DCode = drainage.DCode
```

The SQL statement lists StandID, Species, Height and Condition as attributes, and includes only those rows from both tables that have identical values in the DCode columns. The result is shown in Table 2.4.

SQL is the standard database language widely implemented in relational database management systems (RDBMSs). Most GIS systems use built-in or external RDBMSs to manage attribute data, but often users need not know any SQL to perform database query and manipulation tasks. Query builders in GIS generally provide intuitive interfaces that make most queries simple (see Section 4.2).

Geo-relational databases

Relational databases allow attribute data in GIS to be organised in a network of interrelated tables, each having its own attributes specific to the type of data being stored. However, it is inefficient to store geographical objects as rows and their coordinates and other geometric properties in columns in the form of tables. A relational database requires each row of a table to be composed of the same attributes and the intersection of a row and a column to be a single value (called the first normal form constraint). It is very rare to use the same number of points to approximate the shape of different geographical objects. Some geographical objects may be approximated using fewer than ten points, while others may use a few hundred or more points. Therefore, it is difficult to define a fixed number of columns for a table to store coordinates. In addition, the table structure is too restrictive for many geographical features that have complex structures, such as a polygon with island polygons inside. Decomposing a complex geographical object into simple geometric objects that can be stored in tables will lead to many relations and unnatural joins, which, as mentioned above, is inefficient. Due to these weaknesses of a relational database, GIS systems generally store and manage spatial data in geo-relational databases, in which the attribute component of spatial data is stored in a conventional relational database managed by an RDBMS, and the locational component (describing feature geometry) of spatial data is stored in separate files based on one of the spatial data structures discussed in the previous section and managed by specialised GIS data management software (Yeung and Hall 2007).

In a geo-relational database, spatial data are conceptualised as a series of independent data layers covering the same geographical space (Figure 2.38). Each layer represents a single theme of a particular feature class, such as soil, land use, geology and vegetation. Each layer is a dataset, consisting of files storing geometric data defining the positions and shapes of geographical features, and relational tables containing attribute data associated with the geographical features.

Table 2.4 Result table for the join operation with SQL on the two relations in Figure 2.37

StandID	Species	Height	Condition
1002	Oak	7.5	Moderately well
1040	Silver birch	15.8	Well
1245	Banksia	2.2	Moderately well
2005	Red gum	8.5	Imperfect
. . .	. . .	. . .	. . .

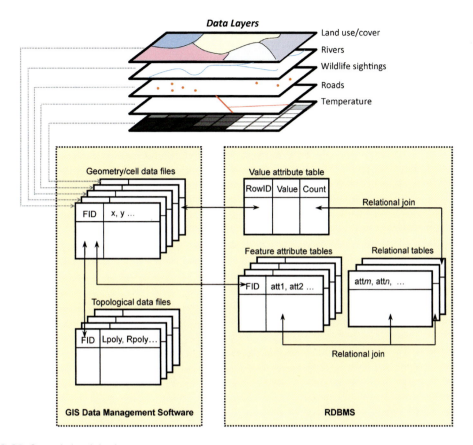

Figure 2.38 Geo-relational database.

When the spatial data are structured using a vector data structure, each feature is assigned a system-generated unique ID number called a feature ID, or FID. The attribute table associated with a vector data layer is called the feature attribute table. The FID is the primary key of a layer's feature attribute table, which is used internally by the GIS system to link a feature's locational data with its attributes. When the spatial data are structured using a raster data structure, the cell data are stored separately from the attribute data associated with each unique cell value, which are stored in a table, often called the value attribute table. A value attribute table uses the unique cell value as the primary key to link it with the associated raster data layer. ArcInfo coverages, shapefiles and grids are all managed with the geo-relational database model in ArcGIS. Many existing spatial databases are geo-relational databases. Box 2.7 provides such an example.

Box 2.7 African Water Resource Database   **CASE STUDY**

The African Water Resource Database (AWRD) was developed by the Food and Agriculture Organization (FAO) of the United Nations to facilitate responsible inland aquatic resource management, planning and development with an overarching goal of promoting food security. It was designed based on recommendations from the Committee on Inland Fisheries and Aquaculture for Africa. The database is both

(continued)

(continued)

an expansion and an update of an earlier project, the Southern African Development Community Water Resource Database, led by the Aquatic Resource Management for Local Community Development Programme.

AWRD is a spatial database developed using ArcView GIS. It consists of twenty-eight thematic map data layers drawn from over twenty-five data sources, containing 156 datasets. The core data layers include surface water bodies, watersheds, aquatic species (766 species), rivers, climate, political boundaries, population density, soils, satellite imagery and other physiographic data. The vector data layers are all in shapefiles. The raster data layers are mostly in ESRI Grid. Figure 2.39 shows the distribution of *Astatotilapia calliptera* and the catchments with the threatened and endangered aquatic species in Southern Africa mapped using the data from AWRD.

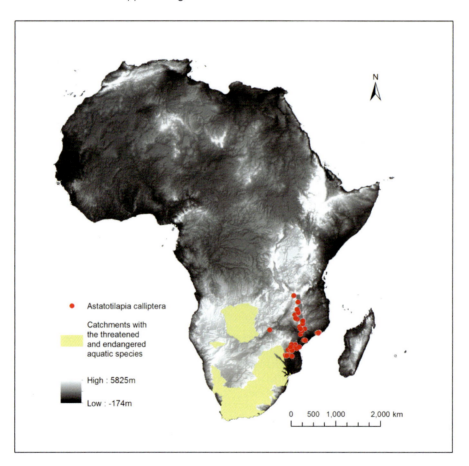

Figure 2.39 The distribution of *Astatotilapia calliptera* in Southern Africa.

AWRD also includes custom applications and tools which allow users to integrate the data and perform analyses. The database can be used to address key inland aquatic resource management issues, such as the status of fishery resources and transboundary movements of aquatic species. It has been applied to create surface water body and inland fishery habitat inventories, predict potential fish yield, perform preliminary hydrological analyses, examine the potential distribution of exotic fish species, analyse the potential spread of invasive fish species, assess the potential impact of invasive and introduced aquatic species on vulnerable or endangered fish species, and to predict the potential distribution of fish species within preferred spawning temperatures and flow regimes.

The database with the interface, toolsets and data are distributed on two DVDs accompanying the technical manual and workbook. Individual data layers can be downloaded through the GeoNetwork (http://www.fao.org/geonetwork/srv/en/main.home) or GISFish portals (http://www.fao.org/fi/gisfish).

Source: Jenness et al. (2007).

Object-relational databases

Object-relational databases extend relational databases to allow objects to be stored in the columns of relational tables. They generally support the ability to declare a new data type that is basically a class, and to set this new type as the data type for a table column. From an object-oriented point of view, the new data types correspond to class names, which are software modules that encapsulate state and behaviour.

When spatial data are stored in an object-relational database, feature attributes are stored in columns as traditional tabular attributes, while feature geometries represented as either vector features or rasters are stored using an extended spatial type. For example, the geodatabase data model presented in Box 2.6 is used to design and

manage object-relational databases in ArcGIS. In such an object-relational database, a feature class is stored as a table. Figure 2.35 illustrates what an object-relational table storing a feature class looks like. Each row in the feature class table represents one feature. The shape column, which is a spatial type (here polygon), stores the coordinates and geometry that define one feature in each row. Therefore, this column does not store single values, but has internal structure, which is encapsulated in binary large objects. The BLOB is an unconventional data type that contains binary information representing an image, a video, a procedure or any large unstructured object. In an object-relational database in GIS, it is often used to store a raster or binary data about the geometry or locational component of a data layer as an object. Figure 2.40 shows

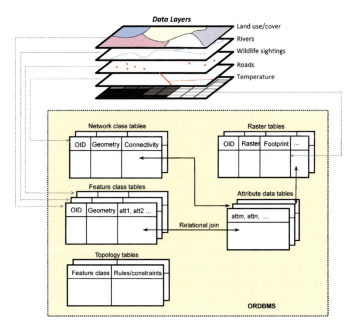

Figure 2.40 Object-relational database.

the general structure of an object-relational database, in which spatial data are stored using tables and managed by the object-relational database management system (ORDBMS).

SQL has been extended to incorporate spatial types and functions (Shekhar and Chawla 2003). Therefore, the contents of relational tables containing feature geometries, when stored as an SQL spatial type, can be accessed through SQL. However, different GIS systems may employ specific database management systems, which may implement SQL differently, and may have different definitions of spatial data types and different storage structures and access methods. A GIS system may also support different types of databases. Selecting the appropriate types of databases to work with will depend on the specific requirements of the GIS project and/or application. Box 2.8 describes four types of spatial databases in ArcGIS.

Box 2.8 Types of spatial databases in ArcGIS   **TECHNICAL**

ArcGIS supports four types of spatial databases: geo-relational databases, file geodatabases, personal geodatabases and ArcSDE geodatabases. Figure 2.41 provides a folder-tree view of examples of the first three types of spatial databases displayed in ArcCatolog.

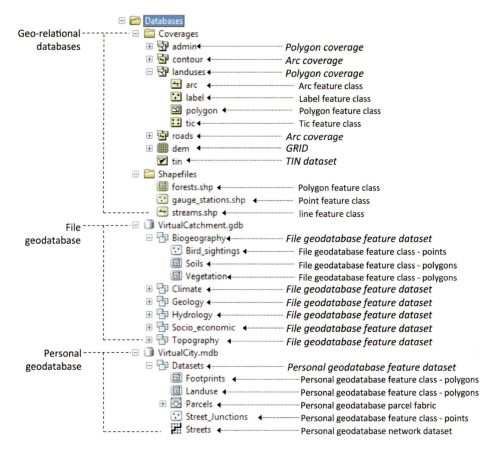

Figure 2.41 A folder-tree view of spatial databases displayed in ArcCatalog.

Geo-relational databases

A geo-relational database in ArcGIS is a workspace containing ArcInfo coverage (including TIN) and grid files, or a collection of shapefiles held in a file system folder. A shapefile contains a single feature class, and a coverage may contain multiple feature classes. All the files are managed by the computer's file management system, but some file management functions, including deleting, renaming, copying, pasting and creating a shapefile or coverage, should be conducted in ArcCatalog. The locational or geometric data and attribute tables of a shapefile or a coverage or a grid are linked, related or joined in ArcGIS via its data management software tools. Geo-relational databases were conventional spatial databases in ArcGIS, and have served the purposes of many environmental GIS applications very well where spatial data management is not a central task.

Figure 2.41 shows two example geo-relational databases, one named Coverages and the other Shapefiles, which appear in ArcCatalog with a folder icon 📁 . The Coverages database contains six data layers: admin, contour, landuses, roads, dem and tin. Different types of coverages and coverage features are represented with distinctive icons. The features classes included in the coverage – landuses – are shown. Note that the landuses coverage contains polygon, arc and label features and the tic feature class (control points for georeferencing). The actual files storing coverages as listed in Figure 2.22 and Table 2.3 are invisible in this view. The Shapefiles database contains three layers: forests, gauge_stations and streams. Only .shp files appear in ArcCatalog, and other files associated with each shapefile, as listed in Box 2.3, are not shown. Shapefiles of different feature classes are represented with different icons.

File geodatabases

A file geodatabase is a collection of spatial datasets of various types (feature classes, topologies, networks, tables, raster datasets and so on) held in a file folder structured based on the geodatabase object-oriented data model and managed by the computer's file management system and ArcGIS's own data management system. Each dataset is stored as a separate file on disk, which can contain more than 300 million features and be up to 1 terabyte in size. It uses a more efficient data structure than shapefiles and coverages in terms of performance and storage. File geodatabases are designed for a single user, and recommended for GIS applications involving very large datasets and requiring fast performance. A file geodatabase is named with the .gdb extension and appears in ArcCatalog with a database icon 🗄 .

An example of a file geodatabase called VirtualCatchment is shown in Figure 2.41, which contains six feature datasets, each consisting of a group of feature classes with the common georeferencing system. For example, Biogeography is a geodatabase feature dataset. It consists of three geodatabase feature classes: Bird_sightings (point feature class), Soils and Vegetation (polygon feature class). Icons are used to differentiate different types of feature classes in ArcCatolog, and actual data files are not visible.

Personal geodatabases

A personal geodatabase is based on the geodatabase object-oriented data model, but stored and managed using Microsoft Access – an RDBMS. All the datasets in a personal geodatabase are held in a single Microsoft Access file. However, the geodatabase is limited to 2GB in size. Personal geodatabases are designed for a single user working with smaller spatial datasets, and only run on Microsoft Windows. A personal geodatabase is named with the .mdb extension and appears in ArcCatalog with a database icon 🗄 .

Figure 2.41 provides an example of a personal geodatabase named VirtualCity. It has one feature dataset, containing five feature classes and datasets. Among them, Parcels is a cadastral or parcel fabric dataset, and Streets is a network dataset (see Box 2.6).

(continued)

(continued)

ArcSDE geodatabases

ArcSDE geodatabases are large, multi-user enterprise spatial databases. They are based on the geo-database object-oriented data model and managed using an ORDBMS such as Oracle, SQL Server, PostgreSQL, Informix or DB2. ArcSDE geodatabases can be used to manage extremely large spatial datasets, primarily used for data-centric spatial database applications in departmental and enterprise settings.

2.5 SUMMARY

- Spatial data have three components: location, attribute and time. The locations recorded in spatial data are defined according to a particular georeferencing (coordinate) system that has a fixed relationship to the Earth's surface and takes the Earth's shape into account.
- Geographical space can be conceptualised in the object or field view. In the object view, the geographical space consists of a collection of discrete objects or entities. Each object can be accurately located and has a well-defined boundary. In the field view, the geographical space constitutes continuous surfaces representing variables whose values change continuously across the space.
- Map scale determines the amount of detail about geographical features captured and represented in spatial data.
- Geographical features represented in spatial data are selected and generalised through cartographic generalisation. The methods of cartographic generalisation include selection, simplification, exaggeration, combination, displacement and classification.
- Geographical coordinate systems uniquely define locations of geographical features on the surface of the Earth using latitude and longitude. They are based on a specific datum.
- Map projections transform the Earth's spherical surface to a flat surface, which converts the latitudes and longitudes of locations on the Earth's surface into locations on a plane. This process always produces some distortions in shape, distance, direction, scale and area. There are three families of map projection: conic, cylindrical and azimuthal.
- Projected coordinate systems are plane coordinate systems established based on a particular map projection. The UTM system, based on the cylindrical transverse Mercator projection, is a popular projected coordinate system widely used for large-scale topographic maps, remote sensing images and GIS databases.
- A spatial data model defines how spatial data represent geographical features, and how spatial data are structured, connected, organised and stored inside a GIS system. Three types of spatial data models have evolved: vector, raster and object-oriented. The majority of the existing GIS systems are based on either vector or raster models, or both. The vector and object-oriented models conform to the object view of geographical space, while the raster model is based on the field view.
- A spatial database is a shared collection of logically related spatial data structured and organised based on one or several spatial data models. Geo-relational and object-relational databases are two typical types of spatial databases. A spatial database is designed to meet the information needs of an organisation.

REVIEW QUESTIONS

1. Discuss the nature of spatial data.
2. What is the object view of geographical space? How does vector representation construct point, line and area features in the computer?
3. What is the field view of geographical space? How does raster representation represent geographical phenomena?
4. Why do all spatial data in an environmental project have to be georeferenced?
5. What is map scale? What impact does map scale have on how we model geographical space using maps?
6. Discuss why spatial data cannot perfectly represent geographical features, and what implications this has for users of spatial data.
7. What is a map projection? Describe the characteristics of the three families of map projection.

8. What are eastings and northings?
9. Describe the UTM zoning system.
10. What is a spatial data model? Outline the importance of spatial data models in GIS.
11. What are vector and raster data models? Discuss their advantages and disadvantages.
12. Discuss the main characteristics of geo-relational and object-relational databases.

REFERENCES

Corbett, J.P. (1979) *Topological Principles in Cartography*, Technical Paper 48, Washington, DC: United States Department of Commerce, Bureau of the Census.

de Berg, M., Cheong, O., van Kreveld, M. and Overmars, M. (2008) *Computational Geometry: Algorithms and Applications*, 3rd edn, Berlin, Germany: Springer.

Ebdon, D. (1992) 'SPANS: a quadtree-based GIS', *Computers and Geosciences*, 18(4): 471–475.

ESRI (1998) *ESRI Shapefile Technical Description: An ESRI White Paper*, Redlands, CA: ESRI Press.

ESRI (2003) *ArcGIS: Working with Geodatabase Topology – An ESRI White Paper*, Redlands, CA: ESRI Press.

Gillenson, M.L. (2012) *Fundamentals of Database Management Systems*, 2nd edn, Hoboken, NJ: Wiley.

Haining, R. (2009) 'The special nature of spatial data', in A. Fotheringham and P. Rogerson (eds) *The Sage Handbook of Spatial Analysis*, Los Angeles, CA: Sage, 5–23.

Jenness, J., Dooley, J., Aguilar-Manjarrez, J. and Riva, C. (2007) *African Water Resource Database, GIS-Based Tools for Inland Aquatic Resource Management, 1. Concepts and Application Case Studies*, CIFA Technical Paper no. 33, Part 1, Rome, Italy: FAO.

Lo, C.P. and Yeung, A.K.W (2007) *Concepts and Techniques in Geographic Information Systems*, 2nd edn, Upper Saddle River, NJ: Pearson Prentice Hall.

Maher, M.M. (2010) *Lining up Data in ArcGIS: A Guide to Map Projections*, Redlands, CA: ESRI Press.

Marx, R.W. (1990) 'The TIGER system: automating the geographic structure of the United States census', in D.J. Peuquet and D.F. Marble (eds) *Introductory Readings in Geographic Information Systems*, London: Taylor and Francis: 120–141.

Peucker, T.K. and Chrisman, N. (1975) 'Cartographic data structure', *The American Cartographer*, 2(1): 55–69.

Peuquet, D.J. (1984) 'A conceptual framework and comparison of spatial data models', *Cartographica*, 21: 66–113.

Robinson, A.H., Morrison, J.L., Muehrcke, P.C., Kimerling, A.J. and Guptill, S.C. (1995) *Elements of Cartography*, 6th edn, New York: Wiley.

Samet, H. (1990a) *Applications of Spatial Data Structures: Computer Graphics, Image Processing and Geographical Information Systems*, Reading, MA: Addison-Wesley.

Samet, H. (1990b) *Design and Analysis of Spatial Data Structures*, Reading, MA: Addison-Wesley.

Shekhar, S. and Chawla, S. (2003) *Spatial Databases: A Tour*, Upper Saddle River, NJ: Prentice Hall.

Theobald, D.M. (2001) 'Topology revisited: representing spatial relations', *International Journal of Geographical Information Science*, 15(8): 689–705.

Torge, W. and Müller, J. (2012) *Geodesy*, 4th edn, Berlin, Germany: De Gruyter.

Yeung, A.K.W. and Hall, G.B. (2007) *Spatial Database Systems: Design, Implementation and Project Management*, Dordrecht, The Netherlands: Springer.

Zeiler, M. (2010) *Modeling Our World: The ESRI Guide to Geodatabase Concepts*, 2nd edn, Redlands, CA: ESRI Press.

CHAPTER THREE

Spatial data input and manipulation

3

This chapter introduces the major means of spatial data capture and input, outlines the principles and methods of spatial data transfer, presents the methods and techniques for spatial data editing and transformation, and discusses the issues of spatial data quality and the concept of metadata.

LEARNING OBJECTIVES

After studying this chapter, you should be able to:

- understand different spatial data capture techniques;
- describe the principles and methods of spatial data transfer;
- recognise different spatial data formats and apply appropriate mechanisms for spatial data transfer;
- grasp basic skills for spatial data editing and manipulation;
- understand the concepts of spatial data quality;
- describe potential errors in spatial data.

3.1 SPATIAL DATA CAPTURE

GIS projects for environmental applications are as varied as the places which they are designed to investigate. We may want to trace pollution in a stream, plan a wildlife reserve or identify land use conflicts, for example. Whatever the case may be, we need to study information about a geographical location to gain a better understanding of that area's environment. The key to the success of such projects depends on having the necessary spatial data. The data needed for a project (such as vegetation, soils and species locations) often do not exist, or they are only available on paper and we need to create a digital version. We may have digital data, but they are not spatially enabled or they are not structured in the data model we need. Spatial data for use in GIS may come from different sources and can be collected through data capture or data transfer. Spatial data capture is the process of generating new digital spatial data. It involves direct measurement or analogue to digital conversion. Spatial data transfer is the process of converting or translating existing digital spatial data and importing them into a GIS. This section focuses on spatial data capture, and Section 3.2 will discuss spatial data transfer.

Vector data capture

There are five main methods for vector data capture depending on the source of data:

1. keyboard entry;
2. digitising;
3. field surveying;
4. photogrammetry;
5. LiDAR.

Keyboard entry

Keyboard entry is mainly used to input the coordinates of a number of point features into a GIS. It involves reading or obtaining the coordinates of point features from maps or other source documents, typing the coordinates into a file in a tabular format and reading the file into the GIS. In some GIS systems, point features are also called XY events. This method is used where the number of point features is small. Box 3.1 illustrates how a set of point features are added from a table of coordinates in ArcGIS.

ArcGIS allows the (*x*, *y*) coordinates of point features stored in tables to be imported and converted to a shapefile or geodatabase feature class. There are several table formats that ArcGIS can read directly, including Microsoft Excel files (.xls and .xlsx), dBase files (.dbf) and plain text files (.csv and .txt). Tables should contain at least an X field and a Y field, storing *x* and *y* coordinates respectively. When using Excel or dBase, the X and Y field should be formatted as NUMBER with a few of decimal places. Figure 3.1 shows an Excel table of wildlife sightings in a region with three fields, where each row represents a sighting site, and POINT_X and POINT_Y are X and Y fields representing UTM coordinates in metres. These coordinates were entered via keyboard and saved as **sightings.xlsx**. It can be found in **C:\Databases\GIS4EnvSci\others**. Here is the procedure for importing the table into ArcGIS and converting it into a shapefile or geodatabase feature class:

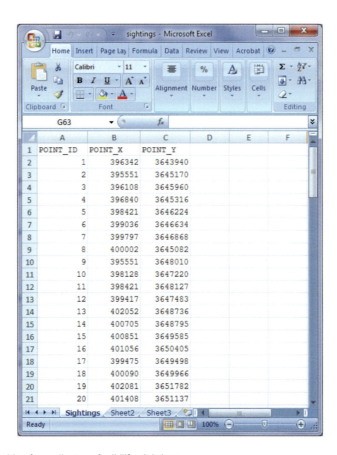

Figure 3.1 Excel table of coordinates of wildlife sightings.

1. Start ArcMap with a new, empty map.
2. In the main menu bar, click **File > Add Data > Add XY Data. . ..**
3. In the **Add XY Data** dialog box:

1) Click the **Browse** button 📂 to browse to the directory where the table resides, double-click the Excel spreadsheet, 'sightings.xlsx', and select the 'Sightings$' sheet containing the XY coordinates.
2) Click the **X Field** drop-down arrow and click 'POINT_X' (containing *x* values).
3) Click the **Y Field** drop-down arrow and click 'POINT_Y' (containing *y* values).
4) Click the **Edit** button to open the **Spatial Reference Properties** dialog box (see Figure 2.7).
5) In the **Spatial Reference Properties** dialog box, define the coordinate system as **WGS_1984_UTM_Zone_46N**. Click **OK**. The *x*, *y* coordinates will be automatically transformed to match the coordinate system of the data frame.
6) Click **OK**. The layer name **Sightings$Events** appears in the table of contents, and the points are drawn as an event theme in the data view.

4. Right-click the file name, select **Data > Export Data. . ..** The points can be exported to a geodatabase feature class or to a shapefile.

Digitising

Digitising is the process of interpreting and converting paper maps or images (aerial photos and satellite images) to digital vector data. It involves choosing geographical features from the map or image and manually tracing the features along their boundaries or shapes one by one using a digitiser while recording the coordinates of the points selected to approximate the boundary or shape of each feature. There are two digitising methods: tablet digitising and on-screen digitising.

Tablet digitising uses a digitising tablet to trace points, lines and polygons on a hard-copy map or image fixed on the tablet. A digitising tablet consists of an electronic tablet and a magnetic pen or puck (Figure 3.2). A puck is similar to a computer mouse, but it has a number of control buttons and a window with crosshairs for pinpoint placement of a point. A magnetic pen (also called a stylus) is similar to a simple ballpoint pen, but uses an electronic head. The tablet has a built-in electronic mesh that can detect movement of the puck, translate its position into digital signals, and send the digital signals to the attached computer. The units of measurement on the digitising tablet are in centimetres or inches. Tablet digitising results in strings of points with (*x*, *y*) values in the digitiser's unit of measurement. These (*x*, *y*) coordinates are later to be converted to real-world coordinates based on the georeferencing system of the map or image. This method was the most popular method of vector data capture from hard-copy maps or aerial photos in the past. It has now been gradually replaced by on-screen digitising and other more advanced methods.

On-screen digitising uses the computer mouse and digitising software to manually trace the features on a scanned map or image displayed on the computer screen to capture vector data. It removes the need for a digitising tablet. On-screen digitising is often called heads-up digitising, because the source document is viewed with the head up on the computer screen, rather than down on the digitising tablet. The source documents may be scanned images (from paper maps and aerial photographs) or digital aerial and satellite images. Most GIS software systems support on-screen digitising. It generally follows the steps below:

1. Scan the source document into a digital image if a paper map or aerial photo is used.
2. Add the image into a GIS.
3. Georeference the image.
4. Trace features on the image using the mouse.
5. Edit features.
6. Add attribute data.

Hard-copy documents are converted into digital images with a scanner. Three types of scanners can be used: flat-bed scanners, rotating drum scanners and large-format feed scanners. Flat-bed scanners are generally small and less accurate, mainly used for small format maps and aerial photos. Rotating drum and large-format feed scanners are more accurate, and are used for large-format maps and aerial photos. Scanners may output digital images in different resolutions. For on-screen digitising, a resolution of 300dpi (dots per inch) is mostly sufficient. Scanned maps do not normally contain georeference information. Therefore, after a scanned map or image is added to a GIS, it has to be georeferenced to the map projection or projected coordinate system of the source map or aerial photograph before digitising begins.

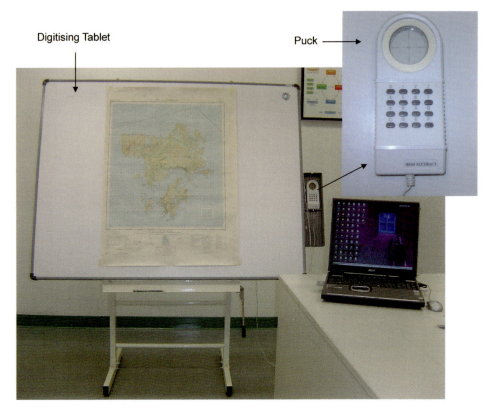

Figure 3.2 Digitising tablet (source: the author).

Georeferencing an image in GIS involves selecting at least three control points that can be accurately identified on the image with known real-world coordinates, using the selected control points to mathematically establish linear or non-linear relationships between image coordinates (rows and columns) and real-world coordinates (x and y coordinates in the projected coordinate system of the source document), and transforming the image so that each location on the image can be defined in real-world coordinates. This process is completed by using the georeferencing tools available in GIS software. Box 3.2 shows how to georeference an image in ArcGIS. After the image is georeferenced, the subsequent digitising process will capture coordinate data in the projected coordinate system of the source map or aerial photograph.

Box 3.2 Georeferencing an image in ArcGIS	PRACTICAL

To follow this example, start a Web browser (such as Internet Explorer, Google Chrome or Mozilla Firefox) on your computer, and then download the scanned 1:250,000 geological map (250dpi) of Southwest Tasmania from the Australian Geoscience Portal at http://www.geoscience.gov.au/geo portal-geologicalmaps/ and save it as **sk5505_sk5507.jpg** to the directory or folder of your choice. All the scanned geological maps available from the portal are not georeferenced. They can be downloaded free of charge.

1. Start ArcMap with a new, empty map.
2. Click on the **Add Data** button ✛ to add **sk5505_sk5507.jpg** to the view. Click **Yes** to build pyramids when prompted.

Note that for a raster or image, pyramids can be built with a reduced resolution layer that copies the original image in decreasing levels of resolution. The lowest level of resolution is used to quickly draw the entire image. When zoomed in, layers with higher resolutions are drawn rapidly, as fewer pixels are needed to represent the successively smaller areas. Pyramids allow rapid display of an image at varying resolutions.

3. Click **OK** when warned about missing spatial reference information. The scanned map is added.

The georeferencing system of the map is **MGA Zone 55**, based on **GDA94** (see Section 2.2). When georeferencing an image in ArcMap, the coordinate system will be taken from the data frame; hence the data frame's coordinate system needs to be set:

4. In the table of contents, right-click **Layers**, and select **Properties**.
5. In the **Data Frame Properties** dialog box:
   1) Click the **Coordinate System** tab.
   2) Expand **Projected Coordinate Systems > National Grids > Australia**, and then select **GDA 1994 MGA Zone 55**.
   3) Click **OK**.
6. Zoom in the scanned map enough so that the coordinates in the lower left corner can be clearly read.

Notice that the MGA *x* and *y* coordinates are marked at regular intervals along the map border: *x* coordinates (eastings) along the top and bottom, and *y* coordinates (northings) down the left and right sides (Figure 3.3). MGA is a grid reference system. The + marks on the map represent the grid

Figure 3.3 Four corners of the geological map of Southwest Tasmania and their MGA coordinates.

(continued)

(continued)

reference lines every 10,000m, which can be used to assist in finding the *x* and *y* coordinates of a location. By using the grid reference lines, the *x* and *y* coordinates of the lower left corner are calculated as 270,000m (east) and 5,150,000m (north). Similarly, the *x* and *y* coordinates of the lower right corner are obtained as 460,000m and 5,150,000m; those of the upper right corner are 460,000m and 5,380,000m; and the coordinates of the upper left corner are 270,000m and 5,380,000m. The four map corners are numbered respectively as 1, 2, 3 and 4 as indicated in Figure 3.3. They are used as control points for georeferencing below.

7. In the table of contents, right-click **sk5505_sk5507.jpg**, and select **Zoom To Raster Resolution**. This will give you the optimal display to locate a control point.
8. Click **Customize** on the main menu bar, point to **Toolbars**, and click **Georeferencing**.
9. On the **Georeferencing** toolbar, click **Georeferencing**. In the pull-down menu, make sure that the box for **Auto Adjust** is checked. The auto adjust function automatically moves the display of the image to its real-world location based on the coordinates you enter.
10. Click the **Add Control Points** tool ⚹ on the **Georeferencing** toolbar. The shape of the mouse pointer changes to a crosshair when you move it over the map display.
11. Pan the map so that you can see its lower left corner – the number 1 control point. Position the crosshair over the point, then click. This enters a **from** control point.
12. Right-click anywhere in the view, and select **Input X and Y. . . .**
13. In the **Enter Coordinates** dialog, enter **270000** for **X** and **5150000** for **Y**. This is the **to** control point. Click **OK**. The image display is moved to real-world coordinate space.
14. In the table of contents, right-click **sk5505_sk5507.jpg**, and click **Zoom To Layer** to redisplay the scanned map.
15. Zoom in on the lower right corner of the map – the second control point. Click the **Add Control Points** tool ⚹ if the shape of the mouse pointer is changed. Position the crosshair over the second control point, then click.
16. As in steps 12 and 13 above, enter **460000** for **X** and **5150000** for **Y** in the **Enter Coordinates** dialog.
17. In a similar way, add the third and fourth control points using the *x*, *y* coordinate values provided above.
18. Click the **View Link Table** button ⊞ on the **Georeferencing** toolbar. In the **Link** window, a table of links (between source and georeferenced control points) and errors is shown. Each row contains the **from** (source)–**to** (map) coordinates for one control link and residuals (errors).The residual error is computed for each link by comparing the actual location of the map coordinate to the position in the source image. The average discrepancy for all links is expressed as the Total RMS (root mean square) Error. The scanned map is transformed using the default **1st Order Polynomial (Affine)** transformation method. The concepts of residual errors, RMS and transformation methods are discussed in detail in Section 3.3. Figure 3.4 shows the scanned map after georeferencing and a link table. The values in your link table will likely differ from those shown in this figure. The smaller the residuals, the better the outcome.
19. On the **Georeferencing** toolbar, click **Georeferencing**. In the pull-down menu, select **Rectify**. Note that after georeferencing the scanned map by following the steps above, only the display of the image is adjusted to real-world coordinate space, not the data itself. To permanently apply the transformation to the data, you must use **Rectify**.
20. In the **Save As** dialog:

    1) For **Output Location**, click the **Open File** button 📂.
    2) In the **Select Workspace** dialog, browse to **C:\Databases\GIS4EnvSci\results**. Click the **Tasmania** folder, then click **Add**.
    3) For **Format**, change the file type to **IMAGINE Image**.

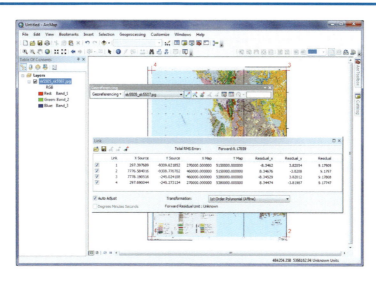

Figure 3.4 Example of a link table resulted from georeferencing.

4) Name the new raster **sk5505_sk55072.img**. Be sure to include the .img extension.
5) Click **Save**. The georeferenced image is saved as **sk5505_sk55072.img** in the selected workspace.

21. Exit ArcMap without saving the map document.

On-screen digitising is performed using the digitising tools provided within GIS software. The georeferenced image is used as the backdrop map. It allows users to use drawing or sketching tools to trace (actually redraw) geographical features on the screen and digitise the features into sequences of point coordinates. There are two modes of digitising: point and stream mode. Point mode digitising is the most common method, using the mouse to select points as vertices of a feature on the image and clicking the mouse to record their individual locations one by one; the GIS then connects the vertices to create a digital feature. As a general rule, for a feature represented as a straight line, only two end-points need to be digitised. For a feature represented as a curved line, more points need to be digitised to make sure its shape is best approximated. The number and density of points selected to digitise for a feature depend on the complexity of the feature's geometric shape (Figure 3.5). The point mode is used particularly when precise digitising is required. Stream mode digitising places a point automatically every time a pre-defined distance or time threshold (for example, every 0.5mm or 0.2 seconds) is passed as

the user moves the mouse, and the user does not need to click to add each individual vertex. Stream mode digitising is quicker than point mode digitising, but may produce larger files with many redundant points.

Digitising is labour-intensive and error-prone. Figure 3.6 shows typical errors that may occur during the digitising process. Digitising errors need to be removed by using editing tools in GIS, which are introduced in Section 3.3.

After errors are corrected and the coordinate data are saved into a data layer, attribute data associated with each digitised feature can be added. Usually, when a new data layer is created after digitising, its associated feature attribute table is automatically generated, in which each row has a unique feature ID, but no attribute data. Attributes and attribute data can be added to the feature attribute table directly through manual keyboard entry. Attribute data can also be entered into a spreadsheet or a database table with common feature IDs, which can be used to join or relate with the feature attribute table (see Figure 2.38). Box 3.3 illustrates how to capture vector data through digitising in ArcGIS.

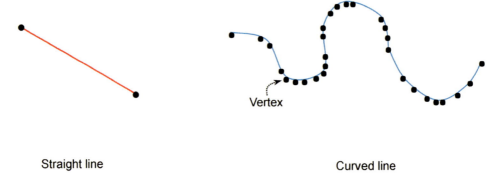

Straight line   Curved line

Figure 3.5 Selection of points in digitising.

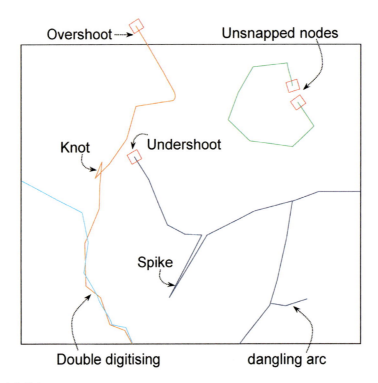

Figure 3.6 Typical digitising errors.

To follow this example, start ArcMap and load **sk5505_sk55072.img** from **C:\Databases\GIS4EnvSci\ results\Tasmania**. This is the geology map of Southwest Tasmania georeferenced and rectified in Box 3.2. Follow the procedure below to extract the lake features from that map using on-screen digitising.

1. Open ArcToolbox by clicking ![icon] on the **Standard** toolbar. Expand **Data Management Tools > Feature Class** in the ArcToolbox window, then double-click the **Create Feature Class** tool. This creates a new feature class to store the features to be created by digitising.
2. In the **Create Feature Class** dialog box:

   1) For **Feature Class Location**, browse to **C:\Databases\GIS4EnvSci\results**, click the folder **Tasmania**, then click **Add**.
   2) Name the **Feature Class** as **lakes**.
   3) Set the **Geometry Type** to **POLYGON** (the default).
   4) For **Coordinate System**, click the **Spatial Reference Properties** button ![icon]. In the **Spatial Reference Properties** dialog, browse to **Projected Coordinate Systems > National Grids > Australia**, and select **GDA 1994 MGA Zone 55**. Click **Add**.
   5) Click **OK**. The new, empty feature class **lakes** is created and added to the table of contents. Close the progress window when the processing completes.

3. Zoom in on **Lake St Clair** in the upper-right part of the map.
4. Click ![icon] on the **Standard** toolbar to open the **Editor** toolbar.
5. Click **Editor** on the **Editor** toolbar. In the pull-down menu, click **Start Editing** to begin an edit session.
6. Click the **Create Features** button ![icon] on the **Editor** toolbar. The **Create Features** window appears beside the data view.
7. In the **Create Features** window, first click the layer name **lakes** in the upper section of the window. A number of construction tools appear in the lower section of the window. Click the **Polygon** tool.
8. Click the sketch tool **Straight Segment** ![icon] on the **Editor** toolbar.
9. Select points along the boundary of **Lake St Clair** which can best approximate the shape of the lake. Click once to add each vertex while tracing the boundary of the lake, as shown in Figure 3.7. Each green dot represents a vertex. To create smoother curves, add more vertices. After the final vertex is added, double-click at the start point. A polygon representing the lake is created and shown as the symbol of a **lakes** feature.

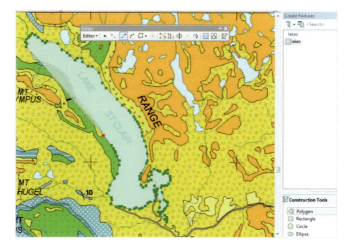

Figure 3.7 Digitising a lake as a polygon.

(continued)

(continued)

10. Click the **Attributes** button ▦ on the **Editor** toolbar. The **Attributes** window opens and appears beside the data view. Two attributes are listed in the window for the digitised lake: **FID** and **Id**. They are two fields in the attribute table created by the system for the layer **lakes**. **FID** is not editable by the user. **Id** can be used to store user-defined feature ID numbers. Now, change the **Id** value from 0 to 1 so that **Lake St Clair** has an ID number of 1. Section 3.4 describes how to add new attributes and edit attribute values.
11. Select some other lakes to digitise by following steps 8–10 above.
12. After finishing digitising, in the pull-down menu of **Editor** on the **Editor** toolbar, click **Stop Editing** and save the edits.
13. Uncheck **sk5505_sk55072.img** in the table of contents and display **lakes** only in the data view. You can see now all the lakes extracted from the geology map.
14. Exit ArcMap without saving the map document.

Field surveying

When there are no existing spatial data (analogue or digital) to meet the requirements of a GIS project, spatial data may need to be collected in the field through surveying and mapping based on direct measurements. Traditionally, precision triangulation, trilateration, traversing and geometric levelling with plane tables, transits and theodolites were used to measure and map geographical locations and altitudes of ground objects (Ghilani and Paul 2012). Today, these conventional surveying instruments are being replaced by total stations and GNSS.

A total station is an electronic digital theodolite integrated with an electronic distance meter (Ghilani and Paul 2012). It can read and record horizontal and vertical angles, measure slope distances and determine (x, y, z) coordinates of surveyed points relative to the total station position. Total stations can measure with an accuracy of about one part per million (for example, 1mm per kilometre or 30mm over a 30 kilometre distance). The data collected by total stations are digital, and can be directly downloaded or read into a computer and imported into a GIS. Ground survey with theodolites or total stations is typically used for capturing ground elevations, shorelines, property boundaries, building footprints, roads, tunnels and other ground objects that need to be located with high accuracy.

GNSS satellite navigation systems are a revolutionary tool for accurate measurement of a position almost anywhere on the surface of the Earth twenty-four hours a day. A GNSS system can give a position (latitude, longitude and altitude), based on a particular geodetic datum, directly and almost instantly without the need to establish control and measure angles and distances between intermediate points between the surveyed point and the

control points as with surveys using theodolites and total stations. Portable hand-held GNSS receivers allow surveyors or data collectors to walk or drive around to collect (x, y, z) coordinates at sample points. The accuracy obtainable from GNSS receivers ranges from 100m to a few millimetres. GPS, GLONASS and BeiDou are three GNSS systems that are currently operational and provide navigation and positioning services. Among them, GPS is most widely used.

GPS is composed of a constellation of twenty-four satellites (plus three spares for backup) and receivers (Hsu 2010). GPS satellites orbit 20,200km above the Earth in a period of twelve hours in six orbital planes. This configuration ensures that at least four satellites (the minimum number required for accurate measurements) are always potentially visible anywhere on Earth. Each satellite emits coded radio signals at precisely timed intervals and has four atomic clocks on board to keep accurate time. GPS hand-held receivers calculate latitude, longitude, altitude and time by comparing signals from visible satellites.

Before May 2000, all GPS measurements were affected by selective availability – an error deliberately added to the signals by the US military. This has now been switched off. GPS measured location is now accurate to within 30m. Location accuracy of 15m or better is usually acquired 95 per cent of the time with a single hand-held GPS receiver. While this order of accuracy is acceptable for certain uses, it is insufficient for most surveying applications. Differential GPS or DGPS on the other hand can improve location accuracies to 1m or better. DGPS involves the simultaneous use of two receivers. One receiver, known as a base station, is located at a precisely surveyed location. The other receiver, known as a rover, is used in the field to collect data. The base station and rover must operate concurrently and be relatively close to each other so that they receive the same GPS signals and have virtually the same errors. The base station computes its position at regular intervals based on satellite signals, and compares them to the known coordinates. The difference is then used by the rover to correct its measurements in real time in the field using radio signals or through post-processing after data capture using special processing software.

Real-time differential correction can be used with any rover that is configured to receive the real-time corrections sent directly from the DGPS provider (or one's own base station). It is usually used for navigational purposes, and provides essentially instantaneous sub-metre positioning. Post-processing differential correction involves first obtaining base station files from the base station which contain the coordinates, the time when they were collected and the amount of error in the x and y directions, then loading the base station files and position data files captured by the rover into differential correction software, which is used to match up each location recorded by the rover with the corresponding error offsets calculated by the base station at the same time and revise the rover data accordingly. Many permanent GPS base stations are currently operating throughout the world, which are maintained by different government agencies or private subscription services. They usually run twenty-four hours, seven days a week, and supply the data necessary for differentially correcting GPS through radio broadcasting or via the Internet.

GNSS survey data are digital, and can be directly downloaded into a GIS. GNSS can be used to capture point, line and polygon data as well as their associated attribute values for building GIS databases. A data dictionary can be designed and input into a GNSS receiver so that all the required attributes of geographical features are recorded in the field. The attribute values can be entered through the GNSS receiver's interface menu while collecting the coordinate data.

GNSS has been widely used for recording ground control points for georeferencing, collecting sample points for environmental model building and validation, and for obtaining reference data for ground truthing of spatial data. Coupled with GIS and remote sensing imagery, GNSS data collection provides a means for environmental monitoring and analysis. Box 3.4 provides an example of the use of GPS in tracking the movement of African elephants in a human-dominated land-use mosaic.

Box 3.4   GNSS in environmental applications: tracking African elephant movements   **CASE STUDY**

Collecting accurate, reliable and timely information is crucial for environmental problem solving and decision making. GNSS data can be captured, mapped and analysed very quickly because they are in a digital form available at all times and in all parts of the world. As a result, GNSS has been rapidly adopted not only for field surveying, but also for environmental applications since its advent. For example, GNSS has been used to track environmental disasters such as fires and oil spills, monitor crustal movement and seismic activities, track the movement of wildlife, and to manage logging activities in forests. When position data collected through GNSS are imported into a GIS, they can be easily linked with other information for comprehensive analysis of environmental concerns. Graham et al. (2009) provided a case study of the use of GPS with GIS to track the movement of African elephants in a human-dominated land-use mosaic.

Elephants play significant roles in ecological dynamics in Africa. Understanding their movements is critical to conservation, particularly in human-dominated landscapes outside protected areas, where both biodiversity conservation and human welfare need to be balanced. Traditionally, seasonal or annual aerial surveys and ground transect data collection methods were employed to count the number of elephants in a region and assess interactions between elephants and humans. In contrast, Graham et al. (2009) applied the GPS technology to track elephant movements every hour, based on which they analysed the where, when, how and why of movements of African elephants (*Loxodonta*

(continued)

(continued)

africana) in relation to different types of land use in the human-occupied landscape of Laikipia District in north-central Kenya. In this study, fifteen elephants were fitted with GPS collars and monitored over about two years, from 2004 to 2006. The GPS units used a global system for mobile communication (GSM) modem for two-way data communication through mobile phone network ground stations, which allowed data to be downloaded remotely via the Internet and the settings of the GPS receivers to be programmed remotely.

Four types of land use in the study area were mapped, including smallholder land, ranch (individually owned areas for commercial activities), pastoral land (communally owned and managed areas for livestock grazing) and forest. The extent to which people tolerate the presence of elephants on ranch and pastoral land was assessed through interviews with local land owners, managers and key informants. All GPS collars except one were programmed to take GPS fixes every hour. After eliminating spurious GPS fixes, a total of 137,816 GPS locations were recorded and used for subsequent analyses. About half of the GPS data were recorded during the day, and the other half at night. ArcGIS was used to ascribe land use, human tolerance of elephant presence, time of day and speed of elephant movement to each of the GPS recorded elephant locations, which were then fed into a statistical software package (SPSS).

Statistical analyses were conducted using the SPSS to assess how elephant movement is related to land use and how variable the speed of elephant movement is in different types of land use with different degrees of human tolerance. The proportion of an elephant's tracking locations that fell within a particular type of land use was used as a measure of elephant use of that type of land use. The speed of elephant movement was calculated as the ratio of the distance between consecutive locations and time. The analyses found that the elephants monitored always preferred ranches, and to a lesser extent forest reserves, over smallholder and pastoral land. They spent significantly less time at night in ranches, and much more time at night in smallholder areas than during the day. Elephants moved fastest in smallholder areas, followed by pastoral areas and ranches, and slowest in forests. But they moved more quickly at night on ranches with less human tolerance than on those with more human tolerance. Overall, elephants preferred areas with low levels of human activity. However, the high-temporal resolution GPS tracking data did reveal that some of the elephants monitored used smallholder, pastoral and forest areas where the risk of predation by humans were fairly high, and they used the cover of darkness to avoid risks and exploit resources in human occupied areas. It was demonstrated that the risk avoidance behaviour of elephants helps maintain connectivity between refugia. The results suggested that securing corridors between existing refugia could enhance the persistence of elephants in the study area into the future, so the identification of existing or appropriate elephant corridors between refugia should be a priority for conservation.

Source: Graham et al. (2009).

GNSS is easy to use, less labour-intensive, and more cost-effective and weather-independent. But its accuracy may degrade in areas with tall buildings or under trees. GNSS signals are often lost in forests with dense canopies, under bridges or indoors, and may be subject to interference from power lines and radio towers.

The attributes of air, land, water, soil, vegetation and other ecological and natural resources are measured in the field using specialised equipment in addition to the locational information. These attributes could be specific chemical, physical and electronic properties. They are often measured by field scientists and data collectors with specialised training. Detailed discussion on this topic is beyond the scope of this book.

Photogrammetry and laser scanning

Both photogrammetry and laser scanning are remote sensing techniques for spatial data capture. Photogrammetry is a technique that makes accurate measurements by means of aerial photography (Jensen 2007). It determines (x, y, z) coordinates of objects by measurements made in overlapping pairs of aerial photographs taken from different points of view.

In aerial photogrammetry, the aircraft usually flies in a series of flight lines. The coverage of the neighbouring flight lines often requires a 30 per cent overlap. Photographs are taken in rapid succession along each flight line with the camera looking vertically towards the ground. There is usually a 50–60 per cent overlap between successive photographs. Two successive photographs are called a stereopair. With a stereopair of photographs, common points of the same object are identified on each. A line of sight (or ray) can be constructed from the camera location to the point on the object. The intersection of these rays determines the 3D position of the point. Stereopair aerial photographs are normally processed in a stereoplotter, which is an instrument that allows an operator to see stereopairs of photographs in a stereoscopic view and make 3D measurements. There are three types of stereoplotter: analogue, analytical and digital (Gomarasca 2009). The latter two are commonly used today. Photogrammetry has been used to draw contours for topographic mapping, capture precise (x, y, z) coordinates of ground objects and obtain geometric measurements of geographical features (length, perimeter and area). Digital photogrammetry has also been used to produce digital elevation models and bathymetric models (Jensen 2007). The possibility of acquiring stereoscopic images with the latest satellites offers an alternative to aerial photogrammetry for medium-scale mapping. However, the photogrammetric technique is complex and the equipment is expensive, which limits its use to specialist surveying and mapping organisations.

In recent years, airborne LiDAR (light detection and ranging) has increasingly become an important source of elevation data. LiDAR uses a laser pulse directed at the ground and detects the amount of time between when the pulse is emitted and when its reflection returns to determine the distance between the ground and the sensor, based on which elevations of the ground and features are derived. LiDAR produces (x, y, z) coordinates for multiple discrete points, called mass points or point clouds. It can achieve a vertical accuracy of 10–15cm and horizontal accuracy of 50–100cm. While photogrammetry is more accurate in the x and y directions, LiDAR data are generally more accurate in the z direction. LiDAR has largely replaced the stereoplotter for capturing elevation data. Chapter 6 provides a more in-depth discussion of LiDAR.

Raster data capture

Remote sensing is the major means of raster data capture. It is a technique for deriving environmental information from remotely sensed images, including satellite images and aerial photographs. Digital remote sensing images are themselves raster data. A digital image consists of a 2D array of individual picture elements called pixels – cells arranged in columns and rows. Each pixel has a digital number, representing the scaled and quantised intensity of electromagnetic radiation (EMR) in a certain part of the EMR spectrum (referred to as a band) reflected, emitted or backscattered from ground features (Jensen 2007). To visually display an image, the digital numbers represent grey scales. Multispectral remote sensing produces images in a number of bands. Different ground features tend to reflect, emit or backscatter different amounts of EMR due to differences in their biophysical properties and chemical compositions, and therefore have different pixel values. By analysing digital numbers in an image or multiple images (in different bands) and relating them to different characteristics of ground features using digital image processing techniques (see Chapter 6), useful information can be extracted. Such information can be used, for example, as a basis for mapping land use/cover, understanding environmental processes and estimating biophysical variables (such as biomass and ozone concentration). Remote sensing images and data derived from them through digital image processing can be directly imported as raster data into GIS for further analysis. Some GIS software systems, such as ArcGIS, GRASS and IDRISI, also provide digital image processing functions for deriving new raster data from remote sensing images. Chapter 6 is devoted to this topic.

Raster data can also be produced through vector-to-raster conversion. The methods for vector data capture discussed above can be used to produce point, line or polygon vector data first, and the vector data are then converted to rasters. This process is known as rasterisation (see Section 3.3), which is a standard feature of most GIS systems.

Crowdsourcing

As mentioned in Section 1.2, crowdsourced mapping is a method for collecting spatial data by soliciting contributions from the community or general public. It involves the use of the Web or Internet-enabled and location-aware mobile devices (such as tablets and smartphones) to gather location-specific information from the masses, or 'crowds'. Three key technologies underpin crowdsourcing mapping: the Web, geotagging and mobile computing.

The Web provides a particularly good platform for crowdsourcing. The latest Web technology allows people to interact and collaborate with each other as creators of user-generated content in a virtual community. In other words, people can not only passively view Web content, but also are able to create, share and remix content (Han 2011). Therefore, using the Web as a platform, we can now harness the power of networked users to build content value.

Geotagging is essentially a process of georeferencing a piece of information, such as a Web entry, a short message service (SMS) message, a note about an object or event, an image or a video, by embedding a geotag within the information so that it can be added to or accessed from maps or georeferenced images in websites, blogs, wikis or social networks (Luo et al. 2011). A geotag specifies a geographical location in a particular geographical or projected coordinate system. Geotags are typically embedded in the metadata of an image (in the Exchangeable Image File Format – EXIF, or Extensible Metadata Platform format – XMP), in the body of an SMS message formatted according to the Open GeoSMS Standard (set by the Open Geospatial Consortium; OGC 2012) and in Web pages in HTML format. Geotagging can be performed by using virtual globes (such as Google Earth), photo-sharing systems or Web services (such as Google Picasa and Flickr) and online crowdsource mapping systems (such as Google Map Maker, OpenStreetMap, Geo-wiki and Wikimapia). Geotagging can also be carried out using a tablet or smartphone equipped with a GPS receiver or Wi-Fi Positioning System (WPS, which uses surrounding wireless networks instead of satellites to measure a location). For example, Google Earth combines satellite imagery, aerial photographs and map data to make a virtual world. Not only does this allow users to explore the virtual environment through the world, but Google Earth also lets them geotag their own multimedia information with placemarks, or view geotagged texts, symbols, images and videos taken by others. Google Earth is now a fully fledged crowdsourced system with millions of users contributing their own content, attaching photos, videos, notes and even 3D models to geographical locations across the world. Picasa is a photo-sharing system which allows users to geotag their own photos by inserting the longitude and latitude to the photos' EXIF metadata, and to place them in Google Map or Google Earth.

Crowdsource mapping systems are built by using or combining the Web, geotagging and mobile computing technologies. The best-known crowdsource mapping systems include Google Map Maker, OpenStreetMap, Geo-wiki and Wikimapia. Google Map Maker and Wikimapia allow users to add places, roads and rivers onto base maps. OpenStreetMap is a free worldwide street map created by volunteers. Geo-wiki is designed to enable a global network of volunteers to gather validation data for global land cover datasets. Crowdsource mapping systems can be targeted to a group among a dedicated audience, or to the public at large. They are developed for collecting different types of data, and may vary in terms of data entry methods, data validation methods, data licensing agreements and ease of use. ClimateWatch (see Box 1.2) is a crowdsource mapping system for collecting data on indicator and animal species in Australia. Other examples include eBird, which collects bird observation data submitted by recreational and professional bird watchers (Sullivan 2009), the Global Biodiversity Informatics Facility (GBIF) data portal, which documents and exchanges global biodiversity data (http://www.gbif.org/), the UK Nature's Calendar, which gathers information on the signs of the seasons recorded by volunteers (http://www.naturesca lendar.org.uk/), and Safecast, which is a global sensor network for collecting and sharing radiation measurements to empower people by providing data about their environments (http://blog.safecast.org/). Ushahidi is an open source development tool that enables the easy deployment of crowdsourced interactive mapping applications with Web forms/e-mail, SMS and Twitter support (http://www.ushahidi.com/). It was used successfully to map incidents of violence and peace efforts throughout Kenya based on reports submitted via the Web and mobile phones, and to map Queensland floods and bushfire events based on citizens' reports in Australia.

Crowdsourcing provides a means of large-scale data collection with limited resources and a way to engage with the community. It is also often criticised as a source of poor-quality data. However, a study on the quality of volunteered geographical information by Hacklay (2010) suggested that spatial data captured through crowdsourcing are fairly accurate. This study compared OpenStreetMap (mainly crowdsourced from uploaded GPS tracks, out-of-copyright maps and Yahoo! Aerial imagery) and Ordnance Survey (OS) datasets (considered reliable and authoritative), and found that on average OpenStreetMap data are within about 6m of the position recorded by the OS, and there is an overlap of up to 100 per cent with OS digitised motorways, A-roads and B-roads. Considering a GPS receiver usually captures a location within 6–10m and the Yahoo! image resolution is about 15m,

OpenStreetMap's data quality is very good indeed. Nevertheless, strategies and methods should be developed and applied to identify errors in the crowdsourced data and to collect data in a robust, trusted manner.

3.2 SPATIAL DATA TRANSFER

Spatial data transfer involves importing existing digital spatial data from external sources. Before starting a GIS project, we usually search for the necessary digital data from various sources and acquire them if they exist. Digital spatial data from external sources are often encoded in different formats. Table 3.1 lists some key digital data formats for vector and raster data. If the acquired data are in the format supported by the GIS system used

in the project, they can be read directly into the system. Otherwise, they need to be translated into a compatible format. For example, the native data formats of ArcGIS are shapefiles and coverages for vector data, and GRID for raster data. ArcGis can also directly read many non-native data formats, such as AutoCAD DXF and DWG, ERDAS IMG, TIFF (Tagged Image File Format), GeoTIFF, GIF, JPEG, and transfer MicroStation DGN, Intergraph MGE, MapInfo MID/MIF and TAB, Oracle and Oracle Spatial, Intergraph GeoMedia Warehouse and other data using data conversion and transformation tools or via intermediate formats.

Direct transfer allows the non-native data to be translated on the fly and displayed with other datasets in the GIS while the source dataset remains in its original format. However, direct transfer usually does not

Table 3.1 Digital spatial data formats

Vector	Raster
ESRI ArcInfo Coverage	ESRI GRID
ESRI ArcInfo E00 (ArcInfo Interchange Format)	DTED (Digital Terrain Elevation Data)
ESRI ArcView Shapefile	ERDAS IMG (IMAGINE)
MapInfo MIF (Map Interchange Format)	BMP (Windows Bitmap)
USGS DLG (Digital Line Graphs)	TIFF (Tagged Image File Format)
AutoCAD DXF (Data Exchange Format)	GeoTIFF
AutoCAD DWG (Drawing)	GIF (Graphics Interchange Format)
CGM (Computer Graphics Metafile, ISO)	JPEG (Joint Photographic Experts Group)
Microstation DGN (Microstation drawing file format)	PNG (Portable Network Graphics)
SDTS (Spatial Data Transfer Standard, US)	USGS DEM
NTF (National Transfer Format, UK)	ECW (Enhanced Compressed Wavelet)
TIGER (Topologically Integrated Geographic Encoding and Referencing System, US Bureau of Census)	
DIME (Dual Independent Map Encoding, US Bureau of Census)	
VPF (Vector Product Format, National Imagery and Mapping Agency, US)	
KML (Keyhole Markup Language, Google)	
WFS (Web Feature Service, Open Geospatial Consortium)	
NetCDF (Network Common Data Form, Unidata)	
S-57 (International Hydrographic Organization)	

allow data editing. To edit the data, they must be converted to one of the system's native formats using data conversion tools. Moreover, direct transfer is mainly used for importing data in relatively simple formats. Complex formats, such as SDTS and NTF, require more sophisticated processing, and are often converted using specialised translation software, such as the Feature Manipulation Engine (FME) (http://www.safe.com/). FME supports over 300 formats of spatial and non-spatial data types, including various GIS, CAD, rasters, databases, 3D, XML and point cloud (LiDAR data). FME technology has been incorporated into ArcGIS. Box 3.5 provides some examples of spatial data transfer in ArcGIS.

Box 3.5   Digital spatial data transfer in ArcGIS   TECHNICAL

ArcGIS provides a number of data conversion tools for digital spatial data transfer. Its data interoperability extension can directly read more than a hundred non-native data formats and export to more than seventy data formats (http://www.esri.com/datainteroperability). Below are some data conversion tools available in **ArcToolBox > Conversion Tools** which do not require the data interoperability extension to be enabled.

KML to file geodatabase

KML is a popular XML-based format for online mapping and spatial data sharing across the Internet (OGC 2008). Its compressed version is a KMZ file. Originally developed by Google for the graphic display of spatial data in applications such as Google Earth and Google Maps, it allows GIS users to add their own spatial data to these applications. As KML uses WGS84 as the default geographical coordinate system, spatial data encoded in KML can also be imported and displayed in a variety of GIS applications. ArcGIS provides conversion tools to translate KML or KMZ files into feature classes, and vice versa.

The **KML To Layer** tool in ArcGIS converts a KML or KMZ file into a file geodatabase containing a feature class within a feature dataset and a layer file. A layer file (.lyr) is a file generated by ArcGIS that stores the path to a source dataset and other layer properties, including symbology. It is not actual data because it does not store the features' attributes or geometry. The layer file created by the **KML To Layer** tool, which will be added to ArcMap to draw the features, maintains the symbology found in the original KML file. The feature class may be point, line or polygon, depending on the original features in the KML file. Each feature class produced by the tool will have attributes which maintain information about the original KML file, such as the original folder structure, pop-up information, and fields defining how the features sit on a surface. If the KML file contains rasters, a raster catalogue will be generated inside the file geodatabase with the source rasters stored in their native formats. The output of the **KML To Layer** tool is in WGS84, and can be re-projected to another georeferencing system using the **Project** tool (see Box 2.1).

There are two data conversion tools in ArcGIS for creating KML files from ArcGIS data. The **Layer To KML** tool allows individual feature or raster layers to be converted to KML, and the **Map To KML** tool translates ArcMap map documents (.mxd) (containing multiple layers) into KML. Both create a KMZ file in the output location which can be uploaded to Google Earth for display. Before they can be converted to KML, feature or raster datasets must first be added to the display or made into a layer using the **Make Feature Layer** or **Make Raster Layer** tools (in **ArcToolBox > Data Management Tools > Layer and Table Views**).

GPX to feature class

GPX, the GPS Exchange Format, is an XML data format for GPS data exchange between GPS receivers, desktop and mobile software, and Web-based services. It defines a common set of data tags for describing GPS data (waypoints and tracks) in XML. Waypoints are single unrelated points. Tracks constitute a route defined with a collection of related points. Waypoints and track points are collected in WGS84 and saved in latitude and longitude pairs, usually with time and elevation information. The **GPX To Features** tool in ArcGIS allows GPX files to be translated into feature classes. The output is a point feature class containing feature geometry and attributes, including name, description, type (waypoints or track points), date and time, and elevation. Track points can then be selected to produce a polyline feature class using the **Points To Line** tool (in **ArcToolBox > Data Management Tools > Features**).

WFS to feature class

ArcGIS can also use the Web Feature Service (WFS) as the data source. WFS is a protocol for serving and transferring feature data across the Web developed by the Open Geospatial Consortium. Each dataset has its own base URL (Uniform Resource Locator, used by Web browsers to identify a network resource on the Internet). A WFS is mainly a feature access service, which allows a client to perform data manipulation operations on one or more geographical features, including getting, retrieving, creating, modifying and deleting a feature. The information that is encoded in WFS includes both feature geometry and attribute values. The **WFS To Feature Class** tool in ArcGIS can be used to import a feature type from WFS to a feature class in a geodatabase. With the tool, the user can specify the URL for the WFS server, select a feature type (such as roads and rivers) published from the server, and import it to a feature class for use in ArcGIS.

3.3 SPATIAL DATA TRANSFORMATION

Spatial data acquired from different sources may use different data structures and different georeferencing systems, or may be at different scales or from different map sheets. Before any GIS analysis begins, all the input spatial data have to be georeferenced or converted to the same georeferencing system, may need to be converted from one data structure to another, the data across different map sheets must be matched at the boundaries, and the data may need to be generalised to contain a certain level of detail required for the task at hand. This section focuses on data structure conversion, geometric transformation, line simplification and edge matching.

Conversion between vector and raster

Conversion between vector and raster data structures enables the use of both vector and raster data when solving an environmental problem with GIS, as well as the use of various GIS analysis functions unique to the two spatial data structures. It provides flexibility when designing a GIS solution by taking into account data sources and data analysis methods. When combining vector and raster data for integrated analysis, such as overlay analysis, they have to be converted into the same data structure.

Converting vector data to raster data is called rasterisation. It involves converting points, lines or polygons into grid cells while interpolating per-vertex values across each shape. An assignment rule (the maximum area rule or the centre location rule, as discussed in Section 2.3) is selected to define how a cell value is determined when two or more categories fall in a single cell. There are a number of algorithms that have been developed for rasterisation (Pavlidis 1982). Different GIS software may use different algorithms. The accuracy of the resultant raster data depends on the resolution of the grid – the cell size, often specified by the user according to the requirements of the analysis. It is recommended that all vector data in a project be converted to rasters at the same resolution. Rasterisation is one of the major methods for producing raster data. However, it is always an approximation of the vector representation of geographical features, and often leads to the loss of information. For example, in Figure 3.8, after rasterisation, the input polygon is split into two parts with the loss of connectivity.

Converting raster data to vector features is referred to as vectorisation. It is a process of converting a raster dataset or scanned image to point, line or polygon features. Vectorisation can be done manually or automatically. On-screen digitising provides a manual method for vectorising scanned images, as discussed in Section 3.1. Automatic vectorisation typically involves

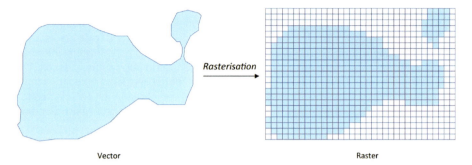

Figure 3.8 Rasterisation and loss of information.

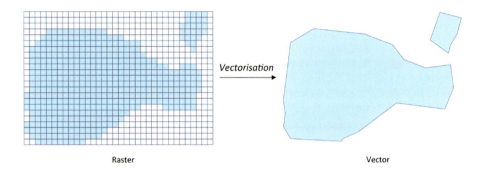

Figure 3.9 Vectorisation and distortion.

line thinning or skeletonisation, line tracing, spike and gap removal, smoothing and topology reconstruction (Peuquet 1981; Teng et al. 2008). Skeletonisation is the process of reducing lines to unit thickness at a given resolution. Line tracing is the process of identifying a specific series of point coordinates that make up a line segment as represented in the input raster layer. The above two processes may produce spikes and gaps due to inaccuracies either present in the input raster data or induced by the specific algorithms used in the processes. After the spikes and gaps are removed, the extracted lines are smoothed. Then the topology is constructed by determining the adjacency relationships among all the lines. The individual line segments may be joined into whole lines or into polygons, depending on the types of geographical features represented in the original raster data.

The resolution of the input raster data is very important for raster to vector conversion as it determines how well the line or area shapes are defined. At a higher resolution, the cells are smaller and the results of vectorisation will approximate the real shapes more closely. As the resolution gets lower, the cells become larger, which leads

to poorer representation of shapes after vectorisation, and may even result in completely false representation of the original shapes. Figure 3.9 shows an example where the raster produced in Figure 3.8 is converted back to the vector representation. Comparing the result of this vectorisation with the original vector map, it is obvious that the vectorised polygons have distorted shapes. Therefore, for raster to vector conversion, we have to make sure that the resolution of the raster data is high enough to adequately represent the essential shapes of the features. Box 3.6 describes several tools available in ArcGIS for data structure conversion.

Geometric transformation

Geometric transformation in GIS is used to adjust a data layer to correct geometric distortions, to register a data layer to a coordinate system (for example, converting a data layer created in digitiser units into real-world units represented on a map) or to transform a data layer by shifting, rotating, scaling or skewing its coordinate system. Transformations are applied uniformly to all features in a data layer.

Box 3.6   Data structure conversion in ArcGIS   TECHNICAL

ArcGIS provides a set of tools for vector to raster and raster to vector conversion. Rasterisation can be done with the **To Raster** toolset, and vectorisation can be accomplished with the **From Raster** toolset and **ArcScan**.

To Raster

The **To Raster** toolset is available in **ArcToolBox > Conversion Tools**. It includes the following vector to raster conversion tools:

- **Feature to Raster** – converts a vector dataset of any feature type into a raster, using the centre location rule to assign the cell value. Any feature class (geodatabase, shapefile or coverage) containing point, line or polygon features can be used as the input vector feature layer. It only requires users to specify the name of the input feature class, the name of the output raster, the field used to assign values to the output raster and the cell size of the output raster.
- **Point to Raster** – converts point features to a raster. With this tool, each cell in the output raster is given the value of the points found within the cell. Cells that do not contain a point are assigned the value of NoData. If more than one point is found in a cell, one of eight rules is used to assign the cell value (the first rule is applicable to both quantitative and qualitative values, and the rest are only applied to quantitative values):

    - MOST FREQUENT – the cell is given the value of the point with the most common attribute.
    - SUM – the cell is given the sum of the attributes of all the points within the cell.
    - MEAN – the cell is given the mean of the attributes of all the points within the cell.
    - STANDARD DEVIATION – the cell is given the standard deviation of attributes of all the points within the cell, or NoData if there are less than two points in the cell.
    - MAXIMUM – the cell is given the maximum value of the attributes of the points within the cell.
    - MINIMUM – the cell is given the minimum value of the attributes of the points within the cell.
    - RANGE – the cell is given the range of the attributes of the points within the cell.
    - COUNT – the cell is given the number of points within the cell.

- **Polyline to Raster** – converts line features to a raster. A cell is given the value of the line that intersects the cell. Cells that are not intersected by a line are given the value of NoData. If more than one line is found in a cell, one of the following two rules can be applied:

    - MAXIMUM LENGTH – the cell is given the value of the line with the longest length that covers the cell.
    - MAXIMUM COMBINED LENGTH – when there is more than one line feature in the cell with the same value, the lengths of these features will be added. The cell is given the value of the combined feature with the longest length within it.

- **Polygon to Raster** – converts polygon features to a raster. The user has three value assignment rules to choose from:

    - CELL CENTER – the centre location rule;
    - MAXIMUM AREA – the maximum area rule;
    - MAXIMUM COMBINED AREA – when two or more polygons in a cell have the same value, these features will be combined and their areas will be added. The cell is given the value of the combined feature with the largest area.

(continued)

(continued)

The **Point to Raster**, **Polyline to Raster** and **Polygon to Raster** tools provide greater control over the conversion process than the **Feature to Raster** tool.

From Raster

The **From Raster** toolset, available in **ArcToolBox > Conversion Tools**, contains the following raster to vector conversion tools:

- **Raster to Point** – converts each cell of the input raster to a point, positioned at the centre of the cell. Cells with the NoData value will not be converted into points.
- **Raster to Polyline** – converts a set of foreground cells to line features. The input raster is viewed as a set of foreground cells and background cells. Background cells can be designated as either the NoData cells or the cells with a zero or negative value. The cells with other valid values are considered as the foreground cells. This tool can also remove dangling lines that are shorter than the pre-set minimum length, and simplify a line by removing small fluctuations or extraneous bends from it while preserving its essential shape.
- **Raster to Polygon** – converts a raster to polygon features. It allows the user to determine whether the output polygons will be smoothed (simplified) or maintain the input raster's cell edges.

The input raster for the above tools may be any valid raster dataset and have any cell size, and the output vector layer will be a feature class (geodatabase or shapefile).

ArcScan

ArcScan is an ArcGIS extension (ESRI 2003). It provides tools for converting scanned images into vector layers. ArcScan supports automatic and interactive vectorisation of scanned images. It provides two methods for automatic vectorisation: centreline and outline. The centreline method produces vector line features along the centre of linear elements in the image, while the outline method creates vector polygon features at the borders of connected raster cells. Interactive vectorisation involves raster tracing and raster snapping. The raster tracing tool allows the user to manually trace raster cells and generate vector features. The raster snapping function allows the user to snap to raster centrelines, intersections, ends and corners while creating vector features with the Editor tools (to be introduced in Section 3.4). ArcScan can only vectorise bi-level images (containing only a background colour and a foreground colour).

Rubber sheeting

Rubber sheeting is a GIS operation for correcting geometric distortions in a data layer. Geometric distortions often occur in source maps, due to low position accuracy (such as in historical maps), inaccurate georeferencing in map compilation, paper shrinkage caused by atmospheric conditions such as humidity and heat, scale distortion or other causes. Rubber sheeting involves mathematically stretching one data layer (the source layer) to comply with another (the target layer, which is often an accurate base map). It geometrically adjusts map features to force the data layer to fit the base map.

For example, in Figure 3.10, the dotted lines represent an existing street layer, and the solid lines represent streets in an updated street layer containing geometric distortion. The updated street layer does not align with the existing street layer. In rubber sheeting, displacement links connecting from- and to- locations (usually control points) are used to define the source and destination coordinates for the adjustment. They are drawn as arrows with the arrowhead pointing towards the destination location (as shown in Figures 3.10b and 3.10d). Displacement links determine where features (consisting of data points) in the source layer are to move. The distance a data point may move depends on its proximity to a displacement link and the length of the link. The further data points are away from displacement links, the less they will move.

In addition to displacement links, identity links are used in rubber sheeting to hold data points in place at specific locations. They represent the locations that are known to be accurate and already match the target layer. Identity links prevent the movement of data points at those locations during an adjustment. After displacement and identity links are specified and created by the user, as shown in Figure 3.10b, rubber sheeting constructs two TINs to interpolate changes in x and y for point coordinates along the displacement links. The two TINs, called x-shift TIN and y-shift TIN respectively, have the same triangulation structure with the from-ends of the displacement links and all identity links used as the triangle nodes. Each node is defined by its (x, y) location and a

z-value. In the x-shift TIN, the z-value of each node is the amount of change in x between the from-end and to-end of a link. In the y-shift TIN, the z-value of each node is the difference in y between the from-end and to-end of a link. Identity links have no change, thus their z-value is zero. The z-value of each node is used to interpolate the amount of x, y adjustment of coordinates for each data point in a triangle. The interpolated z-value from the x-shift TIN is added to the x coordinate of the data point in the triangle, while the z-value interpolated from the y-shift TIN is added to its y coordinate. In this way, rubber sheeting adjusts the coordinates of all the data points in the source data layer to match the target layer.

Two methods are commonly used for interpolation of the TINs in rubber sheeting: linear and natural neighbour (see Section 4.5). Linear rubber sheeting method uses planar facets fitted to each triangle, and is mainly used when many links spread uniformly over the area to be

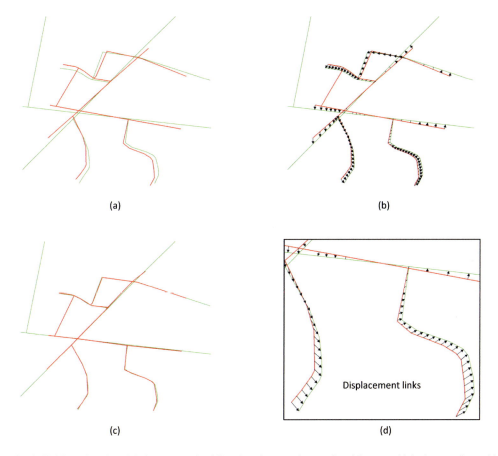

(a)

(b)

(c)

(d)

Figure 3.10 Rubber sheeting: (a) the source (red lines) and target (green lines) layers which do not align with each other, (b) displacement links represented as arrows, (c) the transformed source layer after rubber sheeting, and (d) an enlarged portion of (b).

adjusted. Natural neighbour rubber sheeting computes interpolated values as a weighted average of the nearest neighbour values, and is better applied when there are fewer links and the links scatter across the adjusted region. Rubber sheeting preserves the interconnectivity between features by stretching, shrinking or reorienting their connecting lines. Box 3.7 lists a procedure of rubber sheeting in ArcGIS. While rubber sheeting is a kind of differential transformation, the following three transformations – similarity transformation, affine transformation and projective transformation – are uniform for all data points in a dataset.

Box 3.7 Rubber sheeting in ArcGIS   PRACTICAL

Rubber sheeting is one of the interactive spatial data transformation tools in ArcGIS spatial adjustment toolset. To follow this example, start ArcMap and add the two data layers **mystreets_src** (as the source layer to be transformed) and **mystreets_tgt** (as the target layer) from **C:\Databases\ GIS4EnvSci\others**. It involves the following steps:

1. Set up the editing environment:

   1) Click the **Editor Toolbar** button 🖉 on the main toolbar.
   2) Click the **Editor** menu on the **Editor** toolbar, and click **Start Editing** to start a data editing session. Select the workspace (the directory, database or file folder on disk) that contains the data to be transformed. In this case, the workspace is the directory where **mystreets_src** and **mystreets_tgt** reside – **C:\Databases\GIS4EnvSci\others**.
   3) Click the **Editor** menu, point to **Snapping**, and click **Snapping Toolbar**.
   4) Ensure vertex snapping is enabled. If it is not, click **Vertex Snapping** ☐ on the **Snapping Toolbar** (this is to ensure that each displacement link to be added will snap to the vertices or endpoints of features).

2. Set up the spatial adjustment environment:

   1) Click **Customize** from the main menu bar. Point to **Toolbars. . .**, then click **Spatial Adjustment** to add the **Spatial Adjustment Toolbar** to ArcMap.
   2) Click the **Spatial Adjustment** menu in the **Spatial Adjustment Toolbar**, and click **Set Adjust Data**.
   3) In the **Choose Input For Adjustment** dialog box, click **All features in these layers**, then uncheck **mystreets_tgt**. Make sure **mystreets_src** (the source layer to be adjusted) is checked.

3. Select a transformation method:

   1) Click the **Spatial Adjustment** menu. Point to **Adjustment methods**, then click **Rubbersheet**.
   2) Select **Options. . .** from the **Spatial Adjustment** menu. In the **Adjustment Properties** dialog, click the **Options. . .** button.
   3) In the **Rubbersheet** dialog, click either the **Natural Neighbor** or **Linear** method. Click OK.
   4) Click **OK** to close the **Adjustment Properties** dialog box.

4. Create displacement links (referring to Figures 3.10b and 3.10d)

   1) Click the **New Displacement Link** tool ✦ on the **Spatial Adjustment** toolbar.
   2) Position the mouse pointer at the source location and click once to start adding a link.
   3) Move the pointer to the target location and click once to finish adding the link. An arrow appears, representing a displacement link connecting the source location to the target location.

4) Repeat the above two steps to add the required number of displacement links.
5) For features requiring many links, such as curved features, the **Multiple Displacement Links** tool ◈ can be used.

5. Create identity links:

1) Click the **New Identity Link tool** ⊞ on the **Spatial Adjustment** toolbar.
2) Position the pointer at the source location and click once.
3) Move the pointer to other source locations to add more identity links, to prevent them moving during the transformation.

6. Perform the transformation:

1) Click the **Spatial Adjustment** menu.
2) Click the **Adjust** menu item.

7. Save the transformation results and stop the edit session:

1) Click the **Editor** menu on the **Editor** toolbar, and click **Save Edits**.
2) Click the **Editor** menu on the **Editor** toolbar again, and click **Stop Editing**. The adjusted source layer should look similar to Figure 3.10c.

Similarity transformation

Similarity transformation is to translate, rotate or scale data without changing angle measurements. It maintains the aspect ratios of the features, and preserves their shapes. Only the position and orientation of the features will change. Suppose x and y are the original coordinates (before transformation), and X and Y are the transformed coordinates. The similarity transformation function can be expressed as:

$$X = Ax + By + C \tag{3.1a}$$

$$Y = -Bx + Ay + F \tag{3.1b}$$

where:

$A = s \times \cos \alpha$
$B = s \times \sin \alpha$
$s = $ scale change (same in x and y directions)
$\alpha = $ rotation angle, measured counter-clockwise from the x axis
$C = $ translation in x direction
$F = $ translation in y direction

Figure 3.11 shows three possible similarity transformations of a rectangle with its lower-left corner located at the origin of the input coordinate system.

The values of A, B, C and F can be obtained by solving the simultaneous equations (Equation 3.1) using a minimum of two control points with known coordinates in the original and transformed coordinate systems (that is, two displacement links). When three or more control points are used, they can be determined more accurately using least squares adjustment (Maling 1991). In general, the more control points are used for a transformation, the more accurate the transformation will be.

Affine transformation

The affine transformation is a generalisation of the similarity transformation. It differentially scales, skews, rotates and translates data. Under the affine transformations, feature shapes and line lengths may not be preserved, but straight lines remain unchanged and parallel lines remain parallel. It may involve shear, a transformation in which all points along a given line L remain fixed while other points are shifted parallel to L by a distance proportional to their perpendicular distance from L. The affine transformation uses the following equations:

$$X = Ax + By + C \tag{3.2a}$$

$$Y = Dx + Ey + F \tag{3.2b}$$

where:

$A = s_x \times \cos \alpha$
$B = s_y \times (k \times \cos \alpha - \sin \alpha)$
$D = s_x \times \sin \alpha$
$E = s_y \times (k \times \sin \alpha + \cos \alpha)$

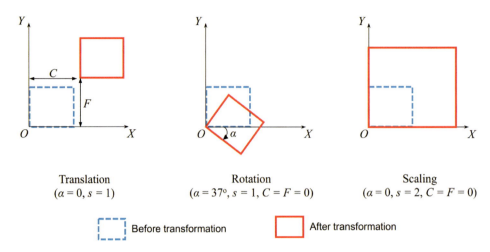

Translation
$(a = 0, s = 1)$

Rotation
$(a = 37°, s = 1, C = F = 0)$

Scaling
$(a = 0, s = 2, C = F = 0)$

☐ Before transformation   ☐ After transformation

Figure 3.11 Similarity transformation.

s_x = scale change in x direction
s_y = scale change in y direction
k = shear factor, i.e. the distance a point moves due to shear divided by the perpendicular distance of the point from the invariant line

$α$, C and F are the same as those given in Equation 3.1. Figure 3.12 uses a rectangle as an example to illustrate differential scaling and skew transformation. The effect of a

skew transformation looks like pushing a feature in one direction parallel to a fixed line (such as a coordinate axis).

The values of A, B, C, D, E and F can be obtained by solving the simultaneous equations in Equations 3.2a and 3.2b using a minimum of three control points with known coordinates in the input and transformed coordinate systems (that is, at least three displacement links). The affine transformation assumes that geometric distortions in an input map or image are uniform. It

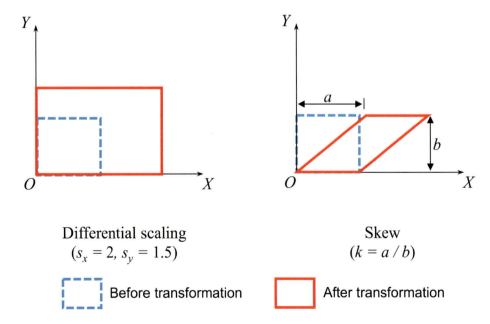

Differential scaling
$(s_x = 2, s_y = 1.5)$

Skew
$(k = a / b)$

☐ Before transformation   ☐ After transformation

Figure 3.12 Affine transformation.

is commonly used for map-to-map transformation (to georeference a digital map to a particular projected coordinate system) and image-to-map transformation (to georeference an image to a particular projected coordinate system). Box 3.8 shows how to implement the affine transformation in ArcGIS.

Box 3.8 Affine transformation in ArcGIS   PRACTICAL

The affine transformation is one of the geometric transformation tools in the ArcGIS spatial adjustment toolset. Like rubber sheeting, all geometric transformations in ArcGIS are based on displacement links, representing the source and destination locations for an adjustment. To follow this example, start ArcMap and add the two data layers **mystreets_src1** (as the source layer to be transformed) and **mystreets_tgt** (as the target layer) from **C:\Databases\GIS4EnvSci\others**. The affine transformation can be implemented by following these steps:

1. Set up the editing environment as in the first step in Box 3.7.
2. Set up the spatial adjustment environment as in the second step in Box 3.7.
3. Select a transformation method
4. Click the **Spatial Adjustment** menu. Point to **Adjustment methods**, then click **Transformation – Affine**.
5. Create displacement links:

   1) Click the **New Displacement Link** tool ➚⁺ on the **Spatial Adjustment** toolbar.
   2) Position the mouse pointer at the source location and click once to start adding a link.
   3) Move the pointer to the target location and click once to finish adding the link. An arrow appears, representing a displacement link connecting the source location to the target location.
   4) Repeat the above two steps to add at least three displacement links.

6. Perform the transformation as in the sixth step in Box 3.7.
7. Save the transformation results and stop the edit session as in the final step in Box 3.7.

The above steps can be used for similarity and projective transformations.

Projective transformation

Projective transformation maps lines to lines, but does not necessarily preserve parallelism. It does not preserve sizes or angles either. Two parallel lines can be transformed to two intersecting lines. For example, a rectangle may be transformed to an irregular quadrilateral using projective transformation. It is based on a more complex mathematical formula, whose general form can be expressed as:

$$X = (Ax + By + C) / (Gx + Hy + 1) \qquad (3.3a)$$

$$Y = (Dx + Ey + F) / (Gx + Hy + 1) \qquad (3.3b)$$

Solving for the transformation coefficients requires at least four control points (or four displacement links). This method is mainly used to transform data captured directly from aerial photography.

Residual errors and root mean square errors in geometric transformation

The accuracy of geometric transformation is usually measured using residual errors and root mean square errors (RMSEs) of the control points. The residual error of a control point is the derivation of the transformed source control points from the destination control points. Assume (x, y) represents the coordinates of a destination control point (the true position) and (x', y') represents the transformed coordinates of the corresponding source control point (the estimated position by transformation). The residual error of the control point is simply calculated as the distance between the true location and the estimated position:

$$e = \sqrt{(x - x')^2 + (y - y')^2} \qquad (3.4)$$

Therefore, the residual error is a measure of the fit between the true locations and the transformed locations of the control points. In ArcGIS, this error is generated for each displacement link. If the residual error exceeds the required tolerance value, then the control point should be removed and a new control point entered.

The RMSE is computed by taking the square root of the mean squared residuals:

$$RMSE_e = \sqrt{\frac{\sum_{i=1}^{n} e_i^2}{n}} \qquad (3.5)$$

where n is the number of control points used in the transformation, and e_i is the residual error of the ith control point. The RMSE measures the accumulation of positional errors of all control points during the transformation. When the RMSE is particularly large, some control points need to be removed, adjusted or added to make sure the error is in the acceptable range.

RMSE is generally used to measure the transformation's overall accuracy. However, a low RMSE value does not mean the transformation is highly accurate. The transformation may still contain significant errors due to a poorly entered control point. The more control points of high quality used, the more accurate the transformation. In addition, control points should be well-defined points whose (x, y) coordinates are known. Significant landmarks, such as road junctions, corner-posts of fenced paddocks, roads or road–railway junctions, lake outlets, stream confluences and the corners of sharply imaged sheds/houses are often selected as control points. They should not be selected in areas that may change, such as shorelines and forested areas. They should also spread out over the entire map rather than being concentrated in one area. Typically, having at least one control point near each corner of the map and a few throughout the interior may produce the best results. In ArcGIS, a minimum of three links are required to produce a transformation that results in an RMSE, which can be viewed together with residual errors in the link table. A more general discussion of RMSE can be found in Section 3.5.

Edge matching

When a study area covers features that extend across the boundaries of adjacent map sheets or data layers, such as soils and contour maps, the map sheets or layers sometimes need to be merged into one. However, in general, individual map sheets are digitised separately. Errors inevitably exist in each map sheet to some degree, introduced by the distortion of the original paper maps or occurring during digitising or georeferencing processes. These errors lead to the features crossing the map sheet boundaries becoming disconnected, as shown in Figure 3.13. Edge matching is a procedure to adjust the position of features at map sheet boundaries from the adjacent layers so that they meet at the join. It mainly involves first identifying the matching features along the edges of two adjacent data layers, then adjusting the data layer with less accurate features so that those features crossing the layer boundaries align with the same features on the adjoining layer (Figure 3.13). After edge matching, the two adjacent layers can be merged into a single layer. If features in the two input layers are polygon features, one extra step is needed to remove the common boundaries of polygons and rebuild topology to form a continuous layer. Box 3.9 illustrates how to perform edge matching in ArcGIS.

Box 3.9   Edge matching in ArcGIS   **PRACTICAL**

Edge matching can be done using the **Edge Match** tool in the ArcGIS spatial adjustment toolset. This tool uses the data layer with more accurate features as the control to adjust the adjoining data layer with less accurate features. Vertices of polygon features or the endpoints of line features are adjusted to the corresponding locations in the control layer using displacement links. To follow this example, start ArcMap and add the two data layers **myrivers_w** (as the source layer) and **myrivers_e** (as the target layer) from **C:\Databases\GIS4EnvSci\others**.

1. Set up the editing environment as in the first step in Box 3.7. Also ensure that end snapping is enabled. If it is not, click **End Snapping** ⊞ on the **Snapping Toolbar**.
2. Set up the spatial adjustment environment:

    1) Click **Customize** from the main menu bar. Point to **Toolbars. . .**, then click **Spatial Adjustment** to add the **Spatial Adjustment Toolbar** to ArcMap.

2) Click the **Spatial Adjustment** menu in the **Spatial Adjustment Toolbar**, and click **Set Adjust Data**.

3) In the **Choose Input For Adjustment** dialog box, choose **Selected features**.

3. Select a transformation method:

1) Click the **Spatial Adjustment** menu. Point to **Adjustment methods**, then click **Edge Snap**.

4. Select a edge snap method:

1) Select **Options. . .** from the **Spatial Adjustment** menu. In the **Adjustment Properties** dialog, click the **Options. . .** button.

2) In the **Adjustment Properties** dialog, click the **General** tab.

3) Click either the **Smooth** or **Line** method. When the **Smooth** edge snap method is used, vertices at the link source point are moved to the destination point. In the meantime, the remaining vertices of the feature are also adjusted to provide an overall smoothing effect. When the **Line** edge snap method is used, only the vertices at the link source point are moved to the destination point, while other vertices of the feature remain unmoved.

4) Check the checkbox if you want to adjust to the midpoint of links.

5. Set the edge snap properties:

1) In the **Adjustment Properties** dialog, click the **Edge Match** tab.

2) Click the **Source Layer** drop-down arrow and select **myrivers_w**, the layer with less accurate features.

3) Click the **Target Layer** drop-down arrow and select **myrivers_e**, the adjoining layer with more accurate features.

4) Check **One link for each destination point** if desired.

5) Check **Preventing duplicate links** if desired.

6) Check **Use Attributes** to ensure links are connected to features that share common attribute values.

7) If you chose to use attributes, click the **Attributes** button. Match the source and target layer fields, and add them to the matched fields.

8) Click **OK** to close the **Adjustment Properties** dialog.

6. Create displacement links:

1) Select (by holding down the SHIFT key) features in a small area to be adjusted (using the **Edit** tool ▶ on the **Editor** toolbar to select features).

2) Click the **Edge Match** tool ⫶⫶⫶ on the **Spatial Adjustment** toolbar.

3) Use the mouse to drag a box around the endpoints or vertices of the selected features. The **Edge Match** tool places multiple displacement links between the closest source and target features located in the box.

4) Repeat the above three steps to select features in another small area to add displacement links until all features to be matched have been selected.

7. Perform the transformation as in the sixth step in Box 3.7.

8. Save the transformation results and stop the edit session as in the last step in Box 3.7.

Line simplification

Data layers captured at different scales and from different sources often contain different levels of detail about the geographical features they represent. The locational accuracy of GIS outputs largely depends on input layers with the lowest accuracy of representation. If data layers used for a GIS analysis were from maps or images at very different scales, it is necessary to simplify and generalise

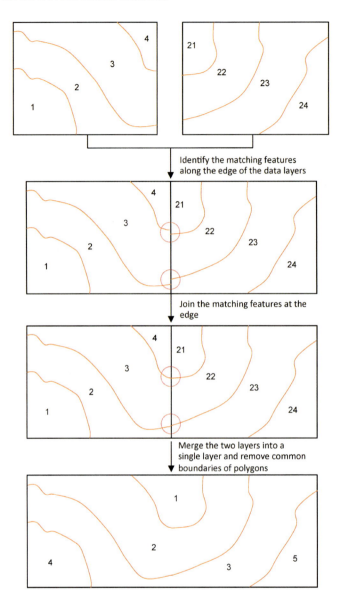

Figure 3.13 Edge matching.

features in the data layers captured from larger-scale maps or images to remove unnecessary geometric detail in their representation. In GIS, this is mainly achieved through line simplification.

Line simplification, also called weeding, is used to remove insignificant bends and redundant points or vertices from a line or a polygon while preserving its essential shape. There are two methods commonly used for line simplification in GIS: point-based and bend-based.

The point-based approach reduces the number of points required to represent a digitised line. It is often implemented using the well-known Douglas-Peucker algorithm (Douglas and Peucker 1973). This algorithm starts by connecting the endpoints of a line with a straight line, called a trend line, then calculates perpendicular distances from the trend line to all intervening points. If these perpendicular distances are all smaller than a user-specified simplification tolerance (a threshold distance), then the trend line is used to represent the whole line in simplified form, and the process stops. Otherwise, the vertex farthest from the trend line is connected to the endpoints of the line to form two new trend lines. For each new trend line, perpendicular distances to its intervening points are measured. If the perpendicular distances are all smaller than the simplification tolerance, the two trend lines are linked to represent the simplified line feature. If not, the vertices with the longest distance from the trend lines (greater than the threshold distance) are used to form new trend lines, and the above process is repeated until the distances from the vertices to the corresponding trend lines are within the simplification tolerance. The final trend lines are connected to form the simplified line. Figure 3.14 illustrates the process. This algorithm is easy to implement, and has been widely used as a scale-independent method for generalising

line features, including boundaries of polygon features. However, it may produce simplified lines with sharp angles and spikes. In some cases, it may self-cross.

The bend-based approach involves identifying and eliminating insignificant bends along a line or polygon boundary. For example, Wang (1996) proposed an algorithm which applies shape recognition techniques to identify bends along a line or polygon, analyse their geometric properties and remove extraneous ones. A line feature can be viewed as a series of bends. Each bend is defined as a fraction of a line composed of a number of subsequent vertices, with the inflection angles on all vertices included in the bend being either positive or negative, and the inflection of the bend's two end vertices being in opposite signs (Wang 1996; Wang and Müller 1998). The algorithm detects bends according

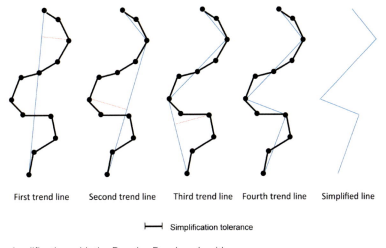

First trend line   Second trend line   Third trend line   Fourth trend line   Simplified line

├──┤ Simplification tolerance

Figure 3.14 Line simplification with the Douglas-Peucker algorithm.

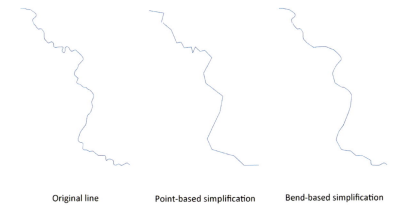

Original line   Point-based simplification   Bend-based simplification

Figure 3.15 Comparison of line simplification with the point- and bend-based approaches.

to the inflection angles at vertices, then measures several geometrical properties of each bend, such as size, shape and the relationship with its neighbouring bends. These properties determine whether a bend is kept or removed. Bend elimination is carried out by replacing the bend curve with its baseline, the line connecting the endpoints of the bend. The iteration for generalisation is carried out until the baseline length of every bend on

the line is larger than the length of the reference bend baseline, which is a simplification tolerance specified by a user. The bend-based approach generally eliminates fewer points, but preserves a geometry that is closer to the original (the main shape and topology of the original line) than the point-based approach does, as shown in Figure 3.15. Both methods can be implemented in ArcGIS (Box 3.10).

Box 3.10 Line simplification in ArcGIS   **PRACTICAL**

To follow this example, start ArcMap and add **myriver** from **C:\Databases\GIS4EnvSci\others**.

1. Click on the **ArcToolbox** button to open the ArcToolbox window.
2. In the **ArcToolbox** window, expand **Cartography Tools > Generalization**, then double-click **Simplify Line**.
3. In the **Simplify Line** dialog box:

    1) Select **myrivers_e** as the line feature class or layer to be simplified.
    2) Type in your preferred name for the output feature class.
    3) Select a simplification algorithm. The **point remove** method implements the Douglas-Peucker algorithm, while the **bend simplify** method applies Wang's algorithm (Wang 1996).
    4) Specify the simplification tolerance as **150** metres.
    5) Optionally, enable the **Check for topological errors** parameter to identify and resolve topological errors introduced by the simplification process.
    6) Click **OK**.

4. When the process is complete, the simplified line features are added to the data view.

3.4 SPATIAL DATA EDITING

Spatial data editing aims to remove the errors that arise during the encoding of spatial data, including location, topological and attribute data errors. Location errors are positional inaccuracies of digitised features, including locational displacement of features, doubling digitising, omission errors (missing features) and commission errors (features erroneously included in the data when they should have been excluded, possibly because the original sources are outdated). Topological errors are those that violate the topology rules defined by either a GIS system or a user – for example, only one point may exist at a given location, lines must intersect at nodes, polygons are bounded by lines, overlapping polygons do not exist and so on. Undershoot, overshoot, dangling arc and unsnapped node, depicted in Figure 3.6, are all topological

errors. Attribute data errors may include incorrect assignment of feature unique identifiers during manual key-in or digitising, missing data records or too many data records. Different from spatial data transformation (discussed in Section 3.3), which corrects systematic errors with regular patterns, spatial data editing is an interactive process involving identification of random spatial and non-spatial errors and graphic editing of these errors.

The easiest way to identify location and topological errors is to plot the digital map at the same scale as its source map, and then overlay it with the source map for comparison. Missing features can be added through digitising, and errors can be corrected using the editing operations described below.

Before using the editing tools to edit features, the editing environment needs to be set up, which

Type of Errors	Before Correction	Edit Function	After Correction
Unclosed Polygon	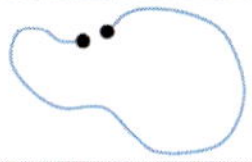	Move a vertex	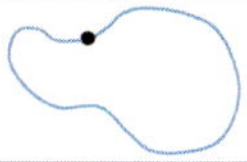
Missing Arc	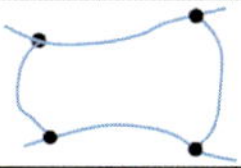	Split a feature	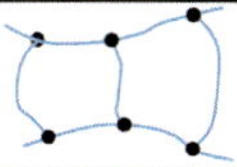
Overshoot	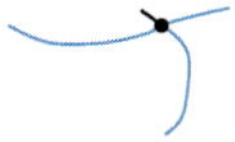	Delete a feature	
Undershoot		Extend a feature	

Figure 3.16 Examples of basic editing operations.

involves defining the tolerances used to snap points and lines. The snap tolerance is a specified distance within which points or features are moved to match or coincide exactly with each other. It can be measured using either map units (for example, metres) or pixels. When the snap distance is defined, in an editing session, the pointer or a feature will snap to another location (vertex, edge, endpoint, intersection and so on) within that distance. Steps 1–7 in Box 3.11 show how to start an editing session and set up the snap environment in ArcGIS.

Editing individual features

Individual features can be modified using graphic editing tools. Basic editing operations available in GIS include extend (lines), trim (lines), delete or erase (vertices or features), add (vertices), copy (features), create (features), rotate (features), move (vertices or features), reshape (features), and split and merge (features). Figure 3.16 provides some examples of the basic editing operations. Box 3.11 illustrates the basic operations for editing individual features in ArcGIS.

Box 3.11   Editing individual features in ArcGIS   **PRACTICAL**

To follow this example, start ArcMap and add **myriver** (a line feature layer) and **mylake** (a polygon feature layer) from **C:\Databases\GIS4EnvSci\others**. First edit **myriver** according to the following steps, then repeat the steps to edit **mylake**.

Set up the editing environment

1. Click the **Editor Toolbar** button 🖉 on the **Standard** toolbar.
2. Click the **Editor** menu on the **Editor** toolbar, and select **Start Editing**.
3. Click the **Editor** menu on the **Editor** toolbar, point to **Snapping**, then click **Snapping Toolbar**.
4. Click the **Snapping** menu on the **Snapping** toolbar, and check **Use Snapping**, if it is not already checked, to enable snapping.
5. On the **Snapping** toolbar, click **End** ⊞ , **Vertex** ☐ and **Edge** ⬜ , if these buttons are not already highlighted. By making them active, points or features will snap to the closet endpoint, vertex and edge (boundary) of another feature within the snap tolerance.

(continued)

(continued)

6. Click the **Snapping** menu on the **Snapping** toolbar, and click **Options**. In the **Snapping Options** dialog box:

   1) Enter the snap tolerance. As a rule of thumb, the snap tolerance should be at least 10 pixels.
   2) Check the boxes for **Show Tips**, **Layer Name**, **Snap Type** and **Background**, if they are unchecked. A snap tip is a piece of pop-up text that indicates the layer to be snapped to and with which snap type (edge, end, vertex and so on).

7. Click **OK** to close the **Snapping Options** dialog box.

Selecting features for editing

8. Click the **Edit tool** 🏴 on the **Editor** toolbar, then click the feature in the data layer to be edited. The selected feature is highlighted in light blue. Hold down the SHIFT key while clicking features to select additional features.

Moving features

9. Simply drag the selected feature or features to the desired location.

Rotating features

10. Click the **Rotate tool** 🔄 on the **Editor** toolbar, click anywhere on the map, and drag the pointer to rotate the selected feature to the desired orientation.

Reshaping a feature

11. Select only one feature to reshape.
12. Click the **Reshape Feature tool** 🔲 on the **Editor** toolbar.

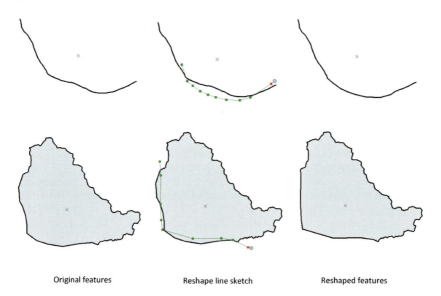

| Original features | Reshape line sketch | Reshaped features |

Figure 3.17 Reshaping a feature.

13. Click on the map to create a line according to the way the feature should be reshaped. The line sketch must cross or touch the edge of the selected feature two or more times so that it can be reshaped (Figure 3.17).
14. After drawing the line sketch, double-click to reshape the feature.

Reshaping a feature by editing vertices

15. Double-click the feature to be edited. All the vertices along the boundary of the feature appear as small dark green squares.
16. To move and delete a vertex, position the pointer over a vertex until it changes to the move pointer ◈.
    1) To move the vertex, drag it to the desired location.
    2) To delete the vertex, right-click, then click **Delete Vertex**.

17. To add a vertex, move the pointer to a location on the boundary of the feature where the vertex is to be added, right-click, then click **Insert Vertex**.

Splitting a feature

18. To split a line feature, select the line, then click the **Split** tool ✂ on the **Editor** toolbar. Move the pointer over the line. Click a place on the line to split the line at that location.
19. To split a polygon feature, select the polygon, then click the **Cut Polygons** tool ⬚ on the **Editor** toolbar. Draw a line sketch that cuts completely through the original polygon as desired. Right-click anywhere on the map, then click **Finish Sketch**.

Extending lines

20. Select the line to which other lines are to be extended.
21. Click the **Editor** menu on the **Editor** toolbar, point to **More Editing Tools**, then click **Advanced Editing**.
22. Click the **Extend** tool ⇢| on the **Advanced Editing** toolbar.
23. Click each line to be extended to meet the selected line.

Deleting features

24. Select the features to be deleted.
25. Click the **Delete** button ✖ on the **Standard** toolbar, or click the **Delete** key on the keyboard.

Editing features with shared geometry

Many vector data layers contain features that share geometry – that is, they have common boundaries or vertices. For instance, a river may be a border between two states, and soil polygons may share borders with land use polygons and shorelines of water bodies. When editing these layers, features that are coincident should be updated simultaneously so that they continue to share geometry.

The operations for editing individual features discussed above do not change the shapes and positions of other features with shared geometry. For example, each polygon is treated as separate from other polygons in the data layer. When a polygon is moved, its adjacent, connected polygons remain unchanged. Therefore, the relocated polygon will overlap with some other polygons and leave its original location an empty or void place, which leads to a change in topological relationships among features. In order to maintain the topology (such as coincident borders and nodes), topological associations among the features need to be built, with which boundaries and nodes/vertices shared by multiple features can be edited, moved or reshaped simultaneously. In other words, when a node is moved, all the edges (which are line segments that define lines or polygons) that connect to it are stretched to remain connected to the node. For example, as shown in Figure 3.18,

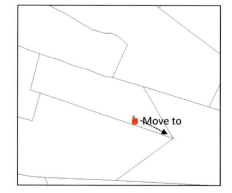

Figure 3.18 Editing features with shared geometry.

by moving a corner vertex shared by three land lots, the boundaries of the three lots are all updated at the same time. Similarly, when an edge is moved, all the adjacent edges are stretched to stay connected.

In ArcGIS, editing features with shared geometry requires first building a map topology, then defining a cluster tolerance. A map topology is a set of topological relationships between coincident parts of features in a data layer or several data layers. For example, a map topology can be constructed between a stream (line) layer and a forest (polygon) layer to ensure that forest boundaries are coincident with streams. With the map topology, coincidental features in all participating data layers can be identified according to a specified cluster tolerance (defined below). When edited with the map topology, features that are coincident are updated simultaneously so that they continue to share geometry. Coincident edges are called topology edges. The points

where edges are connected are referred to as topology nodes. In topological editing, they are collectively called topology elements. A cluster tolerance is the distance within which lines (edges) and vertices are close enough to be considered coincident. Lines or vertices that fall within the specified cluster tolerance will be snapped together. The default cluster tolerance in ArcGIS is 1mm. Increasing the cluster tolerance may cause more features to be snapped together unintentionally, leading to poor data accuracy and feature distortion. A typical cluster tolerance is at least an order of magnitude smaller than the accuracy of the data. For example, if the features are accurate to 1m, the cluster tolerance should be no more than 0.1m. If the data accuracy is unknown, a good strategy is to use the default cluster tolerance and increase it only to handle more severe but localised errors (Chang 2014). Box 3.12 provides a procedure for topological editing with shared geometry in ArcGIS.

Box 3.12   Editing features with shared geometry in ArcGIS   TECHNICAL

1. Start ArcMap, load the data layer(s) (shapefiles and geodatabase feature classes) to be edited, start an editing session, and set up the editing environment as described in Box 3.11.
2. Click the **Editor** menu on the **Editor** toolbar, point to **More Editing Tools**, then click **Topology**. The **Topology** toolbar appears.

Selecting data layers to participate in a map topology

3. Click **Select Topology** button 🗗 on the **Topology** toolbar.
4. In the **Select Topology** dialog box:

    1) Check all the data layers to participate in the map topology.
    2) Optionally, click **Options** to view the cluster tolerance. The default is 0.001m. Change the value if necessary.

3) Click **OK**. A temporary set of topological relationships between the parts of features that are coincident is created.

Selecting topology elements to edit

5. Click the **Topology Edit** tool 🔧 on the **Topology** toolbar.
6. Click a topology edge or node, or drag a box to select several topology elements. The selected topology elements appear in magenta. To clear the selection, simply click away from the element(s). When a topology element is selected, a topology cache is created, which stores the topological relationships between edges and nodes of the features that are located in the current display extent. If the map is zoomed out, the topology cache needs to be rebuilt in order to include other features in the display extent.

Moving a topology element

7. Use the **Topology Edit** tool to select a topology edge or node to move.
8. Drag the edge or node to a new location. Edge segments stretch so that the edge's endpoints (shared with other edges) are connected to their previous positions.

Modifying a topological edge

9. Click the **Modify Edge** tool 🔧 on the **Topology** toolbar, then click the topological edge to be edited, or double-click the topological edge with the **Topology Edit** tool. The edge turns into an edit sketch with small squares representing vertices.
10. Follow steps 15–17 in Box 3.11 to reshape the edge by moving, deleting and adding vertices.

Reshaping a topological edge

11. Select the topological edge to be edited with the **Topology Edit** tool.
12. Click the **Reshape Edge** tool 🔧 on the **Topology** toolbar.
13. Click on the map to create a line according to the way the feature should be reshaped, as shown in Figure 3.17.
14. After drawing the line sketch, double-click to reshape the feature.

Aligning an edge to match another edge

15. Click the **Align Edge** tool 🔧 on the **Topology** toolbar.
16. Click the edge that needs to be aligned (shown in solid magenta).
17. Click the edge to which the above edge should be aligned. The first edge is adjusted to match the second edge.

Rebuilding the topology cache

18. With the **Topology Edit** tool, right-click the map, and click **Build Topology Cache**. When editing features in a map topology, the topology cache generally needs to be rebuilt after making edits that change the topological connections among features (for example, after modifying an edge to cross another).

Editing with topology rules

Beyond editing shared geometry, some GIS systems support the use of topology rules to discover and remove topological errors in spatial data. Topology rules can be defined by the GIS system or the user. They control the acceptable topological relationships among features in a data layer (or feature class) or several data layers (or different feature classes): for example, land cover boundaries must not overlap, roads must not intersect with building footprints, or address points must be properly inside property boundaries. The features are checked against the rules to identify those in violation – that is, topological errors.

Geodatabase in ArcGIS supports spatial data editing with thirty-one topology rules, ten rules for polygons, fifteen rules for lines and six rules for points. Key topology rules in Geodatabase can be found in Perencsik (2004). Table 3.2 gives some examples of topology rules for polygons, lines and points. The topology rules in ArcGIS can only be built and applied to feature classes in a feature dataset based on the geodatabase data model.

Like editing features with shared geometry, editing with topology rules in a geodatabase first requires the construction of a topology for all participating feature classes in a feature dataset. It involves specifying a cluster tolerance, topology rules and ranks. Ranks are assigned to the participating feature classes in the topology. They determine which feature classes are more likely to be moved when fixing topological errors. Edges or vertices of lower-ranking features within the cluster tolerance will

be snapped to nearby edges or vertices of higher-ranking features. Edges or vertices of features of equal rank that lie within the cluster tolerance will be geometrically averaged together. Generally, lower-ranking features are known to have less reliable or less accurate positions. Figure 3.19 shows a set of feature classes and the associated ranks and rules that apply to those participating in a topology. Geodatabase topology is created and stored as a separate data layer, not with features, in a geodatabase.

Once a topology is built, it is validated: (1) to evaluate the features against the topology rules and identify topological errors where the rules are violated, and (2) to identify coincident features that fall within the cluster tolerance and snap them together automatically. The automatic editing of adjacent features with shared geometry in the validation process involves cracking and snapping features (Figure 3.20). In the cracking stage, new vertices are created at the intersection of feature edges. That is, when a vertex of one feature in the topology is within the cluster tolerance of an edge of another feature in the topology, a new vertex is automatically created on the edge so that the features can be geometrically linked

Table 3.2 Examples of geodatabase topology rules for polygons, lines and points

Feature type	Rule	Usage
Polygon	Must not overlay	An area cannot belong to two or more polygons.
	Must not have gaps	All polygons must form a continuous surface, and the data must cover the entire area.
	Contains one point	Each polygon must contain exactly one point. Each point must fall within a polygon.
	Must be covered by	Area features of a given type must be located within features of another type.
Line	Must not self-intersect	Line features cannot cross or overlap themselves.
	Must not self-overlap	Line features can cross or touch themselves, but not overlap themselves.
	Must not have dangles	A line feature must touch lines from the same feature class at both endpoints.
	Must be covered by boundary of	Line features must be covered by the boundaries of area features.
Point	Must be covered by boundary of	Points must fall on the boundaries of area features
	Must be properly inside polygons	Points must fall within area features
	Must be covered by endpoint of	Points in one feature class must be covered by the endpoints of lines in another feature class.
	Must be covered by line	Points in one feature class must be covered by lines in another feature class.

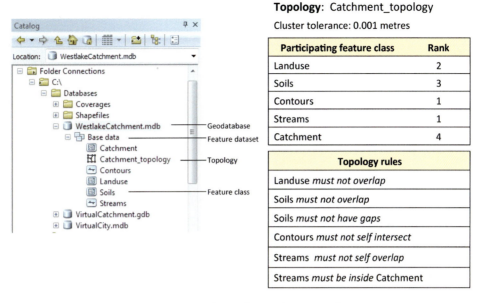

Topology: Catchment_topology

Cluster tolerance: 0.001 metres

Participating feature class	Rank
Landuse	2
Soils	3
Contours	1
Streams	1
Catchment	4

Topology rules
Landuse *must not overlap*
Soils *must not overlap*
Soils *must not have gaps*
Contours *must not self intersect*
Streams *must not self overlap*
Streams *must be inside* Catchment

Figure 3.19 Participating features classes, rules and ranks in a catchment topology.

in the snapping stage. When snapping features during topology validation, vertices that fall within the cluster tolerance are snapped together. The rank of a feature class determines whether or not its vertices will move when they fall within the cluster tolerance of vertices of another feature. The results of validation are saved into a topology layer (such as the Catchment_topology layer in Figure 3.19).

After topology validation, the identified topological errors can be viewed, fixed or accepted as exceptions using a set of topology error fixing tools. These editing tools allow the user to select errors and apply predefined topology fixes of various types to them. After removing errors, the topology is validated again to ensure the edit is correct. Box 3.13 illustrates how to edit features with topology rules in the geodatabase.

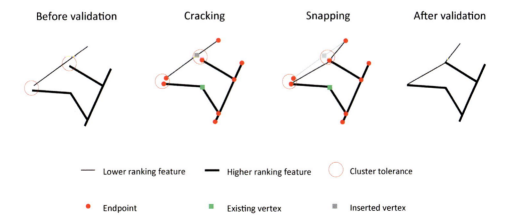

Figure 3.20 Topology validation process.

Box 3.13 Editing with geodatabase topology in ArcGIS   **TECHNICAL**

Geodatabase topology is created, validated and managed in ArcCatalog. Before creating a topology, feature classes and topological rules need to be defined. Non-topological data, such as shapefiles, need to be migrated into a geodatabase. Suppose a geodatabase contains several feature datasets, each having a number of feature classes.

Building and validating a topology

1. Start ArcCatalog, and navigate to the geodatabase.
2. Right-click the feature dataset for which the topology will be built. Point to **New**, and click **Topology**. The **New Topology** wizard begins.
3. Click **Next** on the first panel.
4. On the second panel, enter a name for the topology, and specify a cluster tolerance. Click **Next**.
5. On the next panel, check the feature classes that will participate in the topology, then click **Next**.
6. On the fourth panel, key in the number of ranks allowed in the topology. Click in the **Rank** column, and assign a specific rank to each feature class. Click **Next**.
7. Click the **Add Rule** button on the fifth panel. In the **Add Rule** dialog:

    1) Select a feature class to participate in a topology rule.
    2) Select a topology rule.
    3) Select the second feature class if the rule relates the feature class to the first one.
    4) Click **OK**. One rule is added.

8. Add additional rules by repeating the previous step. After all the rules are added, click **Next** in the fifth panel.
9. Review the parameters and rules that have been defined for the topology. Click **Finish**. The wizard starts to build the new topology for the participating feature classes.
10. After the topology has been created, click **Yes** when prompted to validate it. The topology is now validated and saved in the feature dataset.

Modifying the topology

11. Right-click the topology saved in the previous step, and click **Properties**.
12. In the **Topology properties** dialog box:

    1) To rename the topology, click the **General** tab. Click in the **Name** text box, and type a new name.
    2) To change the cluster tolerance, click the **General** tab. Click in the **Cluster Tolerance** text box, and enter a new one.
    3) To add a feature class from the same feature dataset to the topology, click the **Feature Classes** tab. Click the **Add Class** button, then select a feature class to add.
    4) To remove a feature class from the topology, click the **Feature Classes** tab. Select a feature class, and click the **Remove** button.
    5) To change the number of ranks or the rank of a feature class, click the **Feature Classes** tab. Click the **Number of Ranks** text box, and type a number of ranks, or change the current rank of a feature class to a new one.
    6) To modify the rules, click the **Rules** tab. From here, rules can be added, removed or changed.
    7) To summarise topological errors, click the **Errors** tab, then click the **Generate Summary** button.

13. Click **OK** to finish the modification.
14. Right-click the topology, and click **Validate** to validate the topology.

Setting up the environment for editing with the topology

15. Start ArcMap, and load the feature dataset with the topology built above. Start an editing session, and set up the editing environment as described in Box 3.11.
16. Click the **Editor** menu on the **Editor** toolbar, point to **More Editing Tools**, then click **Topology**. The **Topology** toolbar appears.

Viewing topological errors

17. Click **Error Inspector** on the **Topology** toolbar.

    1) To view errors for all rules, click the **Show** drop-down arrow, and click **Errors from all rules**.
    2) To view errors for a specific topology rule, click the **Show** drop-down arrow, and click the rule.

18. Click the **Search Now** button. A report appears, which lists the errors.

Fixing topological errors

19. Click the error in the **Error Inspector** list, or use the **Fix Topology Error** tool on the **Topology** toolbar to click the error on the map. The error is drawn in black on the map.
20. Right-click the error in the **Error Inspector** list or on the map with the **Fix Topology Error** tool. Select one of the available fixes, which will depend on the type of error. For example, to fix a Must Not Have Dangles error, right-click the error, click **Snap**, then select a snap tolerance. This will snap the node to a vertex of another feature within the tolerance, and remove the error.
21. Validate the topology again by clicking the **Validate Topology In Specified Area** tool on the **Topology** toolbar and dragging a box around the area of interest to validate, or by clicking the **Validate Topology In Current Extent** tool on the **Topology** toolbar to validate the areas that are currently visible on the map.
22. Continue fixing topology errors as necessary.

Editing attribute data

Attribute data errors are more difficult to identify than locational errors as they are usually not apparent until later on in the data processing and analysis. They may result from observation or measurement errors, data entry errors or outdated data. Simple data entry errors, such as missing or duplicate data records, may become evident when linking locational and attribute data. Other errors may be hard to detect, particularly when they are syntactically correct. Attribute data editing in GIS mainly includes adding, deleting or updating attributes associated with features and their values.

Editing attributes

When a new data layer is created through digitising, the feature attribute table is generated automatically with system-generated attributes or fields, including feature or object unique identifiers assigned to each feature and linked to the feature geometry. The user needs to add new fields to store attribute values for each feature. A feature attribute table may be the only table associated with a data layer. For some data layers, several attribute tables, storing related attribute data, may be used and linked to the feature attribute table. When a new table is created or new attributes are added to an existing table, fields need to be defined in terms of name, data type (text, number, date and so on) and property (for example, the length of the field and the number of decimal points). In some databases, attribute domains (rules that indicate valid values for a field) can also be defined.

Deleting an attribute simply involves deleting a field from an attribute table. Adding an attribute involves defining a new field. Box 3.14 shows how to create a table and add and delete a field in ArcGIS.

Box 3.14   Editing attributes in ArcGIS   **PRACTICAL**

ArcGIS uses tables in many formats and sources to store attribute data, including INFO tables, dBASE tables, Excel tables, tables from text files, those from geodatabases and other databases such as Access, Oracle and Informix. A new table can be created in ArcGIS as a dBASE table or a geodatabase table.

Creating a new table

1. Start ArcMap, and open **ArcToolBox**.
2. In **ArcToolbox**, expand the **Data Management Tools** > **Tables** toolset, then click **Create Table**.
3. In the **Create Table** dialog, specify the name of the geodatabase in which the new geodatabase table is to be created, or specify the name of the output workspace (file folder) in which a new dBASE table is to be created.
4. Enter the name of the new table and an optional table whose fields can be used as a template for the fields in the new table.
5. Click **OK**. A new table is added to the table of contents.

Adding a field

6. Right-click the table in the table of contents, and click **Open Attribute Table**.
7. Click the **Table Options** button [icon] in the **Table** window, then click **Add Field**.
8. Enter the name of the field.
9. Click the **Type** pull-down arrow, and select the field type. A dBASE table supports six field types, including short integer (numeric values without fractional values within the range of -32,768 to 32,767), long integer (numeric values without fractional values within the range of -2,147,483,648 to 2,147,483,647), float (numeric values with fractional values within the range of about -3.4E38 to 1.2E38), double (numeric values with fractional values within the range of about -2.2E308 to 1.8E308), text and date. In addition to these field types, a geodatabase table supports BLOB (a binary large object storing annotation, images, multimedia and so on), GUID (a globally unique identifier, commonly used as the primary key of database tables) and raster (storing the raster data within or alongside the geodatabase).
10. Set any other field properties as necessary. For example, when numeric fields are defined, two field properties need to be specified: the precision, which is the maximum length of the field, and scale, which is the maximum number of decimal places.
11. Click **OK**.

Deleting a field

12. Right-click the table in the table of contents, and click **Open Attribute Table**.
13. Right-click the field header in the **Table** window of the field to be deleted, then click **Delete Field**.
14. Click **Yes** to confirm the deletion.

Editing attribute values

Attribute values in an existing table can be copied, pasted, added, deleted and updated. Some GIS software packages also allow users to enter attribute values immediately after creating a new feature, to apply the same attribute values to multiple features in a data layer and to edit the attributes of related features. Box 3.15 shows a procedure for editing attribute values in ArcGIS.

Box 3.15   Editing attribute values in ArcGIS   **TECHNICAL**

Editing attribute values for selected features

1. Start ArcMap, and load the data layer to be edited.
2. Start an editing session, and set up the editing environment as described in Box 3.11.
3. Click the **Edit** tool ▶ on the **Editor** toolbar, and click a feature in the data layer whose attribute values are to be edited.
4. Click the **Attributes** button ▦ on the **Editor** toolbar. The top part of the **Attributes** window lists the selected feature, and the bottom part of the window lists the fields and their values.
5. To change an attribute value, click a cell on the right side in the bottom part of the **Attributes** window, and enter the value for that field.
6. To copy an attribute value, select the contents of the cell, right-click, then click **Copy**. Then right-click, and click **Paste** to paste the attribute value into a different cell.
7. To delete an attribute value, select the contents of the cell, right-click the cell, and click **Delete**.

Adding attribute values upon creating a new feature

8. Click the **Editor** menu, and click **Options**.
9. Click the **Attributes** tab.
10. Check the **Display** box.
11. Specify whether to show the **Attributes** window for all layers, or just certain ones.
12. Click **OK**. The user will be prompted to enter attribute values into the **Attributes** window when a new feature is added.

3.5 SPATIAL DATA QUALITY

As discussed in Section 2.1, spatial data are generalised and simplified representations of real-world phenomena based on a particular way in which the geographical space is conceptualised. They are also usually observational and collected under non-controlled conditions. Therefore, spatial data may contain inherent errors that are insignificant at certain map scales and for some applications, but significant at other map scales and for other applications (Morris 2008). For spatial data to be useful, their quality must be compatible with their intended applications. For example, a 30m resolution digital elevation model is good enough for hydrological modelling in a large catchment, but it is of poor quality for predicting coastal flooding in flat terrain.

Data quality refers to the state of accuracy, precision, completeness, consistency and timeliness of the data that makes them appropriate for a particular use. To solve environmental problems successfully, spatial data must be accurate and in agreement with the real world they represent at certain scales, must be complete, consistent and precise with the minimum acceptable level of uncertainty, and must be sufficiently current or timely for the intended use.

Measures of spatial data quality

Spatial data quality is largely determined by accuracy, precision, completeness and consistency measured for each of the locational, attribute and temporal components of spatial data (Veregin 2005). However, time is not dealt with explicitly in conventional spatial data models. The discussions in this section focus on the quality assessment of locational and attribute components.

Accuracy

Accuracy is a measure of the closeness of the data values to the true values or values accepted as being true. The difference between observed and true values is error. In other words, accuracy is a measure of the degree to which a data value is free from error. Errors may be single and random departures from reality, or may be widespread, systematic deviations throughout a dataset. It is impossible for spatial data to be 100 per cent accurate, but possible to have data that are accurate to within specified tolerances. For instance, a sample point coordinate may be accurate to within ±5m. Therefore, accuracy is always a measure relative to the specification.

Positional accuracy is a measure of how closely the coordinate descriptions of features represented in the data compare to their true location. Measurement of positional accuracy depends on dimensionality (Veregin 2005). Positional accuracy of a single point feature is defined as the distance between the encoded location and the actual location, usually composed of horizontal and vertical accuracy. Suppose the actual location of a point is (x, y, z), and its encoded location is (x', y', z'). Its horizontal accuracy is calculated as $\sqrt{(x-x')^2+(y-y')^2}$ – that is, the horizontal distance between the actual and encoded locations. Its vertical accuracy is the height difference, $|z-z'|$. To assess the positional accuracy of a set of points in a data layer, the RMSE is often used.

Here the RMSE is defined as the square root of the average of the squared differences between each data value and its corresponding true value. The general equation for calculating the RMSE can be expressed as:

$$RMSE = \sqrt{\frac{\sum_{i=1}^{n}\left(X_i - X_i'\right)^2}{n}} \quad (3.6)$$

where X_i' is the ith data value, X_i is its corresponding true value, and n is the number of data values. The closer the RMSE is to zero, the more accurate the data are. For n points whose true coordinates are (x_1, y_1, z_1), $(x_2, y_2,$ $z_2), \ldots, (x_n, y_n, z_n)$, and whose encoded coordinates are (x'_1, y'_1, z'_1), $(x'_2, y'_2, z'_2), \ldots, (x'_n, y'_n, z'_n)$, the RMSE of their horizontal positions is

$$RMSE_{(x,y)} = \sqrt{\frac{\sum_{i=1}^{n}\left(x_i - x_i'\right)^2 + \sum_{i=1}^{n}\left(y_i - y_i'\right)^2}{n}} \quad (3.7)$$

and the RMSE of their vertical positions is

$$RMSE_z = \sqrt{\frac{\sum_{i=1}^{n}\left(z_i - z_i'\right)^2}{n}} \quad (3.8)$$

Assessment of positional accuracy of a line or polygon feature is more complex, as error is a mixture of positional error (error in locating points along the line or polygon) and generalisation error (error in selecting points to represent the line or polygon). There are various methods for measuring the positional accuracy of line or polygon features. One method is to measure equally spaced perpendicular offsets along the encoded line or polygon feature to their intersection with the true line or polygon, and then to calculate the RMSE (Figure 3.21). This method is recommended by the Australian map and spatial data horizontal accuracy standard established by the Intergovernmental Committee on Surveying and Mapping. Other methods include Hausdorff distance, epsilon band, single buffer overlay and double buffer overlay (Ariza-López and Mozas-Calvache 2012).

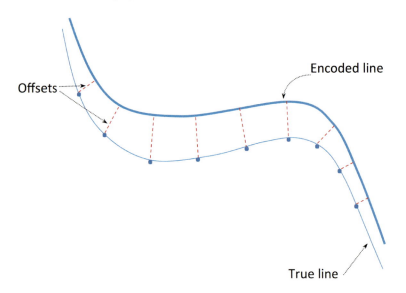

Figure 3.21 Measuring horizontal positional accuracy of a line.

Table 3.3 An error matrix for a land cover layer

Classified data	Reference data					Row total	User's accuracy
	Wetland	Forest	Pasture	Glacier/snow	Dryland		
Wetland	74	0	0	0	0	74	100%
Forest	9	398	30	18	28	483	82%
Pasture	1	21	65	4	2	93	70%
Glacier/snow	2	22	5	300	16	345	87%
Dryland	4	14	3	11	29	61	48%
Column total	90	455	103	333	75	1056	
Producer's accuracy	82%	87%	63%	90%	39%		
Overall accuracy = 82%				**Kappa = 0.74**			

In practice, positional accuracy of a test dataset is assessed against an independent dataset of higher accuracy by comparing the coordinates of sample locations in the test dataset with coordinates of reference locations that can be assumed to be the same in the independent source. Possible sources for higher-accuracy information include geodetic ground surveys, GPS ground surveys, photogrammetric surveys and spatial databases of substantially higher accuracy. In addition, a minimum of twenty sample locations must be selected for assessment, which should be evenly distributed over the geographical area of interest and reflect the distribution of error in the dataset.

Attribute accuracy is a measure of how closely the attribute values of features represented in the data compare to their true values. It is as important as positional accuracy. Attribute accuracy may be measured in different ways, depending on the nature of the data. For numerical attributes such as elevation, precipitation and temperature, accuracy may be measured in terms of measurement error (for example, precipitation accurate to 1mm) or RMSE. For categorical attributes such as land use and soil types, accuracy is commonly assessed using classification error matrices. A classification error matrix, also known as an error matrix or confusion matrix, is a cross-tabulation of encoded and actual classes at sample locations. In practice, error matrices compare on a category-by-category basis the relationship between known reference data from an independent source of higher accuracy and the test data. For example, suppose we have a land cover data layer containing five types of land cover: wetland, forest, pasture, glacier/

snow and dryland. By comparing the land cover types classified in the data layer with the known reference data of higher accuracy at 1,056 sample locations, an error matrix is produced as shown in Table 3.3. The element in row i and column j of the matrix is the number of sample locations assigned to class i but actually belonging to class j. The sum of row i is the total number of sample points assigned to class i. The sum of column j is the total number of sample points actually belonging to class j.

Four metrics have been developed for assessing attribute accuracy based on an error matrix: overall accuracy, producer's accuracy, user's accuracy and kappa index of agreement. The overall accuracy is defined as the sum of the diagonal values divided by the number of sample locations. For the above example, the overall accuracy for this land cover dataset is (74 + 398 + 65 + 300 + 29) / 1,056 = 82 per cent. This indicates that overall, 82 per cent of the sample locations are classified correctly.

The producer's accuracy indicates how well the sample locations of a given class are classified in the dataset. It is calculated by dividing the number of correctly classified sample locations in each category (on the major diagonal) by the number of sample locations actually belonging to that category (the column total). For example, the producer's accuracy of forest is 87 per cent. This means that 13 per cent of the forested locations are wrongly classified – that is, missing. Therefore, the producer's accuracy is a measure of omission error.

The user's accuracy is a measure of commission error. It indicates the probability that a sample location classified into a given category actually represents that category on the ground. The user's accuracy is calculated by dividing the number of correctly classified pixels in each category by the number of sample locations classified for that category (the row total). For example, the user's accuracy of forest is 82 per cent. This means that 82 per cent of the sample locations classified as forest are actually forested locations, but 25 per cent of the sample locations classified as forest are not the forested locations on the ground – they are wrongly classified and included as forest.

Kappa index of agreement, or simply kappa, is a statistic measuring the agreement on classification by taking into account the agreement occurring by chance. It is the proportion of the correctly classified locations after accounting for the probability of chance agreement. The kappa statistic of an error matrix, $\hat{k}$, is computed as:

$$\hat{k} = \frac{N\sum_{i=1}^{r} e_{ii} - \sum_{i=1}^{r}\left(e_{i+} \times e_{+i}\right)}{N^2 - \sum_{i=1}^{r}\left(e_{i+} \times e_{+i}\right)} \qquad (3.9)$$

where r is the number of rows in the matrix, e_{ii} is the element in row i and column i, x_{i+} and x_{+i} are the row and column totals for row i and column i respectively, and N is the total number of sample locations. This statistic describes the relative strength of agreement using a scale represented in Table 3.4.

For the error matrix listed in Table 3.3, its kappa statistic is calculated as 0.74, indicating substantial agreement. While the overall accuracy uses only the data along the diagonal of the error matrix and excludes the omission and commission errors, kappa incorporates the non-diagonal elements.

As a general guideline, a minimum of fifty sample locations for each category should be included in an error matrix. A large area with a large number of categories requires more sample locations. More sample locations should be allocated to more important or more variable categories (for example, more sample locations

Table 3.4 Kappa scale of relative strength of agreement

Kappa statistic	Strength of agreement
< 0.00	Poor
0.00–0.20	Slight
0.21–0.40	Fair
0.41–0.60	Moderate
0.61–0.80	Substantial
0.81–1.00	Almost perfect

for wetlands and fewer for open water) in the accuracy assessment.

Precision

Precision refers to the recorded level of detail, or level of measurement and exactness of description in spatial data. Positional precision is the number of significant digits used to record coordinates. For example, a coordinate specified in metres to the nearest ten decimal places (for example, 2340.3822474087) is more precise than one recorded to the nearest three decimal places (for example, 2340.382). Attribute precision is the level of detail of the data, measured in terms of the number of significant digits for numerical attributes, or the number or fineness of categories in a classification for categorical attributes. Precise attribute data may specify the characteristics of geographical features in great detail. For example, a precise description of a person living at a particular address might include gender, age, income, occupation, level of education and many other characteristics. An imprecise description might include just income or gender. General land use/land cover classes like urban are less precise than specific classes like residential and commercial. The precision of spatial data must match the level of detail required for the application.

However, it is important to note that high precision does not indicate high accuracy, or vice versa. Precise data, no matter how carefully measured, may be inaccurate. For example, a wrong reading in a distance measurement made to 0.01m is very precise, but inaccurate. The level of precision required for different applications varies greatly. Engineering projects such as road and utility construction demand precise information measured to 0.01m. It is possible to create a global database at a precision of less than 1m. However, such precision is unnecessary, and far exceeds the requirement for any typical analysis at the global scale. In addition, high-accuracy and high-precision data are both time-consuming and expensive to collect. Not every GIS application requires highly accurate and precise data. Excessive accuracy and precision are not only costly, but can give rise to considerable details which are not useful and helpful.

Consistency

Consistency refers to the lack of apparent contradictions between datasets or within a dataset. It is a measure of the internal validity of a database. Positional consistency mainly specifies conformance of spatial data with certain topology rules, such as those listed in

Table 3.2. The topological editing functions in a GIS can be used to detect topological inconsistencies and eliminate these errors, as described in Section 3.4. Attribute consistency refers to the absence of contradictions in thematic attribute values. For example, attribute values for bird population, density and size of the area must agree for area units. If the bird population observed in an area is forty and the size of the area is 1,000m², but the recorded bird density for the area in the dataset is 0.4 birds per m², this is inconsistent, as the bird density in the area should be 0.04 birds per m². The absence of inconsistencies does not necessarily imply that the data are accurate.

Completeness

Completeness is an assessment of the extent and range of a dataset with regard to completeness of coverage, classification and verification. Completeness of coverage refers to the proportion of the dataset available in its entirety (including its locational and attribute components). For example, is the spatial data coverage complete for the entire dataset? If not, what amount of spatial data are incomplete? Are attribute data available for the entire dataset? If not, what amount of attribute data are incomplete?

Consider a dataset of buildings in Melbourne. If it contains only buildings in Clayton (one suburb in Melbourne, rather than all suburbs of Melbourne), it is considered spatially incomplete. If the dataset contains only residential buildings, with no commercial, institutional or other private and public buildings, it is thematically incomplete.

Completeness of classification assesses how well the chosen classification method is able to represent real-world features. For example, is the adopted classification method exhaustive? Are there minimum area or minimum width rules (for example, roads less than 30m wide to be represented as a single line, lakes smaller than 0.25km² to be omitted) used to represent features?

Completeness of verification is an assessment of the amount of work (field work or other methods) carried out to validate the correct representation of real-world features contained within the dataset. For example, what are the extent and method of field verification carried out to validate both locational and attribute data? Are the positions of any geographical features in the dataset inferred? If so, what is the method of inference?

Completeness is application-dependent. A highly generalised spatial database can be complete if it contains all of the data and information required for particular applications.

Major sources of error in spatial data

There are many sources of error that may affect the quality of spatial data. Here we focus on three main sources: source data, data encoding and transformation, and data processing and analysis.

Errors in source data

Spatial data are produced through a process of selection, generalisation, projection and symbolisation. Subject to the rules of scale, many features and detailed spatial variations in the real world may be removed; shapes, directions, distances and areas of the selected features may be distorted; area features may have to be represented as point features (such as lakes) or line features (such as rivers) due to their small size; and 'vague, gradual, or fuzzy' boundaries in the real world (such as soil boundaries) may be misrepresented as 'crisp' boundaries. Therefore, source data inherently contain errors of an unknown magnitude.

In addition, spatial data are structured based on vector, raster or object-oriented data models. However, all spatial data models have limitations. The vector data model assumes that the world can be conceptualised as being composed of discrete points, lines and polygons, which can be represented using a single coordinate pair or a collection of them. Indeed, the vector model approximates, but does not exactly portray the shape of features using points. The raster model uses individual grid cells or collections of cells to represent features, and assumes that there are no spatial variations in each cell. The accuracy of the raster representation depends on resolution. At a lower resolution, more features and spatial variations are lost.

Source data may have been captured through field survey, photogrammetric measurement, remote sensing, image interpretation and analysis, or crowdsourcing. Measurement errors may be introduced by faulty observation, biased observers or by miscalibrated or inappropriate equipment. For example, a hand-held, non-differential GPS receiver may produce a position error of 30m. An incorrectly calibrated dissolved oxygen meter would give wrong values of oxygen concentration in a stream. Remote sensing images could be misinterpreted, and automatic image classification is rarely 100 per cent accurate (see Chapter 6). Source data may be too old to be useful for current GIS projects. Not only may much of the information base have changed, but also the past data collection standards may be mostly unknown or currently unacceptable. Errors in source data may skew, bias or negate results in spite of the power of GIS.

Errors in data encoding and transformation

Spatial data encoding is the process of digitising and inputting data into a GIS. Digitising is a significant source of error. In addition to the errors in source data passed to the GIS, operator's errors often occur, such as those shown in Figure 3.6. Errors may also be introduced when registering maps inappropriately during digitising. Keyboard attribute data input may also produce erroneous data. Such errors may result from uncertainty regarding characteristics of geographical features, mixing up of attribute values assigned to features, or incorrect entry.

Moreover, errors may occur when data are translated from one data structure to another or from one georeferencing system to another. Figures 3.8 and 3.9 illustrate information loss and distortion produced in the rasterisation and vectorisation process. As indicated in Section 3.3, errors are incurred in geometric transformation. The result of geometric transformation is considered to be accurate if the RMSE is within the acceptable tolerance value. The RMSE represents the magnitude of the average error, but many errors may be larger than the RMSE, and others may be smaller. It does not give any indication of the spatial variation of the errors.

Errors in data processing and analysis

Errors may arise during data processing and analysis. In particular, numerical rounding, overlay analysis and spatial interpolation are most likely to incur errors. Rounding errors occur when the results of data processing are truncated to the nearest integers, which may cause points near boundaries to be rounded off inside or outside an area (Burrough and McDonnell 1998).

Overlay analysis is a process of placing two or more data layers on top of each other to create a composite layer for integrated analysis (to be discussed in Section 4.4). It is important to note that data representing the same features in different layers may not be uniform and are subject to variation. Overlaying multiple data layers may result in problems such as slivers (thin polygons formed where the common boundaries overlap; see Figure 3.6). In addition, overlay analysis uses locational information to construct new features (points, lines or polygons) from input layers. As a result, the positional and attribute errors contained in the input layers will be passed to the combined output layer. The output from overlay analysis is only as good as the worst input layer (Heywood et al. 2011, p. 327).

Spatial interpolation is used to estimate the values at unknown locations based on a number of point samples. As will be discussed in Section 4.5, there are many spatial interpolation techniques. However, different interpolation techniques may produce different results based on the same sample data. Every interpolation technique involves errors. The output of GIS may contain an accumulation of errors from all the above three sources.

Metadata standards

Errors in spatial data are mainly managed through tracking and documentation. Quality of spatial data is often described in the metadata document accompanying the dataset. The term metadata refers to data about data, describing the content, quality, condition and other characteristics of a dataset. Several metadata standards have been developed and ratified by national or international standards bodies, such as the ISO 19139 Metadata Implementation Specification (by the International Organization for Standardization, or ISO), ISO 19115 Geographic Information – Metadata (by ISO), FGDC CSDGM Metadata (used in the USA), the North American Profile of ISO 19115 (used in USA and Canada), INSPIRE Metadata Directive (the European Profile of ISO 19115 and ISO 19119 Geographic information – Services) and ANZLIC Metadata Guidelines (the Australian/New Zealand Profile of ISO 19115). Each standard or profile identifies the content the metadata should provide to describe spatial data and other GIS resources, and describes the format in which the metadata should be stored. The key metadata elements for a spatial dataset generally include dataset title, abstract, geographical elements such as geographical extent, georeferencing system and spatial resolution, data quality measures, lineage, and database elements such as attribute label definitions and attribute domain values.

Lineage is a history of both the source data and the processing steps used to produce the spatial dataset. The source data may be from one or more data sources. Lineage details the names, descriptions, map scales and projections, production dates, producers and references of the source data. The processing steps are the sequence of operational steps performed on the source data to produce the final dataset. The history of the processing steps described in lineage usually includes the data capture methods, the intermediate processing methods, the production methods and the software and hardware used. Lineage provides useful information about how the dataset was derived. It can help trace the source datasets that may contain errors, assess the reliability and validity of the data and data processing methods, and build confidence in the use of the data.

Several GIS software systems and tool kits offer the capability to view, edit and validate metadata based on particular metadata standards or profiles. Examples include ArcGIS, AutoCAD Map 3D, GeoMedia, GeoNetwork (http://geonetwork-opensource.org/), ISO Metadata Editor (IME) (http://www.crepad.rcanaria.es/metadata/en/index_en.htm) and CatMDEdit (http://catmdedit.sourceforge.net/). In ArcGIS, the metadata editor is designed to create metadata documents for multiple metadata standards. The user selects a metadata style to implement a specific metadata standard. The metadata style determines how the metadata are viewed, edited, exported and validated for that metadata standard or profile. It can change how much metadata content is viewed and how it is displayed, and which pages are included in the editor and how they work. The standard ArcGIS Desktop supports five metadata styles: Item Description, ISO 19139 Metadata Implementation Specification, North American Profile of ISO 19115, INSPIRE Metadata Directive and FGDC CSDGM Metadata. The Item Description metadata style is the default. It provides only one page of information, describing a simple set of metadata properties without adhering to any metadata standards. Other metadata styles are used to create metadata that comply with the corresponding metadata standard. All metadata created using the ArcGIS metadata editor are stored in ArcGIS metadata format no matter which metadata style is used. Metadata for a shapefile are stored in the same location on disk in its accompanying XML file. For geodatabase datasets, metadata are stored in the geodatabase tables. Box 3.16 shows how to view and edit metadata in ArcGIS.

Box 3.16 Viewing and editing metadata in ArcGIS   **TECHNICAL**

Choosing a metadata style

1. Start ArcCatalog.
2. In **Catalog** main menu bar, click **Customize > ArcCatalog Options**. The **ArcCatalog Options** dialog box appears.
3. In the **ArcCatalog Options** dialog:

   1) Click the **Metadata** tab.
   2) Click the drop-down arrow, and select the style of metadata.
   3) Click **OK**.

Viewing metadata

4. In **ArcCatalog**, click the data layer whose metadata is to be viewed.
5. Click the **Description** tab, and the dataset's metadata is displayed in the ArcGIS metadata format.

Editing metadata

6. In **ArcCatalog**, click the data layer whose metadata is to be edited.
7. View the data layer's metadata.
8. Click the **Edit** button 📝 in the **Description** tab. The pages used to edit metadata appear in the **Description** tab.
9. Edit the metadata elements page by page. Metadata elements with a red background are mandatory for the selected metadata style.
10. Stop editing by clicking the **Save** button 💾 to save the edits, or by clicking the **Exit** button ✖ to stop editing without saving.

3.6 SUMMARY

- Spatial data are collected through data capture or data transfer. The main methods employed to capture spatial data include keyboard entry, digitising,

field surveying and remote sensing (including photogrammetry and LiDAR).

- Spatial data may come from different sources and exist in different formats. They are often required to be translated into a format compatible with the GIS system in use by using data conversion and transformation tools or via intermediate formats.

- Spatial data collated for a GIS project may need to be transformed prior to any GIS analysis. This may involve transforming all the data so that they are in the same georeferencing system and the same data structure, and contain the required level of details. It may also involve edge matching of data covering multiple map sheets.

- Rubber sheeting is a kind of spatial data transformation operation that aims to correct geometric distortions in the spatial data. Similarity, affine and projective transformations are used to transform data by shifting, rotating, scaling or skewing the coordinate system.

- Line simplification is used to remove the excessive vertices along line features or boundaries of area features in order to generalise their shapes.

- Spatial data may contain locational, topological and attribute errors resulting from data entry, encoding and digitising processes. They need to be examined and edited before being analysed and presented as information. Certain edits are performed on almost all spatial data once they are captured. Editing tools in GIS allow users to edit both individual features and features with shared geometry, with or without topology rules. These actions ensure that the data are accurate, complete and consistent.

- Spatial data quality is an essential characteristic that determines the reliability of the data for deriving insights to support problem solving and decision making. Spatial data are considered of high quality if their accuracy, precision, completeness, consistency and timeliness are suitable for their intended uses.

- Positional accuracy of spatial data is usually measured using RMSE, and attribute accuracy is assessed using error matrices.

- The major sources of error in spatial data include source data, data encoding and transformation, and data processing and analysis. Careful description of sources of error allows future improvements in data collection and processing techniques. It is also important for users to be aware of these errors so that they can assess the likely impacts of their use on the analysis or modelling outcomes. Spatial data quality is often described in metadata.

REVIEW QUESTIONS

1. What are the main methods used for spatial data capture?

2. Discuss the main problems that may occur in the process of conversion between vector and raster.

3. What is rubber sheeting? What is its purpose?

4. What are the main methods for transforming the coordinate system of the data layers?

5. Describe the edge matching process.

6. Describe the point- and bend-based approaches to line simplification.

7. Discuss potential errors arising from the digitising process, and appropriate edits that should be undertaken.

8. Give some examples of topology rules, and describe how they can be applied in spatial data editing.

9. What is data quality? What are the measures of spatial data quality?

10. How is positional accuracy of spatial data assessed?

11. What is an error matrix? How is it used to assess attribute accuracy?

12. What is the difference between accuracy and precision?

13. Discuss the major sources of error in spatial data.

14. What is metadata? What are the key elements of metadata?

15. Discuss the importance of metadata for spatial data.

REFERENCES

Ariza-López, F. and Mozas-Calvache, A. (2012) 'Comparison of four line-based positional assessment methods by means of synthetic data', *GeoInformatica*, 16(2): 221–243.

Burrough, P.A. and McDonnell, R.A. (1998) *Principles of Geographical Information Systems*, New York: Oxford University Press.

Chang, K. (2014) *Introduction to Geographic Information Systems*, 7th edn, New York: McGraw-Hill.

Douglas, D.H. and Peucker, T.K. (1973) 'Algorithms for the reduction of the number of points required to represent a digitized line or its caricature', *The Canadian Cartographer*, 10(2): 112–122.

ESRI (2003) *Introducing ArcScan for ArcGIS: An ESRI White Paper*, Redlands, CA: ESRI.

Ghilani, C.D. and Paul R.W. (2012) *Elementary Surveying: An Introduction to Geomatics*, 13th edn, Upper Saddle River, NJ: Pearson Education.

Gomarasca, M.A. (2009) *Basics of Geomatics*, Dordrecht, The Netherlands: Springer.

Graham, M.D., Douglas-Hamilton, I., Adams, W.M. and Lee, P.C. (2009) 'The movement of African elephants in a human-dominated land-use mosaic', *Animal Conservation*, 12: 445–455.

Hacklay, M. (2010) 'How good is volunteered geographical information? A comparative study of OpenStreetMap and Ordnance Survey datasets', *Environment and Planning B: Planning and Design*, 37(4): 682–703.

Han, S. (2011) *Web 2.0*, New York: Routledge.

Heywood, I., Cornelius, S. and Carver, S. (2011) *An Introduction to Geographical Information Systems*, 4th edn, Harlow, England: Pearson Education.

Hsu, J.M. (2010) 'Introduction to global satellite positioning system (GPS)', in C.M. Huang and Y.S. Chen (eds) *Telematics Communication Technologies and Vehicular Networks, Wireless Architectures and Applications*, Hershey, PA: IGI Global, 108–118.

Jensen, J.R. (2007) *Remote Sensing of the Environment: An Earth Resource Perspective*, 2nd edn, Upper Saddle River, NJ: Prentice Hall.

Luo, J., Joshi, D., Yu, J. and Gallagher, A. (2011) 'Geotagging in multimedia and computer vision: a survey', *Multimedia Tools and Applications*, 51(1): 187–211.

Maling, D.H. (1991) 'Coordinate systems and map projections for GIS', in D.J. Maguire, M.F. Goodchild and D.W. Rhind (eds) *Geographical Information Systems: Principles and Applications*, Vol. 1, Harlow, England: Longman Scientific & Technical, 135–146.

Morris, A. (2008) 'Uncertainties in spatial databases', in J.P. Wilson and A.S. Fotheringham (eds) *The Handbook of Geographic Information Science*, Malden, MA: Blackwell, 80–93.

OGC (2008) *OGC KML*, accessed 12 October 2013 at http://www.opengeospatial.org/standards/kml/.

OGC (2012) *Open GeoSMS Standard: Core*, accessed 2 September 2013 at http://www.opengeospatial.org/standards/opengeosms.

Pavlidis, T. (1982) *Algorithms for Graphics and Image Processing*. Rockville, MD: Springer-Verlag.

Perencsik, A. (2004) *ArcGIS 9: Building a Geodatabase*, Redlands, CA: ESRI.

Peuquet, D.J. (1981) 'An examination of techniques for reformatting digital cartographic data. Part 1: the raster-to-vector process', *Cartographica*, 18(1): 34–48.

Sullivan, B.L., Wood, C.L., Iliff, M.J., Bonney, R.E., Fink, D. and Kelling, S. (2009) 'eBird: a citizen-based bird observation network in the biological sciences', *Biological Conservation*, 142: 2,282–2,292.

Teng, J., Wang, F. and Liu, Y. (2008) 'An efficient algorithm for raster-to-vector data conversion', *Geographic Information Sciences*, 14(1): 54–62.

Veregin, H. (2005) 'Data quality parameters', in P.A. Longley, M.F. Goodchild, D.J. Maguire and D.W. Rhind (eds) *Geographical Information Systems: Principles, Techniques, Management, and Applications*, 2nd edn, Hoboken, NJ: John Wiley & Sons, 177–189.

Wang, Z. (1996) 'Manual versus automated line generalization', in *GIS/LIS '96 Annual Conference and Exposition Proceedings, November 19–21, 1996, Colorado Convention Center, Denver, Colorado*, Bethesda, MD: American Society for Photogrammetry and Remote Sensing, 94–106.

Wang, Z. and Müller, J.C. (1998) 'Line generalization based on analysis of shape characteristics', *Cartography and Geographic Information Systems*, 25(1): 3–15.

Spatial analysis

This chapter introduces and illustrates fundamental GIS operations for spatial analysis, including spatial query, reclassification, geometric and distance measurement, overlay analysis, map algebra and spatial interpolation.

LEARNING OBJECTIVES

After studying this chapter, you should be able to:

- understand the concept of spatial analysis and modelling;
- describe how spatial data can be retrieved based on attributes and locations;
- explain and apply the methods of data classification;
- understand the difference between simple Euclidean distance, cost distance and network distance, as well as how they are measured;
- grasp the principle of overlay analysis and its usage in spatial data integration;
- understand the principle of map algebra, and apply map algebra in model building with spatial data;
- explain the assumptions and techniques of spatial interpolation, and apply them for environmental data analysis and modelling.

4.1 SCOPE OF SPATIAL ANALYSIS AND MODELLING

Spatial analysis and modelling refers to a set of methods used to examine, summarise, manipulate, analyse and interpret spatial data to discover, understand and predict spatial patterns, spatial relationships, trends and their underlying causes. The distinctive feature of spatial analysis compared with classic statistical analysis and other analytical methods is that its results change when the locations of the geographical phenomena being analysed change (Longley et al. 2011, p. 353). A huge number of spatial analysis and modelling techniques have been developed over the past decades, ranging from simple spatial query for identifying where geographical features of a particular type are and what features are found at particular locations to spatial statistics for investigating spatial dependency and relationships and characterising spatial patterns, spatial predictive modelling for predicting possible spatial patterns and trends, and spatial decision analysis for understanding problem situations and evaluating decision alternatives. Although it is by no means comprehensive, Table 4.1 lists seven broad categories of commonly used spatial analysis and modelling techniques in environmental analysis and modelling. Many of these techniques have been incorporated into GIS to form part of the spatial data analysis component, as described in Section 1.1.

Spatial analysis and modelling provides a means of turning spatial data into useful information and knowledge. It is the core of GIS for environmental problem solving and decision making. As the technology matures and the demand for sophisticated data analysis grows, GIS packages are increasingly equipped with more spatial analysis and modelling tools as standard built-in functions or as optional toolsets, add-ins or extensions. In addition to those provided directly by the original GIS software developers and suppliers, many spatial analysis and modelling tools are developed and provided by third parties. As mentioned in Section 1.1, many GIS software systems offer application programming interfaces or software development kits to enable users to develop their own analytical tools or variants.

Table 4.1 Seven broad categories of spatial analysis and modelling tools

Category	Tools
GIS fundamental operations	Spatial query
	Reclassification
	Geometric measurements
	Buffering and proximity measurements
	Overlay
Network analysis	Finding shortest path
	Finding closest facilities
	Finding service areas
Surface analysis	Slope
	Aspect
	Curvature
	Visibility
	Profiling
Spatial interpolation	Inverse distance weighted
	Spline
	Trend surface
	Kriging
Spatial statistics	Spatial descriptive statistics
	Spatial inferential statistics
	Spatial regression
	Spatial clustering
Spatial decision analysis	Multi-criteria analysis
	Multi-objective optimisation
Spatial modelling	Cellular automata
	Agent-based modelling
	Weights-of-evidence

Due to the vast range of spatial analysis and modelling techniques available and the huge variations in the spatial analysis and modelling functionality of current GIS software packages (including extensions), it is impossible to provide a comprehensive introduction to all available spatial analysis and modelling techniques in this book. Only the tools listed in Table 4.1 are examined. Also, by no means all can be explained in a single chapter of this scope. Therefore, this chapter examines the fundamental GIS operations as building blocks of spatial analysis, introduces network analysis, and discusses spatial interpolation.

Chapter 5 is devoted to spatial statistics, while Chapter 9 introduces some spatial modelling and decision analysis techniques. Surface analysis is included in Chapter 7. All discussions are focused on the basic concepts and uses of these tools, rather than the theoretical issues in spatial analysis, the algorithms behind the tools and their implementations in GIS. For such information, readers may refer to Berry (1993), Fotheringham and Rogerson (1994), Fischer and Leung (2001) and De Smith et al. (2013).

4.2 SPATIAL QUERY AND RECLASSIFICATION

Spatial query

Spatial query is sometimes called spatial selection. There are basically two types of spatial query: query by attribute and query by location. As current GIS systems query vector and raster data in different ways and have no ability to query raster data by location (that is, based on spatial constraints), this section discusses spatial query on vector data only. Raster data query is mainly implemented through map algebra, to be discussed in Section 4.4.

Query by attribute

Query by attribute is used to retrieve all features with particular attribute values from one data layer or feature class in a spatial database and display their locations in a map. For example, consider the following query: 'Retrieve all water mains installed before 2010 with a diameter less than 30cm.' Here water mains are features represented in a data layer, and installation year and diameter are two attributes. After the query is implemented in a GIS, all the water mains with the specified attribute values are selected and displayed in a map.

However, when making a query by attribute in GIS, the user is generally required to first build a query using a specific database query language, then submit the query expression for implementation. Different GIS systems may use different query languages. ArcGIS supports standard SQL (Structured Query Language). For the above query example, suppose the water main data layer is named w_mains, and installation years and diameters (in cm) are stored in two fields in its associated attribute table, named year and diameter respectively. Both fields are of numeric type. In SQL, the query can be expressed as:

```
SELECT * FROM w_mains
WHERE "year" < 2010 AND "diameter" < 30
```

Here, * represents all the features meeting the criteria or condition expressed in the WHERE clause. < and AND are operators. In the general form, an SQL expression for query by attribute can be written as:

```
SELECT * FROM <Layer or dataset>
WHERE <Field name> <Operator>
<Value> <Connector> <Field name>
<Operator> <Value> ...
```

The expression of the query criteria in the WHERE clause specifies the features to be retrieved. It may consist of attributes (fields), operators and functions. There are three types of operators: arithmetic, logical and comparison operators. Arithmetic operators are used to add, subtract, multiply and divide numeric values, logical operators are mainly used as connectors to combine two criteria together and comparison operators are used to compare one expression to another. Functions are one-word commands that return a single value calculated from values in a field. Table 4.2 lists the major SQL operators, and Table 4.3 lists some selected SQL functions for query by attribute available in ArcGIS. Box 4.1 shows how query by attribute is implemented. The rules for evaluating a query expression include the following:

- An expression is evaluated left-to-right.
- Expressions within parentheses assume highest priority.
- If parentheses are nested, the evaluation starts with the innermost parentheses.
- AND is evaluated before OR.
- NOT is evaluated before AND and OR.

Table 4.2 Major SQL operators for query in ArcGIS

Type	Operator	Description
Arithmetic	*	Multiplication
	/	Division
	+	Addition
	−	Subtraction
	%	Modulus
Logical	AND	Combines two conditions together. Returns TRUE if both conditions are true.
	OR	Combines two conditions together. Returns TRUE if at least one condition is true.
	NOT	Returns TRUE if the condition is false.
Comparison	=	Returns TRUE if the values of two operands are equal.
	<>	Returns TRUE if the values of two operands are not equal.
	<	Returns TRUE if the value of left operand is less than the value of right operand.
	<=	Returns TRUE if the value of left operand is less than or equal to the value of right operand.
	>	Returns TRUE if the value of left operand is greater than the value of right operand.
	>=	Returns TRUE if the value of left operand is greater than or equal to the value of right operand.

Table 4.3 Example SQL functions for query in ArcGIS

Type	Function	Description
String	CHAR_LENGTH(s)	Returns the length in characters of string s.
	LOWER(s)	Returns string s in lowercase.
	UPPER(s)	Returns string s in uppercase.
	SUBSTRING(s FROM pos FOR len)	Returns a substring len characters long from string s, starting at position pos.

(continued)

Table 4.3 *(continued)*

Type	Function	Description
Numeric	ABS(x)	Returns the absolute value of numeric expression x.
	LOG(x)	Returns the natural logarithm of numeric expression x.
	POWER(x, n)	Returns the value of numeric expression x to the power of n.
	CEILING(x)	Returns the smallest integer greater than or equal to numeric expression x.
	ROUND(x, n)	Returns x rounded to n decimal places.
	TRUNCATE(x, n)	Return the value of x truncated to n number of decimal places. If n is negative, x is truncated to |n| places to the left of the decimal point.

Box 4.1   Query by attribute in ArcGIS   **PRACTICAL**

To follow this example, start ArcMap, and load the **soils** feature class and the **soil properties** table from **C:\Databases\GIS4EnvSci\VirtualCatchment\Geodata.gdb**.

Search alpine soils

1. In the ArcMap main menu bar, click **Selection > Select By Attributes**. The **Select By Attributes** dialog appears. All the fields or attributes associated with the layer are listed in the dialog box.
2. In the **Select By Attributes** dialog box, enter the query expression by following steps 1)–6) below.

    1)   Choose the layer **soils** to perform the query.
    2)   Choose the **Create a new selection** method.
    3)   Double-click **"Soil_order"** in the field listing window. This field stores soil type names.
    4)   Click the **Get Unique Values** button to see the values for the selected field.
    5)   Click the = operator.
    6)   Double-click the **'Alpine soil'** value. Now, the query expression ″Soil_order″ = 'Alpine soil' is entered, as shown in Figure 4.1. Note that SELECT * FROM forms the first part of the SQL expression and is automatically supplied together with the layer name. The user only needs to enter the WHERE clause.
    7)   Validate the query expression by clicking the **Verify** button.
    8)   Click **OK**. The selected features are highlighted in the data view.

3. Right-click the layer **soils** in the table of contents, point to **Selection**, then click **Create Layer from Selected Features**. A new layer, **soils selection**, is created. This layer shows the query results, which are alpine soils in this case.

Search soils with the pH value within the range 6–7

4. In the table of contents, uncheck **soils selection**.
5. Click the **Clear Selected Features** button ⟨icon⟩ on the **Tools** toolbar to unselect the currently selected features in **soils**.
6. Right-click **soils** in the table of contents, point to **Joins and Relates**, and select **Join**.

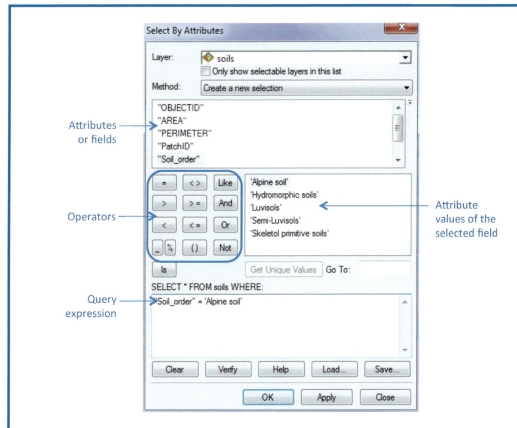

Figure 4.1 The Select By Attributes dialog.

7. In the **Join Data** dialog box:

    1) Make sure **Join attributes from a table** is selected.
    2) Select **PatchID** as the field in the layer that the join will be based on.
    3) **soilProperties** should be automatically set as the table to join to this layer. If not, select it.
    4) Select **PatchID** as the field in the table to base the join on.
    5) Tick **Keep all records** as the join option.
    6) Click **OK**. The table **soilProperties** is joined with the attribute table associated with **soils** through **PatchID**.

8. Right-click **soils** in the table of contents, and click **Open Attribute Table**. Examine the attribute table. All the soil properties and their values from **soilProperties**, including soil depth, texture and pH, should be linked to each feature in the layer.

9. Close the attribute table, and open the **Select By Attributes** dialog.

10. In the **Select By Attributes** dialog box, enter and validate the query expression below by following Step 2 above:

    ```
    soilProperties.pH >= 6 AND soilProperties.pH <= 7
    ```

11. Follow Step 3 above to create a new layer showing the query results – the retrieved features with the pH value of 6–7.

Query by location

Query by location is used to find features from one or multiple data layers based on their locations relative to features from another layer in the same spatial database, or to identify attribute values associated with features at particular locations. An example query could be 'Find all crop fields within flood plains.' Here crop fields are a set of features, and flood plains are another set of features. They are typically represented in two separate data layers. 'Within' indicates the location of crop fields relative to flood plains. In some GIS systems, spatial operators are used in query to retrieve features according to the spatial relationship (relative location) they define. In its OpenGIS Implementation Standard for Geographic Information, OGC defined eleven spatial operators or spatial relationship predicates for SQL. They include Equals, Disjoint, Intersects, Touches, Crosses, Within, Contains, Overlaps, Relate, LocateAlong and LocateBetween. For detailed descriptions of these spatial operators, please refer to OGC (2011). This standard has been adopted by several GIS or database management software systems, including ArcGIS, DB2 Spatial Extender, Informix Spatial Datablade, PostgreSQL, SQL Server and Oracle Spatial. But the choice of spatial operators, their names and syntaxes may differ from system to system. ArcGIS provides thirteen spatial operators for query by location (Table 4.4). Some of them are used in the examples given in Box 4.2.

Table 4.4 Spatial operators for query in ArcGIS

Operator	Description
Intersect	Retrieves all features in the target layers that intersect the feature(s) in the source layer.
Are within a distance of	Retrieves all features from the target layers that are within a user-specified distance from the feature(s) in the source layer.
Are within	Retrieves any feature in the target layers whose geometry falls inside the geometry of a feature in the source layer. Boundaries of selected features and source features can overlap.
Are completely within	Retrieves any feature in the target layers whose geometry falls entirely inside the geometry of a feature in the source layer without touching its boundary.
Contain	Retrieves any feature in the target layers whose geometry contains the geometries of features in the source layer. Boundaries of selected features and source features can overlap.
Completely contain	Retrieves any feature in the target layers whose geometry contains the geometries of features in the source layer without touching their boundaries.
Have their centroid in	Retrieves any feature in the target layers whose centroid of the geometry falls into the geometry of a feature in the source layer or on its boundary.
Share a line segment with	Retrieves any feature in the target layers that shares a line segment with a feature in the source layer.
Touch the boundary of	Retrieves any feature in the target layers whose boundary intersects with the boundary of a feature in the source layer, but their interiors do not intersect.
Are identical to	Retrieves any feature in the target layer whose interior and boundary are identical to those of a feature in the source layer.
Are crossed by the outline of	Retrieves any feature in the target layer whose boundary is crossed by the boundary of a feature in the source layer, but they do not share a line segment.
Contain (Clementini)	Same as Contain, but the source feature must not be entirely on the boundary of the target feature with no part of the source feature inside the target feature.
Are Within (Clementini)	Same as Are Within, but the target feature must not be entirely on the boundary of the source feature with no part of the target feature inside the source feature.

Query by location generally involves two or more data layers. The layers from which features are to be searched or selected are called target layers. The layer whose feature geometries are used to select features from the target layers is called the source layer. Another case of query by location involves retrieving attribute values associated with a feature or features at a particular location, which is also demonstrated in Box 4.2.

Box 4.2   Query by location in ArcGIS   **PRACTICAL**

To follow this example, start ArcMap, and load the **landcover**, **pheasant** and **rivers** feature classes from **C:\Databases\GIS4EnvSci\VirtualCatchment\Geodata.gdb**. The **pheasant** layer is a point feature class, representing sightings of blood pheasant in the area. The **sightings** field in the attribute table of **pheasant** records the number of sightings at each observation point.

Display and summarise pheasant sightings within 200m from rivers

1. In the ArcMap main menu, click on **Selection**, then choose **Select By Location**. The **Select By Location** dialog appears.
2. In the **Select By Location** dialog box:

    1) Choose **select features from** as the selection method.
    2) Select (check) **pheasant** as the target layer.
    3) Select **rivers** as the source layer.
    4) Choose **are within a distance of the source layer feature** as the spatial operator.
    5) Enter **200 meters** as the search distance.
    6) Click **OK**. Fourteen sighting locations within 200m from rivers are highlighted in the data view.

3. Right-click **pheasant** in the table of contents, then click **Open Attribute Table**. The attribute table of the layer opens.
4. In the table window, right-click the field heading **sightings**, and click **Statistics**. The descriptive statistical summary for this field is displayed, which shows that the sum of **sightings** is 180. This means there are 180 sightings of blood pheasant within 200m from rivers.

Identify land covers in which pheasants were sighted

5. Open the **Select By Location** dialog box.
6. In the **Select By Location** dialog box:

    1) Choose **select features from** as the selection method.
    2) Select **landcover** as the target layer.
    3) Select **pheasant** as the source layer.
    4) Choose **contain the source layer feature** as the spatial operator.
    5) Click **OK**. The land cover polygons containing pheasant sightings are highlighted in the data view.

7. Right-click **landcover** in the table of contents, click **Open Attribute Table**. The attribute table of the layer opens. You may notice from the status bar at the bottom of the table that twenty-six out of 4,250 features are selected.
8. In the Table window, click the **Show selected records** button ▤ on the status bar. The records of attributes associated with the twenty-six selected land cover polygons are displayed in the table.

(continued)

(continued)

9. Right-click the header **Type** in the table, and select **Summarize**.
10. In the **Summarize** dialog, specify the name of the output table, and make sure **Summarize on the selected records only** is checked. Click **OK.** The summarised table is created. Add it to the table of contents, then open the summarised table to examine in which types of land cover pheasants were sighted.

Identify the number of pheasant sightings at a particular location

11. Click the **Identify** tool 🛈 on the **Tools** toolbar.
12. Click on a location in the data view where pheasants were sighted.
13. In the **Identify** window, specify **pheasant** as the layer to identify. The attributes associated with the selected observation point in the **pheasant** layer are then listed in the bottom panel. The number of sightings at the selected location is the value of the field **sightings**.

Reclassification

Reclassification is used to recode the attributes of features in the feature attribute table or reclassify the grid cell values to produce a new raster data layer. Reclassification can be used for data simplification and measurement scale change.

Data simplification involves assigning new values to classes or ranges of the existing values to reduce the number of classes or ranges in the original input layer, or to group attribute values into categories in a new classification scheme. For example, to simplify the land use map in Figure 4.2, we may reclassify 'corn' and 'soybean' as 'agriculture', and 'coniferous' and 'deciduous' as 'forest'. Reclassification of vector data may involve removing boundaries between the areas with the same reclassified values, as shown in Figure 4.2. Reclassification of raster data is simpler, only involving changing cell values to new values. Box 4.3 demonstrates the differences in their implementations.

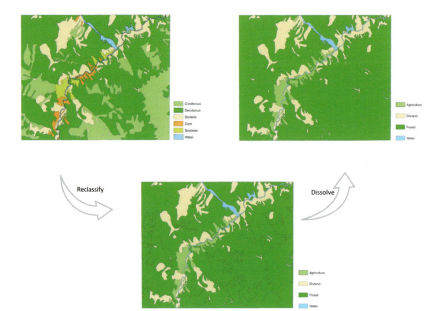

Figure 4.2 Reclassification in vector.

Box 4.3 Reclassification in ArcGIS PRACTICAL

To follow this example, start ArcMap, and load the **landcover** and **degreeSlope** data layers from **C:\ Databases\GIS4EnvSci\VirtualCatchment\Geodata.gdb**. The **landcover** layer is a vector layer, and **degreeSlope** is a raster layer. The **Spatial Analyst** extension is required.

Enable the Spatial Analyst extension

1. Click the **Customize > Extensions** menu from the ArcMap main menu bar.
2. In the **Extensions** dialog box:

    1) Check the **Spatial Analyst** checkbox.
    2) Click **Close**.

Simplify a land cover map

The **landcover** layer contains eight types of land cover: bare rock, cold-temperate coniferous forest, cropland, deciduous broadleaf forest, deciduous broadleaf shrub, meadow, temperate coniferous and broadleaf mixed forest, and water body. It is to be reclassified to have two types: forest and non-forest.

3. Right-click the **landcover** layer in the table of contents, and click **Open Attribute Table**. The feature attribute table of the layer opens.
4. In the table window, click the **Table Options** button ▣ . Select **Add Field**. In the **Add Field** dialog:

    1) Enter **Class** as the name of the field.
    2) Click the **Type** arrow, and select **Text** as the field type.
    3) Change the value of **Length** to 20.
    4) Click **OK**. The field **Class** is added to the table with NULL value for every record.

5. In the table window, click the **Table Options** button. Click **Select By Attributes**.
6. In the **Select By Attributes** dialog box, enter and validate the query expression below in a similar way to that used in Step 2 in Box 4.1:

    ```
    "Type" = 'Cold temperate coniferous forest' OR "Type" = 'Deciduous
    broadleaf forest' OR "Type" = 'Temperate coniferous and broadleaf mixed
    forest'
    ```

Click **Apply**. The records of the three types of forest are highlighted in the table.
Click **Close** to dismiss the **Select By Attributes** dialog box.

7. In the table window, right-click the field heading **Class**, and select **Field Calculator**. When prompted ' . . . Do you wish to continue?', click **Yes**. The field calculator pops up.
8. In the **Field Calculator** dialog, enter 'Forest' in the lower panel, so that the equation is Class = "Forest".

Click **OK**. 'Forest' is assigned to all the selected records (features).

(continued)

(continued)

9. In the table window, click the **Table Options** button. Click **Switch Selection**. When prompted '. . . Do you wish to continue?', click **Yes**. All the records except those linked to the forest features are selected and highlighted in the table.
10. Right-click the field heading **Class**, and select **Field Calculator**. When prompted '. . . Do you wish to continue?', click **Yes**.
11. In the **Field Calculator** dialog, enter 'Non-forest' in the lower panel, so that the equation is:
    Class = "Non-forest".

Click **OK**. 'Non-forest' is assigned to all the selected records.

12. In the table window, click the **Table Options** button. Click **Clear Selection**. Now the table contains the field **Class** with 'Forest' and 'Non-forest' values.
13. Use **Class** as the value field to make a map showing only two land cover classes 'Forest' and 'Non-forest'.

The following step shows how to remove the boundaries between the adjacent polygons with the same class on the map.

14. Open **ArcToolBox**. Navigate to **Data Management Tools > Generalization**, and double-click **Dissolve**.
15. In the **Dissolve** dialog:

    1) Select **landcover** as the input features.
    2) Name the output feature class.
    3) Select or check **Class** as the **Dissolve_Field(s)**.
    4) Click **OK**. The result is added to the data view.

Reclassify slope

The **degreeSlope** layer stores slope values in degree in a raster. It is to be reclassified into three classes: gentle slope (<2°), medium slope (2–20°) and steep slope (>20°). In the resultant raster, 1 represents gentle, 2 medium and 3 steep.

16. Open **ArcToolBox**. Navigate to **Spatial Analyst Tools > Reclass**, and double-click **Reclassify**.
17. In the **Reclassify** dialog box, select **degreeSlope** as the input raster and **Value** as the reclass field. By default, the numerical cell values in the input raster are classified into nine ranges (classes), which are listed in the reclassification table in the dialog box.
18. Click the **Classify** button in the **Reclassify** dialog box.
19. In the **Classification** dialog box:

    1) Change the number of classes from **9** to **3**.
    2) In the lower-right text box, change the first two break values to **2** and **20** respectively, leave the last break value **72.245575** (which is the largest slope value in the input raster) unchanged.
    3) Click **OK**. The **Classification** dialog box is dismissed, and the reclassification table in the **Reclassify** dialog box is changed, in which the old values (slope in degrees) are regrouped to three classes: 0–2, 2–20 and 20–72.245575, and their reclassified new values are 1, 2 and 3.

20. In the **Reclassify** dialog box, name the output raster, then click **OK**. A new raster layer is generated and added to the data view. It represents three classes of slope.

Measurement scale change is used to assign ranks, orders, ratings or weights representing preference, importance, priority, sensitivity, suitability, capability or other criteria to unique values, ranges or categories of values found in the input data layer. For example, we may use the reclassification operation to assign new values to different land use classes based on their ecological importance to help identify areas of high conservation value, and to assign values of 1–10 to different ranges of slope to represent landslide risk. Measurement scale change can be applied to a number of data layers to create a common scale of values for analysis and modelling.

4.3 GEOMETRIC AND DISTANCE MEASUREMENT

Geometric properties (length, area, perimeter and shape) of geographical features and distances between geographical entities are useful parameters for characterising environmental phenomena and their spatial patterns. For example, landscape metrics, which are used to quantify the spatial configuration of landscape and investigate the interaction between spatial pattern and ecological process, are calculated based on geometric properties of landscape units and distances, such as patch size, edge diversity, contiguity, shape complexity, proximity and connectivity (Turner et al. 2001; see also Case Study 9 in Chapter 10). Geometric properties and distances are usually not encoded during spatial data input. GIS offers measurement functions to calculate them.

Geometric measurement

Length

Spatial data are provided either in a projected coordinate system or in a geographical coordinate system. Length and area are calculated differently in the two types of georeferencing systems.

In a projected coordinate system, the length of the shortest line between two points is a straight-line distance, calculated based on the Pythagorean theorem:

$$D = \sqrt{(x_2 - x_1)^2 + (y_2 - y_1)^2} \qquad (4.1)$$

where (x_1, y_1) and (x_2, y_2) are the coordinates of the two points. This is called Euclidean distance.

In a geographical coordinate system, the length of the shortest line between two points is measured as the great circle distance. A great circle is a circle on the spherical surface of the Earth whose centre coincides with the centre of the Earth. The two points separate the great circle into two arcs. The length of the shorter arc is the great circle distance between the points, also called a geodesic. This distance is not straight-line distance, but spherical distance, calculated as:

$$D_{geodesic} = r\cos^{-1}[\sin\varphi_1 \sin\varphi_2 + \cos\varphi_1 \cos\varphi_2 (\lambda_1 - \lambda_2)] \qquad (4.2)$$

where (φ_1, λ_1) and (φ_2, λ_2) are latitude and longitude coordinates of two points, and r is the radius of the Earth. Some GIS systems can select an appropriate equation to calculate length or distance according to the projection applied. For example, in ArcGIS, planar is the default when working in a projected coordinate system, while geodesic is the default when working in a geographical coordinate system. Some systems provide options for users to specify which type of measurement is going to be made. It will produce wrong results when using geographical coordinates to calculate Euclidean distances. It will be equally wrong when using Cartesian coordinates to calculate geodesics.

In vector, the length of a line feature is simply the sum of the lengths of each straight-line segment measured using one of the distance equations, as the line feature is represented as a polyline (see Figure 2.18). It should be noted that the length of a line feature calculated in GIS is almost always shorter than its true length, as the real smoothly curved shape of the feature is replaced by straight-line segments.

Measuring the length of a line feature in raster involves calculating the number of grid cells representing the line and multiplying by the cell resolution. However, the length of a line segment on the diagonal is measured as the number of cells forming the line times the cell resolution times 1.414 ($= \sqrt{1^2 + 1^2}$), as shown in Figure 4.3.

Perimeter and area

In vector, area features are represented as polygons. The perimeter of a polygon is the sum of the lengths of the line segments forming the polygon. The area of a polygon is calculated by summing the areas of a number of trapeziums formed by drawing perpendiculars to the x axis (Longley et al. 2011, p. 382), which can be mathematically expressed as:

$$A = \frac{1}{2}\left|\sum_{i=1}^{n}(x_{i+1} - x_i)(y_{i+1} + y_i)\right| \qquad (4.3)$$

where n is the number of vertices of the polygon, and $(x_1, y_1), (x_2, y_2), \ldots, (x_n, y_n), (x_{n+1}, y_{n+1})$ are the coordinates of the n vertices with $x_{n+1} = x_1$ and $y_{n+1} = y_1$. They

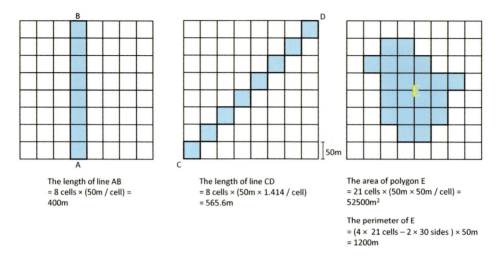

The length of line AB
= 8 cells × (50m / cell) =
400m

The length of line CD
= 8 cells × (50m × 1.414 / cell)
= 565.6m

The area of polygon E
= 21 cells × (50m × 50m / cell) =
52500m²

The perimeter of E
= (4 × 21 cells − 2 × 30 sides) × 50m
= 1200m

Figure 4.3 Measuring length, perimeter and area in raster (the cell resolution is 50m).

are coordinates in a projected coordinate system. Most GIS systems do not provide a measurement function to calculate areas in geographical coordinates.

In raster, area features are represented as groups of contiguous grid cells with a common attribute value. The area of an area feature is simply calculated as the number of cells it is composed of times the cell resolution. The perimeter of an area is computed as:

$$P = \left(4n_a - 2n_s\right)r \tag{4.4}$$

where n_a is the total number of cells the area is composed of, n_s is the number of shared cell sides in the area and r is the cell resolution (Bogaert et al. 2000). Figure 4.3 provides an example.

Box 4.4 shows how to measure length and area in ArcGIS.

Box 4.4   Length and area measurement in ArcGIS   **PRACTICAL**

To follow this example, start ArcMap, and load the **lakes** shapefile from the shapefiles directory and the **landcv_r** raster from **Geodata.gdb** in **C:\Databases\GIS4EnvSci\VirtualCatchment**. The **landcv_r** layer is a raster version of **landcover**.

Measure length and area on the map

1. Click the **Measure tool** on the **Tools** toolbar.
2. In the **Measure** dialog box:

   1) Click the **Measure Line** tool ∿ , and draw a line or polyline to the desired shape on the map in the data view. Double-click to complete the line. The measurements of the lengths of the line and the last segment of the line are displayed in the dialog box.
   2) Click the **Measure An Area** tool ▱ , and draw a polygon to the desired shape on the map in the data view. Double-click to complete the polygon. The measurements of the perimeter and area of the polygon are displayed in the dialog box.
   3) Click the **Measure A Feature** tool ✛ , and click on a lake on the map in the data view. The perimeter and area of the selected lake feature are displayed in the dialog box. As the data frame is in a projected coordinate system, all the measurements are planar.

Calculate geometries of features and adding them to the attribute table

3. Right-click the **lakes** layer in the table of contents, and click **Open Attribute Table**. The feature attribute table of the layer opens.
4. In the table window, click the **Table Options** button [icon] . Select **Add Field**. In the **Add Field** dialog:

   1) Enter **Area** as the name of the field.
   2) Click the **Type** arrow, and select **Double** as the field type.
   3) Enter **20** for **Precision**, and **2** for **Scale**.
   4) Click **OK**. The field **Area** is added to the table.

5. Repeat the previous step to add a new field named **Perimeter** with the same type, precision and scale. Now the attribute table has two new fields.
6. Right-click the field heading **Area**, and click **Calculate Geometry**. When you see a warning message, click **Yes**.
7. In the **Calculate Geometry** dialog box:

   1) Choose **Area** as the geometric **property**.
   2) Check **Use coordinate system of the data source**.
   3) Select **Square Meters** as **Units**.
   4) Click **OK**. When you see a warning message, click **Yes**. The area measurements for each lake are added to the table.

8. Repeat Steps 6 and 7 to calculate perimeters for each feature.
9. Browse the feature attribute table to review the areas and perimeters calculated for each feature. Note that the area, length or perimeter of features can only be calculated when the coordinate system being used is projected. The **Calculate Geometry** tool can only be used for vector data.

Calculate areas in raster

10. Right-click the **landcv_r** layer in the table of contents, and click **Properties**. In the **Layer Properties** dialog, click the **Source** tab. In the list of properties and values of the raster, the linear unit is shown to be **Meter** and the cell size is **40**. Therefore, the cell resolution of the raster is **40** metres.
11. Right-click the **landcv_r** layer in the table of contents, and click **Open Attribute Table**. The feature attribute table of the layer opens.
12. In the table, the **Count** field stores the number of cells each type of land cover comprises.

    1) Click the **Table Options** button [icon] . Select **Add Field**. In the **Add Field** dialog, enter **Area** as the name, and select **Double** as the type. Click **OK**.
    2) Right-click the field heading **Area**, and select **Field Calculator**.
    3) In the **Field Calculator**, enter `Area = [Count]*40*40`. Click **OK**. The areas for each type of land cover are added to the table.

Shape

Shape describes the geometric form of individual features. It is considered a critical factor in defining and analysing habitat regions, assessing species richness, studying the effects of urban growth, determining the impact of deforestation, measuring the parameters of drainage basins and examining the patterns of geological formations (Wentz 2000). There are numerous measures that have been developed for describing and representing shapes. Wentz (2000) groups these measures into four categories: compactness, boundary, form and multiple-parameter measures.

Compactness measures quantify how far a feature deviates from a standard shape, which is the most compact.

A standard shape is usually a circle or square. An example of the shape compactness measure is given as:

$$SI_c = \begin{cases} \dfrac{P}{2\sqrt{\pi A}} & \text{for vector data} \quad (4.5) \\[2ex] \dfrac{P}{4\sqrt{\pi A}} & \text{for raster data} \quad (4.6) \end{cases}$$

where P is the perimeter and A is the area of the feature (Bogaert et al. 2000). When the feature is a circle (in vector) or a square (in raster), $SI_c = 1$. For shapes different from the standard shape, $SI_c > 1$. Higher values mean greater departure from the standard shape and less compactness.

Boundary measures quantitatively describe the roughness or the smoothness of a feature's outer edge. A more uneven boundary has more edges. A boundary measure is an indicator of shape complexity. A useful boundary measure is the fractal dimension of an area, expressed as follows (Moser et al. 2002):

$$SI_f = \frac{2\ln P}{\ln A} \quad (4.7)$$

SI_f is 1 for shapes with very simple edges such as circles and squares, and approaches 2 when shapes become more complex.

Form measures quantify the overall geometric configuration of a feature by comparing it to a standard shape such as a circle. One example is the Boyce-Clark Index (Boyce and Clark 1964), which involves drawing a set of equally spaced radials from a reference point within

a shape to its perimeter. For example, for a circle with the centre as the reference point, all radials would be of equal length. Its mathematical form is:

$$\sum_{i=1}^{n} \left| \frac{r_i}{\sum_{k=1}^{n} r_k} \times 100 - \frac{100}{n} \right| \quad (4.8)$$

where r_i is the length of radial i from a reference point, and n is the number of radials.

Multiple-parameter measures combine measurements of several shape properties, such as compactness, jaggedness, elongation and fragmentation, and involve multivariate analysis. Most of the shape measures are based on length, perimeter and area. Although shape measures are not readily implemented in GIS, they can be calculated by using the basic geometric measurement and attribute calculation functions.

Buffering

Buffering is a distance measurement tool for vector data commonly available in GIS. It creates a zone around a feature or a set of features with a specific width. This zone is called a buffer. As shown in Figure 4.4, a buffer around a point feature is a circle with the buffer width as its radius, a buffer around a line feature is a band with a specific distance (buffer width) on both sides of the line conforming to its curve, and a buffer around an area feature is a zone of a specific distance from the edge of the area conforming to its shape. Any point in a buffer

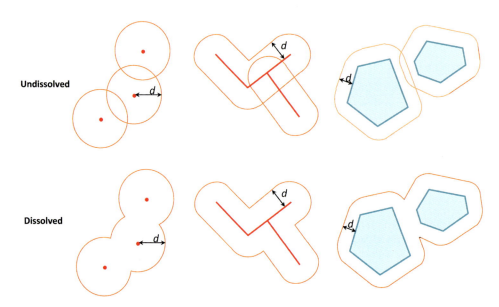

Figure 4.4 Buffers.

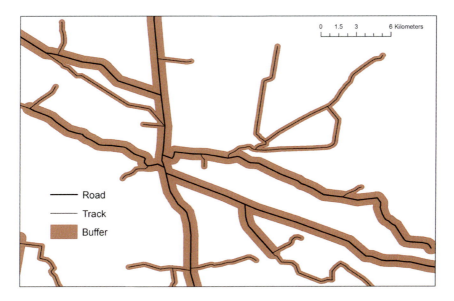

Figure 4.5 Variable buffers.

around a feature is within the distance of the buffer width from the feature.

The buffer width is a user-specified distance. For neighbouring or connected features, their buffers tend to overlap as shown in Figure 4.4. Usually the boundaries of overlapping buffers are dissolved so that a single coherent zone is created around the features. Buffers may be generated with a fixed width or variable widths according to the types or attributes of features. In the example shown in Figure 4.5, buffers of 500m were drawn around roads, and buffers of 200m around tracks.

When working in a projected coordinate system, Euclidean buffers are produced, which measure straight-line distance in a two-dimensional Cartesian plane. When working in a geographical coordinate system, geodesic buffers are created, which accounts for the actual shape of the Earth. Some GIS systems, such as ArcGIS, allow users to produce either Euclidean or geodesic buffers. Buffering has a wide range of applications – for example, defining riparian zones to protect riparian vegetation or creating buffers around a power station to assess environmental and social impacts. Box 4.5 illustrates buffering in ArcGIS.

Box 4.5 Buffering in ArcGIS   **PRACTICAL**

To follow this example, start ArcMap, and load the **lakes** shapefile from the shapefiles directory in **C:\Databases\GIS4EnvSci\VirtualCatchment**. The **Seasonal** field in its feature attribute table indicates whether a lake is ephemeral (with the value of 1) and non-ephemeral (with the value of 0). Buffers will be produced around lakes.

Buffering with a fixed distance

1. Click **Geoprocessing** on the main menu bar, then click the **Buffer** tool.
2. In the **Buffer** dialog box:

   1) Select **lakes** as the input features.
   2) Enter the name of the output feature class for storing the buffers to be produced.

(continued)

(continued)

3) Check **Linear unit**, and enter **200** into the text box below.
4) Change the linear unit to **Meters**.
5) Select **FULL** as the side type, and **ALL** as the dissolve type.
6) Click **OK**, then 200m buffers are created and drawn around the lakes in the layer.

Buffering with variable distances

3. Right-click the **lakes** layer in the table of contents, and click **Open Attribute Table**. The feature attribute table of the layer opens.
4. In the table window, use **Add Field** to add a new field named **Width**. The field type is set to **Float**, the precision to **5**, and the scale to **0**. This field is used to store variable buffer widths for ephemeral and non-ephemeral lakes.
5. In the table, use **Select By Attributes** to select all ephemeral lakes by implementing the query expression: `"Seasonal" = 1`. Then, right-click the field heading **Width,** and use **Field Calculator** to set the value of **Width** for all the selected lakes to **200**.
6. In the table window, click the **Table Options** button. Click **Switch Selection**. All the non-ephemeral lakes are selected. Use **Field Calculator** to set the value of **Width** for all the non-ephemeral lakes to **400**, then clear the selections.
7. Start the **Buffer** tool. In the **Buffer** dialog box:

    1) Select **lakes** as the input features.
    2) Enter the name of the output feature class for storing the buffers to be produced.
    3) Check **Field**, and select **Width**, from which various distances are used to create buffers.
    4) Select **FULL** as the side type, and **ALL** as the dissolve type.
    5) Click **OK**, then 200m buffers around ephemeral lakes and 400m buffers around non-ephemeral lakes are created.

Distance surface

Distance measurement in raster calculates the shortest distances from the location(s) representing the source(s) or origin(s) to every other location and creates a distance surface in raster, which shows a series of concentric rings of equal distance around the source location(s), as shown in Figure 4.6. Distance in raster may be measured based on absolute physical distance between places, or costs incurred while traversing the physical distance. The distance measured by taking into account costs is called cost distance or weighted distance. Here, the cost can be travel time, energy consumption, or other costs or the sum of them, which may be affected by topography such as slope, land cover and other factors. For example, the cost for constructing a highway may include construction costs, land costs and potential costs of environmental and social impacts.

Cost distance is measured as a function of physical distance and costs. Typically, a cost distance measurement tool in GIS uses a source raster layer representing the source locations and a cost raster layer with the same resolution containing the costs associated with each cell to calculate an accumulated cost surface. The value associated with each cell on the cost layer represents the cost per unit distance for moving through the cell. For example, if a cost layer represents travel time, which is in minutes, a cost value will be minutes per metre. Suppose the cost layer has a cell resolution of 50m and a particular cell has a cost value of 0.01 minutes per metre: the time for travelling through the cell is $0.01 \times 50 = 0.5$ minutes. That is, the total cost per cell is the cost value times the cell resolution. Cost is calculated from cell centre to cell centre. Therefore, the cost to move from one cell to its adjacent cell in the cardinal directions is the sum of the cost of each multiplied by half the cell size. If the movement is diagonal, the cost is increased by multiplying the cell size by 1.414. Using the cost layer, cost distance is calculated by totalling the cost while traversing each cell from the source, assigning a least accumulative cost of getting back to the source locations to each cell in a new layer it creates. A number of algorithms have been developed for calculating cost distance, for example the spreading

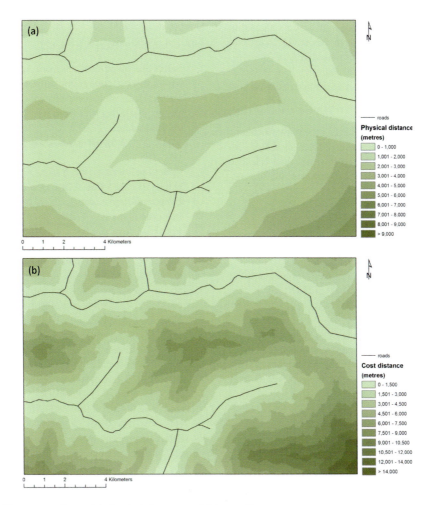

Figure 4.6 Distance surfaces: (a) physical distance and (b) cost distance.

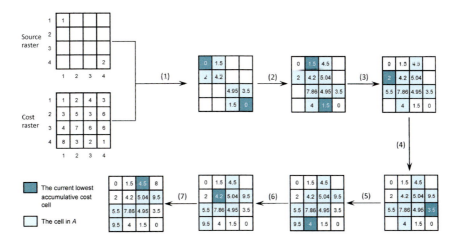

Figure 4.7 Cost distance calculation.

algorithm (Douglas 1994) and the pushbroom algorithm (Eastman 1989). The example in Figure 4.7 is used to illustrate the spreading algorithm used in ArcGIS.

Let $R_{i,j}$ be the cell at row i and column j, $c_{i,j}$ be the accumulative cost value of $R_{i,j}$ and A be the accumulative cost cell list.

1. Assign 0 to the cells representing the source locations (labelled 1 and 2 in the source layer) as there is no accumulative cost to return to them, then calculate the accumulative costs of their adjacent cells like so:

$$c_{1,2} = 0 + (1 + 2) / 2 = 1.5$$

$$c_{2,1} = 0 + (1 + 3) / 2 = 2$$

$$c_{2,2} = 0 + [(1 + 5) / 2] \times 1.414 = 4.24$$

Similarly, calculate $c_{3,3}$, $c_{3,4}$ and $c_{4,3}$. Add all the accumulative cost values to $A = \{(R_{1,2}, 1.5), (R_{2,1}, 2), (R_{2,2}, 4.2), (R_{3,3}, 4.95), (R_{3,4}, 3.5), (R_{4,3}, 1.5)\}$.

2. Select the lowest-cost cells from A, which are $R_{1,2}$ and $R_{4,3}$, and assign the cost values to the cells in the output layer. Calculate the accumulative costs to move into the neighbouring cells (except $R_{1,1}$ and $R_{4,4}$) from $R_{1,2}$ and $R_{4,3}$ like so:

$$c_{1,3} = c_{1,2} + (2 + 4) / 2 = 1.5 + 3 = 4.5$$

$$c_{2,1} = c_{1,2} + [(2 + 3) / 2] \times 1.414 = 1.5 + 3.535 = 5.04$$

$$c_{2,2} = c_{1,2} + (2 + 5) / 2 = 5$$

$$c_{2,3} = c_{1,2} + [(2 + 3) / 2] \times 1.414 = 5.04$$

Similarly, calculate $c_{3,2}$, $c_{3,3}$, $c_{3,4}$ and $c_{4,2}$. Delete $(R_{1,2}, 1.5)$ and $(R_{4,3}, 1.5)$ from A, and add all the newly calculated accumulative cost values to A. If the new accumulative cost for a cell in A is equal to or greater than the value the cell currently has, the value is ignored; otherwise, the old accumulative cost value for the cell is replaced in A with the new value, so $c_{2,1}$, $c_{2,2}$, $c_{3,3}$ and $c_{3,4}$ are not updated.

Now, $A = \{(R_{1,3}, 4.5), (R_{2,1}, 2), (R_{2,2}, 4.2), (R_{2,3}, 5.04), (R_{3,2}, 7.86), (R_{3,3}, 4.95), (R_{3,4}, 3.5), (R_{4,2}, 4)\}$.

3. Select the lowest-cost cell from A, which is $R_{2,1}$, and assign its cost value to the output layer. In a similar way, calculate the accumulative costs to move into the neighbouring cells from $R_{2,1}$. Delete $(R_{2,1}, 2)$ from A, and update A. Now $A = \{(R_{1,3}, 4.5), (R_{2,2}, 4.2), (R_{2,3}, 5.04), (R_{3,1}, 5.5), (R_{3,2}, 7.86), (R_{3,3}, 4.95), (R_{3,4}, 3.5), (R_{4,2}, 4)\}$.

4. Select the lowest-cost cell from A, which is $R_{3,4}$, and assign its cost value to the output layer. Calculate the accumulative costs to move into the neighbouring cells from $R_{3,4}$. Delete $(R_{3,4}, 3.5)$ from A, and update A. Now $A = \{(R_{1,3}, 4.5), (R_{2,2}, 4.2), (R_{2,3}, 5.04), (R_{2,4}, 9.5), (R_{3,1}, 5.5), (R_{3,2}, 7.86), (R_{3,3}, 4.95), (R_{4,2}, 4)\}$.

5. Select the lowest-cost cell from A, which is $R_{4,2}$, and assign its cost value to the output layer. Calculate the accumulative costs to move into the neighbouring cells from $R_{4,2}$. Delete $(R_{4,2}, 4)$ from A, and update A. Now $A = \{(R_{1,3}, 4.5), (R_{2,2}, 4.2), (R_{2,3}, 5.04), (R_{2,4}, 9.5), (R_{3,1}, 5.5), (R_{3,2}, 7.86), (R_{3,3}, 4.95), (R_{4,1}, 9.5)\}$.

6. Select the lowest-cost cell from A, which is $R_{2,2}$, and assign its cost value to the output layer. Calculate the accumulative costs to move into the neighbouring cells from $R_{2,2}$. Delete $(R_{2,2}, 4.2)$ from A, and update A. Now $A = \{(R_{1,3}, 4.5), (R_{2,3}, 5.04), (R_{2,4}, 9.5), (R_{3,1}, 5.5), (R_{3,2}, 7.86), (R_{3,3}, 4.95), (R_{4,1}, 9.5)\}$.

7. Select the lowest-cost cell from A, which is $R_{1,3}$, and assign its cost value to the output layer. Calculate the accumulative costs to move into the neighbouring cells from $R_{1,3}$. Delete $(R_{1,3}, 4.5)$ from A, and update A. Now $A = \{(R_{2,3}, 5.04), (R_{2,4}, 9.5), (R_{3,1}, 5.5), (R_{3,2}, 7.86), (R_{3,3}, 4.95), (R_{4,1}, 9.5)\}$.

8. The process continues to select the lowest-cost cell from A, assign the lowest cost value to the output layer, and update A until A becomes empty or the maximum distance is reached.

Box 4.6 demonstrates how distance surfaces can be produced in ArcGIS.

Box 4.6   Generating distance surfaces in ArcGIS   **PRACTICAL**

To follow this example, start ArcMap, and load the **roads** feature class and **costOnSlope** raster layer from **C:\Databases\GIS4EnvSci\VirtualCatchment\Geodata.gdb**. In ArcGIS, a feature dataset can be used as the data layer representing input source locations with distance measurement tools. The **costOnSlope** raster records costs for travel on different types of slope, measured on a 1–5 scale (1 assigned to < 10°, 2 to 10–20°, 3 to 20–30°, 4 to 30–40° and 5 to > 40°). Enable the **Spatial Analyst** extension.

Least-cost path

Cost distance provides a basis for determining the least-cost path from a destination to a source location. As discussed above, a cell value on a cost distance surface represents the least accumulative cost of returning to the source locations from the cell. Based on the least accumulative cost values, the neighbour that is the next cell on the least accumulative cost path to the nearest source, also known as a back-link, can be tracked. A cost distance function in GIS typically outputs a cost distance surface as well as a direction or back-link raster that records the sequence of movements that created the cost distance surface. On the direction raster, a cell value defines the direction that identifies the next neighbouring cell along the least accumulative cost path from a cell to reach its least-cost source. The directions may be recorded in different ways in different GIS systems. For example, in GRASS, directions are recorded as degrees anti-clockwise from the east. In ArcGIS, directions are represented using numerals 1–8, with 1 for the right neighbouring cell, 2 for the lower right diagonal cell, and continuing clockwise (Figure 4.8). A least-cost path from the destination to the nearest source can be traced by following the given directions recorded in the direction raster. Figure 4.9 shows a direction raster (using ArcGIS direction codes) generated based on the distance surface

resulted from the process depicted in Figure 4.7, and the least-cost path from the cell $R_{3,2}$ (the destination or target location) to the source 2.

Least-cost path analysis is important for corridor planning. It can help select alternative routes for roads, railway lines and utility lines to minimise construction, operational and environmental costs, and assist in the design of wildlife corridors to facilitate wildlife movements and repair landscapes that have become fragmented. Box 4.7 provides an example of finding the least-cost paths to an outlook from villages weighted by slope.

Figure 4.8 Direction values in ArcGIS.

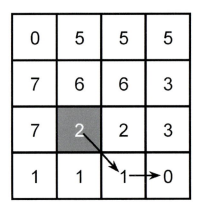

Figure 4.9 Direction raster and least-cost path (in arrows).

Network distance

A network is a set of interconnected line features, such as road, drainage, pipeline and utility networks. Geometrically, a network consists of nodes and links through which resources flow. Nodes represent intersections, interchanges and confluence points in the network. Links denote network segments between nodes. Network distance is the distance involved in passing through a network. It is measured as the shortest distance or least-cost distance between nodes on a network. Several algorithms have been developed for finding the shortest path and calculating the network distance along a network (Cherkassky et al. 1996). The most widely used algorithm is the one proposed by Dijkstra (1959).

Box 4.7   Creating least-cost paths in ArcGIS   PRACTICAL

To follow this example, start ArcMap, and load the **villages** feature class, **outlook** shapefile and **costOnSlope** raster layer from **C:\Databases\GIS4EnvSci\VirtualCatchment**. The **outlook** layer contains a point representing the location of an outlook, which is used as the source. The **villages** feature class represents point locations of several villages, which are used as the destinations. The **costOnSlope** raster records costs for travel on different types of slope, as described in Box 4.6. Enable the **Spatial Analyst** extension.

Generate a cost distance surface and a direction raster

1. Open **ArcToolBox**. Navigate to **Spatial Analyst Tools > Distance**, and double-click **Cost Distance**.
2. In the **Cost Distance** dialog box:

    1) Select **outlook** as the input features.
    2) Select **costOnSlope** as the input cost raster.
    3) Enter the name of the output distance raster.
    4) Enter the name of the output backlink raster.
    5) Click **OK**. A distance surface map and a direction or backlink raster map are created.

Calculate the least-cost paths

3. Open **ArcToolBox**. Navigate to **Spatial Analyst Tools > Distance**, and double-click **Cost Path**.
4. In the **Cost Path** dialog box:

    1) Select **villages** as the input feature destination data.
    2) Input the distance surface and the direction raster generated in Step 2 above as the input cost distance raster and input cost back-link raster.
    3) Enter the name of the output raster.
    4) Select **EACH_CELL** as the path type.
    5) Click **OK**. A raster containing the least-cost paths from the villages to the outlook is created and displayed, as shown in Figure 4.10.

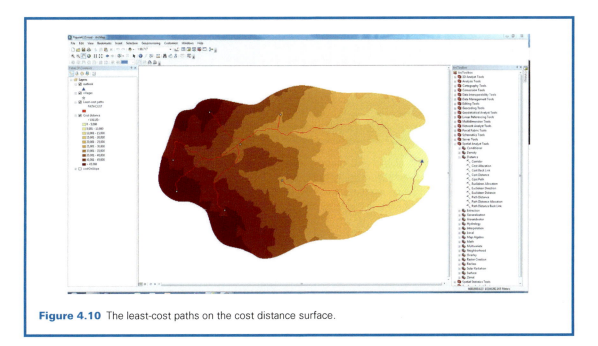

Figure 4.10 The least-cost paths on the cost distance surface.

The classic Dijkstra's algorithm solves the single-source shortest path problem in a network. It selects the unvisited node with the shortest distance, calculates the distance through it to each unvisited neighbouring node and updates the neighbour's distance if smaller. The basic procedure for finding a shortest path from a source node to a destination node with the algorithm can be described as below:

1. Assign zero to the source node and infinity to all other nodes as their tentative network distances to the source.
2. Set the source node as current, and mark all nodes as unvisited.
3. For the current node, calculate the distances from the source through it to all of its unvisited neighbouring nodes (for example, if the current node A has a distance of 10 to the source, and the link connecting it with a neighbour B has a length of 4, then the distance from the source to node B through node A will be 10 + 4 = 14).
4. Compare the calculated distances of the neighbouring nodes with their previously assigned distance values. If the previously assigned distance value for a node is larger than the newly calculated distance, replace the previously assigned distance with the new one as the tentative distance for the node.
5. Mark the current node as visited.

6. If the destination node is marked visited or if the smallest distance among the unvisited nodes is infinity (that is, there is no connection between the source and remaining unvisited nodes), then stop to end the algorithm. Otherwise, go to the next step.
7. Select the unvisited node that has the smallest tentative distance, set it as the current node, then go back to step 3.

After the algorithm is finished, the calculated distance for the destination node is the network distance between the source and destination. When the link length is replaced with travel time or cost over the link, the resulting network distance becomes the cost distance, and the shortest path becomes the quickest or least-cost path. For a full discussion on the algorithm, please refer to Dijkstra (1959). Most GIS systems apply this algorithm directly, or modify it to measure network distances and find the shortest paths in a network. It also provides a basis for other network analysis operations in GIS, including finding the closest facilities and delineating service areas.

However, network distance measurement in GIS requires preparation of a network dataset, which consists of three topologically structured elements: edges (links), junctions (nodes) and, optionally, turns (a transition from one link to another at a node). The network characteristics are recorded as attribute data linked with edges, junctions and turns. Examples of network attributes include link impedance, link demand, flow direction, speed limit, turn restriction and turn impedance. Link impedance is the cost involved in traversing a link, which may be measured as travel time, distance and fuel cost. Turn impedance is usually referred to as the time it takes to complete a turn. Box 4.8 shows how to create a network dataset and measure network distances in ArcGIS.

Box 4.8   Measure network distances in ArcGIS   **PRACTICAL**

To follow this example, start ArcMap, and load the **roads** and **villages** feature class from **C:\Databases\ GIS4EnvSci\VirtualCatchment\Geodata.gdb**. The physical network distances from the villages and two selected locations in the road network are to be measured and saved in an origin–destination (OD) matrix. ArcGIS **Network Analyst** extension is required.

Configure the Network Analyst environment

1. Select **Customize > Extensions** from the main menu bar. The **Extensions** dialog box opens.
2. In the **Extensions** dialog box:

    1) Check the **Network Analyst** checkbox.
    2) Click **Close**. The **Network Analyst** extension is enabled.

3. Click **Customize > Toolbars > Network Analyst**. The **Network Analyst** toolbar is added.

Create a network dataset

4. Click the **Catalog** window button on the **Standard** toolbar.
5. In the **Catalog** window, navigate to the folder where the network dataset is to be stored. Right-click the folder name, point to **New**, and click **Personal Geodatabase** to create a new geodatabase. Change the name of the geodatabase to **MyGeodatabase.mdb**.
6. Right-click **MyGeodatabase.mdb**, point to **New**, and click **Feature Dataset** to create a new feature dataset.
7. In the **New Feature Dataset** dialog:

    1) Name the feature dataset **Infrastructure**.
    2) Select **World Equidistant Conic** as the coordinate system.
    3) Select **none** for the vertical coordinate system.
    4) Take the default values for the tolerances.
    5) Click **Finish**.

8. Right-click **Infrastructure**, point to **Import,** and click **Feature Class (single)**.
9. In the **Feature Class to Feature Class** dialog box:

    1) Select **roads** as the input features.
    2) Make sure the output location is **Infrastructure**.
    3) Enter **myroads** as the output feature class.
    4) Click **OK**. The feature class **myroads** is created and added to the **Infrastructure** feature dataset.

10. Right-click **Infrastructure**, point to **New**, and click **Network Dataset**.
11. In the **Network Dataset** dialog box:

    1) Change the name for the network dataset to **roads_ND**. Click **Next**.
    2) Select **myroads** to participate in the network dataset. Click **Next**.
    3) Click **No** to model turns. Click **Next**.
    4) Take the default connectivity settings. Click **Next**.
    5) Check **None** to model the elevation of the network features. Click **Next**.

6) Keep **Length** as the attribute for the network dataset, and the units as **Meters**. Click **Next**.
7) Click **No** to establish the driving directions. Click **Next**.
8) Review the summary, and click **Finish**.
9) After the network dataset is created, click **Yes** to build the network when prompted.
10) Click **Yes** when prompted, to add all feature classes participating in the **roads_ND** network to the map. Three layers, **roads_ND**, **roads_junctions** and **myroads**, are added to the data view.

Create an OD matrix

12. From the **Network Analyst** toolbar, click **Network Analyst**, then select **New OD Cost Matrix**. OD cost matrix analysis layers appear in the table of contents grouped as a geodatabase, which is named **OD Cost Matrix**. Six feature layers are contained in the database: **Origins**, **Destinations**, **Lines**, **Point Barriers**, **Line Barriers** and **Polygon Barriers**. Each of the layers has default symbology. All the classes are now empty.
13. Click the **Network Analyst** window button on the **Network Analyst** toolbar.
14. In the **Network Analyst** window, right-click **Origins**, then click **Load Locations**.
15. In the **Load Locations** dialog box:

    1) Click the **Load From** drop-down arrow, and select **villages**.
    2) Select **Name** as the sort field.
    3) Set the search tolerance as **5000 meters**.
    4) Click **OK**. All the villages in the **villages** layer are added as the origins and displayed using the default symbols for the **Origins** layer.

16. In the **Network Analyst window**, click **Destination** to make it the active layer.
17. Click the **Create Network Locations Tool** button on the **Network Analyst** toolbar. Click two separate locations of your choice on the road network as the destinations. The two locations are named **Graphic Pick 1** and **Graphic Pick 2** by default. The locations can be adjusted by using the **Select/Move Network Location Tool** on the **Network Analyst** toolbar.
18. Click the **Solve** button on the **Network Analyst** toolbar. The network distances between the villages and two selected locations are calculated. Right-click **Lines** in the **Network Analyst window**, and click **Open Attribute Table** to view the OD matrix. The results are similar to those in Figure 4.11.

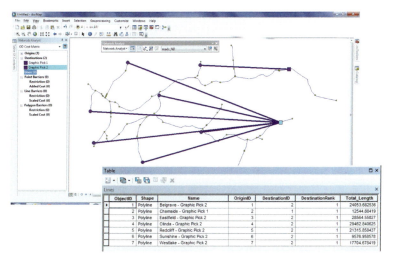

Figure 4.11 OD matrix analysis.

4.4 OVERLAY ANALYSIS AND MAP ALGEBRA

Overlay analysis is a process of stacking multiple data layers registered to a common georeferencing system on top of each other so that the relationships between features at each location can be analysed. It is the most common GIS operation for comparing and analysing multiple data layers simultaneously. As a tool for describing spatial correspondence of different geographical phenomena, overlay analysis has been used in a wide range of applications, such as examining the relationship between rainfall distribution and elevation, calculating the number of properties within a 100-year flood plain, evaluating environmental sensitivity based on slope, surface drainage, bedrock foundation, soil erosion and other environmental factors, and establishing spatial associations between wildlife occurrence and habitat conditions. Overlay analysis can be conducted in both vector and raster.

Vector overlay

Vector overlay involves combining geometries and attributes from the input vector layers into a single vector layer. There are three types of vector overlay: point-in-polygon, line-in-polygon and polygon-on-polygon.

Point-in-polygon overlay identifies in which area a point feature is located. One input layer contains point features, and the other represents polygon features. The result is a point layer representing the same set of point features with additional attributes transferred from the polygon layer. Figure 4.12 illustrates the process.

Line-in-polygon overlay determines which areas a line feature is crossing. One input layer represents line features, and the other represents polygon features. It calculates the geometric intersection of the line features with the polygon features, and splits the line features at polygon boundaries. Each of the resulting line features is assigned the attributes of the polygon it fell within, along with the original attributes from the input line feature layer. The result is a line feature layer composed of the split line features (Figure 4.13).

Polygon-on-polygon overlay combines two polygon layers. It involves calculating the geometric intersection of the polygon features from the two input layers, creating a new set of polygons with the polygon boundaries from the original features split at the intersections, and writing these new polygons as features to the output layer (Figure 4.14). Each new polygon is assigned the attributes from the two input layers.

There are basically three methods for vector overlay: union, intersect and identity (Figure 4.15). The union method performs polygon-on-polygon overlay, and keep all the features from the two input layers. The intersect method performs point-in-polygon, line-in-polygon or polygon-on-polygon overlay, and preserves

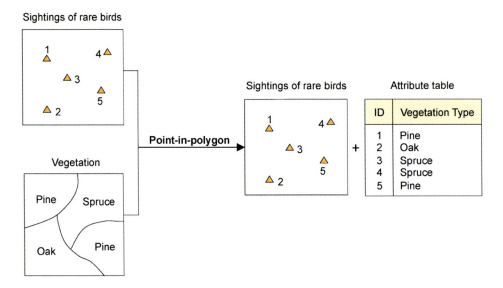

Figure 4.12 Point-in-polygon overlay.

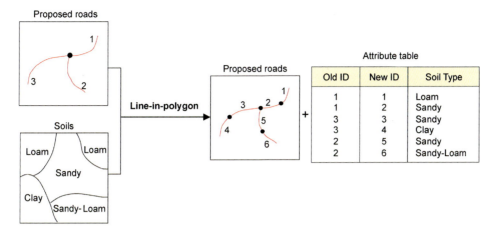

Figure 4.13 Line-in-polygon overlay.

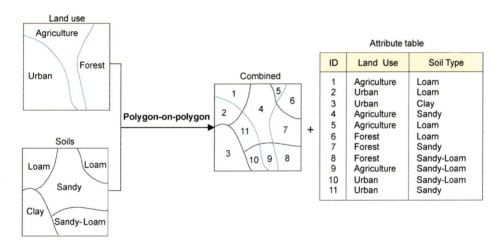

Figure 4.14 Polygon-on-polygon overlay.

the features or portion of the features that fall within the common area of all inputs. The identity method performs point-in-polygon, line-in-polygon or polygon-on-polygon overlay, but keeps all the features in the first input layer and deletes the portions of the second layer features which cross the boundary of the first input layer.

In ArcGIS, the intersect method can be used with inputs with any combination of feature types (point, line and polygon). The output feature type will be of the same type, or a type of the input features with the lowest dimension geometry. For example, if all the inputs are polygons, the output type will be polygon. If one of the inputs is line and none are points, the output type will be line. If any of the inputs are point, the output type will be point. The intersect method can also run with a single input. In this case, it will identify the intersections between features within the single input, which can be useful to find polygon overlap and line intersections. Box 4.9 illustrates vector overlay in ArcGIS.

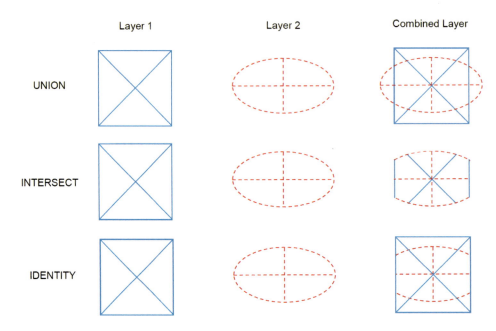

Figure 4.15 Three basic methods for vector overlay.

Box 4.9   Vector overlay in ArcGIS   **PRACTICAL**

To follow this example, start ArcMap, and load the **roads**, **pheasant**, **soils** and **landcover** feature class from **C:\Databases\GIS4EnvSci\VirtualCatchment\Geodata.gdb**.

Point-in-polygon overlay

1. Open **ArcToolBox**. Navigate to **Analysis Tools > Overlay**, and double-click **Intersect**.
2. In the **Intersect** dialog box:

    1) Add **pheasant** and **landcover** as the input features.
    2) Name the output feature class.
    3) Click **OK**. A new point layer is created and added to the table of contents. This layer contains all locations of pheasant sightings with the additional attributes from the **landcover** layer.

3. Right-click the new point layer in the table of contents, and click **Open Attribute Table**. Its attribute table opens, as shown in Figure 4.16. View the land cover type for each sighting location.

Line-in-polygon overlay

4. In **ArcToolBox**, navigate to **Analysis Tools > Overlay**, and double-click **Intersect**.
5. In the **Intersect** dialog box:

    1) Add **roads** and **landcover** as the input features.
    2) Name the output feature class.
    3) Click **OK**. A new line layer is created and added to the table of contents. This layer contains all road features with the additional attributes from the **landcover** layer.

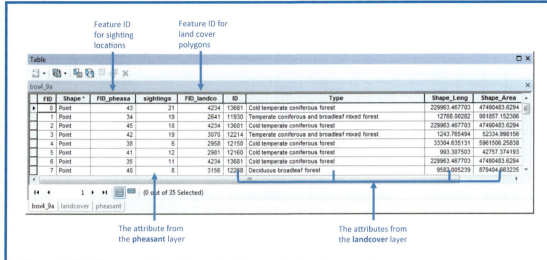

Figure 4.16 A feature attribute table resulted from a point-in-polygon overlay.

6. Right-click the new line layer in the table of contents, and open its attribute table. The table looks similar to Figure 4.16, but contains the attributes from both **roads** and **landcover**. View the land cover type for each road segment.

Polygon-on-polygon overlay

7. In **ArcToolBox**, navigate to **Analysis Tools > Overlay**, and double-click **Union**.
8. In the **Union** dialog box:

    1) Add **soils** and **landcover** as the input features.
    2) Name the output feature class.
    3) Set **NO_FID** for the **Join Attributes** parameter so that all the attributes except the FID from the input features will be transferred to the output layer.
    4) Click **OK**. A new polygon layer is created and added to the table of contents. This layer contains all polygon features with combined attributes from the two input layers.

9. Right-click the new polygon layer in the table of contents, and open its attribute table. Examine how each type of soil corresponds to land cover types.

Raster overlay

Raster overlay combines multiple raster layers into a single raster, in which every cell of each raster references the same geographical location. Therefore, raster overlay does not involve geometric intersection as vector overlay does. It involves operations on individual values of corresponding cells in the input layers to create a new raster layer with new cell values. A value can be assigned to each cell in the output raster based on unique combinations of values from several input rasters. Usually raster overlay is used to mathematically combine the rasters assigned with numerical cell values to calculate a new value for each

cell in the output raster. Figure 4.17 shows an example of raster overlay by addition. Two input rasters are added together to create an output raster with the values for each cell summed.

Raster overlay can be used to perform arithmetical, logical, relational, conditional and statistical operations on the input rasters (see Table 4.5). When applying logical operators in raster overlay, we often first reclassify the input rasters into binary rasters with cell values of either 1 or 0, then perform overlay operations, as shown in Figure 4.18. Box 4.10 shows how to perform raster overlay in ArcGIS.

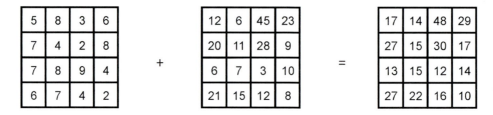

Figure 4.17 Raster overlay by addition.

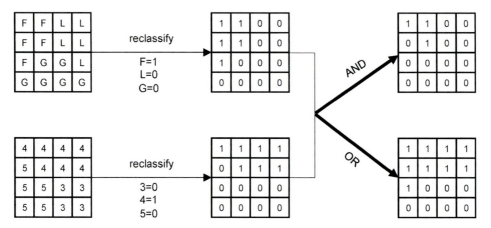

Figure 4.18 Raster overlay by logical operation.

Box 4.10 Raster overlay in ArcGIS   **PRACTICAL**

To follow this example, start ArcMap, and load the **soils** feature class and **landcv_r** raster from **C:\ Databases\GIS4EnvSci\VirtualCatchment\Geodata.gdb**. Enable the **Spatial Analyst** extension.

Raster overlay based on unique combinations of values

1. Use the **Feature to Raster** tool to convert **soils** to a raster layer (see Box 3.6). Name the output raster **soils_r**, select **Soil_order** as the field to assign values to the output raster, and set the output cell size as **40** (metres).
2. Open **ArcToolBox**. Navigate to **Spatial Analyst Tools > Local**, and double-click **Combine**.
3. In the **Combine** dialog box:

    1) Select **soils_r** and **landcv_r** as the input rasters.
    2) Name the output raster.
    3) Click **OK**. The output raster is created and added to the data view. Each cell value in the output raster represents a unique combination of soil and land cover.

4. Right-click the output raster in the table of contents, and click **Open Attribute Table**. In the attribute table, the **SOILS_R** and **LANDCV_R** fields list the values representing soil and land cover types corresponding to each cell value in the output raster.

Raster overlay by mathematical operations

5. Reclassify **landcv_r** by assigning the value **1** to the land cover type of **cold temperate coniferous forest** and **0** to other types of land cover using the **Reclassify** tool (see Box 4.3). Select **Type** as the reclass field. Name the output raster **coldConifer**.
6. Open **ArcToolBox**. Navigate to **Spatial Analyst Tools > Math**, and double-click **Times**.
7. In the **Times** dialog box:

   1) Select **soils_r** as the input raster 1 and **coldConifer** as the input raster 2.
   2) Name the output raster.
   3) Click **OK**. The output raster is created, which shows the soil types (coded 1–5) in the cold-temperate coniferous forest area. The other areas obtain a value of 0.

Map algebra

Raster overlay with mathematical operators is one of the map algebra tools commonly available in GIS packages that support raster data analysis. Map algebra is the application of raster data analysis and mathematical operators and functions to an algebraic equation with raster data layers as input variables to produce a new raster as a solution.

In conventional algebra, real values are represented by symbols which are usually alphabetically based, such as x, y and z, each representing a single numerical value. In map algebra, these symbols may represent single-theme raster layers such as soil type, soil pH value, land use type or wildlife density. For arithmetic and logic operations, each cell in the input raster layers must contain a numeric value representing the attribute of the geographical feature in the cell. These numeric values can be mathematically or logically combined and manipulated. Numbers assigned to symbols in an equation are transformed or combined into new numbers using mathematical operators or functions such as add, subtract, multiply and divide in the same way. Map algebra also incorporates raster data analysis functions. For example:

$$outlayer = inlayer1 * inlayer2 - log10(inlayer3) + slope(inlayer4) \tag{4.9}$$

In this algebraic equation, *outlayer* is the variable representing the output raster; *inlayer1*, *inlayer2*, *inlayer3* and *inlayer4* are variables representing four input rasters; +, – and * are mathematical operators; *log10()* is a mathematical function and *slope()* is a raster data analysis function for calculating slope for each cell in the input raster. Both operators and functions manipulate data items and return a result. Functions differ from operators in that they operate on arguments. The right side of the equation is called a map algebra expression.

Mathematical operators and functions in map algebra perform mathematical computations. They mainly include arithmetic, logical, relational/comparison operators, conditional, trigonometric and statistical functions. Some of the basic operators and functions are listed in Table 4.5.

Table 4.5 Mathematical operators and functions in map algebra

Type	Operators/functions
Arithmetic	+, –, ×, /, power, square root, exponential, logarithm
Logical	AND, OR, NOT
Relational	=, >, ≥, <, ≤
Conditional	IF . . . THEN . . ., IF . . . THEN . . . ELSE
Trigonometric	(inverse) cosine, (inverse) sine, (inverse) tangent
Statistical	maximum, minimum, mean, range, standard deviation, sum, median, mode, majority, minority, variety

There are four common types of raster data analysis functions that can be used in map algebra: local, focal/neighbourhood, zonal and global functions. Local functions perform cell-by-cell analysis. They compute a new value for every cell in the output raster as a function of one or more existing values of the cell at the same location on a single or multiple input rasters. Raster data reclassification and raster overlay operations, discussed earlier, are all local functions. Raster overlay operates on multiple rasters. Reclassification and many mathematical functions listed in Table 4.5 (such as power, logarithmic and trigonometric functions) operate on a single raster. Box 4.11 shows the use of the Raster Calculator tool in ArcGIS for running local functions. Raster Calculator executes a single map algebra expression using Python syntax in a calculator-like interface.

Box 4.11 Examples of local functions in ArcGIS   PRACTICAL

To follow this example, start ArcMap, and load the **dem**, **degreeSlope** and **landcv_r** raster from **C:\Databases\GIS4EnvSci\VirtualCatchment\Geodata.gdb**. Local functions in ArcGIS can be accessed from the **Spatial Analyst**'s **Local** toolset or **Raster Calculator**. In this example, **Raster Calculator** is used. Enable the **Spatial Analyst** extension.

Convert degree slope to percent slope

1. Open **ArcToolBox**. Navigate to **Spatial Analyst Tools > Map Algebra**. Double-click **Raster Calculator**. It appears similar to Figure 4.19.

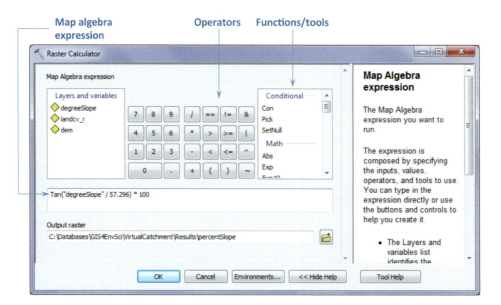

Figure 4.19 Raster Calculator.

2. In the **Raster Calculator** dialog box:

    1) Scroll down the functions/tools list to find the function **Tan**, then double-click it. **Tan()** is added to the expression box with the cursor placed in the parentheses. It is a trigonometric

function that calculates the tangent of every cell value in an input raster. It requires input cell values be measured in radians.

2) Double-click **degreeSlope** in the layers and variables list box. The expression becomes `Tan("degreeSlope")`. However, **degreeSlope** records slope in degrees. The values in **degreeSlope** must first be divided by the radians-to-degrees conversion factor of 180/π, or approximately 57.296.

3) Click the division (**/**) operator button ⬜ , then key in 57.296. The expression becomes `Tan("degreeSlope"/57.296)`.

4) Move the cursor to the end of the expression. Click the multiplication (*****) operator button ⬜ , and key in 100. The expression becomes `Tan("degreeSlope"/57.296)*100`. This is the map algebra expression used to convert degree slope to percent slope.

5) Name the output raster as **percentSlope**. Click **OK**. The output raster is created.

Identify meadows located on a particular type of slope in a certain range of altitude

3. Open **Raster Calculator**.

4. In the **Raster Calculator** dialog box:

   1) Enter the expression `Con("degreeSlope" < 15, 1, 0)`.
   2) **Con** is a conditional function that allows users to control the output value for each cell based on whether the cell value is evaluated as true or false in a conditional statement specified within the parentheses. In the above expression, if the value of a cell in **degreeSlope** is less than 15 (degrees), then 1 will be assigned to that cell location (true) on the output raster; otherwise, cell values greater than or equal to 15 will be assigned 0 (false) on the output raster. By running this function, you actually perform a reclassification operation to create a binary map.
   3) Name the output raster **gentleSlope**. Click **OK**. The output binary raster is created.

5. Re-open **Raster Calculator**, then enter the map algebra expression: `Con((dem > 1000) & (dem < 3000), 1, 0)`.

This expression assigns 1 to the cell locations where the elevation is greater than 1,000 and less than 3,000, and 0 to other cell locations. **&** is the logical **AND** operator. Name the output raster **midElevation**. Click **OK** to create the output raster.

6. Create a new raster named **meadow** using **Raster Calculator** with the following expression: `Con("landcv_r" != 4, 0, 1)`.

In this expression, **!=** is the relational operator **NOT EQUAL**, and **4** is the class code for the land cover class **Meadow**. If the value of a cell in **landcv_r** is not equal to 4, 0 will be assigned to that cell location on the output raster; otherwise, 1 will be assigned to that cell location on the output raster.

7. Re-open **Raster Calculator**, then enter the map algebra expression: `"meadow" * "midElevation" * "gentleSlope"`.

Name the output raster. Execute the above expression. The output raster shows meadows located at 1,000–3,000m in altitude and on slopes less than 15°. This operation is essentially raster overlay.

Focal functions calculate a new value for every cell in the output raster as a function of the existing values, distances and/or directions of the cells in its neighbourhood on the input raster. The cell being processed currently is called the focal cell. The neighbourhood can be defined in different shapes and sizes. The most commonly used shapes include rectangle (usually defined by its width and height in cells centred at the focal cell, for example 3×3), circle (defined by its radius centred at the focal cell), wedge (defined the radius extending from the focal cell, the start and end angles measured counter-clockwise from the east) and annulus (defined by the inner and outer radii centred at the focal cell) (see Figure 4.20). A neighbourhood can also be defined as an irregular shape called a kernel, which can be used to determine which cells in the neighbourhood will be included in the operation or to apply different weights to the cells in the neighbourhood (see Figure 4.21).

Focal functions are mainly used for focal or neighbourhood statistics. Case Study 7 in Chapter 10 provides an example of how to use focal functions in air pollution modelling. They are also very useful for filtering in digital image processing (see Section 6.3). These functions calculate summary statistics of the cell values within the pre-defined neighbourhood of every cell. During the operation of a focal function, the focal cell is shifted from one cell to another following the left-to-right and top-to-bottom direction until all cells are visited. Therefore, the neighbourhood of the focal cell is just like a moving window (Figure 4.22). The statistical functions listed in Table 4.5 can all be used in focal functions. Figure 4.23 provided some examples,

in which a 3×3 rectangular neighbourhood is used, but only the output values in yellow shaded cells are shown. The focal function 'mean' calculates the average of the cell values in the neighbourhood, effectively reducing location variation and smoothing the data. It is often used as a low-pass filter to smooth a digital image to reduce noises from the image. Figure 4.24 shows a part of a digital elevation model before and after smoothing using mean. The focal function 'range' computes the difference between the maximum and minimum cell values in the neighbourhood. A large range value indicates a likely boundary between different classes or categories. The range function can thus be used as an edge detection filter for edge enhancement in digital image processing, particularly useful for making line features and boundaries of area features more distinctive. The focal function 'majority' determines the cell value that occurs most often in the neighbourhood. Section 6.3 will provide a full discussion on spatial filters in digital image processing.

In addition to focal statistics, slope, aspect and surface curvature functions (see Section 7.4) are usually grouped into the focal functions, as their calculations are based on neighbourhood cell values. Box 4.12 provides some examples of the use of focal functions in ArcGIS.

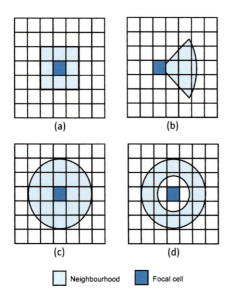

(a)   (b)

(c)   (d)

▢ Neighbourhood ▪ Focal cell

Figure 4.20 Neighbourhood types: (a) rectangle, (b) wedge, (c) circle and (d) annulus.

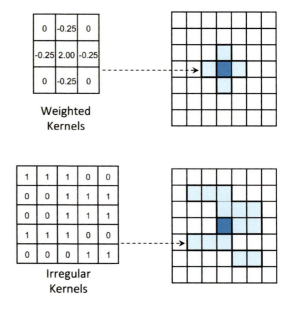

0	-0.25	0
-0.25	2.00	-0.25
0	-0.25	0

Weighted Kernels

1	1	1	0	0
0	0	1	1	1
0	0	1	1	1
1	1	1	0	0
0	0	0	1	1

Irregular Kernels

Figure 4.21 Kernels.

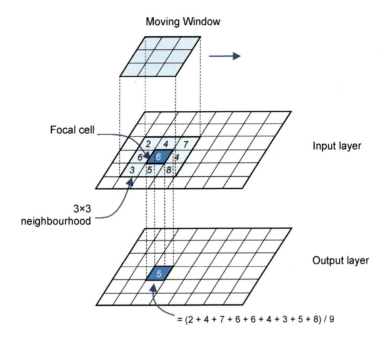

Figure 4.22 Moving window.

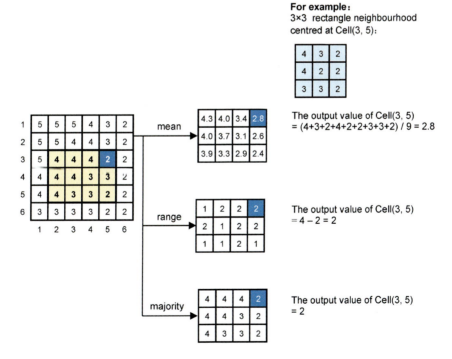

Figure 4.23 Examples of neighbourhood statistics.

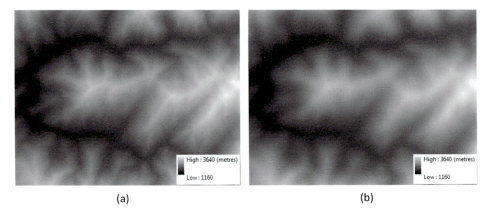

(a)   (b)

Figure 4.24 Part of a DEM: (a) before and (b) after smoothing.

Box 4.12   Examples of focal functions in ArcGIS   **PRACTICAL**

To follow this example, start ArcMap, and load the **dem** and **landcv_r** raster from **C:\Databases\ GIS4EnvSci\VirtualCatchment\Geodata.gdb**. Enable the **Spatial Analyst** extension.

Calculate the variety of land cover within the specified neighbourhood

1. Open **ArcToolBox**. Navigate to **Spatial Analyst Tools > Neighborhood**. Double-click **Focal Statistics**.
2. In the **Focal Statistics** dialog box:

    1) Select **landcv_r** as the input raster.
    2) Name the output raster.
    3) Select **Rectangle** as the shape of the neighbourhood, and set both its height and width to **5** cells – that is, the neighbourhood size is 5×5 cells.
    4) Select **Variety** as the statistics type, which calculates the number of unique values of the cells in the neighbourhood.
    5) Click **OK**. The output raster is created. The raster shows the areas with more than one land cover type. Open the attribute table. It lists how many cells there are that contain multiple types of land cover.

Smooth the DEM with the focal function MEAN

3. In **ArcToolBox**, navigate to **Spatial Analyst Tools > Neighborhood**. Double-click **Focal Statistics**.
4. In the **Focal Statistics** dialog box:

    1) Select **dem** as the input raster.
    2) Name the output raster.
    3) Select **Rectangle** as the shape of the neighbourhood, and set both of its height and width to **10** cells. In general, the larger the size of the neighbourhood, the smoother the result.
    4) Select **Mean** as the statistics type.
    5) Click **OK**. The output raster is created. Part of **dem** and its smoothed output are shown in Figure 4.24.

Zonal functions compute a new value for every cell in the output raster as a function of the existing values of the cells belonging to the same zone. A zone is a group of cells of the same value. It could make up a single region, or several disconnected regions that have the same value. Zonal functions calculate summary statistics of all the cell values in each zone and write the statistics into each cell of the zone in the output raster. The summary statistics include all those listed in Table 4.5. Zones are usually identified from another layer separate from the raster whose cell values are to be summarised statistically. Figure 4.25 shows two examples of zonal statistics, which calculate the mean and maximum elevation in each type of forests. Some GIS packages have zonal geometry functions that can be used to calculate area, perimeter, thickness and other geometric characteristics for each zone in a raster. Box 4.13 demonstrates how to access the zonal functions in ArcGIS.

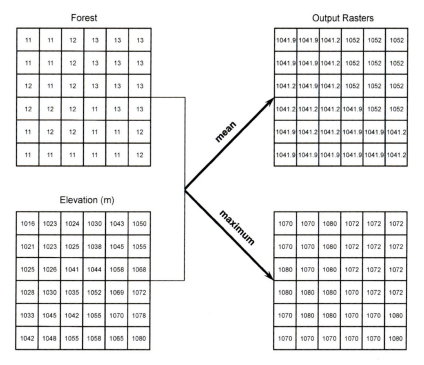

Figure 4.25 Examples of zonal statistics.

Box 4.13   Examples of zonal functions in ArcGIS   **PRACTICAL**

To follow this example, start ArcMap, and load the **dem** and **landcv_r** raster from **C:\Databases\ GIS4EnvSci\VirtualCatchment\Geodata.gdb**. Enable the **Spatial Analyst** extension.

Calculate the average elevation in each type of land cover

1. Open **ArcToolBox**. Navigate to **Spatial Analyst Tools > Zonal**. Double-click **Zonal Statistics**.
2. In the **Zonal Statistics** dialog box:

    1) Select **landcv_r** as the input raster, and **Value** as the zone field. Here, each type of land cover is used to define a zone.

(continued)

(continued)

    2) Select **dem** as the **input value raster**.
    3) Name the output raster.
    4) Select **Mean** as the statistics type.
    5) Click **OK**. The output raster is created. It shows the average elevation for each land cover zone.

3. In **ArcToolBox**, navigate to **Spatial Analyst Tools > Zonal**. Double-click **Zonal Statistics as Table**.
4. In the **Zonal Statistics as Table** dialog box:

    1) Select **landcv_r** as the input raster, and **Type** as the zone field.
    2) Select **dem** as the **input value raster**.
    3) Name the output table.
    4) Select **ALL** as the statistics type, and click **OK**. A table is created which contains seven summary statistics of elevation in each type of land cover, including minimum, maximum, range, mean, standard deviation and sum.

Calculate the geometry of land covers

5. In **ArcToolBox**, navigate to **Spatial Analyst Tools > Zonal**. Double-click **Zonal Geometry**.
6. In the **Zonal Statistics** dialog box:

    1) Select **landcv_r** as the input raster, and **Value** as the zone field.
    2) Name the output raster.
    3) Select **Area** as the geometry type, and **40** as the output cell size.
    4) Click **OK**. The output raster is created. Each cell in the output raster is assigned an area value for the type of land cover it belongs to.

7. In **ArcToolBox**, navigate to **Spatial Analyst Tools > Zonal**. Double-click **Zonal Geometry as Table**.
8. In the **Zonal Geometry as Table** dialog box:

    1) Select **landcv_r** as the input raster and **Value** as the zone field.
    2) Name the output table.
    3) Set the processing cell size as **40**.
    4) Click **OK**. The output table is generated, which lists all the major geometric characteristics of the eight types of land cover, including area, perimeter, thickness and the characteristics of ellipse.

Global functions calculate a new value for every cell in the output raster as a function of the existing values of all cells in the input raster. Distance surface and least-cost path functions, discussed earlier, and visibility analysis functions (see Section 7.5) are global functions.

In ArcGIS, all its Spatial Analyst extension tools, operators and functions can be used with map algebra. In addition to local, focal, zonal and global functions discussed above, map algebra in ArcGIS can use and implement application-specific functions in Spatial Analyst, including hydrology, groundwater, surface analysis and solar radiation functions. There are two ways to use map algebra in ArcGIS. One is to use the Raster Calculator tool, as shown in Box 4.11. The second is to use the geoprocessing ArcPy site package, which provides an integrated Python environment enabling users to write map algebra expressions in Python scripts and take advantage of Python and third-party Python modules and libraries. Using ArcGIS's integrated Python environment requires familiarity with Python, which is beyond the scope of this book.

4.5 SPATIAL INTERPOLATION

Spatial interpolation is the procedure of estimating values of an environmental variable at locations whose values are unknown based on a sample of locations with known values. Environmental data are often collected as discrete observations at points or along transects, such as soil cores, soil moisture, vegetation transects and meteorological station data. Through spatial interpolation, spatial distributions of the environmental phenomena can be approximated. This is an important technique in creating continuous environmental datasets from sampling networks of observation points. For example, it can be used to estimate rainfall based on the data recorded at a network of rain gauge stations, to estimate concentrations of air pollutants such as CO_2, SO_2, NO_2 and particulate matter (PM) using samples from monitor stations, and to assess mining potential based on drilling data.

Basically, spatial interpolation involves applying an interpolation method to a set of points with known values to create a continuous surface. These points are also known as sample points or observations. They can be regularly or irregularly distributed. Many interpolation methods have been developed. Li and Heap (2008) provide a comprehensive review of more than forty spatial interpolation methods. The current volume only introduces the most commonly used methods available in GIS. These include inverse distance weighted (IDW), natural neighbour, trend surface, spline and kriging.

IDW

IDW estimates a value of each location by taking the distance-weighted average of the values of sample points in its neighbourhood (Figure 4.26). The closer a sample point is to the location being estimated, the more influence or weight it has in the averaging process. In other words, each sample point has a local influence that diminishes with distance. The rationale behind IDW is Tobler's first law of geography: 'everything is related to everything else, but near things are more related than distant things' (Tobler 1970, p. 236).

Mathematically, IDW can be expressed as:

$$Z_0 = \sum_{i=1}^{n} w_i Z_i \tag{4.10}$$

$$w_i = \frac{d_i^r}{\sum_{j=1}^{n}\left(1/d_j^r\right)} \tag{4.11}$$

where Z_0 is the interpolated value of the point being estimated, n is the number of sample points in the

neighbourhood, Z_i ($i = 1, \ldots, n$) is the value of the ith sample point in the neighbourhood, d_i ($i = 1, \ldots, n$) is the distance from the point being estimated to the ith sample point in the neighbourhood, and r is the power.

The power parameter determines the significance of the sample points on the interpolated value. By defining a higher power, more emphasis is placed on the nearby points, producing a more varying and less smooth surface. Specifying a lower power will give more influence to the distant points, resulting in a smoother surface. A power of 2 is most commonly used with IDW. Figure 4.27 shows two surfaces interpolated using IDW with different r values, based on the same set of sample points. An optimal r value is considered to be where the minimum mean absolute error or RMSE is at its lowest. Some GIS systems offer tools to find the optimal r value, as demonstrated in Box 4.14 later in this chapter.

The characteristics of the interpolated surface can also be controlled by the shape and size of neighbourhood. Usually, a neighbourhood is defined as a circle, with a fixed or variable radius, which limits the number of sample points that can be used for interpolation (see Figure 4.26).

IDW is easy to understand and use, which makes it a very popular method for interpolation in GIS. The interpolated surface passes through the sample points. However, the interpolated values are always within the range of the measured values of sample points and will never be beyond the maximum and minimum values of the samples. If the original sample points do not capture the underlying true maximum and minimum surface values (for example, mountain tops and valley bottoms), they will be missing from the interpolated surface.

Natural neighbour

Natural neighbour estimates the value of an unsampled location by finding the closest subset of sample points to the location being estimated, then applying weights to them based on proportionate areas (Sibson 1981). It is based on Voronoi (Thiessen) tessellation.

Voronoi tessellation builds a network of polygons based on a set of sample points, called Voronoi or Thiessen polygons. Each polygon contains only one sample point, and any location within a polygon is closer to the polygon's sample point than to any other sample point contained in other polygons (Figure 4.28). Voronoi tessellation was originally used for computing areal estimates from rainfall data. It now has ecological applications, such as determining territories and influence zones

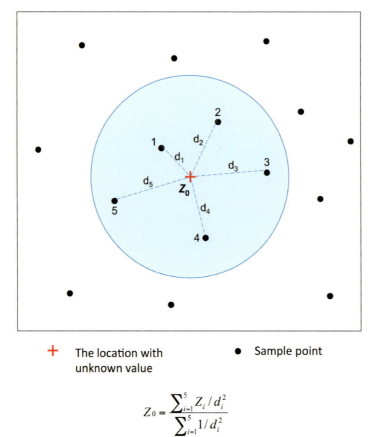

+ The location with unknown value

● Sample point

$$Z_0 = \frac{\sum_{i=1}^{5} Z_i / d_i^2}{\sum_{i=1}^{5} 1 / d_i^2}$$

where

$d_1, ..., d_5$ are distances

$Z_1, ..., Z_5$ are the values of Point 1, Point 2, ..., Point 5

Figure 4.26 Sample points and the circular neighbourhood used in IDW.

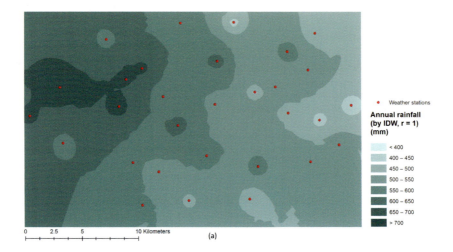

(a)

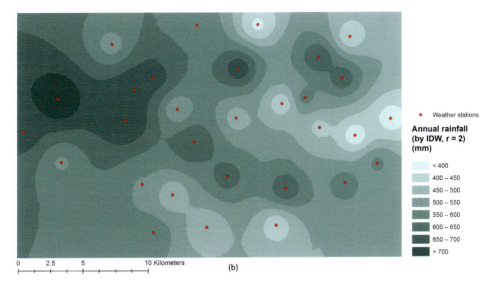

Figure 4.27 Rainfall surfaces interpolated using IDW with different power values.

of wild animals, delineating spaces available for the roots of individual plants and spaces around flowering orchid stems, studying the growth patterns of forests and forest canopies, and predicting forest fires.

For natural neighbour interpolation, initially a network of Voronoi polygons is created of all the sample points. Then a new Voronoi polygon network is constructed of all the sample points plus an unsampled point being estimated. The addition of the unsampled point changes the shape and size of Voronoi polygons of the surrounding sample points. For example, in Figure 4.28, the dashed lines represent the edges of the initial Voronoi network before the unsampled point P_0 is added, and the solid lines represent the edges of the Voronoi network after P_0 is inserted. Only those sample points whose Voronoi polygons have been altered by the insertion of P_0 are included in the subset of sample points for interpolation of a value at P_0. In this case, sample points 1, 2, 3, 4 and 7 are used. The weights of these sample points are calculated as the proportion of overlap between the new Voronoi polygon around P_0 and their own initial polygons created before the addition of P_0. These weights are then used in Equation 4.10 to estimate the value of P_0. Therefore, the basic equation used in natural neighbour interpolation is identical to the one used in IDW interpolation. Also, just like IDW, interpolated values are within the range of the samples, and it will not produce peaks or valleys that are not already represented by the samples.

Trend surface

A trend surface interpolation fits a smooth surface defined by a polynomial function to a set of sample points, then uses the polynomial function to estimate the values of unsampled locations. It rarely passes through original sample points. The simplest form of a trend surface is a planar surface with no curvature, which is defined by a linear or first-order polynomial:

$$f(x, y) = b_{0,0} + b_{1,0}x + b_{0,1}y \tag{4.12}$$

where x and y are coordinates, $b_{i,j}$ $(i,j = 0, 1)$ are polynomial coefficients, and $f(x, y)$ is the value of an environmental variable at the location (x, y). Suppose there are n sample points, whose values are $z_1, z_2, \ldots, z_n$, and whose coordinates are $(x_1, y_1), (x_2, y_2), \ldots, (x_n, y_n)$. The coefficients of the polynomial function can be determined from the sample points by minimising:

$$\Sigma_{i=1}^n \left\{ z_i - f(x_i, y_i) \right\}^2 \tag{4.13}$$

This is called a least-squares method, which ensures that the sum of the squared deviations of the observed values at the sample points from the trend surface is a minimum.

In many circumstances, the modelled surface will not correspond to a plane, but a curved surface. A second-order (quadratic) polynomial can be used, which is expressed as:

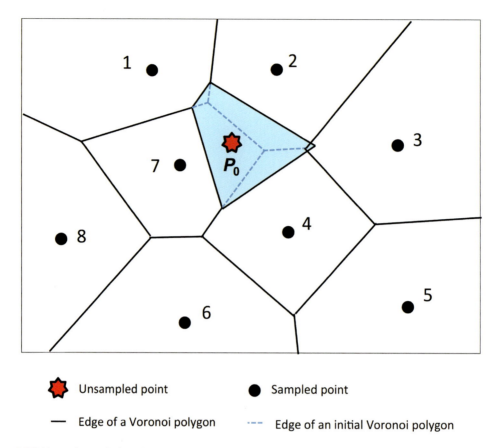

Figure 4.28 Voronoi tessellation (the shaded polygon containing the unsampled point).

$$f(x,y) = b_{0,0} + b_{1,0}x + b_{0,1}\ y + $$
$$b_{2,0}x^2 + b_{1,1}\ xy + b_{0,2}y^2 \qquad (4.14)$$

or a third-order (cubic) polynomial can be used:

$$f(x,y) = b_{0,0} + b_{1,0}x + b_{0,1}\ y + b_{2,0}x^2$$
$$+ b_{1,1}\ xy + b_{0,2}y^2 + b_{3,0}x^3 \qquad (4.15)$$
$$+ b_{2,1}x^2y + b_{1,2}xy^2 + b_{0,3}y^3$$

The general equation of a trend surface is:

$$f(x,y) = \Sigma_{i,j=0}^{p} b_{i,j}x^i y^j \qquad (4.16)$$

where p is the degree of the polynomial. As the polynomial order is increased, the surface being fitted has an increasing number of curvatures and becomes progressively more complex, as shown in Figure 4.29. Although polynomial orders as high as 10 are accepted, numerical instability in the analysis often creates artifacts in trend surfaces of orders greater than 5. This is mainly caused by the limited number of sample points compared with the rapid increase in number of coefficients attached to high order polynomials. Once the coefficients have been estimated using the least-squares method described above, the polynomial function can be evaluated at any point within the area of interest. Because of the least-squares fitting procedure, no other polynomial equation of the same order can provide a better approximation of the data.

Trend surface interpolation creates a smooth surface, representing the global trend of the distribution of an environmental variable. Therefore, it is often used to fit a surface to the sample points when the value of the variable changes slowly across the area. It is also used as a way of removing broad features of the data prior to using some other interpolation methods. Low-order polynomials can be used to characterise slowly varying physical process, such as pollution and wind direction. However, the higher the order of the polynomial, the more difficult it is to ascribe physical meaning to it. The most common

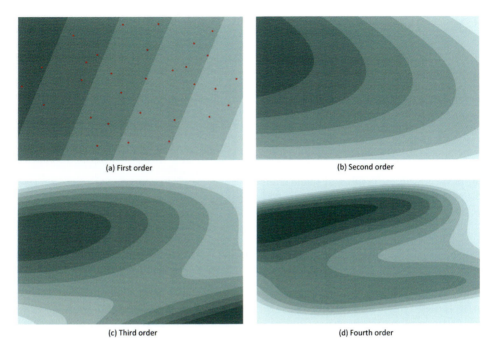

(a) First order

(b) Second order

(c) Third order

(d) Fourth order

Figure 4.29 Rainfall surfaces interpolated by trend surfaces of different orders: (a) first order, (b) second order, (c) third order and (d) fourth order (using the same dataset and same legend as in Figure 4.27).

order of polynomials is 1–4. Furthermore, the interpolated surfaces are highly susceptible to outliers (that is, extremely high and low values), especially at the edges.

Spline

The spline method estimates values at unsampled locations using a mathematical function that minimises overall surface curvature. There are several different types of splines. The most common splines used in GIS for spatial interpolation are thin-plate splines. Such a type of spline produces a surface that passes exactly through the sample points while ensuring the surface is as smooth as possible. The general form of a spline function can be written as:

$$f(x,y) = t(x,y) + \sum_{i=1}^{n} b_i r(d_i) \tag{4.17}$$

where x and y are coordinates, $t(x, y)$ is a trend function, n is the number of sample points, d_i is the distance from the ith sample point to the point (x, y), $r(d)$ is a basis function designed to obtain a minimum curvature surface and b_i are coefficients (Mitas and Mitasova 1999). For thin-plate splines, a basis function minimises the cumulative sum of the squares of the second derivative terms of the surface taken over each point on the surface

(Mitas and Mitasova 1999). Different thin-plate splines take different forms of $t(x, y)$ and $r(d)$. There are two most commonly used forms of thin-plate splines: regularised spline and spline with tension.

Regularised spline

For the regularised spline, the trend and basis functions take the following forms:

$$t(x,y) = a_1 + a_2 x + a_3 y \tag{4.18}$$

$$r(d) = \frac{1}{2\pi} \left\{ \frac{d^2}{4} \left[\ln \frac{d}{2\pi} + c - 1 \right] + \tau^2 \left[K_0 \left(\frac{d}{\tau} \right) + c + \ln \frac{d}{2\pi} \right] \right\} \tag{4.19}$$

where τ is the weight, c is a constant equal to 0.577215 and K_0 is the modified Bessel function (Mitas and Mitasova 1988). Typical values of τ are 0, 0.001, 0.01, 0.1 or 0.5. The higher the weight, the smoother the interpolated surface. The b_i coefficients in Equation 4.17 and a_1, a_2 and a_3 in Equation 4.18 are found by a system of linear equations:

$$f(x_i, y_i) = z_i \; (i = 1, 2, \ldots, n) \qquad (4.20a)$$

$$\sum_{i=1}^{n} b_i = 0 \qquad (4.20b)$$

$$\sum_{i=1}^{n} b_i x_i = 0 \qquad (4.20c)$$

$$\sum_{i=1}^{n} b_i y_i = 0 \qquad (4.20d)$$

where z_i is the value at the ith sample point, and n is the number of sample points.

A major problem with the regularised spline is the steep gradients in data-poor areas (where there are very few sample points), often known as overshoots. A weight τ of higher than 0.5 tends to produce a higher number of overshoots in data-poor areas. These overshoots need to be corrected. Thin-plate splines with tension are designed to mitigate the problem.

Spline with tension

The spline with tension allows users to control the tension to be applied to the edges of the surface. The trend and basis functions of spline with tension have the following forms:

$$t(x, y) = a \qquad (4.21)$$

$$r(d) = -\frac{1}{2\pi\varphi^2}\left[\ln\frac{d\varphi}{2} + c + K_0(d\varphi) \right] \qquad (4.22)$$

where φ is the weight. The higher the weight, the coarser the interpolated surface, and the more the estimated values conform to the sample data range. In other words, a higher weight reduces the range of interpolated values, or the stiffness of the surface. Typical values of φ are set as 0, 1, 5 or 10.

Both regularised splines and splines with tension create smooth, gradually changing surfaces with estimated values that may lie outside the range of the maximum and minimum values of sample points. For full discussions of the two methods, please refer to Mitas and Mitasova (1988). Figure 4.30 shows two examples of rainfall surfaces interpolated with a regularised spline and a spline with tension.

Kriging

Kriging is a geostatistical method for spatial interpolation. It was named after the South African mining engineer Danie G. Krige. This technique is similar to IDW

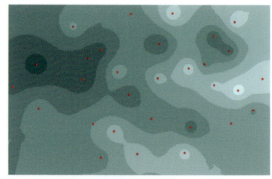

(a) Regularised spline

(a) Spline with tension

Figure 4.30 Rainfall surfaces interpolated using splines: (a) regularised spline and (b) spline with tension (using the same dataset and same legend as in Figure 4.27).

in that it estimates the value of a variable at an unsampled location by computing a weighted average of the known values of the sample points in its neighbourhood. However, weights in kriging are dependent on the spatial variability in the values of sample points.

Kriging assumes that in most cases spatial variations observed in environmental phenomena, such as variations in soil qualities and changes in the grade of ores, are random yet spatially correlated, and the data values characterising such phenomena conform to Tobler's first law of geography (data values at locations which are close to each other generally exhibit less variability than data values at locations which are farther away from each other). This is called spatial autocorrelation. The exact nature of spatial autocorrelation varies from dataset to dataset, and each set of data has its own unique function of variability and distance between sample points. This variability is represented by a semivariogram.

A semivariogram is a graph of the semivariance on the y-axis and the distance between sample points, called the lag, on the x-axis (Figure 4.31). The semivariance measures the variability of the observed values at the sample points that are separated by a certain distance. It is calculated by:

$$\gamma(h) = \frac{1}{2n}\sum_{i=1}^{n}\left[z(x_i + h) - z(x_i)\right]^2 \qquad (4.23)$$

where $\gamma(h)$ is the semivariance for a distance, h, separating two sample points $z(x_i)$ and $z(x_i+h)$, and n is the number of pairs of sample points separated by h. In order to estimate the semivariance at any given distance, the data points in the semivariogram are fitted with a continuous curve, as shown in Figure 4.31. The curves, which are mathematical functions called semivariogram models, can be in different forms. Figure 4.32 provides some examples of semivariogram models, including spherical, circular, exponential and Gaussian models. Each model is designed to fit different types of phenomena. Different models may have different effects on the estimation of the unknown values, particularly when the shape of the curve near the origin differs significantly. Models should be selected based on the spatial autocorrelation of the data and on prior knowledge of the phenomenon.

As illustrated in Figure 4.31, $\gamma(h)$ is expected to increase with increasing distance between two data points. The increase typically tapers off at a certain distance, known as the range. Beyond the range, changes in h have no significant influence on the semivariance, thus there is no longer any spatial autocorrelation. The range determines the spatial scale over which data are correlated. Smaller ranges indicate that data values change more rapidly over space. The range is used in kriging for defining the size of the neighbourhood so that spatially correlated sample points are selected for interpolation.

The semivariance at the range is called the sill. Frequently, the sill is equal to the statistical variance of the data. Theoretically, at zero separation distance, the semivariance is zero. However, at an infinitely small separation distance, the semivariogram often exhibits a nugget effect. A nugget value represents a degree of randomness attributed to measurement errors and/or spatial variations at distances smaller than the sampling interval.

Kriging uses the semivariance values obtained from the fitted semivariogram to estimate the weights used in interpolation and the variance of the interpolated values (as a measure of predication errors). There are many forms of kriging (see Burrough et al. 2015, Chapter 9). Two main types of kriging are introduced below.

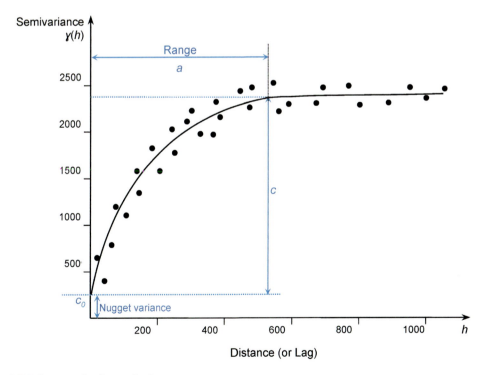

Figure 4.31 An example of a semivariogram.

Spherical

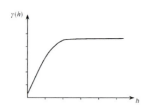

$$\gamma(h) = c_0 + c\left(\frac{3h}{2a} - \frac{1}{2}\left(\frac{h}{a}\right)^3\right) \qquad 0 < h \le a$$

$$\gamma(h) = c_0 + c \qquad h > a$$

$$\gamma(0) = 0$$

Circular

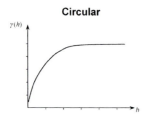

$$\gamma(h) = c_0 + c\left(1 - \frac{2}{\pi}\cos^{-1}\left(\frac{h}{a}\right) + \sqrt{1 - \frac{h^2}{a^2}}\right) \qquad 0 < h \le a$$

$$\gamma(h) = c_0 + c \qquad h > a$$

$$\gamma(0) = 0$$

Exponential

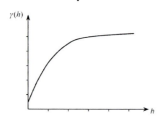

$$\gamma(h) = c_0 + c\left(1 - \exp\left(\frac{-h}{a}\right)\right) \qquad h > 0$$

$$\gamma(0) = 0$$

Gaussian

$$\gamma(h) = c_0 + c\left(1 - \exp\left(\frac{-h^2}{a^2}\right)\right) \qquad h > 0$$

$$\gamma(0) = 0$$

Figure 4.32 Examples of semivariogram models.

Ordinary kriging

Ordinary kriging assumes that there is no trend in the data and that the mean of the dataset is unknown. The weights are derived by solving the system of linear equations, which minimizes the expected variance of data values:

$$\sum_{j=1}^{k} w_j \gamma\left(h_{ij}\right) + \mu = \gamma\left(h_{i0}\right) \text{ for all } i = 1, \ldots, n \qquad (4.24a)$$

$$\sum_{i=1}^{k} w_i = 1 \qquad (4.24b)$$

$$\sigma^2 = \sum_{i=1}^{k} w_i \gamma\left(h_{i0}\right) + \mu \qquad (4.25)$$

where k is the number of sample points within the neighbourhood, w_i is the weight for the ith sample point to be estimated, $\gamma(h_{ij})$ is the semivariance between sample points i and j, $\gamma(h_{i0})$ is the semivariance between sample point i and the point to be estimated, and λ is a Lagrange multiplier, which is added to ensure the minimum possible estimation error. Once w_i $(i = 1, \ldots, k)$ are found, they are used in Equation 4.10 to estimate the values at unsampled locations.

The error variance for each interpolated point is estimated by:

The square root of the variance gives the standard error at the interpolated point, which provides an error estimate and confidence interval for the unknown point. Suppose the interpolated value is z_0. If the interpolation errors have a normal distribution, the real value at the interpolated point is within $z_0 \pm (\sqrt{\sigma^2} \times 2)$ with a probability of 95 per cent. Figure 4.33 shows the rainfall surface interpolated using ordinary kriging on the same dataset as in Figure 4.27 and the map of the standard errors of the surface.

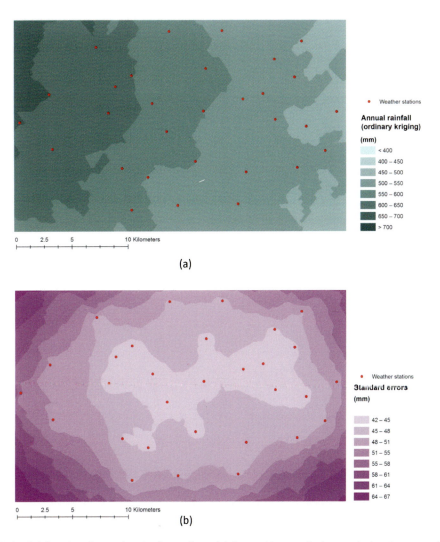

Figure 4.33 A rainfall surface interpolated using ordinary kriging and its standard errors (using the same dataset as in Figure 4.27).

Universal kriging

Universal kriging assumes that there is an overriding trend in the data in addition to spatial autocorrelation among the sample points, and this trend can be modelled by a polynomial function. If the trend is defined using a first-order polynomial

$$f(x, y) = ax + by \qquad (4.26)$$

then the weights for interpolation can be derived by solving the following set of simultaneous equations:

$$\sum_{j=1}^{k} w_j \gamma\left(h_{ij}\right) + \mu + ax_i + by_i = \gamma\left(h_{i0}\right)$$
$$\text{for all } i = 1, \ldots, n \qquad (4.27a)$$

$$\sum_{i=1}^{k} w_i = 1 \qquad (4.27b)$$

$$\sum_{i=1}^{k} w_i x_i = x_0 \qquad (4.27c)$$

$$\sum_{i=1}^{k} w_i y_i = y_0 \qquad (4.27d)$$

where (x_i, y_i) are the coordinates of sample point i, and (x_0, y_0) are the coordinates of the point to be estimated. The variance for each interpolated point is estimated by:

$$\sigma^2 = \sum_{i=1}^{k} w_i \gamma\left(h_{i0}\right) + \mu + ax_0 + by_0 \qquad (4.28)$$

Higher-order polynomials could be handled in a similar way, with a larger set of simultaneous equations to be solved. Figure 4.34 shows the rainfall surface interpolated using universal kriging with the first-order linear trend and the distribution of its standard errors. Case Study 8 in Chapter 10 provides an example of the use of universal kriging in climate trend analysis.

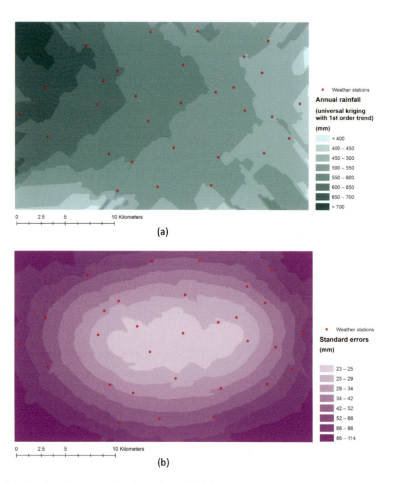

Figure 4.34 A rainfall surface interpolated using universal kriging and its standard errors (using the same dataset as in Figure 4.27).

In the use of kriging, more sample points will produce a more accurate semivariogram model, hence a more accurate interpolated surface. The standard errors, as one of the outputs of kriging, provide an indicator of the reliability of the interpolated values. This proves to be very useful for the future improvement of the surface by identifying where more samples are required. Kriging produces a surface which passes through the sample points. But the interpolated values are not bound by the maximum and minimum of the sample data.

Accuracy assessment and sources of error

Spatial interpolation routines are commonly available in GIS nowadays. They are easy and simple to use. However, it is important to realise that every spatial interpolation method involves errors. They all approximate mathematically the continuous distribution of a particular environmental variable across an area based on a set of discretely distributed point data. Mathematical approximation itself contains uncertainty. From the interpolated rainfall surfaces presented above, it is also not difficult to see that different interpolation methods produce different results with the same dataset. Some methods may yield more accurate results than others. The accuracy of a surface interpolated by a particular method is often evaluated through cross-validation.

Cross-validation evaluates the performance of an interpolation method in two steps. It first removes each sample location one at a time and estimates its value based on the remaining sample points using the interpolation method. Then, for all sample points, it compares the observed and estimated values to calculate estimation errors. Errors are commonly measured in terms of the mean error and RMSE (see Section 3.5). For kriging, additional measures are used, including the standard error and standardised RMSE, which is calculated as the ratio of the RMSE to the standard error.

The mean error is the averaged difference between the measured and the estimated values. The RMSE serves to aggregate the magnitudes of the errors in interpolated values for all sample points into a single measure of interpolation accuracy. It is a measure of magnitude of errors. The smaller the RMSE, the more accurate the interpolated surface is. A better interpolation method produces a smaller RMSE. The mean error and the RMSE can be used together to diagnose the variation in the errors. The RMSE will always be larger or equal to the mean error, and the greater difference between them, the greater variance in the individual errors. If the RMSE is equal to the mean error, then all the errors are of the same magnitude.

The standardised RMSE is a measure of the goodness of the assessment of the standard error. As mentioned before, the standard error in kriging provides an error estimate and confidence interval for interpolated values. If the standard error is accurate or correctly assessed, the ratio of the RMSE to the standard error should be close to 1. Therefore, a better kriging method should produce a smaller RMSE as well as a standardised RMSE closer to 1.

In addition to errors inherent in interpolation methods, there are other two common sources of error in spatial interpolation, including sample data uncertainty and edge effects. Uncertainty in sample data mainly results from too few sample points, limited or clustered distributions of sample points, and uncertainty about locations and/or values of sample points. In general, the larger the number of sample points, the more accurate the estimates. However, clustered sample points may not add much information compared with those that are more evenly spread out. Indeed, clustered sample points may bias the interpolation results. In order to get a good result of interpolation, it is important to have a sufficient number of sample points that are well distributed. When it is impractical to obtain an extensive and evenly distributed coverage of sample data in an area with complex topography and high variability in land use and other environmental variables, it is necessary to incorporate spatial variations of the related environmental phenomena in interpolation. Case Study 6 in Chapter 10 provides such an example.

Edge effects refer to distortions of the interpolated values near the boundary of the study area due to the lack of sample data outside the area. As shown in Figures 4.33b and 4.34b, the border areas have the highest standard errors. In fact, in the border areas, the interpolation method is no longer interpolating – that is, estimating unknown values within a region; rather, it is now extrapolating – that is, predicting values in areas where there are no sample data. To remove the edge effects, an easy solution is to collect sample data outside the area of interest, include them in the interpolation, then clip the area of interest out of the interpolated surface. In this way, most of the inaccuracy is relegated to the area outside the sample points, producing more accurate results within the area of interest.

An understanding of the major spatial interpolation methods and their error sources can help not only to improve the accuracy of the interpolated surfaces, but also to raise awareness of the errors involved in any subsequent analysis that is based on the interpolation results. Box 4.14 demonstrates how to perform spatial interpolation and compare different spatial interpolation methods in ArcGIS.

Box 4.14 Spatial interpolation in ArcGIS   **PRACTICAL**

To follow this example, start ArcMap, and load the **gauges** shapefile and **boundary** feature class from **C:\Databases\GIS4EnvSci\VirtualCatchment** The shapefile contains meteorological records at a set of weather stations in the virtual catchment. The feature class depicts the boundary of the catchment. ArcGIS **Geostatistical Analyst** extension is required.

Start the Geostatistical Analyst extension

1. Select **Customize > Extensions** from the main menu. The **Extensions** dialog box opens.
2. In the **Extensions** dialog box:

    1) Check the **Geostatistical Analyst** checkbox.
    2) Click **Close**. The **Geostatistical Analyst** extension is enabled.

3. Click **Customize > Toolbars > Geostatistical Analyst**. The **Geostatistical Analyst** toolbar is added.

Interpolate an average annual rainfall surface using IDW

4. From the **Geostatistical Analyst** pull-down menu, select **Geostatistical Wizard**.
5. In the **Geostatistical Wizard** dialog box:

    1) In the **Methods** list box on the left of the dialog, click **Inverse Distance Weighting**.
    2) In the **Input Data** list box on the right of the dialog, click the **Source Dataset** arrow, and select **gauges**.
    3) Click the **Data Field** arrow, and select the **rainfall** field. This field stores the average rainfall data for every weather station.
    4) Click the **Next** button. Notice that the default power value is 2 in the second dialog panel.
    5) Click the **Optimize Power Value** button ⊾ to the right of the power box. Now notice that the power changes from 2 to 1. This means that the optimal power value for this case is 1. Other power values will produce more and larger errors.
    6) Use the defaults for other settings, then click **Next**.

The results of **cross-validation** are listed, including a table of the measured and predicted values for all sample points, a cross-validation diagram showing how well the interpolation method predicts the values at the unknown locations, and the **Mean** (the mean error) and **Root-Mean-Square** (RMSE) values. In this case, the mean error is 3.26, and the RMSE is 114.19.

    7) Click the **Finish** button, then click **OK** in the **Method Report** dialog. The **Inverse Distance Weighting** layer is added to the data view.

6. In the table of contents, right-click **Inverse Distance Weighting**, and click **Properties**.
7. In the **Layer Properties** dialog box:

    1) Click the **Extent** tab, and set the extent to **the rectangular extent of boundary**.
    2) Click the **Symbology** tab. Click **Classify**. In the **Classification** dialog, change the classification method to **Manual**, set the number of classes to **8**, and change the class breaks respectively to 400, 450, 500, 550, 600, 650 and 700, then click **OK**.
    3) Click **OK**. The interpolated surface now covers the entire area of the catchment, and should look like Figure 4.27a. You may try 2 for the power value, and the result should be similar to Figure 4.27b.

Interpolate an average annual rainfall surface using a trend surface model

8. Start the **Geostatistical Wizard**.
9. In the **Geostatistical Wizard** dialog box:

    1) In the **Methods** list box, click **Global Polynomial Interpolation**.
    2) In the **Input Data** list box, click the **Source Dataset** arrow, and select **gauges**.
    3) Click the **Data Field** arrow, and select the **rainfall** field.
    4) Click the **Next** button. In the second dialog panel, set the order of polynomial to **3**. The third-order trend surface will be used.
    5) Click **Next**. View the results of cross validation. The mean error is 6.14, and the RMSE is 144.69.
    6) Click **Finish** button, then click **OK** in the **Method Report** dialog. The **Global Polynomial Interpolation** layer is added to the data view.

10. Follow Steps 6 and 7 above to change the symbology of the **Global Polynomial Interpolation** layer. The layer should look similar to Figure 4.29c. You may try the first-, second- and fourth-order trend surface to create other rainfall surfaces similar to those shown in Figure 4.29.

Interpolate an average annual rainfall surface using a spline

11. Start the **Geostatistical Wizard**.
12. In the **Geostatistical Wizard** dialog box:

    1) In the **Methods** list box, click **Radial Basis Functions**.
    2) Select **gauges** as the source dataset, and **rainfall** as the data field. Click **Next**.
    3) Select **Completely Regularized Spline** as the kernel function (that is, the spline function).
    4) Click the **Optimize** button ⊾ to optimise the kernel parameter value. The optimisation process evaluates several models and chooses the **Kernel Parameter** value for the selected kernel function, which produces the lowest RMSE.
    5) Use the defaults for other settings, then click **Next**.
    6) View the results of cross-validation. The mean error is 3.19, and the RMSE is 118.2.
    7) Click the **Finish** button, then click **OK** in the **Method Report** dialog. The **Radial Basis Functions** layer is added to the data view.

13. Follow Steps 6 and 7 above to change the symbology of the **Radial Basis Functions** layer. It should look similar to Figure 4.30a. In a similar way, a rainfall surface can be created using **Spline with Tension**, as shown in Figure 4.30b.

Interpolate an average annual rainfall surface using ordinary kriging

14. Start the **Geostatistical Wizard**.
15. In the **Geostatistical Wizard** dialog box:

    1) In the **Methods** list box, click **Kriging/CoKriging**.
    2) Select **gauges** as the source dataset, and **rainfall** as the data field. Click **Next**.
    3) Click **Ordinary** (kriging). Select **Prediction** as the output surface type.
    4) Click **Next**. The **semivariogram** and **semivariogram model** are displayed. Expand **Model #1**, and change the type to **Exponential**. Click the **Optimize** button ⊾ to optimise the **semivariogram model**'s parameters. Use the defaults for other settings, then click **Next**.
    5) The preview pane on the left in the dialog panel shows the sample points that are used to calculate the value at an unknown location (marked by a crosshair). The red points in the

(continued)

(continued)

   preview are going to be weighted more than the green points since they are closer to the location being estimated.
   6) Expand the **Weights** panel on the right. The number of sample points is tallied in the list, along with an approximate scale of their weight in the calculation.
   7) Click in a few different places on the preview pane on the left. Notice that the neighbourhood moves and new sample points are selected.
   8) Use the defaults for other settings, then click **Next**.
   9) View the results of cross-validation. The mean error is 2.65, the RMSE is 109.16, the standardised RMSE is 0.96 and the average standard error is 114.23.
   10) Click the **Finish** button, then click **OK** in the **Method Report** dialog. The **Kriging** layer is added to the data view.

16. Follow Steps 6 and 7 above to change the symbology of the **Kriging** layer. It should look similar to Figure 4.33a.
17. Start the **Geostatistical Wizard** again. Use ordinary kriging and the same parameters as set in Step 15 above, but select **Prediction Standard Error** as the output surface type. The result is a standard error map, as shown in Figure 4.33b.

Interpolate an average annual rainfall surface using universal kriging

18. Start the **Geostatistical Wizard**.
19. In the **Geostatistical Wizard** dialog box:

   1) Select **Kriging/CoKriging** as the interpolation method, **gauges** as the source dataset, and **rainfall** as the data field. Click **Next**.
   2) Select **Universal** as the type of kriging and **Prediction** as the output surface type. On the right-hand panel, set **Order of trend removal** as **First**. Click **Next**.
   3) Select **Exponential** as the kernel function. Click **Next**.
   4) Expand **Model #1**, and change the type to **Exponential**. Click the **Optimize** button ⇙ to optimise the **semivariogram model**'s parameters. Use the defaults for other settings, then click **Next**. Click **Next** again.
   5) View the results of cross-validation. The mean error is 0.47, the RMSE is 121.1, the standardised RMSE is 1.076 and the average standard error is 114.11.
   6) Click the **Finish** button, then click **OK** in the **Method Report** dialog. The **Kriging_2** layer is added to the data view.

20. Follow Steps 6 and 7 above to change the symbology of the **Kriging_2** layer. It should look similar to Figure 4.34a. Similarly, the standard error map can be generated, which is similar to Figure 4.34b.

Compare different interpolation methods

The performance of different interpolation methods can be assessed in terms of the mean error, RMSE, standardised RMSE and average standard error. These statistics for the tested interpolation methods are listed in Table 4.6.

   From Table 4.6, ordinary kriging produced the smallest magnitude of errors as it has the smallest RMSE. Although universal kriging has the smallest mean error, the difference between its RMSE and mean error is greater than that of ordinary kriging, which indicates that it has the larger variance in individual errors. The standardised RMSE values also show that the standard errors estimated in ordinary kriging are more reliable than those estimated in universal kriging. Therefore, among the methods tested, ordinary kriging is the best, producing the most accurate results.

Table 4.6 Estimation errors of the interpolation methods tested in Box 4.14

Method	Mean error	RSME	Standardised RMSE	Average standard error
Inverse Distance Weighting ($r = 1$)	3.26	114.19		
Global Polynomial Interpolation (third-order)	6.14	144.69		
Radial Basis Functions (Completely regularised spline)	3.19	118.2		
Ordinary kriging	2.65	109.16	0.96	114.23
Universal kriging (with the first-order trend)	0.47	121.10	1.08	114.11

4.6 SUMMARY

- Spatial analysis derives information and knowledge about spatial patterns of geographical features and the spatial relationships between them from spatial data. The results of spatial analysis are dependent on the locations of the geographical features being analysed.
- Spatial query in GIS is an operation for retrieving information from existing spatial datasets or databases, but will not produce new information.
- Reclassification is a GIS operation for regrouping features into different classes or assigning new values to the features. It is mainly used for simplification, generalisation and measurement scale change.
- Distance, length, area, perimeter and shape can be calculated in a GIS and used to describe geometric characteristics of geographical features and derive landscape metrics.
- Buffering tools draw either Euclidean or geodesic buffers (in a vector format) around point, line and area features with fixed or variable widths, while distance surfaces (in a raster format) calculate the physical or cost distances to a particular feature or a group of features. Least-cost paths from one location to another are found based on distance surfaces.
- Network distance is measured as the shortest distance along a network using a network dataset, which consists of links, nodes and turns. It may take into account link and turn impedance.
- Overlay involves superimposing two or more map layers registered to a common georeferencing

system for the purpose of showing the relationships between environmental phenomena that occur in the same geographical space. Overlay analysis involves combining information from the input layers to derive or infer new information. It can be conducted in vector or raster.
- Map algebra is implemented through raster overlay with mathematical operators and functions (including arithmetic, logical, comparison operators, conditional, trigonometric and statistical functions) and raster data analysis functions (including local, focal, zonal and global functions).
- Spatial interpolation is the process of using a sample of locations with known values to estimate values at other unmeasured locations. It creates surfaces of continuous values. IDW, natural neighbour, trend surface, spline and kriging are spatial interpolation techniques commonly available in GIS. The accuracy of spatial interpolation is often evaluated through cross-validation.

REVIEW QUESTIONS

1. What is spatial analysis?
2. Using Query by Attributes in GIS, given the data in Table 4.7 and the following series of queries:

   First, create a new selection – site quality < 3.

   Then select from current selection – slope < 10.

   Then add to current selection – 800 < elevation < 1200.

Table 4.7 Forest stand characteristics.

Forest stand no.	Elevation	Slope	Site quality
1	800	5	1
2	1,200	25	3
3	950	11	2

Finally, select from current selection – site quality > 2.

Which forest stand is selected?

3. Provide three examples of the use of the reclassification function.
4. What is a buffer? What are typical applications of buffering?
5. What is cost distance? Describe the spreading algorithm for cost distance measurement.
6. What is network distance? Describe Dijkstra's algorithm for calculating network distance.
7. What is overlay analysis? Why must all the input data layers in overlay have the same georeferencing system?
8. How are vector and raster overlay implemented?
9. Discuss the three methods of vector overlay: union, intersect and identity.
10. What are local, focal, zonal and global functions?
11. What is spatial interpolation? What is its underlying principle?
12. What are the common problems with spatial interpolation techniques? How can they be overcome?

REFERENCES

Berry, J.K. (1993) *Beyond Mapping: Concepts, Algorithms and Issues in GIS*, Fort Collins, CO: GIS World.

Bogaert, J., Rousseau, R., Hecke, P.V. and Impens, I. (2000) 'Alternative area-perimeter ratios for measurement of 2D shape compactness of habitats', *Applied Mathematics and Computation*, 111: 71–85.

Boyce, R.R. and Clark, W.A.V. (1964) 'The concept of shape in geography', *The Geographical Review*, 54: 561–572.

Burrough, P.A., McDonnel, R.A. and Lloyd, C.D. (2015) *Principles of Geographical Information Systems*, 3rd edn, Oxford: Oxford University Press.

Cherkassky, B., Goldberg, A. and Radzik, T. (1996) 'Shortest paths algorithms: theory and experimental evaluation', *Mathematical Programming*, 73(2): 129–174.

De Smith, M., Goodchild, M.F. and Longley, P.A. (2013) *Geospatial Analysis: A Comprehensive Guide to Principles, Techniques and Software Tools*, 4th edn, Winchelsea, England: The Winchelsea Press.

Dijkstra, E.W. (1959) 'A note on two problems in connexion with graphs', *Numerische Mathematik*, 1: 269–271.

Douglas, D.H. (1994) 'Least cost path in GIS using an accumulated cost surface and slope lines', *Cartographica*, 31: 37–51.

Eastman, J.R. (1989) 'Pushbroom algorithms for calculating distances in raster grids', in *Proceedings of Autocarto 9*, 288–297, accessed 20 February 2016 at http://mapcontext.com/autocarto/proceedings/auto-carto-9.

Fischer, M. and Leung, Y. (eds) (2001) *GeoComputational Modelling: Techniques and Applications*, Berlin, Germany: Springer-Verlag.

Fotheringham, A.S. and Rogerson, P. (eds) (1994) *Spatial Analysis and GIS*, London: Taylor & Francis.

Li, J. and Heap, A.D. (2008) *A Review of Spatial Interpolation Methods for Environmental Scientists*, Record 2008/23, Canberra, Australia: Geoscience Australia.

Longley, P.A., Goodchild, M.F., Maguire, D.W. and Rhind, D.W. (2011) *Geographical Information Systems and Science*, 3rd edn, Hoboken, NJ: John Wiley & Sons.

Mitas, L. and Mitasova, H. (1988) 'General variational approach to the interpolation problem', *Computers and Mathematics with Applications*, 16(12): 983–992.

Mitas, L. and Mitasova, H. (1999) 'Spatial interpolation', in P. Longley, M.F. Goodchild, D.J. Maguire and D.W. Rhind (eds) *Geographical Information Systems: Principles, Techniques, Management and Applications*, Hoboken, NJ: John Wiley & Sons.

Moser, D., Zechmeister, H.G., Plutzar, C., Sauberer, N., Wrbka, T. and Grabherr, G. (2002) 'Landscape patch shape complexity as an effective measure for plant species richness in rural landscapes', *Landscape Ecology*, 17(7): 657–669.

OGC (2011) *OpenGIS Implementation Specification for Geographic Information – Simple Feature Access, Part 1: Common Architecture*, accessed 2 September 2013 at http://www.opengeospatial.org/standards/sfa.

Sibson, R. (1981) 'A brief description of natural neighbor interpolation', in V. Barnett (ed.) *Interpreting Multivariate Data*, New York: John Wiley & Sons, 21–36.

Tobler, W. (1970) 'A computer movie simulating urban growth in the Detroit region', *Economic Geography*, 46(2): 234–240.

Turner, M.G., Gardner, R.H. and O'Neill, R.V. (2001) *Pattern and Process: Landscape Ecology in Theory and Practice*, New York: Springer-Verlag.

Wentz, E.A. (2000) 'A shape definition for geographic applications based on edge, elongation, and perforation', *Geographical Analysis*, 32: 95–112.

Spatial data exploration with statistics

5

This chapter introduces the spatial statistical techniques for describing, exploring and interrogating observational or sample data in order to measure spatial distributions, detect spatial patterns, identify spatial clusters and explore the spatial associations or spatial relationships among environmental variables.

LEARNING OBJECTIVES

After studying this chapter, you should be able to:

- understand what exploratory spatial data analysis is about;
- describe the differences between spatial and non-spatial statistics;
- select and use spatial sampling methods;
- comprehend the issues in spatial sampling design;
- know how to describe and measure spatial distributions;
- understand how to detect spatial patterns;
- grasp the spatial statistical techniques for identifying spatial clusters;
- explain the concept of spatial autocorrelation;
- build models to describe spatial relationships;
- understand the assumptions for different spatial statistical techniques;
- interpret results of spatial statistical analysis.

5.1 EXPLORATORY SPATIAL DATA ANALYSIS

Spatial data provide information on where an environmental event occurs and how an environmental phenomenon is distributed. The location where an environmental event occurs may provide some indication regarding why that particular event happens. Exploratory spatial data analysis is the process of applying spatial statistical methods and tools to investigate spatial data in order to detect and quantify patterns in the data and to establish spatial associations between a given set of environmental events or phenomena (Anselin 1998). Spatial statistics supports both exploratory and confirmatory data analysis. Exploratory spatial statistical methods are used to describe and summarise the characteristics of a spatial dataset and quantify the relationships between environmental variables. They help develop hypotheses or suggest ideas for further investigation. Confirmatory spatial statistical methods are applied for hypothesis testing – that is, making inferences about a population from observations and analyses of a sample. Here, a population is the total set of all elements (plants, animals, buildings and so on) to be studied. A sample is a subset of the member selected from a population. Most environmental or spatial data are regarded as samples of observations on some environmental process operating in space. This chapter only covers the exploratory spatial statistical methods that have been integrated into GIS.

Spatial statistics and non-spatial or traditional statistics have similar concepts and objectives. However, spatial statistics is used to detect, characterise and make inferences about spatial patterns, as well as to assess the significance of the patterns. For example, traditional statistics can be used to calculate the total count, mean age and median height of trees of different species in a region, but cannot describe how the different species of trees are distributed spatially in the area. In contrast, spatial statistics can be used to answer such questions as 'Are different species of plants across a region distributed in patches or along a particular direction?' and 'How is the distribution related to soils, drainage and other environmental factors?'

Unlike traditional statistics, spatial statistics incorporates location, distance, direction or other spatial properties into statistical measures. These measures describe how environmental phenomena manifest themselves in geographical space in terms of location and/or characteristics. More importantly, spatial statistics assumes that observations are spatially dependent due to spatial autocorrelation, whereas traditional statistics assumes that observations are independent (Getis 2005). This chapter focuses on the basic concepts of the fundamental spatial statistics, and their use and interpretation in exploratory spatial data analysis, including spatial sampling methods. The formal mathematics of these methods can be found in several advanced spatial statistics books (for example, Ripley 2004; Chun and Griffith 2013).

5.2 SPATIAL SAMPLING

Spatial sampling methods

Spatial data in environmental science are mostly observations or measurements made for individual objects, along a line transect or over an area. Spatial sampling is the process of selecting representative locations (sample points) to make observations or take measurements. The method used to collect spatial data is critically important. If sample data are not collected in an appropriate way, statistical analysis will produce biased or misleading results. A sample taken from a population is intended to be representative of that population. The list of all elements in the population is referred to as the sampling frame, or framework. The sampling frame for spatial sampling consists of all the points located in a geographical region of interest. There are many spatial sampling methods. The most frequently used methods include simple point random sampling, systematic point sampling, stratified point sampling, clustered point sampling and transect sampling.

Simple random point sampling involves obtaining a certain number of sample points in the study area by randomly selecting x and y coordinates. Every location within the study area has an equal chance of being chosen. The selected sample points may be unevenly distributed (Figure 5.1a), resulting in some areas appearing under-represented and others over-represented. The spatial unevenness is a natural result of the randomisation process.

Stratified point sampling divides the study area into a number of mutually exclusive and collectively exhaustive strata (homogeneous sub-regions), then takes a random sample within each stratum (Figure 5.1b). Separate

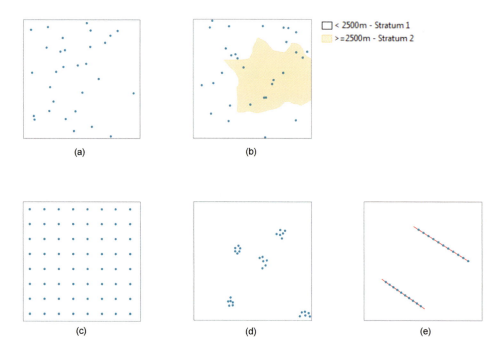

Figure 5.1 Spatial sampling methods.

samples from each stratum are pooled together to form a single stratified random sample. Stratified point sampling may be either proportionate or disproportionate. Proportionate stratified point sampling uses a sampling fraction (the proportion of sample points) in each of the strata, which is the percentage of the area of the stratum compared to the entire study area. For instance, suppose 120 sample points are needed for an environmental study in a region that is broken into two strata, a vegetated area and a non-vegetated area. If the vegetated area occupies 40 per cent of the study area and the non-vegetated area comprises 60 per cent, forty-eight ($=120\times40$ per cent) sample points should be taken in the vegetated area, and seventy-two ($=12\times60$ per cent) sample points in the non-vegetated area. Disproportionate stratified point sampling uses a sampling fraction in each of the strata that is not proportional to its area size. Suppose particular attention needs to be paid to the vegetated area in the environmental study. In this case, 70 per cent of the 120 sample points need to be taken from the vegetated area and 30 per cent from the non-vegetated area, so that the vegetated stratum has more observations while maintaining an adequate number of sample points from the non-vegetated area. Stratified point sampling ensures adequate coverage of all parts or sections of the area.

Systematic point sampling takes a sample according to some regular pattern, usually a regularly spaced grid (Figure 5.1c). For this type of sampling, a distance is defined as the regular sampling interval. It first selects a starting point randomly, then locates other sample points using the sampling interval. The result is a set of sample points spaced evenly across the study area. This method overcomes the problem of spatial unevenness in simple random point sampling, and is often used when

dealing with continuously distributed environmental phenomena. It is simple, and provides sampling designs which are spatially well balanced. However, it may fail to detect the true extent of heterogeneity in the spatial pattern of the phenomenon being investigated.

Clustered point sampling first selects a number of sites randomly, then takes a random sample in the nearby area surrounding each site (Figure 5.1d). This method is convenient, and may reduce travel time and costs. But it excludes substantial parts of the study area, and it is hard to tell whether the sample is representative.

Transect sampling involves taking samples at fixed intervals, usually along lines (Figure 5.1e). This method involves doing transects, where a sampling line is set up across areas with clear environmental gradients. For example a transect can be set up to study the changes of plant species in a transition zone from woodland to grassland, or to investigate the impact on species composition of a pollutant spreading out from a particular source. The position of the transect line depends on the direction of the environmental gradient being studied. This method has the advantages of being efficient, cost-effective and representative.

Different spatial sampling methods can be used in combination. The choice of sampling method depends on the spatial variability of the environmental phenomenon to be investigated. Because of spatial autocorrelation, simple random and clustered point sampling may produce redundant information when sample points are close to each other, in which cases systematic point sampling and stratified point sampling tend to be better (Rogerson 2015, p. 153). All the above sampling methods can be implemented in GIS. Box 5.1 shows some examples with the use of ArcGIS.

Box 5.1   Spatial sampling using ArcGIS   **PRACTICAL**

To follow the examples, start ArcMap, and load the **sampledarea**, **altband** and **transect** shapefiles from **C:\Databases\GIS4EnvSci\VirtualCatchment\Shapefiles**. **sampledarea** depicts the extent of the sampled area, **altband** represents two elevation bands, < 2,500m and ≥ 2,500m, and **transect** contains two transect lines. In these examples, a total of thirty sample points will be obtained using different spatial sampling methods (except systematic point sampling and transect sampling).

Simple random point sampling

1. Open **ArcToolBox**. Navigate to **Data Management Tools > Feature Class**, and double-click **Create Random Points**.

(continued)

(continued)

2. In the **Create Random Points** dialog:

    1)   Enter the output location (a directory name), and name the output point feature class **random**.
    2)   Select **sampledarea** as the constraining feature class.
    3)   Enter **30** as the number of points.
    4)   Click **OK**. Thirty randomly selected sample points within the sampled area are created and shown in the data view, which is similar to Figure 5.1a.

Proportionate stratified point sampling

3. Right-click **altband** in the table of contents, and open its attribute table.
4. In the attribute table, add a new field named **Sample**. The field type is set to **Short Integer** and the precision is **2** (refer to Step 4 in Box 4.3).
5. In the table window, right-click the column heading **Sample**, and select **Field Calculator**.
6. In the **Field Calculator** dialog, enter `Sample = 30*([Area]/(5071327+11324689))`. Here, `(5071327+11324689)` is the total area of the sampled area. Click **OK**. The result indicates that the elevation band of $\geq 2,500$m is allocated nine points and the other band twenty-one points.
7. Start the **Create Random Points** tool.
8. In the **Create Random Points** dialog:

    1)   Enter the output location (a directory name), and name the output point feature class **stratified**.
    2)   Select **altband** as the constraining feature class.
    3)   Check **Field**, and select **Sample**.
    4)   Click **OK**. Thirty sample points are created in two elevation bands, which look similar to Figure 5.1b.

Systematic point sampling

9. In **ArcToolBox**, navigate to **Data Management Tools > Feature Class**, and double-click **Create Fishnet**.
10. In the **Create Fishnet** dialog:

    1)   Name the output feature class **systematic**.
    2)   Select **Same as layer sampledarea** as the template extent.
    3)   Change the fishnet origin coordinates to **6675900** for X Coordinate and **10344800** for Y Coordinate.
    4)   Change the Y-Axis Coordinate to **6676000** for X Coordinate and **10344800** for Y Coordinate, which makes sure the fishnet will not rotate.
    5)   Assign **500** as both the width and height of a cell.
    6)   Assign **8** as the number of rows and the number of columns.
    7)   Tick **Create Label Points**.
    8)   Click **OK**. Two layers are created, **systematic** (a line feature class) and **systematic_label** (a point feature class). **systematic_label** contains the sample points resulting from systematic point sampling, as shown in Figure 5.1c. The sampling interval is 500m.

Clustered point sampling

11. Start the **Create Random Points** tool to select five sites randomly as the centres around which sample points are to be selected.

12. In the **Create Random Points** dialog:

    1) Enter the output location (a directory name), and name the output point feature class **clusteredsites**.
    2) Select **sampledarea** as the constraining feature class.
    3) Enter **5** as the number of points.
    4) Enter **800** metres as the Minimum Allowed Distance.
    5) Click **OK**. Five centres are selected and shown in the data view.

13. Use the **buffer** tool to create 200m buffers around each centre as the neighbourhood for sampling (refer to Steps 1–2 in Box 4.5). Name the buffer layer **clusteredsites_buffer**.
14. After **clusteredsites_buffer** is created, open its feature attribute table, and add a new field named **Sample** with the field type set to **Short Integer** and the precision to **2** (see Step 4 above). Then use **Field Calculator** to set the **Sample** value for all the buffers as **6** by entering Sample = 6 in the **Field Calculator** dialog.
15. Do stratified point sampling using each buffer as a stratum, **clusteredsites_buffer** as the constraining feature class and the **Sample** field value as the number of points by repeating Steps 7 and 8 above. Name the output feature class **clustered**. The result looks similar to Figure 5.1d.

Transect sampling

16. Open **ArcCatalog** to create a new point shapefile.
17. In the **ArcCatalog** window, highlight the folder that will contain the new shapefile.
18. Right-click on the folder, and select **New > Shapefile**.
19. In the **Create New Shapefile** dialog box:

    1) Name the shapefile **transectpoints**, and select **Point** as the feature type.
    2) Click the **Edit** button. In the **Spatial Reference Properties** dialog, click the **Add Coordinate System** button 🌐 and click **Import**. Navigate to **sampledarea** to import its coordinate system for the new shapefile. Click **OK**.
    3) Click **OK** again. **transectpoints** is created and added to the table of contents. This layer will hold the sample points to be created along the two transect lines in **transect**.

20. Show **transect** in the data view.
21. Open the **Editor Toolbar**, and start editing (refer to Step 1 in Box 3.7).
22. Click the **Edit** tool ► on the **Editor Toolbar**, and click the upper transect line to highlight it.
23. Click the **Editor** menu, then **Construct Points**.
24. In the **Construct Points** dialog, check **Distance**, and enter **200** as the sampling interval. Click **OK**. Eleven sample points separated by 200m are created along the line.
25. Click the **Edit** tool ► on the **Editor Toolbar**, and click the lower transect line to highlight it.
26. In the **Construct Points** dialog, check **Number of Points**, and enter **10**. Click **OK**. Ten equally spaced sample points are created along the line.
27. Click the **Editor** menu, then **Stop Editing**. Click **Yes** when prompted to save the edits. Twenty-one sample points are selected along the two transect lines and represented in the layer **transectpoints**.

The x, y coordinates of sample points can be identified using the **Calculate Geometry** tool (see Box 4.4).

Sample size

The size of the sample determines the accuracy of statistical and spatial analysis. A sample should be large enough so that the real characteristics of spatial patterns and the true nature of spatial relationships can be derived. The minimum sample size depends on which sampling method is to be used, what the subsequent statistical and spatial analysis is to measure or estimate, and what level of confidence is expected to be obtained for the results of analysis based on the sample. Non-spatial sampling theory has been used to compute the sample size (Rogerson 2015, pp. 153–156). Some methods were developed to estimate the sample size by considering spatial autocorrelation (Griffith 2005; Wang et al. 2012). As rules of thumb, a minimum of thirty sample points is required for detecting significant spatial autocorrelation, and 100 or more for obtaining reliable measurements of spatial pattern or structure (for example, the presence of environmental gradients or patches) (Stein and Ettema 2003; Fortin and Dale 2005, p. 18).

5.3 MEASURES OF SPATIAL DISTRIBUTIONS

Spatial statistics provide a number of measures to describe and summarise the characteristics of spatial distributions. They include measures of central tendency that describe the centre of a spatial distribution, and measures of dispersion that describe the degree of dispersion of geographical features around the centre of their spatial distribution.

Measures of central tendency

There are several different ways to determine the centre of a spatial distribution, including central feature, mean centre, median centre and linear directional mean.

Central feature identifies the geographical feature that has the shortest accumulative distance to all other features in the dataset. It is used to find the most centrally located or the most accessible feature. This measure is useful for locating the centre to minimise distance for all features to the centre.

Mean centre is the most commonly used measure of central tendency. For point feature data, the mean centre represents the point whose location is defined by the means of the x and y coordinates of all point features. For area or line feature data, the mean centre is the point that lies at the intersection of the means of the x and y coordinates of the centroids or geometric centres of the area or line features. In general, the mean centre's coordinates $(\bar{x}, \bar{y})$ can be calculated as:

$$\bar{x} = \frac{\sum_{i=1}^{n} x_i}{n}; \bar{y} = \frac{\sum_{i=1}^{n} y_i}{n} \tag{5.1}$$

where n is the number of features, and (x_i, y_i) are the coordinates of the ith point or the centroid of the ith feature. Weights, representing the magnitude of a variable or frequency count, can be added to find a weighted mean centre. Let w_i be the weight for the ith point or the centroid of the ith feature. The weighted mean centre's coordinates $(\bar{x}, \bar{y})$ are defined as:

$$\bar{x} = \frac{\sum_{i=1}^{n} w_i x_i}{\sum_{i=1}^{n} w_i}; \bar{y} = \frac{\sum_{i=1}^{n} w_i y_i}{\sum_{i=1}^{n} w_i} \tag{5.2}$$

The mean centre may be considered as the centre of gravity of a spatial distribution. It is useful for tracking changes in the spatial distribution or for comparing the distributions of different types of features. For example, Figure 5.2 shows the distributions of takin observations in the virtual catchment taken in 1975 and 1995, and their respective mean centres. It can be seen that the mean centre of takin population shifted towards the west over twenty years. It is most likely a response to changes in habitat conditions. The mean centre has found many other applications – for instance, to identify the possible origin of a disease epidemic and to show where wildlife animals congregate in different seasons.

Median centre finds the location that minimises the sum of distances to all features. The coordinates of the median centre are calculated using an iterative algorithm (Rogerson 2015, p. 41). The algorithm starts with an initial location, usually the mean centre, as a candidate median centre. It then updates the candidate median centre's coordinates using the equations:

$$\dot{x} = \frac{\sum_{i=1}^{n} w_i x_i / d_i}{\sum_{i=1}^{n} w_i / d_i}; \dot{y} = \frac{\sum_{i=1}^{n} w_i y_i / d_i}{\sum_{i=1}^{n} w_i / d_i} \tag{5.3}$$

where d_i is the distance from the candidate median centre to the ith point (or centroid), and w_i is the weight (for example, frequency count) associated with ith point (or centroid). The process is repeated using the same equations with redefined d_i to calculate the location of a new candidate median centre until the newly calculated location does not differ significantly from the previously computed candidate location.

The median centre is often used for finding the most accessible place or locating services and facilities in terms of accessibility. For example, it can be used to locate a site for an electronics waste (e-waste) recycling facility based on a predicted pattern of potential e-waste distribution in a region in order to minimise the travel cost or distance involved in reaching the recycling facility. In addition,

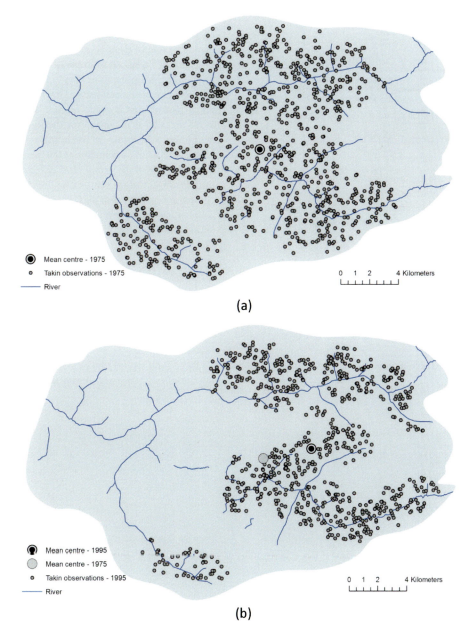

(a)

(b)

Figure 5.2 Mean centres of takin distributions.

the median centre is less sensitive to spatial outliers (rare events or features that are spatially distant from the rest of the features) than the mean centre. For some applications, the median centre is a more representative measure of central tendency than the mean centre.

Linear directional mean identifies the general direction or orientation for a set of line features (such as fault lines, cyclone paths and wildlife migration routes) by calculating the average angle of the features using the following equations:

$$\overline{LD} = \begin{cases} \tan^{-1} \dfrac{\sum_{i=1}^{n} \sin\theta_i}{\sum_{i=1}^{n} \cos\theta_i}, \\ \quad when \ \sum_{i=1}^{n} \sin\theta_i \geq 0 \ and \ \sum_{i=1}^{n} \cos\theta_i > 0 \\ 360 - \tan^{-1} \dfrac{\sum_{i=1}^{n} \sin\theta_i}{\sum_{i=1}^{n} \cos\theta_i}, \\ \quad when \ \sum_{i=1}^{n} \sin\theta_i < 0 \ and \ \sum_{i=1}^{n} \cos\theta_i > 0 \\ 180 - \tan^{-1} \dfrac{\sum_{i=1}^{n} \sin\theta_i}{\sum_{i=1}^{n} \cos\theta_i}, \ otherwise \end{cases}$$ (5.4)

where θ_i is the angle of the ith line feature measured based on its start and end points. Some GIS systems also produce the circular variance as a measure of directional variation among the set of line features, calculated as:

$$CV = 1 - \frac{\sqrt{\left(\sum_{i=1}^{n} \sin\theta_i\right)^2 + \left(\sum_{i=1}^{n} \cos\theta_i\right)^2}}{n}$$ (5.5)

The circular variance ranges from 0 to 1. It is close to 0 when all line features have about the same direction, and approaches one when the line features have distinctly different directions.

The linear directional mean can be used to compare the directional trend of migration routes for different species of animals, track the movement of storms, cyclones and hurricanes, study wind patterns and directions, analyse the movement of glaciers, and so on.

Figure 5.3 gives an example of the linear directional means measuring the orientation of two groups of geological faults, measured using ArcGIS. The linear directional mean for each group of fault lines is represented as an arrowed line centred on the mean centre of the fault lines in the group, whose length equals to the mean length, and whose orientation is the mean orientation of the fault lines in the group. The table in the figure lists the following attribute values: CompassA (compass angle, clockwise from due north), DirMean (directional mean, counterclockwise from due east), CirVar (circular variance), AveX and AveY (the mean centre's x and y coordinates) and AveLen (mean length). As indicated in the table, the general orientation of faults in Group 1 is 115.9° (compass angle), and the general orientation of faults in Group 2 is 38.6° (compass angle). Since the circular variances are all close to zero, the fault lines in each group have quite similar orientations, which is evident from the fault line map. There are slightly more directional variations in Group 2.

Box 5.2 illustrates some tools in ArcGIS for measuring the central tendency of spatial distributions.

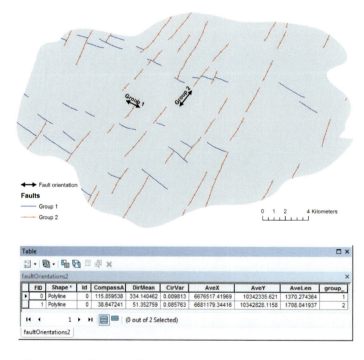

Figure 5.3 Linear directional means of geological faults.

Box 5.2   Measuring the central tendency of spatial distributions in ArcGIS
PRACTICAL

To follow this example, start ArcMap, and load the **takin75** (takin observations in 1975), **takin95** (takin observations in 1995) and **faults** feature classes from **C:\Databases\GIS4EnvSci\VirtualCatchment\ Geodata.gdb**.

Mean centre

1. Open **ArcToolBox**. Navigate to **Spatial Statistics Tools > Measuring Geographic Distributions**, and double-click **Mean Centre**.
2. In the **Mean Centre** dialog box:

    1) Select **takin75** as the input feature class.
    2) Name the output feature class.
    3) Click **OK**. The output point layer is created, which contains the identified mean centre. It looks similar to Figure 5.2a.

3. Repeat the previous two steps to calculate the mean centre for **takin95**. The result should be similar to Figure 5.2b. See the text for the interpretation of the two mean centres.

Linear directional mean

4. In **ArcToolBox**, navigate to **Spatial Statistics Tools > Measuring Geographic Distributions**, and double-click **Linear Directional Mean**.
5. In the **Linear Directional Mean** dialog box:

    1) Select **faults** as the input feature class.
    2) Name the output feature class.
    3) Tick **Orientation Only**.
    4) Select **group_** as the case field to group features for separate directional mean calculations. The **group_** field in **faults** contains two values, 1 and 2, representing two groups of fault lines. Faults in the same group have similar orientation.
    5) Click **OK**. Two lines are created in the output layer, as shown in Figure 5.3.

6. Right-click on the output layer name in the table of contents, and open its attribute table. The interpretation of the attribute values is also given in the text.

You may try the central feature and median centre tools in the spatial statistics toolset using the same datasets in a similar way.

Measures of dispersion

Standard distance and standard deviational ellipse are two main measures of dispersion of a spatial distribution. Standard distance measures dispersion around the mean centre. It is calculated as:

$$S_d = \sqrt{\frac{\sum_{i=1}^{n}(x_i - \bar{x})^2}{n} + \frac{\sum_{i=1}^{n}(y_i - \bar{y})^2}{n}} = \sqrt{\frac{\sum_{i=1}^{n} d_{ic}^2}{n}} \quad (5.6)$$

where ($\bar{x}$, $\bar{y}$) are the coordinates of the mean centre, and d_{ic} is the distance from the ith point or centroid to the mean centre. Therefore, standard distance is the square root of the average squared distance from the points or centroids to the mean centre. It is expressed in the unit in which distance is measured. A larger S_d indicates that the features are more spread out. When the weighted mean centre is required, the weighted standard distance can be computed using the equation:

$$S_{wd} = \sqrt{\frac{\sum_{i=1}^{n} w_i d_{ic}^2}{\sum_{i=1}^{n} w_i}} \qquad (5.7)$$

The standard distance can be used to answer questions like 'Which species has broader territory?', 'Are forest fires more widespread now than ten years ago?' and 'Are gorals more dispersed than serows?' In GIS, the standard distance is represented as a circle around the mean centre with S_d or S_{wd} as the radius. Figure 5.4 shows the weighted standard distances of pheasant distribution in three areas in the virtual catchment calculated using ArcGIS by following the procedure depicted in Box 5.3. The weight is the number of sightings. Area 3 has a weighted standard distance of 2,343.2m, which is the largest among the three, thus pheasants are more dispersed in this area than in the other two areas.

Standard deviational ellipse measures the directional trend of a spatial distribution by computing an ellipse

centred on the mean centre of the distribution. The major axis of the ellipse shows the direction of maximum dispersion of the features, and the minor axis shows that of the minimum dispersion. The area of the ellipse indicates the concentration of the features. The compass angle of the major axis θ is calculated as:

$$\tan\theta = \frac{\sum_{i=1}^{n} \tilde{x}_i^2 - \sum_{i=1}^{n} \tilde{y}_i^2 + \sqrt{\left(\sum_{i=1}^{n} \tilde{x}_i^2 - \sum_{i=1}^{n} \tilde{y}_i^2\right)^2 + 4(\sum_{i=1}^{n} \tilde{x}_i \tilde{y}_i)^2}}{2\sum_{i=1}^{n} \tilde{x}_i \tilde{y}_i} \qquad (5.8)$$

$$\tilde{x}_i = x_i - \bar{x}, \; i = 1, 2, \ldots, n$$

$$\tilde{y}_i = y_i - \bar{y}, \; i = 1, 2, \ldots, n$$

where $(\bar{x}, \bar{y})$ are the coordinates of the mean centre. θ represents the orientation of the spatial distribution. The above formula is derived by shifting the coordinate

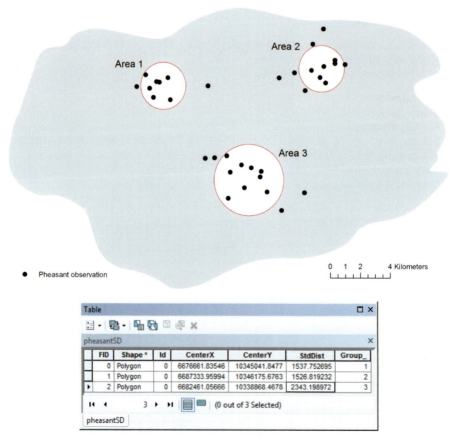

Figure 5.4 Standard distances of pheasant observations.

system's origin to the mean centre and then rotating it through similarity transformation (see Section 3.3) until the maximum and minimum standard deviations of the transformed y coordinates, y_i' $(i = 1, 2, \ldots, n)$, are obtained (Figure 5.5). θ is the rotated angle of the y axis corresponding to the maximum standard deviation.

According to Equation 3.1b in Section 3.3, when the standard deviation of the transformed y coordinates is maximised, the deviation of y_i' from the mean centre is:

$$y_i' - \bar{y} = -x_i \sin\theta + y_i \cos\theta \qquad (5.9)$$

The average of squared deviations between each individual data value and the mean of the dataset is called variance. Standard deviation is the square root of the variance (the standard deviation is the most important and useful measure of variation of values about the mean). Therefore, the maximum standard deviation of the transformed y coordinates is:

$$\sigma_{max} = \sqrt{\frac{\sum_{i=1}^{n}\left(y_i' - \bar{y}\right)^2}{n}}$$
$$= \sqrt{\frac{\sum_{i=1}^{n}\left(-x_i \sin\theta + y_i \cos\theta\right)^2}{n}} \qquad (5.10)$$

Accordingly, the minimum standard deviation is calculated as:

$$\sigma_{min} = \sqrt{\frac{\sum_{i=1}^{n}\left(\begin{array}{c}-x_i \sin(\theta + 90) + \\ y_i \cos(\theta + 90)\end{array}\right)^2}{n}}$$
$$= \sqrt{\frac{\sum_{i=1}^{n}\left(\begin{array}{c}-x_i \cos(\theta) + \\ y_i \sin(\theta)\end{array}\right)^2}{n}} \qquad (5.11)$$

The standard deviational ellipse is defined by σ_{max} as the semi-major axis, σ_{min} as the semi-minor axis and θ as the orientation of the major axis. For the full mathematical description, see Yuill (1971).

The standard deviational ellipse provides information about the dispersion of features and the orientation and core area of their distribution. For example, Figure 5.6 shows the dispersion patterns of pheasant observations in three different areas, created in ArcGIS as described in Box 5.3. It can be seen that the three ellipses are small compared with the entire area of the catchment. Obviously, pheasant distributions are clustered, not widely distributed. Using some GIS measurement tools and spatial query functions, we can calculate the area and

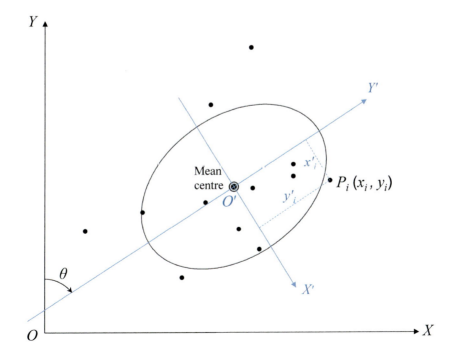

Figure 5.5 Standard deviational ellipse.

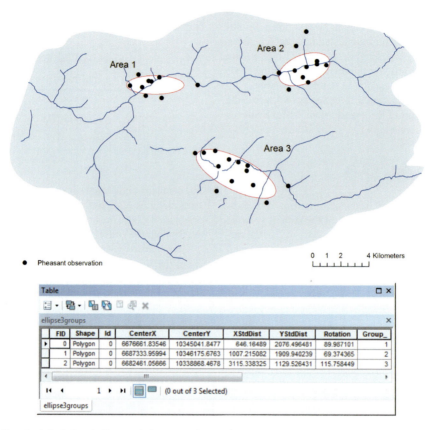

Figure 5.6 Standard deviational ellipses of pheasant observations.

count of data points in each ellipse, then compute the densities of data points inside the ellipses. In this example, the density of pheasant sightings for each ellipse is calculated. The ellipse in Area 1 has a density of 13.53 sightings/km², the ellipse in Area 2 has a density of 16.88 sightings/km² and the density for the ellipse in Area 3 is 9.14 sightings/km². Therefore, pheasants are more concentrated in Area 2. In addition, the orientations of these ellipses indicate that the distributions of pheasant in the catchment are largely affected by the river courses.

Box 5.3   Measuring the dispersion of spatial distributions in ArcGIS

PRACTICAL

To follow this example, start ArcMap, and load the **pheasant** feature class from **C:\Databases\ GIS4EnvSci\VirtualCatchment\Geodata.gdb**. The **Group_** field in the attribute table of **pheasant** groups the observations into three different areas in the virtual catchment.

Standard distance

1.   Open **ArcToolBox**. Navigate to **Spatial Statistics Tools > Measuring Geographic Distributions**, and double-click **Standard Distance**.

2.   In the **Standard Distance** dialog box:

1)   Select **pheasant** as the input feature class.
2)   Name the output standard distance feature class.
3)   Select **1_STANDARD_DEVIATION** as the circle size (it is actually the standard distance).
4)   Select **sightings** as the weight field (the number of sightings at each point).
5)   Select **Group_** as the case field.
6)   Click **OK**. The output standard distance feature class is created to show three weighted standard distance circles, as shown in Figure 5.4. Open its attribute table, which lists the coordinates of the mean centres and weighted standard distance values for the three areas.

Standard deviational ellipse

3.   In **ArcToolBox**, navigate to **Spatial Statistics Tools > Measuring Geographic Distributions**, and double-click **Directional Distribution (Standard Deviational Ellipse)**.
4.   In the **Directional Distribution (Standard Deviational Ellipse)** dialog box:

1)   Select **pheasant** as the input feature class.
2)   Name the output ellipse feature class.
3)   Select **1_STANDARD_DEVIATION** as the ellipse size.
4)   Select **sightings** as the weight field, and **Group_** as the case field.
5)   Click **OK**. The output ellipse feature class is created and shows three weighted standard deviational ellipses, as illustrated in Figure 5.6. Open its attribute table, which lists the coordinates of the mean centres, the orientations of the major axes (in the field **Rotation**) and the standard distances in the x and y directions for the three areas.

The standard deviational ellipse has many other applications, such as investigating how a pollutant is spreading based on groundwater well samples, identifying the dominant wind direction based on the dispersion pattern of seeds from a tree source and detecting the directional trends of a disease outbreak in different seasons. Box 5.4 presents a case study of using the standard deviational ellipse to investigate the distribution of malaria in Sukabumi, Indonesia.

Box 5.4   The use of standard deviational ellipses to understand the geographical distribution of malaria in Sukabumi, Indonesia   **CASE STUDY**

Malaria is caused by a parasite called *Plasmodium*, transmitted to humans by the bite of infected female *Anopheles* species mosquitoes. The disease is widespread in the tropical and subtropical regions of the world, including much of Sub-Saharan Africa, Asia and Latin America. Sukabumi District, located in the southern area of West Java province, Indonesia, has been a malaria endemic area, where an outbreak of malaria occurred in 2004. In order to understand the spatial distribution of malaria and associated environmental factors and support malaria surveillance in the district, Eryando et al. (2012) conducted research on the use of the standard deviational ellipse to characterise the spatial pattern of malaria distribution.

In this research, malaria data were collected through GPS plotting, surveys and interviews based on positive malaria cases in the period 2011–2012. The malaria cases were recorded as point features. The mean centre of the cases was calculated, which was located at 106.603° longitude and −7.118°

(continued)

(continued)

latitude. The standard deviational ellipse was established, which covered an area of about 146.1km². The orientation of its major axis was 148.52°, indicating that the distribution skewed towards the northwest–southeast. The semi-major axis was 7,968.41m in length, while the semi-minor axis had a length of 5,836.57m.

The standard deviational ellipse was then used to identify the physical environmental factors that might trigger epidemics through overlay analysis. It was found that rainfall and temperature anomalies were two major triggering factors, and the highest malaria incidence mainly occurred in the southern hills and mountainous areas with ambient temperatures of 22–25°C and altitude of 500–1,000m, in the northern areas with annual rainfall of 3,453–3,846mm and in the areas with a distance from breeding places of less than 500m.

The findings from the research provided an initial understanding of the spatial variation of malaria cases and associated environmental factors, which could help formulate various strategies and methods of intervention.

Source: Eryando et al. (2012).

5.4 ANALYSIS OF SPATIAL PATTERNS

Spatial pattern analysis aims to measure the degree to which geographical features or their attribute values are clustered, dispersed or randomly distributed across a region. For example, are takins distributed randomly, or concentrated in clusters in particular areas? Does the spatial pattern of the green catbird mirror that of the strangler fig? GIS offers several spatial statistical tools which not only provide quantitative measures of spatial pattern, but also assess their statistical significance.

Significance testing of spatial patterns

Statistical significance is the probability that the observed spatial pattern is not likely due to just chance alone. The observed spatial pattern is considered significant if it has been predicted as unlikely to have occurred by chance. In spatial pattern analysis based on sample data, it is important to determine whether the spatial pattern quantified by a statistical measure could have arisen by chance alone or through sampling error, or whether it might be attributed to statistically significant manifestations of a spatial process at work. This process is called significance or hypothesis testing. It often starts by defining a null hypothesis and an alternative hypothesis.

The null hypothesis, H_0, is a default position that, for example, there is no spatial pattern or there is no relationship between two observed spatial patterns, depending on the research question under investigation. The

alternative hypothesis, H_1, is an opposite position to H_0. H_0 and H_1 are mutually exclusive, and only one of them is considered valid. Spatial pattern analysis frequently adopts the H_0 that the observed spatial pattern is random, and the H_1 that the observed spatial pattern is not random. Significance testing applies a specific statistic test to find the observed test statistic and use it to determine whether H_0 should be rejected or accepted. If H_0 is rejected, H_1 must be accepted.

The test statistic is found by converting the sample statistic (such as the mean and standard deviation) to a score (such as z, F, t or χ^2 score). Several different test statistics are used to test different types of claims. This section only discusses the z test statistic as it is used in spatial pattern analysis. F and t test statistics will be discussed in Section 5.6. Comprehensive discussion of significance testing and other test statistics can be found in any statistics textbook.

The z test statistic, or z score, is used to test the null hypothesis about a mean (or proportion) and determine whether the observed (sample) mean (or proportion) is different from the expected mean (or proportion) of the population under the null hypothesis when the sample size is large (thirty or more). The z test statistic for the mean is found by the following equations:

$$z = \frac{\bar{x} - \mu}{\sigma_{\bar{x}}} = \frac{\bar{x} - \mu}{\frac{\sigma}{\sqrt{n}}} \tag{5.12}$$

where $\bar{x}$ is the observed mean, μ is the expected mean, σ is the population standard deviation, and n is the sample size. $\sigma_{\bar{x}}$ is the standard error of the mean. Therefore, when testing the null hypothesis about a mean, the z scores are units of the standard deviation. The z test statistic for the proportion is calculated differently as:

$$z = \frac{\hat{p} - p}{\sigma_{\hat{p}}} = \frac{\hat{p} - p}{\sqrt{\dfrac{p(1-p)}{n}}} \qquad (5.13)$$

where $\hat{p}$ is the observed proportion, p is the expected proportion and $\sigma_{\hat{p}}$ is the standard error of the proportion.

For example, suppose we have a sample of 120 farms from a rural area, which has a mean size of 130ha. The population of farms in the area has a mean size of 65ha and a standard deviation of 114ha. H_0 is that there is no difference between the sample mean and the population mean. The z score for this example is:

$$z = \frac{130 - 65}{\dfrac{114}{\sqrt{120}}} = 6.246 \qquad (5.14)$$

The z test statistic is assumed to have a normal probability distribution, which can be depicted as a bell-shaped curve. As shown in Figure 5.7, the curve shows the probabilities associated with different z scores. It is peaked when $z = 0$, which is the mean of the distribution, and its tails approach, but never reach, the horizontal axis. The total area under the curve equals 1.0. The proportion of the area under the curve between two z scores is the probability of any feature or value falling between the two

z scores. For example, the area under the curve between the two z values of -0.5 and 0.5 (as shaded in Figure 5.7) is 38.29 per cent of the total area covered by the curve, thus there is a 38.29 per cent probability of any feature or value falling between the two z scores. The probability of a feature or value not occurring between the two scores (that is, outside the shaded area) is 100.0 per cent $-$ 38.29 per cent $= 61.71$ per cent.

The probabilities associated with z scores form the basis of decisions about whether to reject or accept the null hypothesis. The probability at which such a decision is made is known as the significance level, denoted by α. This is also the probability of making the mistake of rejecting H_0 when it is actually valid. The typical choices for α are 0.05, 0.01 and 0.10, with 0.05 being most common. The absolute z score at the chosen significance level is called the critical value. The critical value separates the area under the normal curve of the z test statistic into critical and non-critical regions (Figure 5.8). The critical region encompasses the test statistics whose absolute values are equal to or larger than the critical value. Any z score in the critical region leads to rejection of H_0. All z scores in the non-critical region lead us to accept H_0. For the previous example of 120 farms in a rural area, a z score of 6.246 is exceptionally large, which is located in the critical region of $\alpha = 0.01$. Indeed, the probability of getting a z score of 6.246 is less than 0.0000000006, which is extremely small. Therefore, the mean size of 130ha of the farm sample is significantly different from the population mean, which is very unlikely to occur by chance, and H_0 can be confidently rejected.

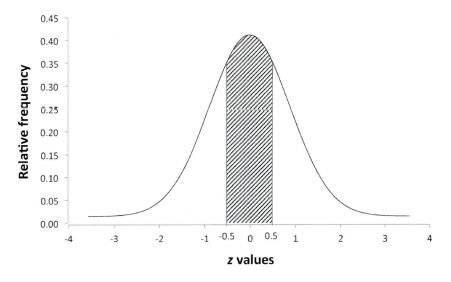

Figure 5.7 Distribution of the z test statistic.

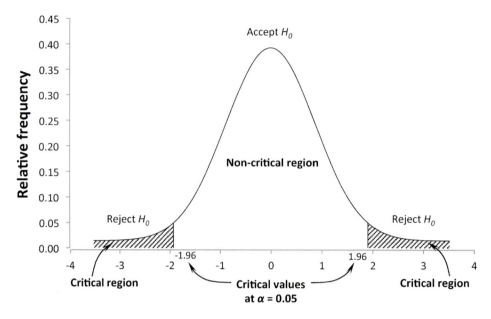

Figure 5.8 Critical values and critical regions of the distribution of the z scores.

The exact probability of getting a value of the test statistic of a given magnitude if the null hypothesis is true is called the *p*-value. The *p*-value is twice the area of the extreme region (tail area) bounded by the test score under the normal curve in the probability distribution of the test statistic (Figure 5.9). If the *z* score calculated for a sample has a *p*-value of equal to or less than the significance level, H_0 should be rejected; otherwise,

H_0 cannot be rejected. The significance level is usually specified before doing a significance testing. The spatial pattern analysis tools in GIS commonly report the *p*-value associated with the calculated *z* score. Reject the null hypothesis of a random spatial pattern or no spatial pattern if the *p*-value $\leq \alpha$ (the significance level, for example 0.05); accept the null hypothesis if the *p*-value $> \alpha$.

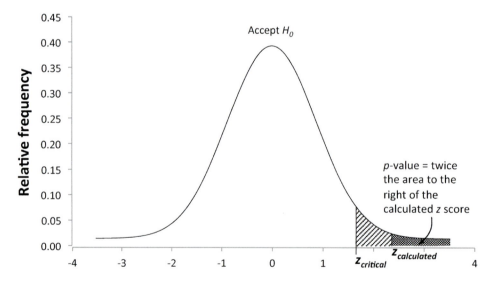

Figure 5.9 Finding *p*-values in the probability distribution of the z scores.

Nearest neighbour analysis

Nearest neighbour analysis involves measuring the distance of each feature to its nearest neighbour, then comparing the observed distances with that expected from a random spatial pattern of features. It was originally developed to analyse the spatial distribution of plant species (Clark and Evans 1954), and is now one of the standard tools in GIS for measuring point patterns.

Nearest neighbour analysis involves calculating the ratio between the observed and expected average or mean distance between points and their nearest neighbours to produce the nearest neighbour statistic R. The mean distance expected from a random pattern of points, r_e, is calculated as:

$$r_e = 0.5\sqrt{A/n} \tag{5.15}$$

where A is the size of the study area, and n is the number of points. The mean observed distance, r_o, is calculated as:

$$r_o = \Sigma_{i=1}^{n} d_i / n \tag{5.16}$$

where d_i is the distance between the ith point to its nearest neighbour. R is then calculated as $R = r_o/r_e$. This gives R values within the range 0.0–2.1491. A value of $R = 0$ represents perfect clustering, where all points occupy exactly the same location. Thus, a clustered pattern tends to have an R value towards 0. A value of 2.1491 is obtained when a regularly dispersed pattern is present, in which points form a triangular lattice so that every point has an equal distance from its six nearest neighbours and the mean distance to the nearest neighbour is maximised. A value of 1.0 indicates a random pattern, where the observed mean distance is equal to the expected mean distance. The following rules are used to describe a spatial pattern using R:

IF $R < 1$, THEN it shows a tendency towards clustering.

IF $R = 1$, THEN it shows a random pattern.

IF $R > 1$, THEN it shows a tendency towards dispersion.

However, an R value other than 1.0 does not by itself suggest that the features have some form of spatial preference, and it may occur by chance or as a result of sampling error. The probability of an R value occurring by chance is established through significance testing using the z test statistic. H_0 states that the sample R does not equal to 1.0 due to sampling error or as the result of a random process. The z test statistic is calculated as:

$$z = \frac{r_o - r_e}{\sigma_{\bar{x}}} \tag{5.17}$$

$$\sigma_{\bar{x}} = \frac{0.26136}{\sqrt{n\left(\dfrac{n}{A}\right)}} \tag{5.18}$$

where $\sigma_{\bar{x}}$ is the standard error. The p-value associated with a z score can then be assessed for testing H_0.

For example, the pheasant distribution depicted in Figures 5.4 and 5.6 has an R value of 0.500566, therefore it shows a clustered pattern. The z score for the R value is −5.652536, and the p-value is less than 0.000001. The p-value indicates that the probability of getting this z score is very small. There is less than 0.0001 per cent likelihood that this clustered pattern could be the result of random chance. Therefore, H_0 is rejected and the conclusion can be made that pheasants in the catchment tend to be clustered, and this clustered pattern does not occur by chance and is statistically significant. Box 5.5 provides another example of nearest neighbour analysis of the mainland serow distribution in the virtual catchment.

Box 5.5 Nearest neighbour analysis in ArcGIS **PRACTICAL**

To follow this example, start ArcMap, and load the **serow** feature class from **C:\Databases\ GIS4EnvSci\VirtualCatchment\Geodata.gdb**. **serow** represents the distribution of the mainland serow in the virtual catchment, which is displayed as in Figure 5.10.

(continued)

(continued)

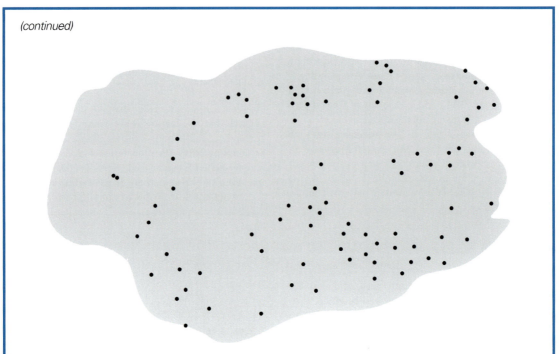

Figure 5.10 The distribution of the mainland serow.

1. Open **ArcToolBox**. Navigate to **Spatial Statistics Tools > Analyzing Patterns**, and double-click **Average Nearest Neighbor**.
2. In the **Average Nearest Neighbor** dialog box:

   1) Select **serow** as the input feature class.
   2) Set the distance method as **EUCLIDEAN_DISTANCE**.
   3) Check **Generate Report** in order to create a graphical summary of results.
   4) Enter **401778312** as the size of the area (which is the area of the catchment).
   5) Click **OK**. Graphical and text summaries of the results are produced.

3. In the main menu, click **Geoprocessing > Results**.
4. In the **Results** window:

   1) Expand **Current Session**, and then **AverageNearestNeighbor**.
   2) Double-click **Report File: NearestNeighbor_Result.html**. The HTML file is opened in the default Internet browser, which shows the graphical summary of results.
   3) Right-click **Messages** in the **Results** window, and select **View**. A text summary of the results is displayed in a message dialog box, as shown in Figure 5.11.

Interpretation

From Figure 5.11, the observed mean distance is 1,030.35m and the expected mean distance is 1,087.06m. The calculated R is 0.95, which is less than but close to 1.0. This value indicates some minor clustering in the distribution of serows. Significance testing resulted in a z score of −0.92, and

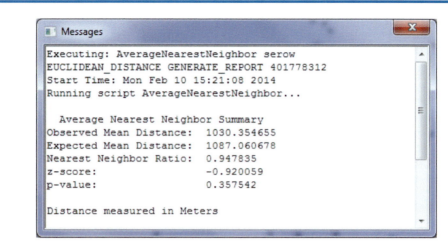

Figure 5.11 Nearest neighbour analysis results.

the p-value associated with the z score is 0.3575. At the significance level $\alpha = 0.05$, we fail to reject H_0 because the p-value $> \alpha$. The conclusion can be reached that the observed spatial pattern could be the result of a random process or sampling error and has no statistical significance. In other words, the distribution of the mainland serow in the reserve exhibits a random pattern.

There are three main problems with nearest neighbour analysis. First, the R statistic is highly dependent on the size and shape of the study area. A long and narrow area may result in a relatively low R value, as the points tend to be close to each other. If the study area is large compared with the extent of the point distribution, a low R value would be obtained. However, if the same points are tightly enclosed by the boundary of the study area, a large R value would be produced. Therefore, the boundary must be carefully drawn by considering the phenomena under investigation, and comparing results of nearest neighbour analysis is most appropriate when the study area is fixed. The second problem is that the same R value could be obtained from very different point patterns, as the R statistic only relates to the distance between points, and does not account for other characteristics of the spatial arrangements of the points, such as angular configuration. Moreover, the statistic only describes a spatial pattern in terms of the number of individual features in a given area and the distribution function of distances between them. It ignores spatial variations in feature attributes and spatial autocorrelation.

Global Moran's *I*

Spatial autocorrelation occurs whenever the values of an environmental variable or feature attributes at one location depend on the values of the same variable or feature at nearby locations. As discussed in Section 4.5, spatial autocorrelation is present at least to some degree in the distribution of many environmental variables or features. The intensity of spatial autocorrelation may vary from location to location and according to direction, producing a variety of spatial patterns. When locations with similar values are nearby forming clusters, or features with similar attributes are concentrated in space, such a clustered spatial pattern as a whole is said to show positive spatial autocorrelation. When locations nearby tend to have dissimilar values, or features closer to each other have more dissimilar attributes than those further apart, which is in opposition to Tobler's first law of geography, such a dispersed spatial pattern exhibits negative spatial autocorrelation. When feature attributes or values of a variable are location-independent, the spatial pattern is random with no spatial autocorrelation.

Moran's I statistic is one of the commonly used measures for characterising a spatial pattern in terms of spatial autocorrelation (Moran 1950). It is calculated as:

$$I = \frac{n\sum_{i=1}^{n}\sum_{j=1}^{n}w_{ij}(x_i - \bar{x})(x_j - \bar{x})}{(\sum_{i=1}^{n}\sum_{j=1}^{n}w_{ij})\sum_{i=1}^{n}(x_i - \bar{x})^2} \quad (5.19)$$

where n is the number of features, x_i is the attribute value of the ith feature, $\bar{x}$ is the mean value of the attribute and w_{ij} is a weight defining spatial contiguity or proximity between features i and j. The matrix $\{w_{ij}\}$ is called a spatial weights matrix. The weights can be defined in several ways. They can be defined based on contiguity between area features, or based on distance between points (point features or centroids of area features).

Contiguity-based spatial weights are basically defined as: $w_{ij} = 1$ if features i and j are contiguous – that is, they share at least one common point and/or a common boundary – and $w_{ij} = 0$ otherwise. Two types of contiguity are often differentiated: queen's case contiguity, which requires features to share at least a common point, and rook's case contiguity, which requires features to share at least one common boundary (or edge).

Distance-based spatial weights are defined as a function of distance between features. Most commonly, they are defined as inverse distance: $w_{ij} = d_{ij}^{-\beta}$, where d_{ij} is the distance between feature i and feature j, and β is the distance decay parameter that is often set as 1 or 2. Sometimes, weights are simply defined as $w_{ij} = 1$ if features i and j are within a pre-defined distance, and $w_{ij} = 0$ if the two features are outside the pre-defined distance. Alternatively, a zone of indifference is defined so that $w_{ij} = 1$ if features i and j are within a pre-defined distance and $w_{ij} = d_{ij}^{-\beta}$ if d_{ij} is larger than the pre-defined distance.

Spatial weights can also be standardised. Standardised weights are usually calculated as $w_{ij}^* = w_{ij} / \sum_{j=1}^{n}w_{ij}$. This is also called row standardisation, because weights are standardised by dividing each weight by its row sum in the weights matrix.

Moran's I statistic ranges from −1 to 1. A spatial pattern that shows strong negative spatial autocorrelation will get a value near −1, which is very rare. Negative values represent negative spatial autocorrelation, indicating dispersed patterns. Positive values correspond to positive spatial autocorrelation, signifying clustered patterns. A strongly clustered pattern will get a value near 1. Values close to zero suggest random patterns. The following rules are used to describe a spatial pattern using I:

IF $I > 0$, THEN it shows a tendency towards clustering – that is, similar values tend to be located nearby.

IF $I = 0$, THEN it shows a random pattern – that is, values are independent of location and they are uncorrelated.

IF $I < 0$, THEN it shows a tendency towards dispersion – that is, dissimilar values are located near one another.

The significance of the calculated value of Moran's I is tested using the z test statistic under the null hypothesis H_0 of a random pattern with zero spatial autocorrelation. The z test statistic is computed as:

$$z = \frac{I - I^e}{\sqrt{I^v}} \quad (5.20)$$

where I^e is the expected I value when H_0 is true, and I^v is the variance of I. I^e is calculated as:

$$I^e = -\frac{1}{n-1} \quad (5.21)$$

If the data are sampled from a population whose distribution is normal, the variance of I is:

$$I^v = \frac{n^2 S_1 - nS_2 + 3S_3^2}{(n+1)(n-1)S_3^2} - (I^e)^2 \quad (5.22)$$

If the data are random samples from a population whose distribution is unknown, the variance of I is:

$$I^v = \frac{n\left[(n^2 - 3n + 3)S_1 - nS_2 + 3S_3^2\right] - b\left[(n^2 - n)S_1 - 2nS_2 + 6S_3^2\right]}{(n-1)(n-2)(n-3)S_3^2} - (I^e)^2 \quad (5.23)$$

where:

$$S_1 = \frac{1}{2}\sum_{i=1}^{n}\sum_{j=1}^{n}\left(w_{ij} + w_{ji}\right)^2, \; i \neq j$$

$$S_2 = \sum_{k=1}^{n}\left(\sum_{i=1}^{n}w_{ik} + \sum_{j=1}^{n}w_{kj}\right)^2$$

$$S_3 = \sum_{i=1}^{n}\sum_{j=1}^{n}w_{ij}, \; i \neq j$$

$$b = \frac{\sum_{i=1}^{n}(x_i - \bar{x})^4 / n}{(\sum_{i=1}^{n}(x_i - \bar{x})^2 / n)^2}$$

Figure 5.12 shows a map of aged population density ($\geq$ 65 years old) in the Melbourne metropolitan area (the areal unit is postcode area). By applying Moran's I to the map and defining spatial weights based on the rook's case contiguity, we produced a Moran's I value of 0.72, a z score of 19.71 and a p-value of less than 0.000001 using

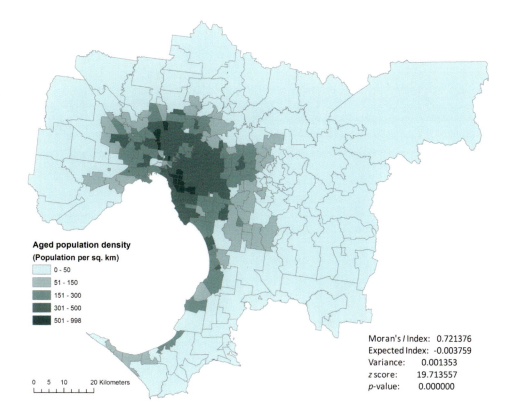

Aged population density
(Population per sq. km)

	0 - 50
	51 - 150
	151 - 300
	301 - 500
	501 - 998

Moran's *I* Index: 0.721376
Expected Index: -0.003759
Variance: 0.001353
z score: 19.713557
p-value: 0.000000

0   5   10   20 Kilometers

Figure 5.12 Aged population distribution in Melbourne and its Moran's *I* (data source: 2006 Census, the Australian Bureau of Statistics).

the Global Moran's *I* tool in ArcGIS. The results indicate that the aged population distribution shows a clustered pattern with quite strong positive spatial autocorrelation. Given the *z* score of 19.71, there is less than 1 per cent likelihood that the clustered pattern could be the result of random chance. Therefore, H_0 is rejected. Box 5.6 uses another example to demonstrate how the Global Moran's *I* tool is used in ArcGIS, which is based on the assumption that the statistical distribution of a population is unknown and might not be a normal distribution.

Box 5.6   Global Moran's *I* in ArcGIS   **EXAMPLE**

To follow this example, start ArcMap, and load the **animalDensity** feature class from **C:\Databases\ GIS4EnvSci\VirtualCatchment\Geodata.gdb**. **animalDensity** is a grid map. Each grid cell is 1km × 1km, represented as an area feature with two main attributes: the pheasant density (in the **pheasant** field) and mainland serow density (in the **serow** field). The density is in the unit sightings/km². Use the mainland serow density as the attribute to make a map. It should be shown as in Figure 5.13.

(continued)

(continued)

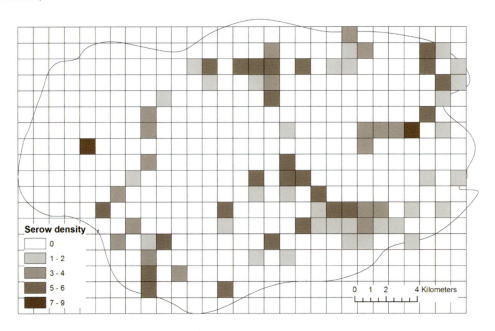

Figure 5.13 Serow density grid map.

1. Open **ArcToolBox**. Navigate to **Spatial Statistics Tools > Analyzing Patterns**, and double-click **Spatial Autocorrelation (Morans I)**.
2. In the **Spatial Autocorrelation (Morans I)** dialog box:
    1) Select **animalDensity** as the input feature class.
    2) Select **serow** as the input field.
    3) Check **Generate Report**.
    4) Select **CONTIGUITY_EDGES_CORNERS** to conceptualise the spatial relationships among features (grid cells) – that is, use queen's case contiguity to define the spatial weights.
    5) Set **None** for standardisation.
    6) Click **OK**. Graphical and text summaries of the results are produced.
3. Open the **Results** window. Expand **Current Session**, then right-click **Messages** and select **View**. A text summary of the results is displayed as below:

```
Moran's Index: 0.115993
Expected Index: -0.001919
Variance: 0.000496
z-score: 5.294792
p-value: 0.000000
```

The graphical summary of results can be viewed by double-clicking **Report File** in the **Results** window under the current session.

Interpretation

The Moran's *I* value of 0.115993 suggests that the mainland serow distribution shows weak positive spatial autocorrelation with a slightly clustered pattern, which is similar to the result of the nearest neighbour analysis described in Box 5.5. However, given the *z* score of 5.294792 and a *p*-value of less than 0.000001, there is a very slim chance that the pattern could be the result of random chance. Therefore, the null hypothesis of a random spatial pattern with zero spatial autocorrelation is rejected. The conclusion can be reached that the mainland serow distribution tends to be slightly clustered although the Moran's *I* value is close to 0, indicating a random pattern. This conclusion is slightly different from that of the nearest neighbour analysis, which concluded that the minor clustered pattern is not statistically significant.

You may also try to use the **animalDensity** layer to analyse the spatial pattern of pheasants using the Moran's *I* statistic based on the pheasant density data by following the steps above.

5.5 DETECTION OF SPATIAL CLUSTERS

The statistics discussed in the previous section provide a single quantity to detect the overall spatial pattern of feature locations and/or feature attributes over the entire study area. They can indicate whether or not there is a spatial pattern, or whether the detected pattern is clustered or dispersed. For example, the Moran's *I* statistic indicates the presence and strength of spatial autocorrelation globally. Such statistics are labelled global spatial statistics. They may detect an overall clustered pattern, but do not identify where spatial clustering occurs. Global spatial statistics assumes homogeneity across the entire region. However, even if there is no global spatial autocorrelation or no clustering, clusters can be still found at a local level using local spatial statistics. Local spatial statistics are designed to evaluate whether there are local patterns of clustering or concentration of high or low values, and if so, where they are. Two local spatial statistics are introduced below for detecting spatial clusters.

Local Moran's *I*

The local Moran's *I* statistic identifies spatial clusters of features with high or low values and spatial outliers (a high value surrounded by low values or a low value surrounded by high values). It is expressed as follows (Anselin 1995):

$$I_i = \frac{\left(x_i - \bar{x}\right)\sum_{j=1}^{n} w_{ij}\left(x_j - \bar{x}\right)}{\sum_{i=1}^{n}(x_i - \bar{x})^2 / n} \tag{5.24}$$

Here, I_i is the local Moran's *I* statistic for feature *i*. Under the null hypothesis of no spatial clustering, the expected I_i is:

$$I_i^e = -\frac{\sum_{j=1}^{n} w_{ij}}{n-1} \tag{5.25}$$

The variance of I_i is found as:

$$I_i^v = \frac{\left(n-b\right)\sum_{j \neq i} w_{ij}^2}{n-1} + \frac{(2b-n)\sum_{k \neq i}\sum_{h \neq i} w_{ik}w_{ih}}{(n-1)(n-2)} - (I_i^e)^2 \tag{5.26}$$

By substituting I_e and I_v respectively with I_i^e and I_i^v in Equation 5.20, the *z* score can be obtained for significance testing of the calculated I_i.

A positive I_i value suggests the presence of positive local spatial autocorrelation, indicating that feature *i* has neighbouring features with similarly high or low attribute values, thus feature *i* is part of a cluster. A negative I_i value indicates that a feature has neighbouring features with dissimilar values (that is, negative local spatial autocorrelation), and the feature is an outlier (for example, a location with high values surrounded by neighbours with low values). In either case, the *p*-value associated with I_i must be smaller than the pre-specified significance level for the cluster or outlier in order to be considered statistically significant.

Figure 5.14 shows spatial clusters identified by applying the local Moran's *I* tool in ArcGIS to the map in Figure 5.12. The spatial weights are defined based on

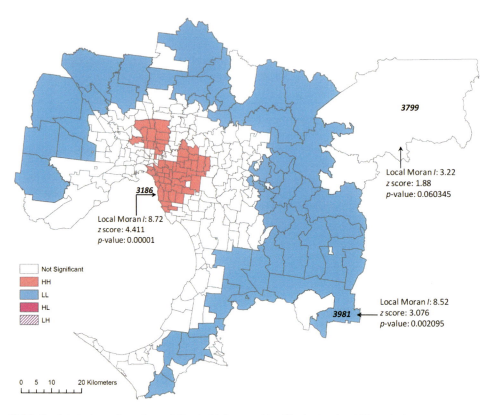

Figure 5.14 Spatial clusters of aged population in Melbourne identified using local Moran's *I*.

rook's case contiguity. Postcode area 3186 has a positive local Moran's *I* value of 8.72 and a *p*-value of 0.00001, indicating that this postcode area is part of a spatial cluster of similar values, and it is statistically significant with a significance level $\alpha = 0.05$ (> the *p*-value). The positive local Moran's *I* and a small *p*-value < 0.05 for postcode area 3981 suggest that it is also part of a significant spatial cluster of similar values at $\alpha = 0.05$. Postcode area 3186 is in a cluster of high aged population density, while postcode area 3981 is part of a cluster of low aged population density. Although postcode area 3799 has a positive local Moran's *I* value,

its associated *p*-value is greater than 0.05, thus spatial clustering around the area is not statistically significant. Typically, GIS outputs a map resulting from the local Moran's *I* test which shows statistically significant (at $\alpha = 0.05$) clusters of high values (HH), spatial clusters of low values (LL), spatial outliers in which a high value is surrounded mainly by low values (HL) and spatial outliers in which a low value is surrounded largely by high values (LH). In Figure 5.14, spatial clusters of HH and LL are detected, but no spatial outliers. Box 5.7 illustrates how to use and interpret the local Moran's *I* statistic in ArcGIS.

Box 5.7   Local Moran's *I* in ArcGIS   **PRACTICAL**

To follow this example, start ArcMap, and load the **animalDensity** feature class from **C:\Databases\GIS4EnvSci\VirtualCatchment\Geodata.gdb**. The density distribution of mainland serow is shown in Figure 5.13.

1. Open **ArcToolBox**. Navigate to **Spatial Statistics Tools > Mapping Clusters**, and double-click **Cluster and Outlier Analysis (Anselin Local Morans I)**.
2. In the **Cluster and Outlier Analysis (Anselin Local Morans I)** dialog box:
    1) Select **animalDensity** as the input feature class.
    2) Select **serow** as the input field.
    3) Name the output feature class.
    4) Select **CONTIGUITY_EDGES_CORNERS** to conceptualise the spatial relationships among cells.
    5) Click **OK**. The output feature class is created, as shown in Figure 5.15.

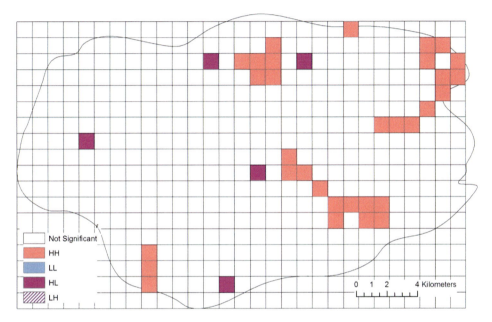

Figure 5.15 Spatial clusters and outliers of mainland serows identified using local Moran's *I*.

3. Open the feature attribute table associated with the output feature class. The table contains the calculated local Moran's *I* value (**LMiIndex**), z-score (**LMiZScore**), p-value (**LMiPValue**) and cluster/outlier type (**COType**) for every cell. The type of spatial clustering (HH or LL) or spatial outlier (HL or LH) is determined at $\alpha = 0.05$. Figure 5.16 shows the first few records of the table.

Interpretation

The local Moran's *I* statistic analysis identified thirty-two out of 522 cells forming five statistically significant spatial clusters of high values (HH), as well as five locations with high density values surrounded by low density values (HL). This is consistent with the results from the global Moran's *I* statistic analysis shown in Box 5.6, which indicate that the overall mainland serow distribution exhibits weak positive spatial autocorrelation with minor spatial clustering. While the global statistic identifies

(continued)

(continued)

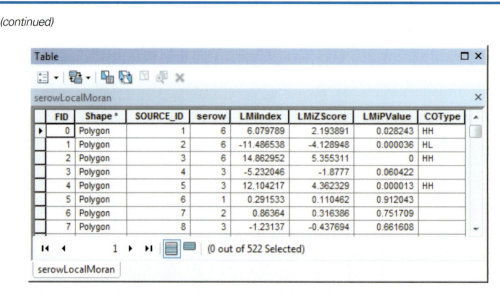

	FID	Shape*	SOURCE_ID	serow	LMiIndex	LMiZScore	LMiPValue	COType
▶	0	Polygon	1	6	6.079789	2.193891	0.028243	HH
	1	Polygon	2	6	-11.486538	-4.128948	0.000036	HL
	2	Polygon	3	6	14.862952	5.355311	0	HH
	3	Polygon	4	3	-5.232046	-1.8777	0.060422	
	4	Polygon	5	3	12.104217	4.362329	0.000013	HH
	5	Polygon	6	1	0.291533	0.110462	0.912043	
	6	Polygon	7	2	0.86364	0.316386	0.751709	
	7	Polygon	8	3	-1.23137	-0.437694	0.661608	

(0 out of 522 Selected)

Figure 5.16 Results of the local Moran's *I* test in the case of the mainland serow.

the overall trend of the distribution, the local statistic uncovers the location and size of clusters and outliers. However, the substantive interpretation of the special nature of the spatial clusters and outliers is beyond the scope of the exploratory data analysis. The role of the analysis here is to point them out, and by doing so, to aid in the suggestion of possible explanations or hypotheses. The indication of the spatial outliers may point to erroneous observations or problems with the choice of the spatial weight definition.

You may also try to use the **animalDensity** layer to analyse the spatial pattern of pheasants using the local Moran's *I* statistic by following the steps above.

Getis-Ord G_i^* (hot spot analysis)

The Getis-Ord G_i^* statistic, or simply G_i^*, is a *z* test statistic used to test for the detection of spatial clusters. Like local Moran's *I*, it enables us to detect spatial clusters that may not show up when using global spatial statistics. G_i^* for feature *i* is calculated as:

$$G_i^* = \frac{\sum_{j=1}^n w_{ij}(d)x_j - \bar{x}\sum_{j=1}^n w_{ij}(d)}{s\sqrt{\left\{\begin{array}{c} n\sum_{j=1}^n [w_{ij}(d)]^2 - \\ [\sum_{j=1}^n w_{ij}(d)]^2 \end{array}\right\}/(n-1)}} \quad (5.27)$$

where $\{w_{ij}\}$ is a symmetric 1 or 0 spatial weights matrix with 1s for all features within distance *d* of feature *i*, including feature *i* itself, and 0s for all other features, x_j is the attribute value associated with feature *j*, and $\bar{x}$ and *s* are the mean and standard deviation of the attribute values respectively (Ord and Getis 1995).

G_i^* tests whether feature *i* and its neighbouring features have higher or lower than average values of a particular attribute or variable. A positive G_i^* value larger than the critical *z* value at a specific significance level indicates that the feature is part of a spatial cluster of particularly high values, while a negative G_i^* value lower than the minus critical *z* value suggests that the feature is located in a spatial cluster of particularly low values at the corresponding significance level. Therefore, given the set of features, G_i^* can be used to identify those clusters of features with values of a particular variable higher or lower in magnitude than it might be expected to find by random chance. The clusters of high values are called hot spots, and the clusters of low values are referred to as cold spots.

Figure 5.17 shows the hot and cold spots of aged population in Melbourne identified using G_i^* at $\alpha =$ 0.05 based on the data in Figure 5.12. The distance *d* used to define the spatial weights was set as 5km. The

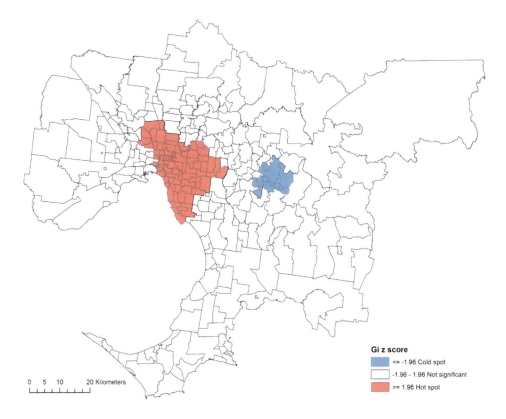

Gi z score

▨	<= -1.96 Cold spot
□	-1.96 - 1.96 Not significant
▨	>= 1.96 Hot spot

0 5 10   20 Kilometers

Figure 5.17 Hot and cold spots of aged population in Melbourne identified using G_i^* at $\alpha = 0.05$.

critical z value at this significance level is 1.96 for hot spots and −1.96 for cold spots. Box 5.8 shows how G_i^* is used in ArcGIS. Case Study 13 in Chapter 10 provides an example of the use of G_i^* to identify hot spots with high landscape values in the Murray River region of Victoria, Australia.

Box 5.8   Getis-Ord G_i^* in ArcGIS   **PRACTICAL**

To follow this example, start ArcMap, and load the **animalDensity** feature class from **C:\Databases\ GIS4EnvSci\VirtualCatchment\Geodata.gdb**. The density distribution of mainland serow is shown in Figure 5.13.

1. Open **ArcToolBox**. Navigate to **Spatial Statistics Tools > Mapping Clusters**, and double-click **Hot Spot Analysis (Getis-Ord Gi*)**.
2. In the **Hot Spot Analysis (Getis-Ord Gi*)** dialog box:

    1) Select **animalDensity** as the input feature class.
    2) Select **serow** as the input field.
    3) Name the output feature class.

(continued)

(continued)

4) Select **FIXED_DISTANCE_BAND** to conceptualise the spatial relationships among cells.
5) Set **EUCLIDEAN_DISTANCE** as the distance methods.
6) Set the distance band or threshold distance as **2000** (metres).
7) Click **OK**. The output feature class is created and displayed in the data view, as shown in Figure 5.18.

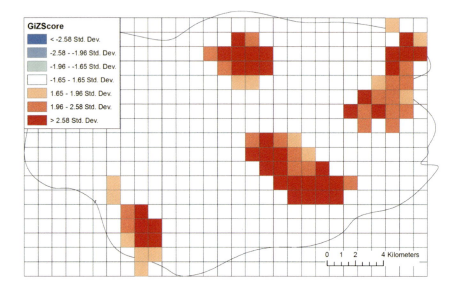

Figure 5.18 Hot spots of mainland serows identified using G_i^*.

3. Open the feature attribute table associated with the output feature class, which lists the calculated G_i^* value (**GiZScore**) and associated *p*-value (**GiPValue**) for each feature (here grid cell). By selecting those cells with a *p*-value ≤ 0.05 or an absolute G_i^* value ≥ 1.96, hot and cold spots of the mainland serow distribution can be identified at $\alpha = 0.05$ and highlighted in the map.

Interpretation

The G_i^* testing identified four statistically significant hot spots at the significant level $\alpha = 0.05$, but no cold spots. As mentioned in Section 5.4, when testing H_0 about a mean, the *z* score is the number of units of standard deviation. Therefore, the ArcGIS G_i^* testing tool displays the G_i^* values as units of standard deviation. Hot and cold spots are dependent on the significant level and the critical values of G_i^* at the significance level. Their size and location are also dependent on the distance used to define the spatial weights or the distance band. The distance band should be large enough to ensure all features have at least one neighbour, but should not include thousands of neighbours. The best distance band should reflect maximum spatial autocorrelation – that is, a distance over which the spatial process are most active or most pronounced.

You may also try to use the **animalDensity** layer to analyse the spatial pattern of pheasants using the G_i^* statistic by following the steps above.

5.6 MODELLING OF SPATIAL RELATIONSHIPS

Spatial patterns result from either exogenous (induced) or endogenous (inherent) processes occurring in space and time. Spatial autocorrelation is largely an effect of endogenous processes. For instance, endogenous processes such as seed dispersal, gene flow and competition tend to make plants within patches resemble each other. Most environmental phenomena exhibit some degree of spatial autocorrelation, which can be measured using Moran's I and other statistics discussed in the previous two sections. A spatial pattern induced by exogenous processes is the function of other factors independent of the variable or characteristics of interest. For example, the precipitation pattern may be a function of altitude and distance from the sea. The spatial pattern of the plants may be the response to soils, as they may only grow on a specific type of soil. The spatial distribution of wildlife animals may be restricted by or actively respond to environmental factors such as rainfall, temperature or habitat type, which themselves are spatially autocorrelated. Such spatial patterns can be investigated and modelled by means of regression.

Regression is a statistical tool used to investigate and estimate the relationships among variables. It has been used as a major means of evaluating spatial relationships in order to explain the factors behind the observed spatial patterns. Traditional simple or multiple regression is used to explore, examine and model spatial relationships by assuming that the relationships are static and uniform across the entire study area. Geographically weighted regression takes into account spatial or regional variations of the relationships.

Simple regression

Simple regression models relationships between two variables using the following linear equation:

$$y = a + bx + \varepsilon \tag{5.28}$$

In this regression equation, y is the dependent variable, x is the independent or explanatory variable, a and b are regression coefficients and ε is the error term. Regression coefficients represent the form and strength of relationship the explanatory variable has to the dependent variable. As a perfect linear relationship is unlikely to be encountered when investigating real-world phenomena, ε always exists. The goal of simple regression analysis is to obtain estimates of the unknown parameters a and b, and to assess the residuals. Once the regression coefficients

are known, the values of y can be estimated from the values of x by:

$$\hat{y} = a + bx \tag{5.29}$$

where $\hat{y}$ is the estimated or predicated y value. For example, suppose a regression of the annual rainfall (y) on the annual number of rain days (x) produced the following model:

$$\hat{y} = -9.172 + 0.4562x \tag{5.30}$$

Assume the total number of rain days next year is forecast to be fifty. Then the predicted annual rainfall is $-9.172 + 0.4562 \times 50 = 13.638$ inches. If the number of rain days next year is revised to seventy, then the annual rainfall estimate is $-9.172 + 0.4562 \times 70 = 22.762$ inches.

Regression coefficients are estimated based on a collection of paired sample data of the dependent and explanatory variables. The form of relationship between two variables can be portrayed graphically by plotting the sample data (paired measurements of x and y for each observation) in a scatterplot, as shown in Figure 5.19. Each point in the plot represents an observation with an x value and a y value. Simple regression involves fitting a straight line to the set of the points so that the sum of squared vertical distances from individual points to the line is smaller than those to any other line – that is, $\sum_{i=1}^{n} e_i^2$ is minimised (Figure 5.19) – where:

$$e_i = y_i - \hat{y}_i \tag{5.31}$$

y_i is the measured y value of the ith observation, and $\hat{y}_i$ is its y value estimated or predicted by the straight line; e_i is termed the residual. The straight line is called the least-squares regression line, or simply the regression line. The regression coefficient a is the intercept of the line on the y axis (the value of y at the point where $x = 0$), and b is the slope of the regression line. They are determined using the sample data as follows:

$$b = \frac{n\sum_{i=1}^{n} x_i y_i - \left(\sum_{i=1}^{n} x_i\right)\left(\sum_{i=1}^{n} y_i\right)}{n\sum_{i=1}^{n} x_i^2 - \left(\sum_{i=1}^{n} x_i\right)^2} \tag{5.32}$$

$$a = \frac{\sum_{i=1}^{n} y_i}{n} - \frac{b\sum_{i=1}^{n} x_i}{n} \tag{5.33}$$

Predicted by the regression line, $\hat{y} = \bar{y}$ when $\hat{x} = \bar{x}$, where $\bar{x}$ and $\bar{y}$ are the mean of x and y respectively.

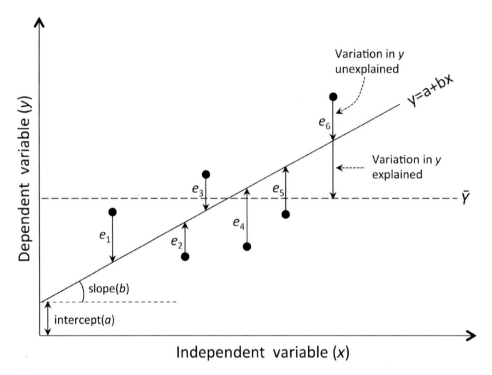

Figure 5.19 Simple regression and related concepts.

This means that the point ($\bar{x}$, $\bar{y}$) is on the regression line. This is true for all linear regression problems. The variation in the observed values of the dependent variable y has two components: the residual that is unexplained or unaccounted for by the independent variable x, and the deviation of the predicted y value from $\bar{y}$ that is explained by x, as illustrated in Figure 5.19. The fraction of the total variation in y that is explained by the regression is called the coefficient of determination. It is calculated by:

$$r^2 = \frac{\sum_{i=1}^{n}\left(\hat{y}_i - \bar{y}\right)^2}{\sum_{i=1}^{n}\left(y_i - \bar{y}\right)^2} \tag{5.34}$$

The value of r^2 is in the range between 0 and 1. A value of 0 indicates that none of the variation in y is explained by x, while a value of 1 suggests that all of the residuals are zero – that is, all the observed or sample points fall exactly on the regression line. A value of 0.82 would mean that 82 per cent of the variation in y is explained by x; and the other $100 - 0.82 = 18$ per cent of the variation is explained by other variables that are not accounted for in the regression model.

The spread of sample points about the regression line is measured by the standard error of estimate, which is defined as:

$$s_e = \sqrt{\frac{\sum_{i=1}^{n}\left(y_i - \hat{y}_i\right)^2}{n-2}} \tag{5.35}$$

S_e may be interpreted as the magnitude of a typical residual. It can be used to compare different samples with respect to the extent to which the observed points are dispersed about their regression line. It can also be used to compare distributions which share the same least-squares regression line.

Significance testing of simple regression

The reliability or statistical significance of the linear relationship represented by a regression equation and the estimated regression coefficients needs to be tested because they are likely to be influenced by sampling errors.

The reliability or overall significance of the regression equation is tested using the F test statistic. The F test is used to compare variations in two samples and test whether a disparity between the two sample variances has arisen by chance. Here it is used to test whether the regression equation can explain a significant portion of the variation in y with the null hypothesis H_0 that the fraction of the variation in y explained by x is 0. The variance in y explained by the regression, called the regression variance, is:

$$s_y^2 = \frac{\sum_{i=1}^{n} \left(\hat{y}_i - \bar{y} \right)^2}{k} \quad (5.36)$$

where k is the number of independent variables. For simple regression, $k = 1$. The variance in y unexplained by the regression, called the residual variance, is:

$$s_e^2 = \frac{\sum_{i=1}^{n} \left(y_i - \hat{y}_i \right)^2}{n-k-1} \quad (5.37)$$

The test statistic F for regression is given by:

$$F = \frac{s_y^2}{s_e^2} \quad (5.38)$$

When a large proportion of the total variance is associated with the regression variance, a high F ratio is obtained, indicating that the regression equation provides a good prediction or estimation of y. Conversely, a large residual variance may result in a low F, suggesting that the regression equation may be unreliable as other variables unaccounted for in the equation may play an important part in explaining the variation of y. Therefore, the magnitude of the F test statistic helps determine whether the amount of the variability in y accounted for by the regression equation is significant compared to that associated with the residuals or errors. Regression analysis tools in GIS provide the p-value for the calculated F ratio. If the p-value is equal to or less than the significance level α, H_0 is rejected and the regression equation is significant overall. Thus, the p-value associated with the F ratio is usually used as a measure of how well the regression equation fits the sample data.

The significance of the regression coefficients is tested using the t test statistic. The null hypothesis H^0 is that the true value of the regression coefficient is zero. So, if the t test for b fails to reject H_0, then the independent variable x has no significant influence on the dependent variable y, thus x is not important in understanding y. If the t test for a fails to reject H_0, we could conclude that a generated from the sample data is the result of chance or sampling error, therefore the predicted y value by the regression

equation is not reliable. The standard errors of a and b are used to carry out a t test. The mathematical formulae for calculating the t test statistics for a and b can be found in Rogerson (2015). We can use p-values associated with the t test scores to make a decision on whether or not to reject H_0.

Multiple regression

Many environmental problems are of a multivariate nature. For example, the concentration of fungal toxin in nuts at growing sites is dependent on rainfall, temperature and wind speed; the intensity of rainfall is related to sea surface temperature, sea level pressure, wind speed and other environmental factors. Multiple regression offers an extension to simple regression for addressing multivariate problems. It establishes a linear relationship between a dependent variable y and two or more independent variables $(x_1, x_2, x_3, \ldots, x_k)$. The general form of a multiple regression equation is:

$$\hat{y} = a + b_1 x_1 + b_2 x_2 + \ldots + b_k x_k \quad (5.39)$$

where k is the number of independent variables, $\hat{y}$ is the predicted value of the dependent variable y, and a and $\{b_i \mid i = 1, 2, \ldots, k\}$ are the regression coefficients estimated from sample data. As with simple regression, the regression coefficients are found by minimising the sum of the squared residuals. How well the multiple regression fits the sample data is measured by the multiple coefficient of determination, R^2, which is the ratio of the explained variation and total variation in y, conceptually identical to r^2 in simple regression. However, R^2 increases as more variables are included, whereas the best multiple regression equation does not necessarily use all of the available variables. To address this flaw, R^2 is adjusted according to the number of independent variables and the sample size as follows:

$$adjusted\ R^2 = 1 - \frac{n-1}{n-k-1}\left(1-R^2\right) \quad (5.40)$$

where n is the sample size. R^2 and adjusted R^2 are interpreted in the same way as r^2.

F and t tests are also used to assess the statistical significance of the multiple regression equation and individual regression coefficients. The p-value is used to determine the statistical significance according to the significance level α.

Both simple regression and multiple regression assume that the data are numerical; the relationship between the dependent and independent variables is linear, but only partial relationships are dealt with, not precise mathematical ones, and the residuals have a normal distribution and are independent of each other with no autocorrelation. Multiple regression also assumes that the independent variables are not significantly correlated with one another – that is, there is no multicollinearity among the independent variables. If there is significant multicollinearity, the variance of the estimated regression coefficients will become inflated, which may make insignificant independent variables to appear significant. The severity of multicollinearity can be quantified using the variance inflation factor (VIF). VIF is calculated for each independent variable as follows:

$$VIF_i = \frac{1}{1 - R_i^2} \qquad (5.41)$$

where VIF_i is the VIF for independent variable x_i, and R_i^2 is the multiple coefficient of determination associated with the regression of x_i (taken as the dependent variable) on all other independent variables. As a rule of thumb, a VIF of greater than 7.5 suggests high multicollinearity – an indication of potential multicollinearity problems. The square root of VIF also indicates how much larger the standard error is, compared with what it would be if that variable was uncorrelated with the other independent variables.

Whether or not the residuals from a regression model are normally distributed is tested using the Jarque-Bera test statistic (Jarque and Bera 1987). The null hypothesis for this test is that the residuals are normally distributed.

When the p-value for the test is less than the significance level, H_0 is rejected and the residuals are not normally distributed, indicating the regression model is biased. If the residuals are also spatially autocorrelated, the bias may result from missing one or more key independent variables. The spatial autocorrelation of the residuals can be tested using global Moran's I.

A multivariate environmental problem may involve a large number of variables. It is not necessary to include all possible variables in a multiple regression equation. Which variables should be included or excluded is decided mainly based on domain knowledge, practical considerations and sometimes common sense. Determination of the best multiple regression equation relies on the assessment of the statistical significance of the regression equation and individual coefficients, adjusted R^2, VIF and some other statistical measures. Generally, a regression equation can be selected when it has overall significance determined by the p-value. For a given number of independent variables, the equation with the largest value of adjusted R^2 should be selected. Insignificant variables and variables with high multicollinearity (VIF > 7.5) need to be weeded out. Therefore, a multiple regression equation can be constructed step by step, by testing one independent variable at a time and including it in the regression model if it is statistically significant, or by including all potential independent variables in the equation and removing those that are not statistically significant. Box 5.9 shows the use of these principles to find a regression equation to model the relationship between annual average temperature, elevation and aspect with ArcGIS.

Box 5.9 Regression analysis in ArcGIS   PRACTICAL

To follow this example, start ArcMap, and load the **gauges** shapefile from **C:\Databases\GIS4EnvSci\ VirtualCatchment\Shapefiles** In this example, you are going to explore the relationships between average annual temperature (the dependent variable), elevation and aspect (the independent variables) using multiple regression. The observed data for the three variables at each weather station are stored in the fields **temp**, **elevation** and **aspect** in the attribute table of **gauges**.

Regression of temperature on elevation and aspect

1. Open **ArcToolBox**. Navigate to **Spatial Statistics Tools > Modeling Spatial Relationships**, and double-click **Ordinary Least Squares**.
2. In the **Ordinary Least Squares** dialog box:

   1) Select **gauges** as the input feature class.
   2) Select **Id** as the unique ID field.

   3) Navigate to your output directory and enter the name of the output feature class.
   4) Select **temp** as the dependent variable.
   5) Tick **elevation** and **aspect** as the explanatory (independent) variables.
   6) Navigate to your output directory, and enter the name of the output report file.
   7) Click **Additional Options**.
   8) Navigate to your output directory and enter the name of the coefficient output table.
   9) Navigate to your output directory, and enter the name of the diagnostic output table.
  10) Click **OK**. Wait until the process is complete. The output feature class, report and tables are created. The output feature class stores the temperature values estimated by regression and residuals. It is shown in the data view as a residual map, revealing the over- and under-estimations at each observation by the regression model.

3.   Add and open the coefficient output table, as shown in Figure 5.20a.
4.   Add and open the diagnostic output table, as shown in Figure 5.20b.
5.   Open the output report PDF file.

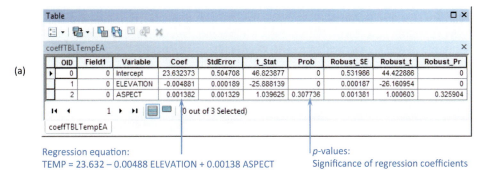

(a)

Regression equation:
TEMP = 23.632 − 0.00488 ELEVATION + 0.00138 ASPECT

p-values:
Significance of regression coefficients

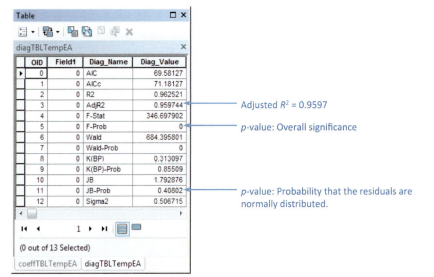

(b)

Adjusted R^2 = 0.9597

p-value: Overall significance

p-value: Probability that the residuals are normally distributed.

Figure 5.20 Regression output tables: (a) coefficient table and (b) diagnostic table.

(continued)

(continued)

Interpretation

A complete analysis of the regression results might include other important elements, but the discussion here is limited to the following components: regression equation, adjusted R^2, overall significance, significance of individual regression variables, multicollinearity and distribution of residuals. They are assessed at the $\alpha = 0.05$ level.

Using the values listed under the heading **Coef** in the coefficient output table in Figure 5.20a, we can express the regression equation as:

$$\hat{y} = 23.632 - 0.00488x_1 + 0.00138x_2$$

where $\hat{y}$ is the estimated average temperature, x_1 is elevation, and x_2 is aspect. The value of the adjusted $R^2 = 0.9597$ in the diagnostic output table in Figure 5.20b indicates that 95.97 per cent of the variation in temperature can be explained by elevation and aspect. The p-value measuring the overall significance of the regression equation listed in the diagnostic output table is 0, indicating that the regression equation has very good overall significance and is reliable for temperature predication or estimation. From the PDF report, the VIF for elevation and that for aspect are both 1.0516, which is < 7.5. Therefore, there is little multicollinearity between the two independent variables. The p-value for the Jarque-Bera test listed in the diagnostic output table is 0.408 (> α), indicating that the residuals are normally distributed. The p-values measuring the significance of individual regression coefficients are listed under the heading **Prob** in the coefficient output table. The p-value for elevation is 0, but the p-value for aspect is 0.3077, which is > α. Thus, aspect is not significant and should be removed.

Regression of temperature on elevation

6. In **ArcToolBox**, navigate to **Spatial Statistics Tools > Modeling Spatial Relationships**, and double-click **Ordinary Least Squares**.
7. In the **Ordinary Least Squares** dialog box:

   1) Select **gauges** as the input feature class.
   2) Select **Id** as the unique ID field.
   3) Navigate to your output directory, and enter the name of the output feature class.
   4) Select **temp** as the dependent variable.
   5) Tick **elevation** as the explanatory variable.
   6) Navigate to your output directory, and enter the name of the output report file.
   7) Click **Additional Options**.
   8) Navigate to your output directory, and enter the name of the coefficient output table.
   9) Navigate to your output directory, and enter the name of the diagnostic output table.

8. Click **OK**. Wait until the process is complete.
9. Interpret the results. According to the coefficient and diagnostic output tables, the simple regression equation can be written as:

$$\hat{y} = 23.7612 - 0.0048x_1$$

The value of the adjusted R^2 is 0.9596, almost the same as that produced by the multiple regression equation above. The p-value for the overall significance of the regression equation and that for elevation are all 0, indicating that both the simple regression equation and the independent variable are highly significant. The Jarque-Bera test is insignificant, suggesting that the residuals have a normal distribution. The simple regression equation appears to be as good as the multiple regression equation, but it requires only one independent variable. You can use the simple regression equation to reliably predict or estimate temperature using elevation values.

Geographically weighted regression

Conventional regression analysis creates a single regression equation to represent the relationship between a dependent variable and a number of independent variables based on the assumption that the relationship is static and consistent across the whole study area. In other words, a conventional simple or multiple regression equation is a global model in which the regression coefficients are considered to be constant over space. However, in environmental analysis these parameters may themselves be a function of geographical location because the form and strength of relationships among the environmental variables could be different across space. The results of conventional regression will not capture the true nature of spatially varying relationships between an environmental phenomenon and the influencing factors. Geographically weighted regression (GWR) was developed to overcome this limitation (Fotheringham et al. 2002).

GWR extends 'global' multiple regression by producing a set of local parameter estimates for each relationship at every data point in a region. Essentially, it estimates a linear regression equation at every data location with varying regression coefficients. A GWR local regression equation can be written as:

$$\hat{y}_i = a_i + b_{i1}x_{i1} + b_{i2}x_{i2} + \ldots + b_{ik}x_{ik} \tag{5.42}$$

where k is the number of independent variables, $\hat{y}_i$ is the predicted value of the dependent variable y at location i, a_i and $\{b_{ij} \mid j = 1, 2, \ldots, k\}$ are the regression coefficients at location i and x_{ij} is the observed value of the independent variable j at location i. Therefore, the regression equation is specific to the location of data point i.

A local regression equation at location i is estimated through weighting all surrounding observations by a function of distance from location i – that is, each observation is assigned a weight representing the influence of the observed value on the value at location i to be estimated. The observations closer to location i get larger weights than those further away. The weights can be determined by several methods. The simplest way is to define the weights as:

$$w_{il} = \begin{cases} 1 \text{ if } d_{il} \leq d \\ 0 \text{ otherwise} \end{cases} \tag{5.43}$$

where w_{il} is the weight assigned to observation or sample point l for estimating a regression equation at location i, d_{il} denotes the distance from location i to sample point l and d is a distance threshold value.

With the above weighting method, any observation further than distance d from location i is considered to have no influence on location i, and is excluded from estimating a regression equation for location i. All the observations within d from location i have equal weights. However, this method suffers from the problem of discontinuity. The regression coefficients at different locations may change considerably, particularly when the density of sample points differs greatly around those locations. One way to address this is to define weights using a continuous weighting function d_{il}, known as a kernel. A typical kernel function (which is used as a fixed radius kernel in ArcGIS) is:

$$w_{il} = e^{-\left(\frac{d_{il}}{h}\right)^2} \tag{5.44}$$

where h is a quantity known as the bandwidth. This function represents a Gaussian curve. The weight at location i is 1, where $i = l$ and $d_{il} = 0$. The weights of other data points decrease according to a Gaussian curve as the distance between i and l increases. When d_{il} increases to a certain distance, the weights will become virtually zero, effectively excluding these observations from the estimation of regression coefficients for location i. The bandwidth in the kernel is expressed in the same units as the coordinates used in the data. When a very large bandwidth is selected, the weights approach 1 and the local GWR model becomes a conventional global regression model.

The Gaussian kernel expressed in Equation 5.44 uses a fixed bandwidth h. It is suitable for use when the sample points are reasonably regularly spaced in the study area. However, when the sample points are not regularly spaced, it is expected that relatively small bandwidths are used in areas where the sample points are densely distributed, and relatively large bandwidths used where the sample points are sparsely distributed. One way of implementing this adaptive bandwidth approach is to specify the exact same number of sample points to be included in the local bandwidth of a kernel for estimation at each location. The following weighting function is such an adaptive kernel, called the bisquare kernel (which is used as an adaptive kernel in ArcGIS):

$$w_{il} = \begin{cases} \left[1 - \left(\frac{d_{il}}{h_i}\right)^2\right]^2 & if \ d_{il} \leq h_i \\ 0 & otherwise \end{cases} \tag{5.45}$$

where h_i is the bandwidth set to include the p observations nearest to location i, assuming p is the specified number of sample points to be included.

The bandwidth may be optimised by minimising the cross-validated sum of squared errors denoted as:

$$CV(h) = \sum_{i=1}^{n} \left[y_i - \hat{y}_{\neq i}(h) \right]^2 \tag{5.46}$$

where y_i is the observed value of y at location i, and $\hat{y}_{\neq i}(h)$ is the predicted value of y at location i made using a specific value of h when observation i is not used in the estimation (Brunsdon et al. 1998). Observation i is omitted from the calculation so that in the area of sparse observations the model is not calibrated solely on observation i.

Another method for choosing the optimal bandwidth is to minimise the corrected Akaike information criteria (AIC_c) (Hurvich et al. 1998). Its mathematic form for GWR is complicated, so we will focus on its interpretation. AIC_c is basically a model selection criterion. It provides a measure of the relative quality of a model, here a regression equation, for a given set of data. The model with the smaller AIC_c value is the better one – that is, it provides a better fit with the observed data. Here, the bandwidth with the lowest AIC_c is the optimal value.

Kernels other than those mentioned above can be used in GWR, and have been implemented in various software packages. Once a kernel or weighting function has been selected and the optimal bandwidth is obtained, all the observations are weighted relative to location i, and the weighted observations are then used to estimate a linear regression at the location based on the least squares principle used in conventional regression. It is clear that the regression coefficient estimates could vary by locations and reflect the spatial heterogeneity. For further details, see Fotheringham et al. (2002). Box 5.10 demonstrates how to implement GWR and interpret the results in ArcGIS. Box 5.11 provides an example of using GWR to model relationships of immature mosquitoes and human densities with the incidence of dengue in Taiwan.

Box 5.10   GWR in ArcGIS   PRACTICAL

To follow this example, start ArcMap, and load the **gauges** shapefile from **C:\Databases\GIS4EnvSci\VirtualCatchment\Shapefiles** In this example, you are going to explore the relationships between average annual temperature (the dependent variable), elevation and aspect (the independent variables) using GWR.

1. Open **ArcToolBox**. Navigate to **Spatial Statistics Tools > Modeling Spatial Relationships**, and double-click **Geographically Weighted Regression**.
2. In the **Geographically Weighted Regression** dialog box:

    1) Select **gauges** as the input feature class.
    2) Select **temp** as the dependent variable.
    3) Select **elevation** and **aspect** as the explanatory (independent) variables.
    4) Navigate to your output directory, and enter the name of the output feature class.
    5) Select **ADAPTIVE** as the kernel type – that is, to use an adaptive kernel.
    6) Select **AIC_c** as the bandwidth method – that is, to automatically find the optimal bandwidth by minimising the AIC_c value.
    7) Click **OK**. Wait until the process is complete. The output feature class is created and added to the data view, shown as a residual map. The GWR tool also creates a DBF table which contains the diagnostic statistics. The DBF table is added to the table of contents, whose name is that of the output feature class with the suffix _supp. Figure 5.21 shows the diagnostic table produced by the GWR regression.

Interpretation

Recall that the bandwidth of the GWR has been estimated for an adaptive kernel by minimising AIC_c. As shown in the GWR diagnostic table in Figure 5.21, twenty-five nearest neighbours have been used in the estimation of each set of coefficients. Let us compare the fit of the GWR and multiple regression models by comparing the diagnostic statistics in Figures 5.20b and 5.21.

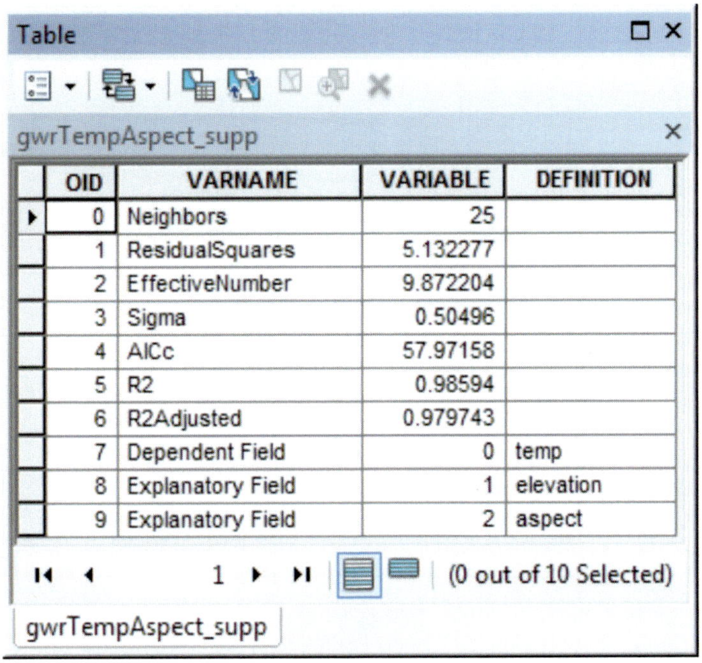
Table

gwrTempAspect_supp

OID	VARNAME	VARIABLE	DEFINITION
0	Neighbors	25	
1	ResidualSquares	5.132277	
2	EffectiveNumber	9.872204	
3	Sigma	0.50496	
4	AICc	57.97158	
5	R2	0.98594	
6	R2Adjusted	0.979743	
7	Dependent Field	0	temp
8	Explanatory Field	1	elevation
9	Explanatory Field	2	aspect

Figure 5.21 GWR output diagnostic table.

The adjusted R^2 for the GWR is 0.9797, while the adjusted R^2 for the multiple regression is 0.9597 – a 2 per cent improvement in model performance for the GWR. The AIC_c for the GWR is 57.97, and that for the multiple regression is 71.18. As a rule of thumb, if the AIC_c difference between two regression models is larger than ten, there is little evidence in support of the model with the larger AIC_c (Burnham and Anderson 2002). Therefore, the difference of 13.21 in the AIC_c in our case suggests that the GWR has a significant improvement in the fit of the model to the data over the multiple regression.

The feature attribute table associated with the output feature class contains the coefficient estimates, their standard errors and a range of diagnostic statistics, as shown in Figure 5.22. GWR fits a linear regression equation at every sample point. The condition number in the table is used to assess whether there is local multicollinearity. A condition number of > 30 suggests that the regression model at that sample point may be unreliable due to the presence of strong local multicollinearity. Our results indicate that all the fitted regression models at the sample points are reliable. Coefficient standard errors measure the reliability of each coefficient estimate. Confidence in those estimates is higher when standard errors are small in relation to the actual coefficient values. Large standard errors may be caused by local multicollinearity. Unlike in conventional multiple regression, GWR will not test whether coefficients are different from zero using a *t* test. There is no significance testing in the current form of GWR. As a result, GWR should be applied to sample data with a large sample size (several hundreds of observations) for best results. It is not appropriate for small samples.

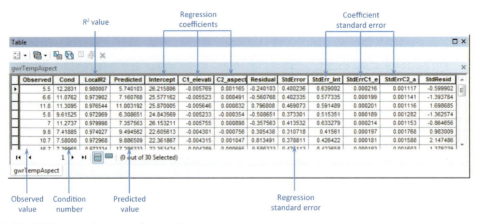

Figure 5.22 GWR output feature attribute table.

Dengue is a viral disease spread by mosquitoes (particularly *Andes aegypti* and *Andes albopictu*) in tropical and subtropical regions. It was once confined to Southeast Asia, but has now spread to Southern China and countries in the Pacific Ocean and the Americas. Scientists have long been studying the relationships between dengue incidence and mosquito abundance, and dengue incidence and human density. However, they are still not well understood. Most studies on these relationships have presented a global view, in which the relationships were assumed to be spatially constant across a region. Lin and Wen (2011) argued that this assumption may not be true, as these relationships could be positively correlated in some areas, but negatively or not correlated at all in other areas. They explored these relationships in the cities of Kaohsiung and Fenhshan in Taiwan using GWR.

Dengue has been observed in southern Taiwan since the late nineteenth century, initially occurring as intermittent epidemics at intervals of up to forty years, but becoming an annual phenomenon in the two cities since the beginning of the twenty-first century. Figure 5.23a shows the spatial distribution of dengue incidence (IR) in 2002 in the two cities. In order to understand the spatial variations in the relationships between dengue incidence and mosquito abundance, and dengue incidence and human density, Lin and Wen (2011) built a GWR model using the disease incidence (measured as cases per 100,000 people) as the dependent variable, and the monthly maximum Breteau index (BImax) and population density (POPden) as independent variables. The Breteau index represents the density of immature *Andes* mosquitoes measured as the number of containers with *Andes* mosquito larvae/pupae per 100 houses. The population density is measured as the number of people per km^2. Figures 5.23b and 5.23c show the spatial distributions of BImax and POPden in 2002 respectively. The GWR model has an adjusted R^2 of 0.59, suggesting that 59 per cent of the total variation could be explained by the model. Figure 5.24 shows the spatial variations in the locally weighted R^2 between the observed and predicted values.

It is evident from Figure 5.24 that the value of local R^2 varies across the study area. The GWR model fits best in districts 1, 5, 10 and 11 (see the key to Figure 5.24), but does not fit well in district 12. This means that other variables would influence the dengue incidence in district 12.

The spatial variations in intercept, BImax and POPden are shown in Figure 5.25. It can be seen that higher intercept values are distributed near the border of the two cities (districts 9, 10 and 11), which

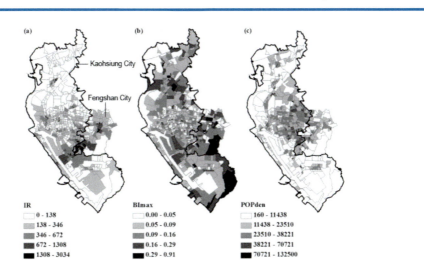

Figure 5.23 Spatial distributions of: (a) dengue incidence, (b) maximum Breteau index and (c) population density in 2002 in Kaohsiung and Fengshan cities (the area unit is Li, the lowest administrative unit in Taiwan) (reproduced with permission of Tzai-Hung Wen).

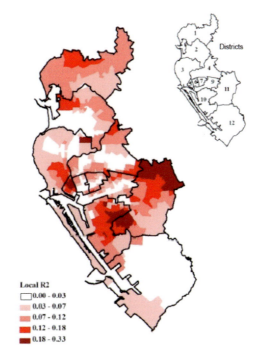

Figure 5.24 Spatial distribution of local R^2 (reproduced with permission of Tzai-Hung Wen).

(continued)

(continued)

implies that besides immature mosquito abundance and human population density, other explanatory variables need to be considered. The relationship between dengue incidence and BImax shown in Figure 5.25 suggests that with other factors the same, in districts 2, 5, 6, 7, 8, 10 and 11 increased dengue incidence were related to increased BImax. However, in the other districts, higher dengue incidence was associated with lower BImax, and vice versa. The distribution of population density parameter shows a more clearly spatial non-stationary pattern. In the northern districts, higher population density tended to be associated with higher dengue incidence, while higher dengue incidence was related to lower population density in the southern districts.

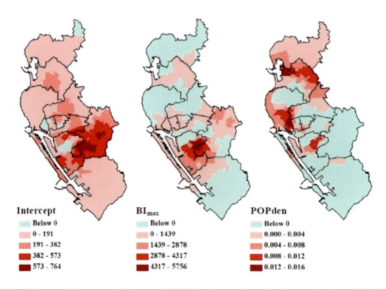

Figure 5.25 Coefficients of intercept, BImax and POPden (reproduced with permission of Tzai-Hung Wen).

The study provided evidence that the relationships between dengue incidence and mosquito abundance, and dengue incidence and human population density were spatially variable in the cities of Kaohsiung and Fengshan. In some areas, mosquito abundance and human population density could be significant predictors of dengue incidence. But in other areas, other environmental and social factors might have more effects on dengue transmission. The findings from the study provided policy makers with useful insights into spatial targeting of intervention against dengue epidemics.

Source: Lin and Wen (2011).

5.7 SUMMARY

- Exploratory spatial data analysis is about detecting and quantifying patterns, trends and associations in spatial data. It can help formulate hypotheses about the causes of observed environmental phenomena, assess assumptions on which statistical inference will be based, determine relationships among the explanatory variables and provide a basis for further data collection. Exploratory spatial data analysis is typically applied before any formal modelling commences, and can help inform the development of more complex models.

- Since it is generally impossible to make observations of environmental phenomena at every location in a region, researchers typically rely on spatial sampling to perform an observational study at a set of selected locations. It is important that the data collected at the set of locations selected is representative, and not biased in a systematic manner. The most common spatial sampling designs are simple point random sampling, systematic point sampling, stratified point sampling, clustered point sampling and transect sampling.

- Spatial distributions of geographical features can be described using statistical measures of central tendency and dispersion. Measures of central tendency find the centre of a distribution, including central feature, mean centre, median centre and linear directional mean. Measures of dispersion describe how geographical features are spread out or dispersed about the centre of their spatial distribution. Standard distance and standard deviational ellipse are two commonly used measures of dispersion.

- Spatial statistics have been used to understand spatial patterns by measuring whether geographical features are clustered, dispersed or randomly distributed across an area. Nearest neighbour analysis and global Moran's I are global spatial statistics used to detect the overall spatial pattern, while local Moran's I and Getis-Ord G_i^* are local spatial statistics for identifying spatial clusters at the local level.

- Spatial relationships between environmental phenomena can be assessed by regression. Conventional regression assumes that the relationships are static and uniform everywhere in a region. Geographically weighted regression assumes the relationships vary spatially across a landscape, and fits a regression equation to any point in space with locally weighted coefficients, which provides a powerful tool for exploring spatial heterogeneity and understanding spatial relationships.

REVIEW QUESTIONS

1. What is the nature of exploratory spatial data analysis?
2. What is the difference between spatial and non-spatial statistics?
3. What is a sample? What is the sampling frame?
4. Describe the following spatial sampling methods: simple point random sampling, systematic point sampling, stratified point sampling, clustered point sampling and transect sampling. Discuss their advantages and disadvantages.

5. Under what conditions are mean centre and median centre most appropriate to be used?
6. To determine the general direction of a group of line features, which measure of central tendency can be used?
7. If you want to measure the directional trend of a spatial distribution, which measure of dispersion should be used?
8. Explain the following terms:
   a) statistical significance;
   b) null hypothesis;
   c) alternative hypothesis;
   d) significance level;
   e) p-value;
   f) spatial autocorrelation;
   g) regression;
   h) residual;
   i) coefficient of determination.
9. What are the differences between global and local spatial statistics?
10. What does nearest neighbour analysis measure, and how is it used to understand spatial patterns?
11. What does global Moran's I measure, and how is it used to understand spatial patterns?
12. Describe how local Moran's I and Getis-Ord G_i^* are used to detect spatial clusters.
13. What are the assumptions of conventional and geographically weighted regression?
14. How does geographically weighted regression model spatial relationships?

REFERENCES

Anselin, L. (1995) 'Local indicators of spatial association: LISA', *Geographical Analysis*, 27: 93–115.

Anselin, L. (1998) 'Exploratory spatial data analysis in a geocomputational environment', in P.A. Longley, S.M. Brooks, R. McDonnell and W. Macmillan (eds) *Geocomputation: A Primer*, New York: Wiley & Sons, 77–94.

Brunsdon, C., Fotheringham, S. and Charlton, M. (1998) 'Geographically weighted regression-modelling spatial non-stationarity', *Journal of the Royal Statistical Society. Series D (The Statistician)*, 47(3): 431–443.

Burnham, K.A. and Anderson, D.R. (2002) *Model Selection and Multimodel Inference: A Practical Information-Theoretic Approach*, 2nd edn, New York: Springer.

Chun, Y. and Griffith, D.A. (2013) *Spatial Statistics and Geostatistics*, London: Sage.

Clark, P.J. and Evans, F.C. (1954) 'Distance to nearest neighbor as a measure of spatial relationships in populations', *Ecology*, 35: 445–453.

Eryando, T., Susanna, D., Pratiwi, D. and Nugraha, F. (2012) 'Standard deviational ellipse (SDE) models for malaria surveillance, case study: Sukabumi district – Indonesia, in 2012', *Malaria Journal*, 11(Suppl. 1): 130.

Fortin, M. and Dale, M.R.T. (2005) *Spatial Analysis: A Guide for Ecologists*, Cambridge: Cambridge University Press.

Fotheringham, A.S., Brunsdon, C. and Charlton, M. (2002) *Geographically Weighted Regression: The Analysis of Spatially Varying Relationships*, Chichester, England: John Wiley & Sons.

Getis, A. (2005) 'Spatial statistics', in P. Longley, M.F. Goodchild, D.J. Maguire and D.W. Rhind (eds) *Geographical Information Systems: Principles, Techniques, Management and Applications*, 2nd abridged edn, Hoboken, NJ: John Wiley & Sons (on CD).

Griffith, D.A. (2005) 'Effective geographic sample size in the presence of spatial autocorrelation', *Annals of the Association of American Geographers*, 95(4): 740–760.

Hurvich, C.M., Simonoff, J.S. and Tsai, C.-L. (1998) 'Smoothing parameter selection in nonparametric regression using an improved Akaike information criterion', *Journal of the Royal Statistical Society, Series B*, 60: 271–293.

Jarque, C.M. and Bera, A.K. (1987) 'A test for normality of observations and regression residuals', *International Statistical Review*, 55(2): 163–172.

Lin, C.H. and Wen, T.H. (2011) 'Using geographically weighted regression (GWR) to explore spatial varying relationships of immature mosquitoes and human densities with the incidence of dengue', *International Journal of Environmental Research and Public Health*, 8(7): 2,798–2,815.

Moran, P.A.P. (1950) 'Notes on continuous stochastic phenomena', *Biometrika*, 37(1): 17–23.

Ord, J.K. and Getis, A. (1995) 'Local spatial autocorrelation statistics: distributional issues and an application', *Geographical Analysis*, 27(4): 286–306.

Ripley, B.D. (2004) *Spatial Statistics*, New York: Wiley.

Rogerson, P.A. (2015) *Statistic Methods for Geography: A Student's Guide*, 4th edn, London: Sage.

Stein, A. and Ettema, C. (2003) 'An overview of spatial sampling procedures and experimental design of spatial studies for ecosystem comparisons', *Agriculture, Ecosystems and Environment*, 94(1): 31–47.

Wang, J., Stein, A., Gao, B. and Ge, Y. (2012) 'A review of spatial sampling', *Spatial Statistics*, 2: 1–14.

Yuill, R.S. (1971) 'The standard deviational ellipse: an updated tool for spatial description', *Geografiska Annaler, Series B, Human Geography*, 53(1): 28–39.

Remote sensing data analysis

Remote sensing is one of the major sources of spatial data. The effective analysis of remote sensing data to derive useful information is of critical importance for environmental analysis and monitoring using GIS. This chapter provides an introduction to remote sensing data, and presents the basics of digital image processing. It introduces commonly used operations of digital image processing in environmental remote sensing, including image preprocessing, transformation and classification.

LEARNING OBJECTIVES

After studying this chapter, you should be able to:

- understand the characteristics of digital remote sensing data;
- comprehend the basic principles of remote sensing;
- identify popular remote sensing systems for environmental monitoring;
- apply digital image analysis operations for processing remote sensing images;
- perform land use/cover classification with satellite images, and assess classification accuracy.

6.1 NATURE OF REMOTE SENSING DATA

Most of the spatial data used in GIS were derived from maps and survey products. As remote sensing technology advances, remote sensing has become an increasingly important source of spatial data for GIS. In particular, remote sensing provides accurate, multiscale, multispectral and rapidly updated environmental data for GIS databases. This chapter focuses on GIS methods

and functions for analysis of remote sensing data. Only the basic concepts of remote sensing are described in this section. For comprehensive and detailed discussions on the theories, techniques and applications of remote sensing, please refer to remote sensing textbooks such as Jensen (2007) and Lillesand et al. (2008).

Remote sensing refers to a group of techniques for collecting environmental data from a distance without coming into direct contact with the object or environmental phenomena of interest. Remote sensing data can take many forms, including variations in electromagnetic energy distributions, acoustic wave distributions or force distributions. This chapter is only concerned with remote sensing data recording the magnitude of electromagnetic energy flux that is reflected or emitted from objects on the Earth's surface or within the atmosphere. Such remote sensing data are normally in some form of imagery, which is further processed, interpreted or analysed to produce useful data and information about the Earth's resources and environment. They are mainly captured by imaging sensors deployed on airborne and spaceborne platforms (aircraft, satellites or other spacecraft).

Characteristics of electromagnetic radiation

Electromagnetic radiation is the only form of energy transfer that can occur in a vacuum such as the space between the Sun and the Earth. It has both wave and particle properties (Jensen 2007). According to its wave properties, electromagnetic radiation propagates through space at the speed of light. This electromagnetic wave consists of an electric and a magnetic field, which are perpendicular to one another, and to the direction of the travel (Figure 6.1). The distance from one wave peak to the next is the wavelength (λ), normally measured in micrometres (μm). The number of wave peaks passing a fixed point in space per unit time is the wave frequency (v). A wave that sends one peak by every second is said to have a frequency of one cycle per second, or 1 hertz (Hz). The wavelength and frequency are inversely related: $c = \lambda v$, where c is the speed of light, which is essentially a constant. Therefore, the longer the wavelength, the lower the frequency, and vice versa.

On the other hand, electromagnetic radiation is composed of many discrete units called quanta or photons, which carry particle properties such as energy and momentum (Jensen 2007). The relationship between the energy of radiation (e) and the wavelength is expressed as: $e = hc/\lambda$, where e is measured in joules and h is the Planck constant (6.6256×10^{-34} joules second). Thus, the longer the wavelength, the lower its energy content.

In remote sensing, it is most common to categorise electromagnetic waves by their wavelength location within the electromagnetic spectrum. We often specify a particular region of the electromagnetic spectrum by identifying a beginning and ending wavelength, then attaching a description. This wavelength interval in the electromagnetic spectrum is called a band, channel or region. As shown in Figure 6.2, the electromagnetic spectrum ranges from the shorter wavelengths (including short-wave gamma and x-rays) to the longer wavelengths (including microwave and long-wave radio). There are several regions of the electromagnetic spectrum which are useful for remote sensing of the environment, including the ultraviolet (0.003–0.4μm), the visible (0.4–0.7μm), the infrared (IR) (0.7–300μm) and

the microwave or radar (0.3–300cm) regions. The visible region of the spectrum can be further broken down into wavelength ranges that correspond to perceived colours of visible light: violet (0.4–0.446μm), blue (0.446–0.5μm), green (0.5–0.578μm), yellow (0.578–0.592μm), orange (0.592–0.620μm) and red (0.620–0.7μm). The visible band is the only portion of the spectrum we can associate with the concept of colours. Two bands in the IR region are important to remote sensing: the reflected IR (0.7–3μm) and thermal IR bands (3–15μm). Thermal IR energy is essentially the radiation that is emitted from the Earth's surface in the form of heat.

Interactions of electromagnetic radiation with matter

The Sun is the most important natural source of electromagnetic radiation for remote sensing. When electromagnetic radiation from the Sun strikes an object, the energy may be transmitted, absorbed, re-emitted, reflected or scattered. Remote sensors observe Earth features mainly by detecting the electromagnetic radiation reflected or emitted from them. An important principle underlying the use of remote sensing data is that different objects or ground materials reflect, absorb, transmit or emit electromagnetic radiation in different proportions, and these differences allow them to be identified.

Before electromagnetic radiation from the Sun reaches the Earth's surface, it has to travel through some distance of the atmosphere, where it interacts with particles and gases. This interaction with the atmosphere

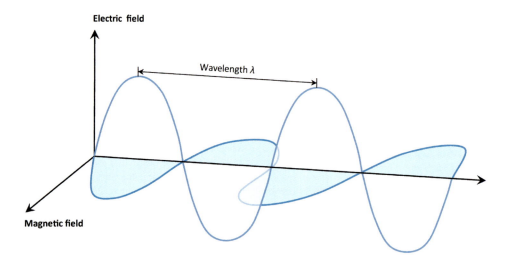

Figure 6.1 Electromagnetic wave.

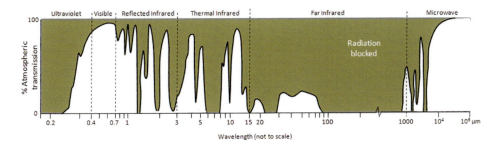

Figure 6.2 Electromagnetic spectrum and atmospheric windows (adapted from Lillesand et al. 2008).

occurs mainly through the mechanisms of scattering and absorption. Atmospheric scattering deflects the radiation from its path, and can severely reduce the information content of remote sensing data to the point that the imagery loses contrast and it becomes difficult to differentiate objects. Atmospheric absorption causes molecules in the atmosphere to absorb energy at various wavelengths. Ozone, carbon dioxide and water vapour are three main atmospheric constituents which absorb electromagnetic radiation. Remote sensing is difficult in those regions of the spectrum that are severely affected by atmospheric scattering and/or absorption. We only use the areas of the spectrum that are relatively, though not completely, unaffected by absorption and scattering for remote sensing data acquisition. These spectral regions are called atmospheric windows. In Figure 6.2, the percentage of incoming electromagnetic radiation through the atmosphere is plotted against wavelength. Atmospheric windows are those spectral regions in which the transmission of radiation through the atmosphere is high.

Electromagnetic radiation that is not absorbed or scattered in the atmosphere can reach the Earth's surface and interact with water, soil, vegetation and other ground features. When electromagnetic radiation is incident on the Earth's surface, it will interact with the surface features in one or more of the three ways: absorption, transmission and reflection. Remote sensing data are mainly measurements of the radiation reflected from targets. The reflectance characteristics of ground features depend on the wavelength of the incident energy, and the material and condition of these features. It may be quantified by measuring the portion of the incoming radiation that is reflected, which is referred to as spectral reflectance.

Spectral reflectance of a particular ground object usually varies with wavelength. A graph of the spectral reflectance of an object as a function of wavelength is called a spectral reflectance curve. Figure 6.3 shows three spectral reflectance curves respectively characterising typical spectral responses of vegetation, soil and clear water over the visible and near infrared bands. It illustrates that healthy vegetation strongly absorbs radiation in the red and blue bands, but reflects green radiation in the visible band, which is why healthy vegetation appears green to human eyes. When we move from the visible to the near IR region at about $0.7\mu m$, the spectral reflectance of healthy vegetation rises sharply. If our eyes were sensitive to near IR, trees would appear extremely bright to us at these wavelengths. Figure 6.3 also shows that the spectral curves of clear water, soils and vegetation have different shapes. This indicates that these features can be separated spectrally, and therefore can be distinguished in remote sensing imagery of certain bands. For example, water and vegetation may reflect somewhat similarly in the visible region, but are almost always separable in the near IR band, thus a near IR image can be effectively used to discern water and vegetation. The spectral differences among different objects or features may be a result of their differences in biophysical properties, chemical composition and surface geometry (such as roughness). Spectral response can be quite variable even for the same feature type, and can also vary with time and geographical location. Spectral reflectance curves give us insight into the spectral reflectance characteristics of the objects under investigation, and provide an indicator of which wavelength regions should be used to acquire remote sensing data for a particular application.

In addition, the electromagnetic energy absorbed by ground features will finally be released again in the form of heat as emitted thermal IR energy. Together with the geothermal energy emitted from the interior of the Earth, they form the source of electromagnetic radiation for thermal IR remote sensing.

Types of remote sensing data

The atmospheric windows shown in Figure 6.2 are utilised by different remote sensing systems. In general, the

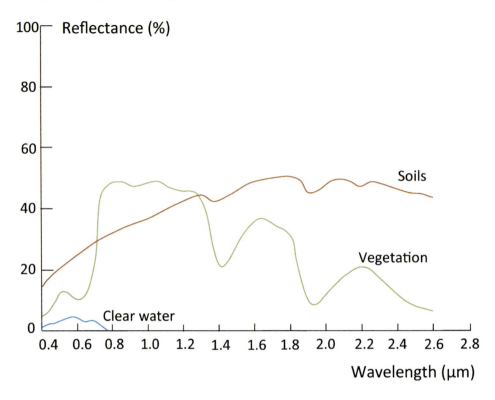

Figure 6.3 Typical spectral reflectance characteristics of vegetation, soils and clear water.

atmospheric windows covering the ultraviolet, visible and near IR bands are used for aerial photography; multiple bands from the visible to thermal infrared region are used by multispectral and hyperspectral remote sensing systems; two atmospheric windows in the thermal IR region, the 3–5µm and 8–14µm wavelength regions, are used for thermal IR remote sensing, and the radar and passive microwave remote sensing systems operate through a window in the region of 1mm–1m. Different remote sensing systems produce different types of remote sensing data. The most commonly used types of remote sensing data include aerial photographs, multispectral and hyperspectral images, thermal IR images, radar images and LiDAR point clouds.

Aerial photographs

Aerial photographs are captured using cameras mounted on aircraft. Cameras are optical sensors that use lenses to form an image at the focal plane. They are usually designed to detect electromagnetic radiation in the ultraviolet (0.3–0.4µm), visible (0.4–0.7µm) and near IR (0.7–0.9µm) regions. Photos or images produced from

cameras (using either photographic films or digital sensors) that are sensitive to the entire visible band are called panchromatic. They are black-and-white images, representing the intensities of visible electromagnetic radiation recorded at the moment of exposure. Cameras that are sensitive to blue, green and red light produce true or normal colour aerial photographs, in which the colours of objects appear to us the same way we see them in the natural environment (for example, a tree appears green). Data captured by cameras that sense green, red and the photographic portion of near IR radiation can be used to produce false colour or colour infrared images, in which objects with a high near IR reflectance appear red, those with a high red reflectance appear green and those with a high green reflectance appear blue. Colour infrared images are used extensively for vegetation analysis and mapping.

There are two types of aerial photos: oblique or vertical photos. Oblique aerial photographs are taken with the camera pointing to the side of the aircraft, but they are not usually used for mapping or spatial data collection as distortions in scale from the foreground to the background make it complicated to measure distance, area and elevation. Vertical photographs are taken with the camera pointing straight down at the ground. When obtaining vertical aerial photographs, the aircraft normally flies in a series of pre-defined flight lines and captures a rapid sequence of photographs while limiting geometric distortion. The camera systems are often linked with navigation systems onboard the aircraft so that accurate geographical coordinates can be immediately assigned to each photograph. Most camera systems for aerial photography also incorporate mechanisms which compensate for the effect of the aircraft's motion relative to the ground in order to minimise distortion.

Aerial photographs are most useful when fine spatial detail is more important than spectral information. Traditionally, aerial photographs are interpreted visually, and these results are then digitised into a GIS. Digital aerial photographs can now be processed directly in a GIS, which makes full use of the spectral detail contained in the photographs for feature enhancement and extraction.

Multispectral and hyperspectral images

Multispectral imagery consists of image data selectively acquired in a number of spectral bands. Although true colour and colour infrared aerial photography can be considered as three-band multispectral data, multispectral images are usually referred to as the image data captured by multispectral scanners in many more spectral bands and over a wider range of the electromagnetic spectrum. A multispectral scanner is a scanning system that uses a set of electronic detectors, each sensitive to a specific spectral band. The electronic detectors detect and measure the energy reflected or emitted from the phenomena of interest for each spectral band. The detected energy is recorded as an electrical signal, which is then converted to a digital value. Multispectral scanners produce digital images.

A digital image is actually a raster dataset. Each cell in the raster is a picture element called a pixel. Each pixel has a brightness value, also called a digital number (DN). A pixel's DN value represents the detected and measured energy in a given wavelength band, which is quantised to an 8-bit or 10-bit or a higher-bit digital number. An 8-bit digital number ranges from 0 to 255 (that is $2^8 - 1$). The detected energy value is scaled and quantised to fit within this range of values, which is termed radiometric resolution or bit resolution. A 10-bit digital number ranges from 0 to 1,023. To visually display an image, the DN values represent grey scales. Thus, the higher the reflected or emitted energy a pixel records, the brighter the pixel is in the image. The size of a pixel is the image's spatial resolution. Digital multispectral imagery is composed of multiple digital images of the same area of the Earth's surface, but for different discrete and narrow bands of the electromagnetic spectrum (Figure 6.4). The ability of a sensor to define fine wavelength intervals is called the spectral resolution. The finer the spectral resolution, the narrower the wavelength range for a particular band.

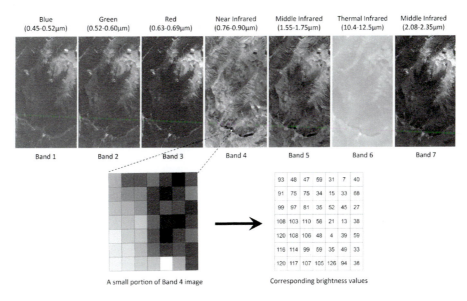

Figure 6.4 Digital multispectral images.

There are many digital multispectral remote sensing systems (Jensen 2007; Lillesand et al. 2008). AVHRR (Advanced Very High Resolution Radiometer) and Landsat are two multispectral remote sensing systems that have played important roles in land use and land cover characterisation. AVHRR systems have been deployed on sixteen satellites maintained by the National Oceanic and Atmospheric Administration (NOAA) since 1978. This scanning radiometer uses six detectors that measure electromagnetic energy in the red, near IR and thermal IR bands, as shown in Table 6.1. The spatial resolution at nadir (directly beneath the satellite) is about 1km. The AVHRR sensors acquire multispectral images of the entire globe twice a day. AVHRR images are widely used for large-area studies of vegetation, soil moisture, snow and ice cover, cloud and surface temperature. The normalised difference vegetation index (NDVI) calculated from reflectance values in the red and near IR bands has been used extensively with AVHRR data for vegetation monitoring on a global, regional or national level. NDVI is a simple transformation based on the following ratio:

$$NDVI = \frac{\rho_{nir} - \rho_{red}}{\rho_{nir} + \rho_{red}} \tag{6.1}$$

where ρ_{nir} and ρ_{red} are the reflectance (brightness) values respectively in the near infrared and red band. It ranges from −1.0 to 1.0. A positive value close to 1.0 indicates green vegetation, and a negative value suggests a non-vegetated surface. NDVI data calculated from AVHRR data gathered on multiple dates can also be composited to provide summary seasonal information.

The Landsat series of satellites were developed to meet the needs of resource managers and Earth scientists in providing global coverage of high-resolution multispectral imagery. The programme is managed by the National Aeronautics and Space Administration (NASA) and the US Geological Survey. Since 1972, eight Landsat satellites have been launched.

The first three Landsat satellites used two sensor systems: the Return Beam Vidicon (RBV) and the Multi-Spectral Scanner (MSS). Landsat 4, 5 (and 6) carried, in addition to the MSS, a new sensor system called the Thematic Mapper (TM). Landsat 7 carries only a single sensor system, the Enhanced Thematic Mapper (ETM+). Landsat 8, launched in February 2013, carries two sensors: the Operational Land Imager (OLI) and the Thermal Infrared Sensor (TIRS). The RBV system is similar to today's digital cameras, and it was discontinued after Landsat 3. The MSS sensor is one of the older generation of multispectral sensors. It senses electromagnetic radiation from the Earth's surface in four spectral bands with a spatial resolution of approximately 80m, ranging from the visible green to the near IR wavelengths. The TM sensor provides several improvements over the MSS sensor, including a higher spatial resolution, finer spectral bands and seven as opposed to four spectral bands. The spatial resolution of TM is 30m for all but the thermal IR band, which is 120m. ETM+ replicates the capabilities of TM, but adds a panchromatic band (0.52–0.90μm) with 15m spatial resolution and increases the spatial resolution of the thermal IR band to 60m. Table 6.2 outlines the characteristics of the individual TM and ETM+ bands and some useful applications. Figure 6.4 provides seven bands of TM images of a mountainous area.

Table 6.1 Characteristics of AVHRR

Band	Wavelength range (μm)	Spatial resolution at nadir (km)	Typical uses
1	0.58–0.68 (red)	1.09	Daytime cloud, snow, ice and vegetation mapping
2	0.725–1.00 (near IR)	1.09	Land–water boundary delineation, snow, ice and vegetation mapping
3A	1.58–1.64 (mid-IR)	1.09	Snow and ice mapping
3B	3.55–3.93 (thermal IR)	1.09	Day/night cloud and sea surface temperature mapping
4	10.3–11.3 (thermal IR)	1.09	Day/night cloud and sea surface temperature mapping
5	11.5–12.5 (thermal IR)	1.09	Sea surface temperature and cloud mapping

Table 6.2 Characteristics of TM and ETM+

Band	Wavelength range (μm)	Spatial resolution at nadir (m)	Typical uses
1	0.45–0.52 (blue)	30	Water body penetration, coastal water mapping and soil/vegetation discrimination
2	0.52–0.60 (green)	30	Vegetation discrimination and vigour assessment
3	0.63–0.69 (red)	30	Photosynthetic activity and vegetation species differentiation
4	0.76–0.90 (near IR)	30	Vegetation types, vigour and biomass assessment, water body delineation and soil moisture discrimination
5	1.55–1.75 (mid-IR)	30	Vegetation water stress, soil moisture and differentiation of snow from clouds
6	10.4–12.5 (thermal IR)	120 for TM 60 for ETM+	Vegetation stress analysis, soil moisture discrimination and thermal mapping
7	2.08–2.35 (mid-IR)	30	Discrimination of mineral and rock types, and vegetation stress analysis
8	0.52–0.9 (panchromatic) (for ETM+)	15	Sharpening multispectral images

The OLI sensor is similar to the ETM+ sensor, but provides enhancements over prior Landsat sensors by adding two more spectral bands: a deep blue visible band (Band 1) designed for water resource and coastal zone investigation, and a new infrared band (Band 9) for cirrus cloud detection. It collects data at a 30m resolution in eight shortwave spectral bands, and in one panchromatic band at a 15m resolution. The TIRS instrument collects data in two thermal IR bands at a 100m resolution for the wavelength covered by a single band on the previous TM and ETM+ sensors. Table 6.3 lists their characteristics. In addition, OLI and TIRS data are quantised over a 12-bit dynamic range and delivered as 16-bit images (scaled to 65,536 grey levels, compared with only 256 grey levels in previous 8-bit images), which enables better characterisation of land cover state and condition.

Table 6.3 Characteristics of OLI and TIRS

Band	Wavelength range (μm)	Spatial resolution (m)	Typical uses
1	0.43–0.45 (deep blue)	30	Coastal/aerosol studies
2	0.45–0.51 (blue)	30	Bathymetric mapping, soil/vegetation discrimination and forest type mapping
3	0.53–0.59 (green)	30	Assessment of vegetation vigour
4	0.64–0.67 (red)	30	Plant species differentiation
5	0.85–0.88 (near IR)	30	Detection of vegetation boundary between land and water, plant vigour and biomass assessment

(continued)

Table 6.3 *(continued)*

Band	Wavelength range (μm)	Spatial resolution (m)	Typical uses
6	1.57–1.65 (mid-IR)	30	Plant drought stress analysis, delineation of burnt areas and fire-affected vegetation and detection of active fires
7	2.11–2.29 (mid-IR)	30	Similar to Band 6
8	0.50–0.68 (panchromatic)	15	Sharpening multispectral images
9	1.36–1.38 (cirrus)	30	Detection of cirrus clouds
10	10.6–11.19 (thermal IR)	100	Thermal mapping of water currents, monitoring of fires and estimation of soil moisture
11	11.5–12.51 (thermal IR)	100	Similar to Band 10

Lansat satellites image the entire Earth every sixteen days. Multispectral images provided by Landsat have been widely and extensively used for global change studies, land use/cover monitoring and assessment and large area mapping. The long lifespan of the Landsat programme has also provided a voluminous archive of Earth resource data, facilitating long-term monitoring, historical records and research. Box 6.1 provides an example of the use of Landsat imagery to assess the impact of an earthquake on the giant panda habitats in China.

Box 6.1 Remote sensing based assessment of the impact of the Wenchuan earthquake on the giant panda habitats in Sichuan, China

CASE STUDY

On 12 May 2008, a devastating 8.0 magnitude earthquake occurred in Sichuan Province, China. The epicentre was in Wenchuan County, where the Wolong Nature Reserve is located. The reserve is the home of about 10 per cent of the world's population of wild giant pandas. The Wolong Nature Reserve and other six nature reserves and eleven scenic parks in the region form the World Nature Heritage Sichuan Giant Panda Sanctuary (WNHSGPS). The Wenchuan earthquake caused not only huge human casualties, but also damage to the panda habitat in the sanctuary. A team of scientists from the Chinese Academy of Sciences conducted an assessment of the damage in the Wenchuan region of WNHSGPS, covering the Wolong Nature Reserve and part of the adjacent Caopo Nature Reserve and Sanjiang Scenic Area by means of remote sensing, aiming to provide scientific evidence to evaluate the impact of the earthquake on the WNHSGPS as a whole. Figure 6.5 shows the study area with a true colour composite of Landsat 5 TM images on 18 September 2007.

Basically, the damage was assessed by comparing changes in land cover before and after the earthquake. The research team first derived land cover maps of the study area for two time points, one before and the other after the earthquake. The Landsat 5 TM imagery captured on 18 September 2007 was used to map the land cover before the earthquake. The Landsat 7 ETM+ imagery taken on 23 May 2008 and Landsat 5 TM imagery on 15 May 2008 were used to produce the post-earthquake land cover map. The two sets of imagery were utilised in order to reduce the impact of clouds. The two land cover maps were derived through supervised classification of the Landsat images (image classification is discussed in Section 6.5). Seven land cover classes were mapped: forest, shrub, meadow, farmland, water body, bare land and snow field, and settlement. The overall accuracy of the

Figure 6.5 The Wenchuan region of the Sichuan Giant Panda Sanctuary (image source: USGS).

post-earthquake land cover classification exceeded 90 per cent, verified through visual inspection and field investigation (the accuracy assessment of image classification is also discussed in Section 6.5). It was assumed that the accuracy of the pre-earthquake land cover classification was at a similar level. The two land cover maps were compared, and earthquake-damaged forests, shrubs and meadows were then identified and mapped.

To quantify the impact of the earthquake, the research team identified the giant panda habitats based on the land cover maps and a DEM by applying the following criteria: (1) within WNHSGPS; (2) with the forest land cover under which bamboo grows; (3) between 1,200 and 3,800m in elevation; and (4) with a slope of <45°. A slope map was derived from the DEM using ArcGIS (see Chapter 7 on terrain analysis). The land cover maps, DEM and slope map were overlaid via map algebra with ArcGIS to produce maps of suitable habitats for giant pandas before and after earthquakes. Four landscape metrics were calculated for the identified habitats using FRAGSTATS software: number of patches (NP), patch density (PD), landscape shape index (LSI) and landscape division index (LDI) (see Case Study 9 in Chapter 10 for more information about landscape metrics). A comparison of the pre- and post-earthquake landscape metrics revealed that the earthquake generally increased the complexity of the identified habitats. NP and PD all increased by about 20 per cent, indicating an increase of the heterogeneity of the landscape. LSI increased by 4.57 per cent and LDI by 3.51 per cent, which suggests that there was an increase in shape irregularity and fragmentation of the habitats.

The size of the damaged vegetated areas was also calculated. About 18.57km² of vegetated area was damaged by the earthquake in the identified panda habitats. This accounts for only 1.76 per cent of the vegetation cover in the habitats. Therefore, the Wenchuan earthquake did not cause severe damage to giant panda habitats. Indeed, most of the bamboo shoots retained their leaves in the habitats after the earthquake. The study concluded that earthquake was not a major factor affecting the welfare of the giant panda in this region.

Source: Yu et al. (2011).

While multispectral remote sensing collects data in multiple spectral bands, hyperspectral remote sensing systems acquire images in hundreds of very narrow, contiguous spectral bands throughout the visible and infrared portions of the electromagnetic spectrum. They can be used to distinguish many surface features that cannot be identified using broadband remote sensing systems such as Landsat ETM+ and OLI. Examples of hyperspectral remote sensing systems include AVIRIS (Advanced Visible/Infrared Imaging Spectrometer, http://aviris.jpl.nasa.gov/), HyMap (Cocks et al. 1998) and MODIS (Moderate Resolution Imaging Spectroradiometer, http://modis.gsfc.nasa.gov/).

Many sources of multispectral and hyperspectral satellite remote sensing data are available, and are covered in most standard remote sensing textbooks. An important and useful source is the USGS Global Visualization Viewer (GloVis), which is an online search and order tool for selected satellite data including the Landsat 8 OLI and TIRS, Landsat 7 ETM+, Landsat 4/5 TM, Landsat 1–5 MSS and MODIS. GloVis also allows users to specify cloud cover limits and date limits in searching for data, and provides access to metadata. Box 6.2 illustrates how to acquire and download Landsat 7 ETM+ and Landsat 8 OLI and TIRS image data from GloVis, and how to composite Landsat data into a multiband image.

Box 6.2   Acquiring Landsat data from GloVis, and compositing and clipping the data in ArcGIS   PRACTICAL

To follow this example, start a Web browser (Internet Explorer, Google Chrome or Mozilla Firefox) on your computer. Landsat imagery is available for download free of charge from GloVis. In this example, Landsat images for the Melbourne region will be downloaded and clipped to the Woori Yallock catchment (see Section 1.4 for a description of the catchment).

Download Landsat data

1. Enter **http://glovis.usgs.gov/** in the Web browser to open GloVis. If you see a Java security alert, click on **Activate Java**.
2. In the left panel of the GloVis viewer, enter the latitude (−**37.81**) and longitude (**144.96**) of Melbourne, Australia, and click the **Go** button beside the latitude/longitude boxes. In the right-hand side view, a set of image scenes laid out in the direction the satellite travelled are shown with a yellow square outlining the scene containing the location.
3. Change the path from 93 to **92**, and click **Go** next to the path/row boxes. The path and row numbers of the scene to be downloaded are respectively **92** and **86**.
4. From the GloVis menu bar, select **Collection > Landsat Archive> L7 SLC-on (1999–2003)** (Landsat 7 ETM+ images).
5. In the left panel of the viewer, change the maximum cloud cover (Max Cloud) to **0%** to limit the search to the cloudless scenes.
6. Set the target month to **Jan** and the target year to **2000**, and click **Go**. A scene is found, and the scene information is displayed, including the scene ID, cloud cover and the date the scene was taken. In this case, the scene ID is LE70920862000031EDC00, taken on 31 January 2000. The quality value 9 indicates a perfect scene with no errors detected.
7. Click **Add**, then **Sent to Cart**. A Web page pops up in a new window. If you have not yet registered, go through the registration process. If you have already registered, enter your user name and password to open the item basket page.
8. In the item basket page, click the **download** icon ⬇ . In the new pop-up window, click the **download** button next to Level 1 Product. Save the compressed file **LE70920862000031EDC00. tar.gz** to a directory of your choice on your computer. It is a large file and may take a few minutes to an hour to download, depending on your Internet connection.
9. After the above file is downloaded, from the GloVis menu bar, select **Collection > Landsat Archive> Landsat 8 OLI**. Repeat Steps 5–6 to search for a cloudless OLI and TIRS scene whose

path/row numbers are 92/86, taken in January 2014. The result should be a scene taken on 14 January 2014. Download the corresponding file **LC80920862014014LGN00.tar.gz** as in Steps 7–8, and save it to the directory where the ETM+ image is stored.

10. Decompress the two downloaded .gz files with WinZip or another compatible software package for creating and extracting ZIP files.
11. Create a folder named **LE012000** and another folder named **LC012014**. Move all the extracted files for the ETM+ scene to the folder **LE012000** and those for the OLI/TIRS scene to the folder **LC012014**. All individual-band images are formatted as georeferenced TIFF (GeoTIFF). An MTL file contains metadata for the scene. Note that all Landsat images from GloVis are radiometrically and geometrically corrected as well as georeferenced.

Composite Landsat data in ArcGIS without sensor property data

Compositing involves combining individual band images into a single multiband image. The following steps will create a composite image without transferring the information on the sensor properties (gains/biases, Sun elevation and so on) to its metadata.

12. Start ArcMap. Click the **Add Data** button ⊕, and navigate to the folder **LE012000**. B1, B2, B3, . . . in the names of the TIFF image files represent Band 1, Band 2, Band 3 and so on. Since ETM+ Band 6 is acquired in both high and low gain (see the radiometric correction in Section 6.2), the gain settings are provided as two separate band files.
13. In the **Add Data** dialog, select Band 1 image **LE70920862000031EDC00_B1.TIF** to add. When prompted to create pyramids for the image for rapid display at different resolutions, click **Yes** and wait for a few moments until the image is added and displayed in the data view.
14. Repeat the previous step to add Band 2, Band 3, Band 4, Band 5 and Band 7 one by one.
15. From the ArcMap **Main** menu, click **Windows > Image Analysis**. The **Image Analysis** window pops up.
16. In the **Image Analysis** window, all the images or raster layers in the table of contents are listed in the layer list box at the top of the window. The layers that are displayed in the data view are ticked.

    1) Click **LE70920862000031EDC00_B1.TIF** in the image or layer list so that it is highlighted in blue. Hold down the **Ctrl** key on the keyboard, and select **LE70920862000031EDC00_B2.TIF**, **LE70920862000031EDC00_B3.TIF**, . . . **LE70920862000031EDC00_B7.TIF** one by one. Release the **Ctrl** key.
    2) Expand the **Processing** section, and click the **Composite Bands** button ⊟. A temporary multiband image is created and named **Composite_ LE70920862000031EDC00_B1.TIF**.
    3) Click this temporary layer in the image list so that it is highlighted in blue.
    4) Click the **Export** button 🖫 in the **Processing** section. In the **Export Raster Data** dialog, enter **C:\Databases\GIS4EnvSci\WooriYallock** as the location, change the name to **le012000n.tif**, and click **Save**. After **le012000n.tif** is created, you are prompted to add it to the view.

17. Repeat Steps 12–16 to composite the first seven bands of the downloaded OLI/TIRS scene, and save the combined image as **C:\Databases\GIS4EnvSci\WooriYallock\c012014n.tif**.

Composite Landsat data in ArcGIS with sensor property data

The following steps will create a composite image with the information on the sensor properties (gains/biases, Sun elevation and so on) transferred to its metadata.

18. Click the **Add Data** button ⊕, navigate to the folder **LE012000**, and click ▦ **LE70920862000031EDC00_MTL.txt**. ArcGIS generates a temporary composited multiband image named

(continued)

(continued)

Multispectral_LE70920862000031EDC00_MTL and displays it in the data view as a natural colour image (see Box 6.5).

19. Right-click **Multispectral_LE70920862000031EDC00_MTL** in the table of contents, and open the **Layer Properties** window. In this window, click on the **Key Metadata** tab to display the information about the sensor and the scenes, including the sensor name, data acquisition date, Sun elevation, wavelength bands, and radian gains and biases for each band.

20. In the **Image Analysis** window, click on **Multispectral_LE70920862000031EDC00_MTL** to highlight it, and click the **Export** button 💾 in the **Processing** section.

21. In the **Export Raster Data** dialog, enter **C:\Databases\GIS4EnvSci\WooriYallock** as the location, change the name to **le012000.tif**, and click **Save**. After **le012000.tif** is created, you are prompted to add it to the view. **le012000.tif** is a multiband image with information on sensor properties.

22. Repeat Steps 18–21 to make a composite image of the downloaded OLI/TIRS scene from 🖼 **LC80920862014014LGN00_MTL.txt**, and save the combined image as **C:\Databases\GIS4EnvSci\WooriYallock\lc012014.tif**.

Clip Landsat data in ArcGIS

An image may not fit a study area. We often clip an image to the study area using its boundary.

23. Add **WooriYallock.shp** from **C:\Databases\GIS4EnvSci\WooriYallock** to ArcMap. This contains the boundary of the Woori Yallock catchment, east to Melbourne.

24. In the **Image Analysis** window, click **le012000.tif** in the image or layer list so that it is highlighted in blue.

25. Move the mouse to the **Tools** toolbar, and click the **Select Features** tool 🖱 .

26. Move the mouse to the data view, and click inside the polygon representing the Woori Yallock catchment boundary. The boundary is selected and highlighted in light blue.

27. Move the mouse back to the **Image Analysis** window. In this window:

    1) Expand the **Processing** section, and click the **Clip** button 🖼 . A temporary image layer is created and named **Clip_le012000.tif**.
    2) Click **Clip_le012000.tif** in the image list so that it is highlighted in blue.
    3) Click the **Export** button 💾 in the **Processing** section. In the **Export Raster Data** dialog, enter **C:\Databases\GIS4EnvSci\WooriYallock** as the location, change the name to **wyle2000.tif**, and click **Save**. After **wyle2000.tif** is created, you are prompted to add it to the view.

28. Follow the same process from Step 24 to Step 27 to extract **lc012014.tif** to the catchment, and save the clipped image as **C:\Databases\GIS4EnvSci\WooriYallock\wylc2014.tif**.

29. Click **File** on the main menu of ArcMap, then click **Save As**. In the **Save As** dialog, navigate to **C:\Databases\GIS4EnvSci\WooriYallock**, enter **chapter6.mxd** as the file name, and click **Save**.

Thermal IR images

Thermal IR images are captured by scanners that use special detectors sensitive to emitted thermal IR radiation. Thermal IR detectors essentially measure the radiant temperature and thermal properties of targets. Many multispectral scanners, such as AVHRR and Landsat TM mentioned earlier, contain thermal IR detectors for producing thermal IR images. Brightness values in a thermal IR image represent relative radiant temperatures. When depicted in grey levels, by convention, a thermal IR image shows warmer objects in light tones and cooler objects in dark tones. For meteorological purposes, this convention is reversed so that clouds (generally cooler than the Earth's surface) appear in light tones.

Due to absorption of thermal IR by atmospheric gases, thermal IR remote sensing is normally restricted to two specific wavelength regions: 3–5μm and 8–14μm. The spatial resolution of thermal images is usually coarse compared to the spatial resolution possible in the visible and reflected IR bands. As the thermal radiation is emitted, not reflected, thermal imagery can be acquired during the day or night. Thermal IR sensing has been used for a variety of applications, such as mapping forest fires, identifying surface and subsurface hydrothermal features, and monitoring water pollution.

Radar and LiDAR data

Radar (radio detection and ranging) and LiDAR (light detection and ranging) are both active remote sensing systems. Unlike aerial photography, multispectral and hyperspectral remote sensing systems that employ electromagnetic radiation naturally reflected or emitted from the Earth's surface and are therefore called passive remote sensing, radar and LiDAR utilise electromagnetic radiation provided by the sensor itself.

Radar systems transmit their own microwave energy at a particular wavelength (in the range 1–30cm) through the atmosphere for a particular duration of time, then measure the energy backscattered from the ground. Most imaging radar systems used for remote sensing of the Earth's resources and environment use side-looking airborne radar (SLAR), which produces radar imagery on one side of the aircraft's flight line. There are two main types of SLAR: real aperture radar and synthetic aperture radar (SAR). Real aperture radar uses an antenna of fixed length to transmit and receive the signal. SAR makes use of a small antenna (for example, 1–2m), but it can simulate a much larger antenna (for example 600m). This is made possible by storing the phases and amplitudes of the returning pulses, which are later synthesised to create a large antenna. SARs are able to achieve very fine resolution from large distances.

A radar image records the intensity of backscattered energy determined by the surface roughness, moisture content and dielectric properties of the ground features. In general, smooth surfaces or dry soils return very little backscatter towards the sensor and show up as dark areas on a radar image, while rough surfaces or moist soils return a lot of backscatter towards the sensor and appear as bright areas. Due to the side-looking viewing geometry, radar images exhibit shadows, which enhance subtle terrain features and can mirror ground surfaces for displaying topography. A major advantage of radar is its all-weather, day and night operation capability, which allows data to be collected at any time. This is because long-wavelength microwave radiation can penetrate through cloud cover, haze, dust and all but the heaviest rainfall.

Different from radar, LiDAR is a vertical- or nadir-looking instrument, and uses electromagnetic radiation in the visible and eye-safe near IR regions. However, it is not an imaging sensor. The LiDAR system emits laser

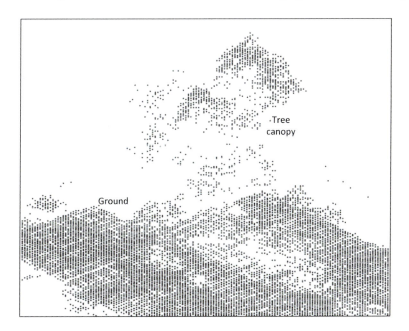

Figure 6.6 LiDAR elevation masspoints.

pulses and measures their travel time from the transmitter to the target on the terrain surface and back to the receiver (Liu 2008). As the velocity of the laser pulse (light) is known, the distance or range between the sensor and the ground can be calculated. When the sensor location can be determined with a high-precision GPS, the range can be converted to absolute coordinates (x, y, z). Here, (x, y) represents the target's location, and z is the absolute elevation on the Earth's surface. The range measurement process produces a collection of elevation data points, commonly referred to as masspoints. Figure 6.6 shows an example of masspoints associated with the ground and tree canopy. Therefore, LiDAR records information at discrete points which is not composed of contiguous pixels.

A laser pulse transmitted from a LiDAR instrument may generate one return or multiple returns (Figure 6.7).

The first returned laser pulse is usually associated with the highest object on the ground, like a treetop or rooftop. The second return may come from tree branches. The third and perhaps last return could be from bare ground. The first return can also represent the ground, in which case only one return will be detected by the LiDAR sensor. Multiple returns can be used to detect the elevations of several objects within a laser footprint. First returns are mainly used to create digital surface models that include features above the ground surface (such as buildings, bridges and trees). Intermediate returns, in general, are used for vegetation structure or for separating vegetation from solid objects among the aboveground features. Last returns are used to build DEMs of the bare ground surface. LiDAR data are often provided as (x, y, z) masspoints (also called point cloud data) or in the LAS format developed by the American Society for

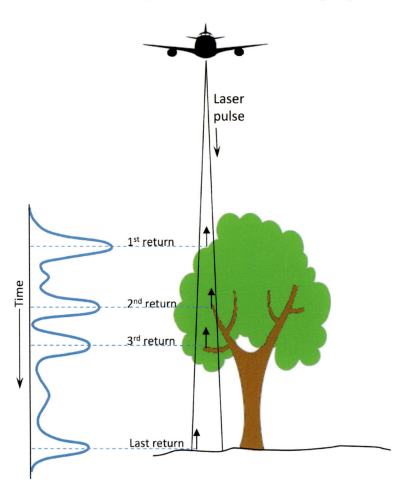

Figure 6.7 Multiple returns of a laser pulse from LiDAR through a tree canopy.

Photogrammetry and Remote Sensing (ASPRS 2013), which may include multiple returns as well as intensities.

6.2 IMAGE PREPROCESSING

Remote sensing images in their raw form as received from remote sensors may be distorted or contain deficiencies. Preprocessing operations are usually required prior to image interpretation and analysis. There are basically two types of image preprocessing operations: geometric and radiometric calibration. Geometric calibration aims to remove geometric distortions in an image caused by sensor–Earth geometry variations (such as those caused by satellites or aircraft departing from their normal altitude, or by sensors deviating from the main focus plane), and georeference the image to a particular projected map coordinate system so that it can be combined with other GIS data layers. Radiometric calibration involves rectifying an image for sensor irregularities, correcting an image for atmospheric effects, or restoring an image so that it accurately represents the radiance measured by the sensor. The type of preprocessing required for an image depends on the quality of the image and the purpose of the use of the data, and varies widely among sensors. This section only introduces some image preprocessing operations available in GIS.

Geometric calibration

Geometric calibration is a process of transforming the distorted image coordinates to geometrically correct image coordinates. It may involve rotating, rescaling, projecting or reprojecting an image. Generally, a pixel in a corrected image does not overlap a pixel in the original distorted image (Figure 6.8). The pixel value in the corrected image must be recalculated based on the pixel values surrounding the transformed position in the original image. This process is called resampling. There are basically three methods of resampling: nearest neighbour, bilinear interpolation and cubic convolution.

The nearest neighbour method determines the value of each pixel in the rectified output image as the value of its nearest pixel (centre to centre) in the original distorted image. For example, the bold blue outline in Figure 6.8 indicates a target pixel in the output image. It is assigned the value of the pixel labelled a on the input image. The nearest neighbour method does not change the values of the pixels, thus it is quick and maintains the integrity of data. But it may create a blocky appearance in the corrected image because features on the output image may be offset spatially by up to half a pixel (Figure 6.9b).

The bilinear interpolation method calculates the value of a rectified output pixel as the average or

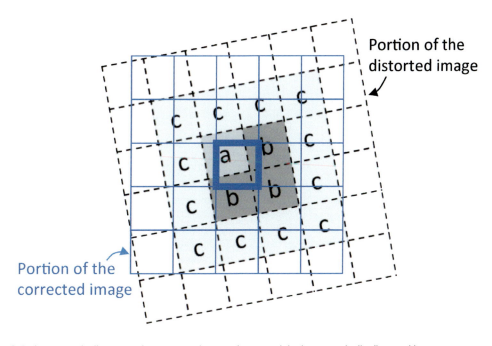

Figure 6.8 A geometrically correct image superimposed on an original geometrically distorted image.

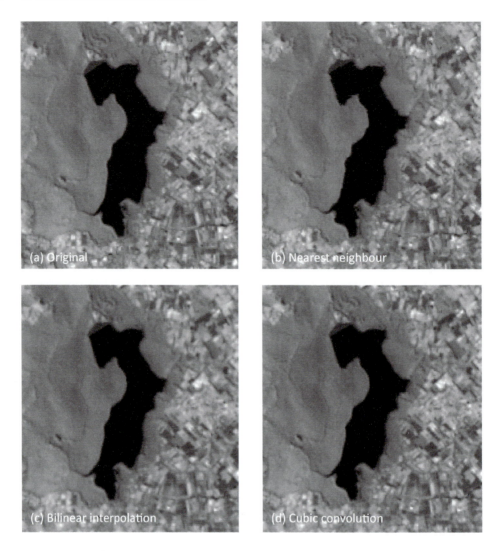

Figure 6.9 Resampling of a Landsat TM image: (a) original, (b) nearest neighbour, (c) bilinear interpolation and (d) cubic convolution.

distance-weighted average of the four nearest input pixels. For example, the target output pixel in Figure 6.8 is determined as the average of the values of the pixels labelled a and b. It produces a smoother image than the nearest neighbour method, but changes the range of brightness values of the original image (Figure 6.9c).

The cubic convolution method computes the value of a rectified pixel as the average or distance-weighted value of the sixteen nearest input pixels. For instance, the target output pixel in Figure 6.8 is determined as the average of the values of the pixels labelled a, b and c. It is similar

to bilinear interpolation, but it is more computationally intensive. The method sharpens an image, and may smooth out noise (Figure 6.9d). Both bilinear interpolation and cubic convolution involve changing the original pixel values, which may cause some problems in subsequent spectral pattern analyses of the image. To avoid those problems, the two methods are often performed after spectral pattern analyses.

Box 6.3 demonstrates how to reproject an image and change its cell size in ArcGIS using one of the resampling methods discussed above.

Box 6.3   Reprojecting, rotating and resampling images in ArcGIS   **PRACTICAL**

To follow this example, start ArcMap and load the **wyle2000.tif** and **wylc2014.tif** images created in Box 6.2. **wyle2000.tif** and **wylc2014.tif** were projected to UTM Zone 55N with WGS84 datum. In this example, both images will be reprojected to Map Grid of Australia (MGA) with the Geocentric Datum of Australia 1994 (GDA94). **wyle2000.tif** will also be resampled from 30m resolution to 20m resolution.

Reproject an image

1. Open **ArcToolBox**. In the **ArcToolBox** window, navigate to **Data Management Tools > Projections and Transformations > Raster**, and double-click **Project Raster**.
2. In the **Project Raster** dialog:

    1) Select **wyle2000.tif** as the input raster.
    2) Enter **C:\Databases\GIS4EnvSci\WooriYallock\wyle2000mga.tif** as the output raster dataset.
    3) Click the **Spatial Reference Properties** [⊞] button. In the **Spatial Reference Properties** dialog, expand **Projected Coordinate Systems > National Grids > Australia**, select **GDA 1994 MGA Zone 55**, and click **OK**. **GDA 1994 MGA Zone 55** is set as the output coordinate system.
    4) Select **NEAREST** (nearest neighbour) as the resampling method.
    5) Click **OK**. The reprojected multiband image is created and added to the data view.

3. Repeat the previous steps to reproject **wylc2014.tif** to **GDA 1994 MGA Zone 55**, and save the reprojected image as **C:\Databases\GIS4EnvSci\WooriYallock\wylc2014mga.tif**.

Resample an image

4. In the **ArcToolBox** window, navigate to **Data Management Tools > Raster > Raster Processing**, and double-click **Resample**.
5. In the **Resample** dialog:

    1) Select **wyle2000mga.tif** as the input raster.
    2) Enter **C:\Databases\GIS4EnvSci\WooriYallock\wyle2000mga20.tif** as the output raster dataset.
    3) Enter **20** in both the X and Y boxes.
    4) Select **CUBIC** (cubic convolution) as the resampling method.
    5) Click **OK**. The input multiband image's pixel size is changed from 30m to 20m.

Radiometric calibration

Environmental monitoring often requires the comparison of images taken at different times or geographical locations. To do this correctly, it is necessary to apply radiometric calibration to the images. This is because the radiance measured by a remote sensor over a particular feature is affected by changes in scene illumination, atmospheric conditions, viewing geometry, sensor response properties and other factors. The type of radiometric correction varies greatly from sensor to sensor, and depends on the specific application. In this section, we will concentrate on DN-to-radiance conversion, Sun elevation correction and radiance-to-reflectance conversion for multispectral imagery.

DN-to-radiance conversion

Sensor systems are normally designed to produce a linear response to incident spectral radiance. As a result, the pixel values or DN values in an image are usually a linear transformation of the physical quantity of spectral radiance measured by the sensor to fit into the range of, for example, 8-bit (0–255) or 12-bit (0–4,095). For a multi-spectral sensor system, each spectral band has its own response function. By applying the response function, the DN values can be converted back to spectral radiance values. In general, a response function for a given band λ can be written as:

$$DN_\lambda = G_\lambda \times L_\lambda + B_\lambda \qquad (6.2)$$

where DN_λ is the digital number recorded for the band λ, L_λ is the spectral radiance measured over the band, G_λ is the slope of the response function referred to as the gain and B_λ is the intercept of the response function called the offset or bias.

Let L_{min} be the spectral radiance corresponding to the minimum quantised and calibrated DN value (usually 0), L_{max} be the spectral radiance at the maximum quantised and calibrated DN value and DN_{max} the maximum DN value for the band λ. L_λ is calculated as:

$$L_\lambda = \left(\frac{L_{max} - L_{min}}{DN_{max}}\right) \times DN_\lambda + L_{min} \qquad (6.3)$$

L_λ, L_{max} and L_{min} are in units of watts/(steradian × square metre × µm). DN_{max} depends on the radiometric resolution designed for the sensor. For an 8-bit sensor like Landsat TM or ETM+, it is 255. One L_{max}/L_{min} set exists for each gain state for each band. Landsat ETM+ has two gain states (low and high gain). The high gain state measures a narrower radiance range with increased sensitivity over areas of low reflectance, while the low gain setting measures a broader radiance range with decreased sensitivity for areas of high reflectance. The goal in switching gain states is to maximise the sensor's 8-bit radiometric resolution without saturating the detectors. L_{max} and L_{min} (or gain and offset) values for each band are typically retrieved from the image's metadata or received from the data provider. The spectral radiance calculated by Equation 6.3 is also called the apparent radiance at sensor or the top of atmosphere (TOA) radiance (measured by a sensor flying higher than the Earth's atmosphere), which includes contributions by the atmosphere.

DN-to-radiance conversion is useful when comparing the actual radiance measured by different sensors, for example Landsat TM and ETM+, or ETM+ versus OLI. It is also useful when establishing quantitative relationships between image data and ground measurements such as water quality and plant biomass data.

Sun elevation correction

Images taken at different times of year or times of day are likely illuminated by the Sun at different angles. As illustrated in Figure 6.10, the solar elevation angle decreases from summer to winter for the same sensor and the same location. To compensate for the different amounts of illumination of scenes captured at different times of day, or at different latitudes or seasons, the Sun elevation correction can be applied. The correction involves normalising the images acquired under different solar elevation angles to the zenith using the following equation:

$$L_\lambda' = \frac{L_\lambda}{\sin \alpha} \qquad (6.4)$$

where L_λ' is the corrected radiance value, and α is the sun elevation angle. This equation ignores topographic and atmospheric effects. The location, the scene centre's Sun elevation angle, and the date and time when a particular scene was captured are usually provided in the image's metadata.

Radiance-to-reflectance conversion

Radiance is the variable directly measured by remote sensing instruments. Reflectance is a property of the target being observed. As reflectance is often used in extracting biophysical information, such as deriving vegetation index values, it is useful to convert the spectral radiance measured by the sensor to the apparent reflectance or TOA planetary reflectance. TOA reflectance is the total spectral reflectance at the sensor from both target and atmosphere. It is also known as albedo. The apparent reflectance is calculated for each individual band using the equation:

$$\rho_\lambda = \frac{\pi \times L_\lambda \times d^2}{E_s \cos \theta} \qquad (6.5)$$

where ρ_λ is the apparent reflectance (unitless) for band λ, d is the relative Earth–Sun distance in astronomical units for the day of image acquisition, E_s is the mean solar exoatmospheric irradiances and θ is the sun's zenith angle in degrees. Information on d and E_s is usually part of the ancillary data provided with the images or supplied in the data users' handbook. If the image has been corrected for atmospheric effects, the value calculated by Equation 6.5 is an estimate of actual target reflectance.

The equations for radiometric calibration presented above can be implemented in GIS via map algebra (see Section 4.4 and Box 4.11). Some GIS packages contain

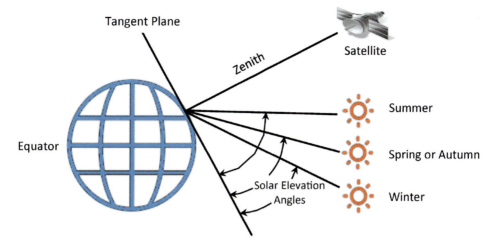

Figure 6.10 Solar elevation angles at different seasons.

specialised functions for radiometric calibration. For example, ArcGIS's Apparent Reflectance function is designed to calibrate radiometrically remote sensing images for some satellite sensors, including all Landsat sensor systems, IKONOS, Quickbird, GeoEye-1, RapidEye, DMCii, WorldView-1, WorldView-2, SPOT 6 and Pleiades. This function performs two calibrations. The first calibration is to convert the DN value to the TOA radiance based on the sensor properties (that is, gain settings). The second one is to convert the TOA radiance to TOA reflectance based on Sun elevation and acquisition date. Box 6.4 illustrates how to use this function.

Box 6.4   The Apparent Reflectance function in ArcGIS   **PRACTICAL**

To follow this example, start ArcMap, and open **chapter6.mxd** from **C:\Databases\GIS4EnvSci\ WooriYallock**.

Radiometrically rectify an ETM+ image

1. From the ArcMap **Main** menu, click **Windows > Image Analysis**.
2. In the **Image Analysis** window:
   1) Click **Multispectral_LE70920862000031EDC00_MTL** in the layer list so that it is highlighted in blue.
   2) Expand the **Processing** section, and click the **Function Editor** button *fx* . In the **Function Template Editor** dialog, right-click **Composite Band Function**, point to **Insert**, and click **Apparent Reflectance Function**. The **Raster Function Properties** dialog pops up. The radiance gain and bias values as well as the Sun elevation angle are automatically read from the scene's metadata and populated in the dialog.
   3) In the **Raster Function Properties** dialog, tick **Albedo**. The results of this function will be expressed as albedo, which is the percentage of the energy reflected by the planetary surface.

(continued)

(continued)

    4) Click the **General** tab in the **Raster Function Properties** dialog, change the **Output Pixel Type** to 32-bit float, then click **OK**.

    5) Click **OK** in the **Function Template Editor** dialog. The apparent reflectance function is implemented on **Multispectral_LE70920862000031EDC00_MTL**. After it is complete, **Func_Multispectral_LE70920862000031EDC00_MTL** is created, which is a temporary image layer. This layer contains the albedo values ranging from 0 to 1.0.

3: Repeat Step 2, but uncheck **Albedo**, and set the **Output Pixel Type** to 8-bit unsigned. The result will be **Func1_Multispectral_LE70920862000031EDC00_MTL**. This image contains TOA values scaled to the specified output pixel type – 8-bit in this example.

4. Use **WooriYallock.shp** to clip **Func_Multispectral_LE70920862000031EDC00_MTL** as you did in Box 6.2, and obtain **Clip_Func_Multispectral_LE70920862000031EDC00_MTL**.

5. Export **Clip_Func_Multispectral_LE70920862000031EDC00_MTL** to the location **C:\Databases\ GIS4EnvSci\WooriYallock**, and save it as **wy00abd.tif**.

6. Repeat Steps 4–5 to clip **Func1_Multispectral_LE70920862000031EDC00_MTL** to the Woori Yallock catchment, and save it to **C:\Databases\GIS4EnvSci\WooriYallock** as **wy00ar.tif**.

Radiometrically rectify an OLI image

7. Repeat Step 2 above to implement the apparent reflectance function on **Multispectral_ LC80920862014014LGN00_MTL** with **Albedo** as the output, and create a temporary albedo image **Func_Multispectral_LC80920862014014LGN00_MTL**. Note that Landsat 8 metadata does not use gain and bias, but instead uses similar concepts under name REFLECTANCE_MULT and REFLECTANCE_ADD. So for Landsat 8 OLI data, the gain and bias values are all set to the fixed numbers 2.0000E-05 and –0.1.

8. Apply the apparent reflectance function on **Multispectral_LC80920862014014LGN00_MTL** with **Albedo** unchecked, and set the **Output Pixel Type** to 16-bit unsigned. The result will be **Func1_Multispectral_LC80920862014014LGN00_MTL**. This image contains TOA values scaled to 16-bit.

9. As in Steps 4–5 above, use **WooriYallock.shp** to clip **Func_Multispectral_LC80920 862014014LGN00_MTL** and **Func1_Multispectral_LC80920862014014LGN00_MTL**, export the clipped images to **C:\Databases\GIS4EnvSci\WooriYallock**, and save them respectively as **wy14abd.tif** (albedo values) and **wy14ar.tif** (scaled TOA values).

10. You may reproject **wy00abd.tif**, **wy00ar.tif**, **wy14abd.tif** and **wy14ar.tif** to **GDA 1994 MGA Zone 55** as you did in Box 6.3.

6.3 IMAGE ENHANCEMENT

Image enhancement is a process of improving the interpretability or human perception of information in images for specific applications. It includes techniques for enhancing the visual distinctions among features in images. Typically, image enhancement is achieved through various band combinations (for increasing the amount of information that can be visually interpreted from an image), pan-sharpening (for increasing the spatial resolution of a multispectral image with a higher-resolution panchromatic image), contrast stretching (for enlarging the tonal distinction between different features in an image) and spatial filtering (for enhancing or suppressing specific spatial details in an image).

Band combinations

Multispectral image data are acquired as a set of individual single-band greyscale images, as shown in Figure 6.4. To facilitate the interpretation and analysis of these data, we often combine single-band images into colour composites. A colour composite can be made by blending three bands at a time, with one band displayed as blue, one band as green and one band as red. Therefore, it is also called an RGB (red–green–blue) colour composite. A true or natural colour composite is generated by assigning the colour blue to the blue band image, the colour green to the green band image and the colour red to the red band image. False colour images are a representation of a multi-spectral image produced using bands other than visible red, green and blue as the red, green and blue components of an image display. A false colour image created by assigning blue, green and red colours to the green, red and near IR band images respectively is also called a colour infrared image. Using bands such as near IR increases the spectral separation and often the interpretability of the remote sensing data.

Different false colour composites may be suitable for identifying different types of objects. For example, a colour infrared composite is a very common false colour composite scheme for displaying multispectral imagery, which allows vegetation to be readily detected. A false colour composite made of mid-infrared, near infrared and green bands may help differentiate different forest types and discern difference in soil moisture content. Box 6.5 shows how to create colour composites in ArcGIS.

Box 6.5 Creating colour composites in ArcGIS   **PRACTICAL**

To follow this example, start ArcMap, and load **lc012014n.tif** created in Box 6.2.

Create a natural colour composite

1. In the table of contents, right-click **lc012014n.tif**, and select **Properties**.
2. In the **Layer Properties** dialog:

    1) Click on the **Symbology** tab. The **RGB Composite** (the red–green–blue colour composite) in the left panel is highlighted. By default, red is assigned to Band 1, green to Band 2 and blue to Band 3, as shown in the right panel. In this case, Band 1 of the OLI imagery is a deep blue band, Band 2 is a blue band and Band 3 is a green band. Therefore, when **lc012014n.tif** is loaded, it is displayed as a false colour composite.
    2) Click the **Band** drop-down arrow next to the **Red** colour channel, and click **Band 4** (the red band) to display for the red colour.
    3) Click the **Band** drop-down arrow next to the **Green** colour channel, and click **Band 3** (the green band).
    4) Click the **Band** drop-down arrow next to the **Blue** colour channel, and click **Band 2** (the blue band).
    5) Click **OK**. A natural colour composite is displayed in the data view, which is close to what we would expect to see in a normal colour photograph.

Natural colour composites tend to have low contrast and be slightly hazy in appearance. This is mainly because blue light is more susceptible to atmospheric scattering than longer wavelengths. Natural colour composites are useful for some applications as land cover is associated with familiar colours – for example, forests with green and roads with grey.

Create a colour infrared image

3. In the table of contents, right-click **lc012014n.tif**, and open the **Layer Properties** dialog.
4. In the **Layer Properties** dialog:

    1) Click on the **Symbology** tab.
    2) Click the **Band** drop-down arrow next to the **Red** channel, and click **Band 5** (the near IR band).

(continued)

(continued)

    3) Click the **Band** drop-down arrow next to the **Green** channel, and click **Band 4**.
    4) Click the **Band** drop-down arrow next to the **Blue** channel, and click **Band 3**.
    5) Click **OK**. The image is displayed as a colour infrared composite.

This band combination makes vegetation appear in different shades of red depending on the types and conditions of the vegetation, with brighter reds indicating more vigorously growing vegetation (higher near IR reflectance). Bare soils, roads and buildings may appear in different shades of blue, yellow or grey, depending on their composition. Water bodies appear blue. Deep, clear water appears dark-bluish (higher green reflectance), while turbid water appears cyan (higher red reflectance due to sediments). Clouds and snow appear bright white.

Create a false colour composite of green, near IR and mid-IR bands

5. In the table of contents, right-click **lc012014n.tif**, and open the **Layer Properties** dialog.
6. In the **Layer Properties** dialog:

    1) Click on the **Symbology** tab.
    2) Click the **Band** drop-down arrow next to the **Red** channel, and click **Band 7** (the mid-IR band).
    3) Click the **Band** drop-down arrow next to the **Green** channel, and click **Band 5**.
    4) Click the **Band** drop-down arrow next to the **Blue** channel, and click **Band 3**.
    5) Click **OK**.

In this false colour composite, healthy vegetation appears bright green and grasslands appear light green, forested coniferous areas are in darker green than deciduous ones, dry soils appear in pink while wet soils appear blue, urban areas appear in varying shades of magenta and water is in dark blue. This combination also penetrates atmospheric particles and smoke, and therefore can detect fires, which would appear red. It is useful for post-fire analysis of burned and non-burned forested areas.

Pan-sharpening

Pan-sharpening – shorthand for panchromatic sharpening – is a process of combining a higher-resolution panchromatic image (in greyscale) with a lower-resolution multispectral image to create a fused colour image with the spatial resolution of the panchromatic image. Lower-resolution multispectral images and higher-resolution panchromatic images of the same scenes are commonly bundled in satellite image datasets. For example, Landsat 7 ETM+ and Landsat 8 OLI include 30m resolution multispectral bands plus a 15m resolution panchromatic band. Pan-sharpening uses spatial information in the higher-resolution panchromatic band and spectral information in the lower-resolution multispectral bands to produce a high-resolution multiband image, which essentially increases the spatial resolution of the multispectral dataset and provides a better visualisation of the multispectral image, as shown in Figure 6.11. As a panchromatic

image usually covers a wavelength range from the visible to near IR region, pan-sharpening is mostly applied to sharpen a colour composite made of red, green, blue and near IR band images.

A pan-sharpened image is a fusion between the multispectral and panchromatic images which gives the best of both types of image: high spectral resolution and high spatial resolution. However, pan-sharpening may result in spectral distortions due to the nature of the panchromatic band. For instance, the Landsat ETM+ panchromatic band is not sensitive to blue light. Consequently, the spectral characteristics of the pan-sharpened colour image may not exactly match those of the corresponding low-resolution colour image, resulting in distorted colour tones. Several methods have been developed for pan-sharpening (Nikolakopoulos 2008; Amro et al. 2011). The popular ones include IHS, Brovey, Gram-Schmidt and simple mean transformations. They are all subject to trade-offs between the final

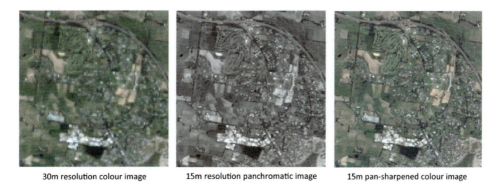

| 30m resolution colour image | 15m resolution panchromatic image | 15m pan-sharpened colour image |

Figure 6.11 Pan-sharpening of a multiband image: (a) 30m resolution colour image, (b) 15m resolution panchromatic image and (c) 15m pan-sharpened colour image.

spatial and spectral resolutions, and there is no method which is clearly the best overall.

IHS (intensity-hue-saturation) transformation is the most basic and popular pan-sharpening technique. It involves the following steps (see Figure 6.12):

1. georeferencing the source images to a standard map projection or projected coordinate system (not necessary when the data come from the same sensor);
2. resampling the multispectral image to the same resolution as the panchromatic band using a nearest neighbour, bilinear or cubic convolution technique;
3. transforming the resampled RGB colour composite of multispectral bands to HSI colour space (see Section 8.2);
4. replacing the intensity component in the transformed space with the panchromatic image, or when a near IR band is included, calculating the intensity values as:

$$Intensity = Pan - NearIR \times w_{nir} \qquad (6.6)$$

where *Pan* refers to the pixel intensity value in the panchromatic image, *NearIR* is the pixel value in the near IR image and w_{nir} is the weight assigned to the near IR band;

5. performing a reverse transformation using the substituted intensity component to convert back to the original RGB colour space.

IHS is very fast, and can quickly process large volumes of data and produce enhanced images. However, the fused images suffer from spectral distortion compared to the original multispectral image.

Brovey transformation assumes that the spectral range spanned by the panchromatic image is the same as that covered by the multispectral bands. It first multiplies each resampled multispectral band by the panchromatic band minus the weighted near IR band, then divides each product by the weighted sum of all the multispectral bands, as expressed below:

$$I^{\lambda}_{out} = I^{\lambda}_{in} \times \frac{Pan - NearIR \times w_{nir}}{R \times w_{red} + G \times w_{green} + B \times w_{blue}} \qquad (6.7)$$

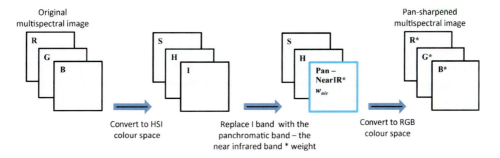

Figure 6.12 IHS pan-sharpening process.

where I_{out}^{λ} is the output intensity value for band λ, I_{in}^{λ} is the original intensity value for band λ, R is the red band, G is the green band, B is the blue band and w_{red}, w_{green} and w_{blue} are the weights assigned to the red, green and blue band respectively. λ could be red, green, blue or near IR. This method can increase contrast in the low- and high-reflectance areas in the image. As a result, it should not be used if the original scene radiometry needs to be preserved.

Gram-Schmidt transformation is based on an algorithm for vector orthogonalisation used in linear algebra and multivariate statistics (Farebrother 1974). Basically, this method is performed in the following steps:

1. computing a simulated low-resolution panchromatic band as a weighted mean of the multispectral bands;
2. applying the Gram-Schmidt orthogonalisation algorithm on the simulated panchromatic band (used as the first vector that is not transformed) and the multispectral bands (each treated as a high-dimensional vector) to make all bands orthogonal, which de-correlates all the bands;
3. replacing the simulated panchromatic band by the gain- and bias-adjusted high-resolution panchromatic band, and resampling all multispectral bands to the same resolution as the high-resolution panchromatic band;

4. applying the inverse Gram-Schmidt transform using the same transform coefficients, but on the high-resolution bands to form the pan-sharpened spectral bands.

The low-resolution spectral bands used to simulate the panchromatic band must fall within the range of the high-resolution panchromatic band. The weights used in the process depend on the spectral response of these bands. It uses the spectral response function of a given sensor to estimate what the panchromatic data look like, so that the sharpness is maximised and at the same time the colour or spectral distortion of the pan-sharpened output image is minimised. For more details about the Gram-Schmidt method, see Laben and Brower (2000).

Simple mean transformation simply calculates the mean between a multispectral band and the panchromatic band as the output for the multispectral band:

$$I_{out}^{\lambda} = \frac{1}{2} \times (I_{in}^{\lambda} + Pan) \qquad (6.8)$$

This method can only be applied to three bands: red, green and blue.

Box 6.6 describes how to pan-sharpen multispectral imagery.

Box 6.6 Pan-sharpening in ArcGIS   PRACTICAL

To follow this example, start ArcMap, and open **chapter6.mxd** from **C:\Databases\GIS4EnvSci\WooriYallock**.

1. Add **LE70920862000031EDC00_B8.TIF** from the folder **LE012000**. This is the panchromatic band of the ETM+ scene downloaded in Box 6.2.
2. Open the **Image Analysis** window if it is closed.
3. In the **Image Analysis** window, click the **Options** button . In the **Image Analysis Options** dialog:

    1) Click the **Pan Sharpen** tab.
    2) Click the **Method** drop-down list, and select **Gram-Schmidt**.
    3) Click the **Sensor** drop-down list, and select **Landsat 7 ETM+**. The weight values for each of the red, green, blue and infrared bands are automatically populated. ArcGIS calculates the weight values for the Gram-Schmidt transformation for a number of popular sensors, including Landsat MSS, TM, ETM+, OLI, SPOT 5, SPOT 6, QuickBird, IKONOS and WorldView-2.
    4) Click **OK**. The above steps defined the pan-sharpening method and weights.

4. In the **Image Analysis** window:

    1) Select **LE70920862000031EDC00_B1.TIF**, **LE70920862000031EDC00_B2.TIF**, **LE7092086 2000031EDC00_B3.TIF** and **LE70920862000031EDC00_B4.TIF**, and use the **Composite**

> **Bands tool** ⬚ to make a four-band composite as you did in Box 6.2. A temporary image layer is created, and named **Composite1_LE70920862000031EDC00_B1.TIF**. Display it as a natural colour image.
>
> 2) Select **Composite1_LE70920862000031EDC00_B1.TIF** in the **Image Analysis** window.
> 3) Hold down the **Ctrl** key, and select **LE70920862000031EDC00_B8.TIF**.
> 4) Click the **Pan Sharpen** button ⬚ . The pan-sharpened image is added to the display as a temporary image layer **Pansharp_Composite1_LE70920862000031EDC00_B1.TIF**. Display it as a natural colour image.
> 5) If you wish to save this image, select it in the **Layers** list box, and click the **Export** button 💾 in the **Processing** section.
>
> 5. Compare the original multispectral composite, the panchromatic band and the pan-sharpened image. Figure 6.11 shows a small portion of the scene in the three images.
> 6. Repeat Steps 3–4 to try the IHS, Brovey and simple mean transformations and set various weights. Save them as you wish.

Contrast stretching

Contrast stretching attempts to improve the contrast in an image by expanding the narrow range of brightness or DN values in the image over a wider range of values or over the entire brightness range of the display medium (such as computer screens). For example, Figure 6.13a is an 8-bit greyscale image with no stretch, which has DN values ranging from 18 to 191. It is rather dark and low in contrast. After contrast stretching by extending the DN values of the image to range from 0 to 255 (the full range of the brightness values for an 8-bit video cathode ray tube display), we obtain Figure 6.13b, which appears brighter and higher in contrast. In other words, in the contrast-stretched image, light-toned areas appear brighter and dark areas appear darker, making visual interpretation much easier. This helps to accentuate the contrast between features of interest and their backgrounds.

Contrast stretching involves adjusting the distribution and range of DN values. The distribution of the DN values that comprise an image is usually represented graphically using a histogram, as shown in Figures 6.13c and 6.13d. In an image histogram, the DN values are displayed along the x-axis of the graph, while the number of pixels in the image at each different DN value is shown on the y-axis. The distribution of DN values in a histogram of remote sensing imagery is often unimodal (with one clear peak). Both Figures 6.13c and 6.13d have a unimodal distribution. Multimodal distributions (with two or more clear peaks) result if a scene contains two or more dominant classes with distinctly different ranges of reflectance.

There are basically two types of methods for contrast stretching: linear and non-linear. Linear contrast stretching expands the range of DN values uniformly to fill the entire range of the output device. It transforms the original image using the formula:

$$DN' = \left(\frac{DN - DN_{min}}{DN_{max} - DN_{min}} \right) \times R \qquad (6.9)$$

where DN' is the stretched value assigned to a pixel in the output image, DN is the original value of the pixel in the input image, DN_{min} and DN_{max} are the minimum and maximum pixel values respectively in the input image and R is the range of DN values that can be displayed on the output device (for example, 255). In the example shown in Figure 6.13, R is 255, DN_{min} 18 and DN_{max} 191. Any pixel with a DN value of 18 in Figure 6.13a would have a DN value of 0 after applying Equation 6.9, while any pixel with a DN value of 191 in Figure 6.13a would now have a DN value of 255. All pixel values between 18 and 191 in the original image would be proportionately stretched between 0 and 255. This method is usually called a min-max contrast stretch.

DN_{min} and DN_{max} in Equation 6.9 could be defined as the minimum and maximum plus/minus one or more standard deviations of the DN values, then a linear stretch could be applied between the two values. Any value that is smaller than DN_{min} becomes 0, and any value that is greater than DN_{max} is changed to R (for example, 255). This approach is commonly referred to as a standard-deviation contrast stretch. Figure 6.13b is a result of this kind of contrast stretching with 2.5 standard deviations. Usually, the DN values falling outside one or two standard deviations may be discarded without serious loss of the most important data.

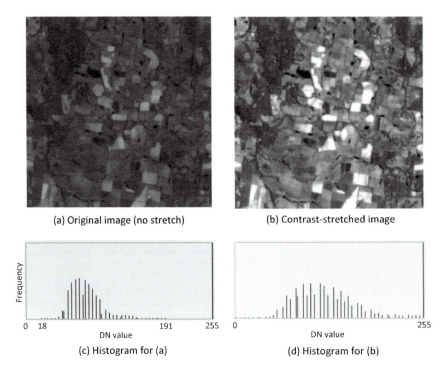

(a) Original image (no stretch)

(b) Contrast-stretched image

(c) Histogram for (a)

(d) Histogram for (b)

Figure 6.13 Images before and after contrast stretching and their histograms.

Another alternative is to define a certain percentage of pixels from the low and high end of the histogram to trim off the extreme values, and use the new lowest DN value as DN_{min} (called percent clip minimum) and the new highest DN value as DN_{max} (called percent clip maximum) to stretch the image linearly. This method is often termed a percent-clip contrast stretch. The clipping percentages can be different at each tail of the histogram. For example, if we set 1 per cent as the clipping percentage at the low end of the histogram in Figure 6.13c and 10 per cent at the high end, DN_{min} would become 42.5, while DN_{max} would be 107.5. The original DN values between 42.5 and 107.5 will spread evenly between 0 and 255, with all values 18–42.5 becoming 0 and 107.5–191 becoming 255. These linear methods may be effective in visualisation of a particular aspect of the information content of an image.

However, linear contrast stretching does not consider the shape of the image histogram. Original DN values between the defined lower and upper limits spread evenly within the full brightness range of the display device no matter whether they are rarely occurring values or frequently occurring values. Non-linear methods can overcome this limitation. The most popular non-linear contrast stretching method is histogram equalisation, which assigns DN values to the output greyscale levels according to their frequency of occurrence. It effectively spreads out the most frequent DN values, thus enhancing the contrast in the most populated range of DN values in the image. Jensen (1996) provides a good explanation of this method with examples. The method is particularly useful for images with backgrounds and foregrounds that are both bright or both dark. A disadvantage of the method is that it is indiscriminate. It may increase the contrast of background noise while decreasing the usable information content.

Box 6.7 shows how to enhance images using various contrast stretching methods in ArcGIS.

Spatial filtering

Spatial filtering is to use spatial filters to detect, sharpen or smooth specific features in an image based on their spatial frequency. The spatial frequency refers to the frequency of change in DN values per unit distance along a particular direction in the image. The areas in an image where changes occur in very close proximity are said to have high spatial frequencies, while those areas with changes which occur over large distances have low spatial frequencies. Therefore, a scene with small details and

To follow this example, start ArcMap, and open **chapter6.mxd** from **C:\Databases\GIS4EnvSci\ WooriYallock**.

1. In the table of contents, check **LE70920862000031EDC00_B4.TIF**, uncheck all other layers, and double-click **LE70920862000031EDC00_B4.TIF**.
2. In the **Layer Properties** dialog:

    1) Click the **Symbology** tab. Notice that the image is stretched using the percent-clip contrast stretch method by default, with the clipping percentage set at 0.5.
    2) Click the **Histograms** button. The histogram for this image is displayed with statistical information, which indicates that the DN values range from 0 to 188 with a standard deviation of 43.63. Click **OK** to close the histogram window.
    3) Click on the stretch type drop-down arrow, change the stretch type to **None**, and click **Apply**. Examine the image in the data view to see what it looks like without contrast stretching.
    4) Change the stretch type to **Standard Deviation**, and click **Apply**. Notice how the image contrast changes. Click the **Histograms** button. Two histograms are shown: the original histogram (displayed in grey) from the percent-clip contrast stretch method, and the stretched histogram displayed with a colour.
    5) Change the stretch type to **Histogram Equalize** then **Minimum-Maximum** to compare views of the image and the DN distributions in histograms as you make changes.
    6) Click **OK** to close the **Layer Properties** dialog.

3. Open the **Image Analysis** window if it is closed.
4. In the **Image Analysis** window:

    1) Select **LE70920862000031EDC00_B4.TIF** to modify its image contrast.
    2) Click the **Interactive Histogram Stretch** button 🔺 . A histogram is displayed.
    3) Change the stretch type to **Percent-Clip**, and click 🔺 . In the histogram, use the sliders (two grey dotted arrows) or type in the values to adjust the minimum and maximum values to set the range limits. Or type in a minimum and maximum clipping percentage value (**Min-Max Percent**) to adjust the image contrast. To return the stretch to the defaults, click the **Undo** button ↶ .
    4) Change the stretch type to **Std-dev** (the standard deviation method), and click 🔺 . In the histogram, use the sliders or type in the values to adjust the minimum and maximum values. Or enter a number of standard deviations for contrast stretching.
    5) Try other contrast stretching methods in the similar way to observe and compare the effects of the image enhancements.

sharp edges contains more high spatial frequency information than one composed of large coarse features. As the ability of the human eyes to discern spatial frequency is limited, it is often desirable to selectively remove certain spatial frequency ranges within an image to make it more interpretable with less noise. Spatial filtering involves examining the spatial variations in DN values of an image and using spatial filters to modify the image by selectively suppressing or separating certain spatial frequency ranges.

A spatial filter is an image operation where each pixel is changed by a function of the intensities or DN values of pixels in its neighbourhood. It essentially calculates the focal sum statistic for each pixel of the input image using a weighted kernel (see Section 4.4). A kernel is an array of coefficients or weights of a few pixels in dimension (for example, 3×3 or 5×5) (see Figure 4.21). It is used as a moving window (see Figure 4.22). A spatial filtering operation moves the kernel along in both the row and column dimensions pixel by pixel, multiplies the pixel values within the neighbourhood by the corresponding coefficients or weights in the kernel, sums all the resulting products and replaces the central pixel by the sum. The calculation is repeated until the entire image is filtered and a new image is generated. This process is also called convolution.

With different types of calculation, spatial filters can be designed to enhance or suppress different types of features. There are basically three types of spatial filters: low pass, high pass and edge detection filters. Low pass filters are used to highlight low spatial frequency features, reduce the local variation in an image and generally serve to smooth the image. Figure 6.14 shows three low pass filters and example outputs. The original image shown in Figure 6.14a is used for all filtering operations depicted in the subsequent figures. Filter A in Figure 6.14 uses a kernel whose coefficients are all set to 0.1111 (1/9). This filter calculates the mean value within each neighbourhood, commonly called a moving average filter. The outcome is that the high and low values within each neighbourhood will be averaged out. However, this low pass filter blurs the image, particularly at the edges of features. To reduce blurring, Filters B and C in Figure 6.14 could be used, whose kernels have unequal coefficients.

High pass filters enhance high spatial frequency features, increase smaller detail in an image and generally serve to sharpen the image. As an image is considered to consist of low- and high-frequency components as well as noise, a high pass filtered image can be derived by subtracting the low pass filtered image from twice the original image (Jensen 1996), for example:

$$I_{high} = 2 \times I - I_{low} \qquad (6.10)$$

where I_{high} is the high pass filtered image, I_{low} is the low pass filtered image, and I is the original image. High pass filters that sharpen edges can also be created using the kernels listed in Figure 6.15. These filters do not keep the low-frequency component of the image, and can be used to enhance boundaries between land and water, and urban structures such as roads and buildings.

Edge detection filters enhance and delineate linear features such as roads, linear geological structures and boundaries of area features. Different from high pass filters, edge detection filters preserve both low- and high-frequency components of an image. Several edge enhancement techniques have been developed. One straightforward method is directional first differencing. This method calculates the difference between each pixel and one of its immediate neighbouring pixels, and uses the difference as the output pixel value. It approximates the first derivative between two adjacent pixels with regard to a certain direction, which can be horizontal, vertical and diagonal. Suppose $DN_{i,j}$ is the brightness value of the pixel at the ith row and jth column in the

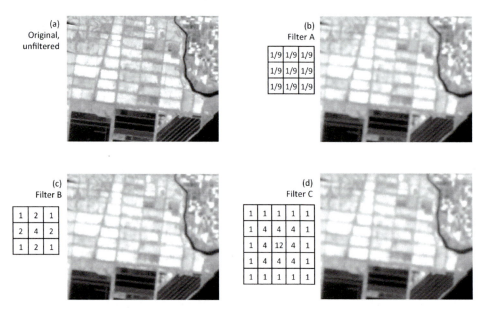

Figure 6.14 Low pass filters.

(a)
Filter D

0	-1/4	0
-1/4	2	-1/4
0	-1/4	0

(b)
Filter E

-1/4	-1/4	-1/4
-1/4	3	-1/4
-1/4	-1/4	-1/4

(c)
Filter F

-1	-1	-1
-1	9	-1
-1	-1	-1

(d)
Filter G

-1	-3	-4	-3	-1
-3	0	6	0	-3
-4	6	21	6	-4
-3	0	6	0	-3
-1	-3	-4	-3	-1

Figure 6.15 High pass filters.

original image, and $DN'_{i,j}$ is the filtered value of the pixel. The horizontal, vertical and diagonal first differences at the pixel would be respectively:

$$\text{Horizontal: } DN'_{i,j} = DN_{i,j} - DN_{i-1,j} + K \quad (6.11)$$

$$\text{Vertical: } DN'_{i,j} = DN_{i,j} - DN_{i,j+1} + K \quad (6.12)$$

$$\text{Northeast diagonal: } DN'_{i,j} = DN_{i,j} - DN_{i+1,j+1} + K \quad (6.13)$$

$$\text{Southeast diagonal: } DN'_{i,j} = DN_{i,j} - DN_{i-1,j+1} + K \quad (6.14)$$

where K is a constant added to make the differences positive. K is usually set as the median of DN values that can be displayed on the output device (for example, 127 for an 8-bit display). The resultant image enhances the edges normal to the direction of differencing, and deemphasises those parallel to the direction of differencing.

Edge enhancement can also be performed by convolving the image with a weighted kernel to produce different effects. Figure 6.16 lists some embossing filters, which emboss edges on a certain side. Embossing filters take a pixel on one side of the centre pixel, and subtract one of the other sides from it. To use the negative pixels as shadow and positive ones as light, an offset of 127

is added to the resultant image. Embossing is achieved by setting pixels black except where there is a change in DN values. The filtered image will represent the rate of change in DN values at each pixel of the original image in a specific direction. Filter H in Figure 6.16 embosses edges in the east direction, Filter I embosses edges in the north direction, and Filters J and K detect changes in the northeast and northwest directions respectively.

Further edge detection filters can be implemented to detect and delineate edges using the kernels in Figure 6.17. Filters L, M, N and O in Figure 6.17 are horizontal, left-diagonal, right-diagonal and vertical directional filters respectively. These filters should be applied before the image is smoothed in order to get better results.

Laplacian filters are non-directional edge detection filters, which emphasise all edges in an image. In other words, any feature with a sharp discontinuity (even noise, unfortunately) will be enhanced by a Laplacian filter. Laplacian filters are a second-derivative method of edge enhancement. One application of a Laplacian filter is to restore fine detail to an image which has been smoothed to remove noise. It generally emphasises points, lines and boundaries in the image, and suppresses uniform and low-frequency areas. Two examples of Laplacian filters are shown in Figure 6.18.

All the filters shown in Figures 6.14–6.18 are available in ArcGIS. Their implementation is demonstrated in Box 6.8.

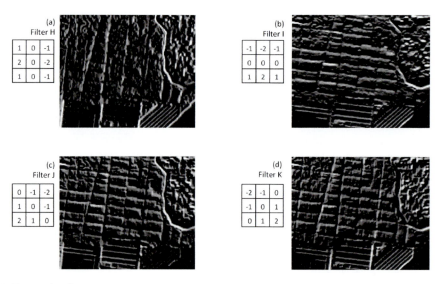

Figure 6.16 Embossing filters.

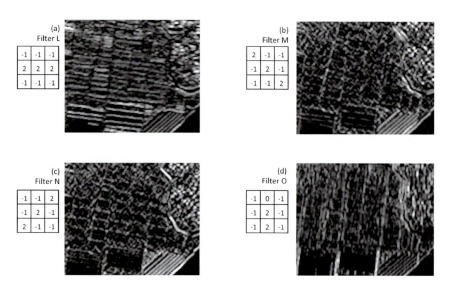

Figure 6.17 Line detection filters.

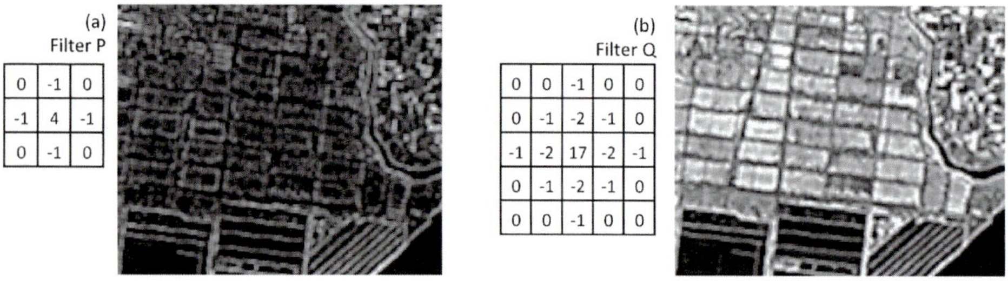

Figure 6.18 Laplacian filters.

Box 6.8   Spatial filtering in ArcGIS   **PRACTICAL**

To follow this example, start ArcMap, and open **chapter6.mxd** from **C:\Databases\GIS4EnvSci\ WooriYallock**.

1. In the table of contents, check **LE70920862000031EDC00_B4.TIF**, and uncheck all other layers.
2. Open the **Image Analysis** window if it is closed.
3. In the **Image Analysis** window:

    1) Select **LE70920862000031EDC00_B4.TIF**.
    2) Click the **Function Editor** button f_x . In the **Function Template Editor** dialog, right-click **LE70920862000031EDC00_B4.TIF**, point to **Insert**, and click **Convolution Function**.
    3) In the **Raster Function Properties** dialog, the input raster is set as **LE70920862000031EDC00_ B4.TIF**. Select the **Smooth Arithmetic Mean filter** (Filter A in Figure 6.14). Click **OK**.
    4) Click **OK**. A temporary image layer **Func_LE70920862000031EDC00_B4.TIF** is created, which is a smoothed image. Change its name to **meanFiltered**. Part of the southwest corner of the smoothed image is shown in Figure 6.14b.

4. Repeat Step 3, and apply the **Sharpen II** filter (Filter E in Figure 6.15) to **LE70920862000031EDC00_ B4.TIF**. Part of the sharpened image looks like Figure 6.15b.
5. As in Step 3, add (insert) **Convolution Function** to **LE70920862000031EDC00_B4.TIF**. In the **Raster Function Properties** dialog, change the filter type to **User Defined**, change the number of rows and columns of the kernel to **5**, and enter the coefficients of the kernel for Filter G shown in Figure 6.15. Apply the filter with this kernel to sharpen the input image. Part of the sharpened image is shown in Figure 6.15d. In this way, users can define their own kernels or filters for convolution.
6. Repeat Step 3 to apply the **Gradient North-East** filter (Filter J in Figure 6.16) to **LE70920862000031EDC00_B4.TIF**. Part of the embossed image looks like Figure 6.16c.
7. Similarly, apply the **Line Detection Horizontal** filter (Filter L in Figure 6.17) to **meanFiltered** (a smoothed image) generated in Step 3 above. Part of the filtered image is shown in Figure 6.17a.
8. Apply the **Laplacian 3x3** filter (Filter P in Figure 6.18) to **meanFiltered**. Figure 6.18a shows part of the filtered image.
9. In a similar way, try the other spatial filters available in ArcGIS and compare the results.

6.4 IMAGE TRANSFORMATION

As digital image data are raster data, all raster data processing and analysis functions available in GIS can be used to transform digital images. Two special image transformations are introduced here: band ratioing and principal components analysis.

Band ratioing

Band ratioing involves dividing pixel values in one spectral band by the corresponding values in another band on a pixel-by-pixel basis. It is a raster overlay by division operation in GIS. The result is a ratio image. Ratio images tend to carry the 'true' spectral characteristics of features

that are not affected by variations in scene illumination conditions caused by topographical slope, aspect, shadows or seasonal changes.

In addition, by ratioing the images from two different spectral bands, unique information about ground features not available in any single band can be derived. For example, healthy vegetation reflects strongly in the near IR spectral region while absorbing strongly in the red band. Other features, such as soil and water, have similar reflectance in both bands. Thus, a ratio image of the Landsat ETM+ red band and near IR band would produce ratios much smaller than 1.0 for vegetation, and ratios around 1.0 for soil and water. This ratio is sometimes used as a vegetation index – a measurement of vegetative amount and condition. Figure 6.19a shows a ratio image of the Woori Yallock ETM+ near IR band (Band 4) and red band (Band 3). The brighter the pixel is, the more photosynthetically active vegetation is present in this example. Other band ratios can be produced with different combinations of bands for the discrimination of different features. For example, a ratio of mid-IR band (for example, ETM+ Band 5) and green band (for example, ETM+ Band 2) may be used to determine moisture contents (Figure 6.19b), a ratio of red band (for example, ETM+ Band 3) and mid-IR band (for example, ETM+ Band 7) may reveal differences in water turbidity, and a ratio of red band and

blue band may help in the detection of ferric iron-rich rocks. In general, the weaker the correlation between the bands, the higher the information content of the ratio image.

However, a simple ratio may become meaningless when the denominator equals zero. Moreover, band ratio values generally vary considerably from one region to another or from one season to another, which makes comparisons across regions or over time rather difficult. Therefore, more complex ratios involving sums of and differences between spectral bands have been developed in order to overcome these difficulties. NDVI, as

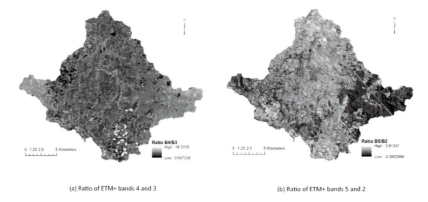

(a) Ratio of ETM+ bands 4 and 3   (b) Ratio of ETM+ bands 5 and 2

Figure 6.19 Ratio images of Woori Yallock: (a) ratio of ETM+ bands 4 and 3, and (b) ratio of ETM+ bands 5 and 2.

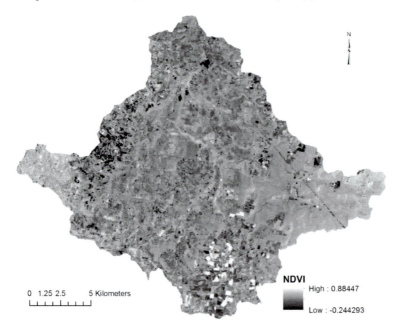

Figure 6.20 NDVI map of Woori Yallock.

expressed in Equation 6.1, is an example. It has been widely used for studying vegetation dynamics, assessing biomass, estimating crop yields, monitoring drought and predicting hazardous fire zones. Figure 6.20 is an NDVI map of Woori Yallock derived from the ETM+ imagery.

Box 6.9 shows how to use the band arithmetic functions in ArcGIS to perform band ratioing.

Box 6.9   Band ratioing in ArcGIS   **PRACTICAL**

To follow this example, start ArcMap, and load the radiometrically corrected Landsat ETM+ image **wy00abd.tif** created in Box 6.4.

Create simple band ratios

1. Open the **Image Analysis** window if it is closed.
2. In the **Image Analysis** window:

   1) Select **wy00abd.tif**.
   2) Click the **Function Editor** button *ƒ* . In the **Function Template Editor** dialog, right-click **wy00abd.tif**, point to **Insert**, and click **Band Arithmetic Function**.
   3) In the **Raster Function Properties** dialog, the input raster is set as **wy00abd.tif**. Select **User Defined** as the method, enter the expression **B4/B3**, then click **OK**.
   4) Click **OK**. A temporary image layer **Func_wy00abd.tif** is created and displayed, as in Figure 6.19a.

3. Repeat Step 2 to try other combinations of bands to perform band ratioing. Interpret the resultant ratio images to extract useful information.

Calculate NDVI

4. In the **Image Analysis** window:

   1) Select and highlight **wy00abd.tif**.
   2) Click the **Function Editor** button *ƒ* . In the **Function Template Editor** dialog, right-click **wy00abd.tif**, point to **Insert**, and click **Band Arithmetic Function**.
   3) In the **Raster Function Properties** dialog, select **NDVI** as the method, and enter the band indices **4 3** (4 refers to Band 4, the near IR band, 3 to Band 3, the red band in the composite image, and 4 and 3 are separated by a space). Click **OK**.
   4) Click **OK**. A temporary image layer is created and displayed, as in Figure 6.20. The NDVI values in this image range from -1 to 1. Generally, clouds, water and snow may have negative values, and the NDVI values of sand, rock and bare soil are close to 0, and generally below 0.1. Moderate values (0.2–0.3) may be generated from shrub and grassland, while high values (0.6–0.8) could represent forests and croplands.

Principal components analysis

Multispectral image data consist of images of various wavelength bands that may appear similar and contain the same information due to similarities of the spectral response of the observed features in those bands. Such multispectral images are said to be highly correlated. Inter-band correlation leads to redundancies in the data.

Principal components analysis (PCA) can be used to reduce such redundancy. It is a mathematical technique developed to identify similarities and differences within a dataset and transform a correlated dataset to a new dataset without correlations. Applied to an n-band multispectral image dataset, this technique transforms the original correlated image dataset to fewer than n (often two or three) bands of new uncorrelated images, called principal component images. These principal component images can then be used for image interpretation and classification in lieu of the original data without much loss of information. Thus, PCA may help produce images that are more interpretable than original ones and increase the computational efficiency of subsequent image analysis, as it reduces the number of bands (or dimensionality) of the image dataset that need to be analysed.

Considering a two-band image dataset, a two-dimensional distribution of DN values from the two bands can be plotted in a scattergram, as shown in Figure 6.21. If all points fall on the line PC_1, the two bands are perfectly correlated. In many cases, the points are scattered around the line with PC_1 as the dominant direction of scatter or variability. PC_1 represents the direction where there is the most variance and where the data are most spread out. It is the direction of the first principal component in the data. PC_2 is the line perpendicular to PC_1 and passing through the mean of the data distribution. This line represents the direction of the second principal component in the data. If the variation in the

direction of PC_2 contains a small fraction of the total variability of the data, it may be disregarded without much loss of information. In order to find the principal component images, PCA translates and/or rotates the Band A–Band B axes onto the PC_1–PC_2 axes, and calculates the new DN value for the pixel at row i, column j in terms of the axes PC_1 and PC_2 as below:

$$DN_{pc1}(i,j) = a_{11}DN_A(i,j) + a_{12}DN_B(i,j) \qquad (6.15)$$

$$DN_{pc2}(i,j) = a_{21}DN_A(i,j) + a_{22}DN_B(i,j) \qquad (6.16)$$

where $DN_A(i,j)$ and $DN_B(i,j)$ are the original values at the pixel (i,j) in the Band A and Band B images, $DN_{pc1}(i,j)$ is the transformed pixel value in the first principal component image, $DN_{pc2}(i,j)$ is the transformed pixel value in the second principal component image, and (a_{11}, a_{12}) and (a_{21}, a_{22}) are coefficients for the transformation known as eigenvectors or principal components. Eigenvectors are computed from the covariance matrix of the pixel values from the two original bands.

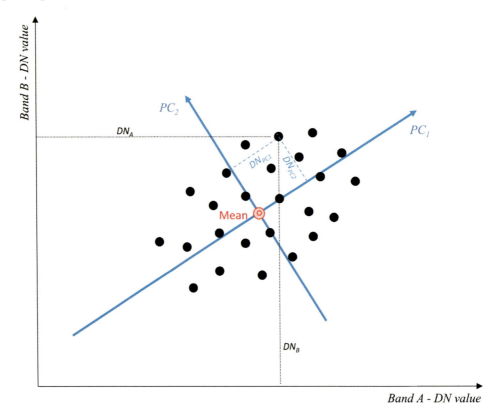

Figure 6.21 First two principal components of a two-band image.

The covariance between Band A and Band B is calculated as:

$$cov(A,B) = \frac{\sum_{i=1}^{p}\sum_{j=1}^{q}\left(DN_A(i,j)-\overline{DN}_A\right)\left(DN_B(i,j)-\overline{DN}_B\right)}{p \times q - 1} \quad (6.17)$$

where p and q are respectively the number of rows and columns of the image, $\overline{DN}_A$ is the mean DN value of Band A, and $\overline{DN}_B$ is the mean DN value of Band B. It is a measure of how much the pixel values of the two bands vary from the mean with respect to each other. If there are more than two bands in the image dataset, there will be more than one covariance measurement to be calculated. In fact, for an n-band image, there will be $n!/\left[(n-2)!\times 2\right]$ different covariance values. The covariance values between the n bands are put in an $n \times n$ matrix in a format as below:

$$C = \begin{pmatrix} cov(B1,B1) & cov(B1,B2) & \dots & cov(B1,Bn) \\ cov(B2,B1) & cov(B2,B2) & \dots & cov(B2,Bn) \\ & \dots & & \\ cov(Bn,B1) & cov(Bn,B2) & \dots & cov(Bn,Bn) \end{pmatrix}$$

$$(6.18)$$

where $B1, B2, \dots, Bn$ are referred to Band 1, Band 2, ... Band n, $cov(a,b)=cov(b,a)$ and $cov(a,a)$ is actually the variance of a. This matrix is called the covariance matrix. It is symmetrical about the main diagonal. The matrix has n eigenvectors, which are perpendicular. Each eigenvector has an associated eigenvalue. The eigenvectors are found by using some linear algebra decomposition techniques, which is beyond the scope of this book.

Further information about eigenvectors and eigenvalues in general and how to derive them can be found in any standard linear algebra textbook. As an example, Table 6.4 lists the covariance matrix derived from wy00abd. tif, the radiometrically corrected six-band ETM+ albedo image of Woori Yallock. Table 6.5 lists the eigenvectors, associated eigenvalues and percentages of the scene variance calculated from the covariance matrix.

Once eigenvectors are found, principal component images can be produced. In general terms, the pixel values for the kth principal component image are found from:

$$DN_{pck}(i,j) = a_{k1}DN_1(i,j) + a_{k2}DN_2(i,j) + \dots + a_{kn}DN_n(i,j) \quad (6.19)$$

where $DN_{pck}(i,j)$ is the DN value of the pixel (i,j) in the kth principal component image, n is the number of bands, $DN_i(i,j)$ is the value of the pixel (i,j) in the ith band image ($i = 1, 2, \dots, n$) and ($a_{k1}, a_{k2}, \dots, a_{kn}$) is the kth eigenvector derived from the covariance matrix of the n-band image pixel values. Figure 6.22 shows six principal component images derived from wy00abd.tif based on the eigenvectors listed in Table 6.5.

The principal component images are ordered by the eigenvalues of their associated eigenvectors from the highest to lowest. This gives the principal component images in order of significance. The first principal component image has the largest eigenvalue, 0.0035 in the case of Woori Yallock, as listed in Table 6.5. It contains the largest portion of the total scene variance, 78.61 per cent in the case of Woori Yallock. The second principal component image contains far less variance, 18.93 per cent in the case of Woori Yallock. The third and succeeding component images have a further decreasing portion

Table 6.4 Covariance matrix of the six-band ETM+ image of Woori Yallock

Band (µm)	1 (0.45–0.52)	2 (0.52–0.60)	3 (0.63–0.69)	4 (0.76–0.90)	5 (1.55–1.75)	6 (2.08–2.35)
1	6.121262e-005	8.644852e-005	1.013835e-004	1.311654e-004	2.309181e-004	1.526053e-004
2	8.644852e-005	1.387132e-004	1.568977e-004	2.961515e-004	4.158296e-004	2.544669e-004
3	1.013835e-004	1.568977e-004	1.976969e-004	2.300761e-004	4.759172e-004	3.137895e-004
4	1.311654e-004	2.961515e-004	2.300761e-004	1.824921e-003	1.152683e-003	4.940394e-004
5	2.309181e-004	4.158296e-004	4.759172e-004	1.152683e-003	1.630471e-003	9.269539e-004
6	1.526053e-004	2.544669e-004	3.137895e-004	4.940394e-004	9.269539e-004	5.958071e-004

Table 6.5 Eigenvectors and eigenvalues calculated from the covariance matrix in Table 6.4

	Eigenvector (component)					
	1	**2**	**3**	**4**	**5**	**6**
Band 1	0.09267	−0.11180	0.45979	0.19860	0.64365	0.56016
2	0.17259	−0.12320	0.48770	0.23683	0.21478	−0.78420
3	0.18154	−0.25690	0.54140	0.13221	−0.72352	0.25878
4	0.61610	0.76734	0.11736	−0.12141	−0.03755	0.04108
5	0.65167	−0.38031	−0.48991	0.43446	0.02437	0.03637
6	0.35272	−0.41578	0.06205	−0.82674	0.11875	−0.03561
Eigenvalue	0.00350	0.00084	0.00008	0.00002	0.00001	0.00000

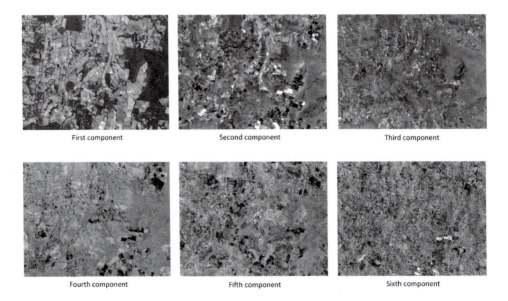

First component   Second component   Third component

Fourth component   Fifth component   Sixth component

Figure 6.22 Principal component images of Woori Yallock (the central part of the catchment).

of the scene variance. Principal component images with high eigenvalues and large percentages of the scene variance provide significant information about the observed features, while those with small eigenvalues and low variances represent noise, and provide little useful information. If the first k ($< n$) principal component images account for most of the variance in the multispectral image dataset, the original n bands of the image data can be set aside and the subsequent image analyses can be performed using just the k principal component images.

In the case of Woori Yallock, the first three principal component images contain 99.35 per cent of the variance in the original image data, which means they explain nearly all of the variance in the data. Components 4, 5 and 6 can be ignored as they account for a very low percentage (0.65 per cent) of the scene variance and are insignificant. The three component images largely represent system noise.

Box 6.10 illustrates the principal components analysis function in ArcGIS.

Box 6.10   Principal components analysis in ArcGIS   **PRACTICAL**

To follow this example, start ArcMap, and load the radiometrically corrected six-band Landsat ETM+ image **wy00abd.tif** created in Box 6.4. Note that the DN values in this multiband image are albedo values in the range 0–1.0.

1. Open **ArcToolBox**. In the **ArcToolBox** window, navigate to **Spatial Analyst Tools > Multivariate**, and double-click **Principal Components**.
2. In the **Principal Components** dialog:
   1) Select **wy00abd.tif** as the input raster bands. The number of principal components is automatically set as **6**.
   2) Enter **C:\Databases\GIS4EnvSci\WooriYallock\wypc** as the output multiband raster.
   3) Enter **C:\Databases\GIS4EnvSci\WooriYallock\wypcp.txt** as the output data file for storing principal component parameters.
   4) Click **OK**. **wypc** is created and displayed as a colour composite made of the first three principal component images.
3. Click the **Add Data** button ✚ , navigate to **C:\Databases\GIS4EnvSci\WooriYallock** and add **wypcc1**, **wypcc2**, . . ., **wypcc6**. These are individual principal component images numbered in order of significance – that is, **wypcc1** is the first principal component image, **wypcc2** is the second principal component image, and so on. Figure 6.22 shows the central part of these images.
4. Use **Notepad** to open **wypcp.txt**. View the covariance matrix for the image data, and eigenvectors, eigenvalues and percentage variances for each principal component images. The key information contained in the text file is presented in Tables 6.4 and 6.5.

6.5 IMAGE CLASSIFICATION

Image classification involves automatically extracting different types of ground features from an image. It is a process of categorising all pixels in an image, normally a multispectral or hyperspectral image, into land cover classes based on the spectral information represented by the DN values in one or more spectral bands. The resulting classified image is a thematic map describing the spatial distribution of various land cover classes (such as water, vegetation and soil).

There are a number of different methods for image classification, including supervised and unsupervised classification, hybrid classification, fuzzy classification and object-oriented classification (Lillesand et al. 2008). This section focuses on multispectral image classification using two of the most common classification approaches: supervised (human-guided) and unsupervised (software-driven) classification.

Supervised classification

Basically, supervised classification involves using pixels of known classes to identify pixels of unknown classes. It requires an image analyst first to select or develop a land cover classification scheme, then to choose sample pixels in the image that are representative of specific classes based on fieldwork, interpretation of aerial photographs and large-scale maps, personal knowledge or a combination of these methods. These samples of known classes are called training sets, also known as testing sets or input classes. The selected training sets are then used by the GIS or image processing software to generate class signatures that describe the spectral characteristics of each class in all spectral bands. Each pixel in the image is then evaluated against each class signature, and assigned to the class it resembles most.

Selection of training sets and generation of class signatures are two very important components of the image classification process. A training set for a particular class is usually drawn from pixels in multiple areas determined by the analyst to represent that class, and should capture both the mean and variability of the class as a whole. The better the training set represents the spectral variation of the class, the more accurate the classification results. For an n-band image, a training set for a given class should contain at least $10 \times n$ pixels so that the spectral response pattern of the class can be reliably characterised. The number of training sets depends on the nature of the classes and the complexity of the study area. It is not unusual to have 100 or more training sets in order to adequately represent the spectral variability in the image. After the training sets are selected and delineated, the pixel values in the training sets are extracted and form the sample for the construction of class signatures.

A class signature is a statistical description of the fundamental spectral characteristics of a class, containing the spectral means, standard deviations, variances, minimum values, maximum values, variances, covariance matrix, correlation matrix and other multivariate statistical measures calculated from its training set. In other words, a class signature is a statistical representation of a particular class. It is used by various decision rules to assign classes to each pixel in the image.

Suppose we have an n-band image. Each pixel has a DN value on each band. These DN values form an n-dimensional vector $X = \left[DN_1(i,j), DN_2(i,j), \ldots, DN_n(i,j) \right]$, where i, j are the row and column numbers of the pixel, and $DN_k(i,j)$ is the DN value of the pixel in band k. This vector is called a measurement vector. This representation allows us to consider each pixel as occupying a point, and each training set as occupying a sub-space (that is, a representative point surrounded by some spread or deviation) within the n-dimensional measurement space. Viewed as such, the classification problem is to determine to which sub-space class each measurement vector belongs.

To simplify the explanation, let us look at a two-band image composed of the red band and near IR band and five training sets representing forest, cropland, grassland, urban and water. Each pixel from the five training sets has an associated 2D measurement vector, which may be plotted on a scattergram, as shown in Figure 6.23. This diagram shows a 2D measurement space with the x-axis representing the DN values in the red band and y-axis representing the DN values in the near IR band. Five sub-spaces or clusters of pixels (enclosed in ovals) can be identified, each corresponding to a particular type of land cover. Each sub-space shows the general spectral response pattern and spectral variability within the class it represents. Class signatures use statistical measures to describe the spectral response pattern and variability of each class exhibited in their own sub-spaces. When a pixel of an unknown class is tested, we must decide how to quantitatively determine which of the sub-spaces (classes) it belongs to based on the class signatures, given the measurement vector of the pixel. There are a number of techniques that can be used to make a decision, but parallelepiped classifier and maximum likelihood classifier are two commonly used methods.

Parallelepiped classifier

This method uses the range of DN values in each training set to define sub-spaces, often called decision regions, for

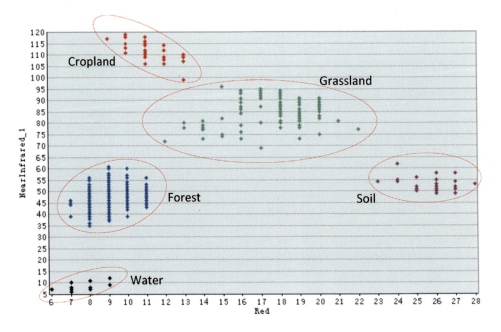

Figure 6.23 Two-dimensional measurement space.

each class. A decision region is usually bounded by the maximum and minimum DN values of each class in each band calculated from the training set. If an unknown pixel lies in a decision region of a particular class, it will be assigned to that class. If it is placed outside all decision regions, it remains unknown. In a 2D measurement space, a decision region is a rectangle, as shown in Figure 6.24. The pixel P is classified to grassland as it is located in its decision region. In general, an unknown pixel at (i, j) is classified to class C if and only if:

$$L_{ck} \leq DN_k(i,j) \leq U_{ck} \; for \, all \; c = 1, 2, \ldots, m;$$
$$k = 1, 2, \ldots, n \quad (6.20)$$

where m is the number of classes, n is the number of bands, L_{ck} is the minimum DN value of the class C training set in Band k and U_{ck} is the maximum DN value of the class C training set in Band k. In some applications, L_{ck} is defined as the mean DN value minus one standard deviation of the class C training set in Band k, and U_{ck} is defined as the mean DN value plus one standard deviation of the class C training set in Band k. A decision region in a multidimensional measurement space is often called a parallelepiped.

The parallelepiped classifier is computationally efficient. However, some decision regions of different classes may overlap. In these cases, unknown pixels falling in the overlap regions are classified as 'not sure' or assigned to one of the overlapping classes. In addition, this method may not be very accurate, as the parallelepipeds are formed based on DN ranges that may not be representative of a class.

Maximum likelihood classifier

This method applies a probability model to determine the decision regions. Each pixel is evaluated and assigned to the class of which it has the highest probability of being a member. It assumes that the DN values of the training set for each class in each band are normally distributed. Under this assumption, a class can be characterised by the mean vector $M_c = \left(\overline{DN}_{c1}, \overline{DN}_{c2}, \ldots, \overline{DN}_{cn} \right)$, where $\overline{DN}_{ck}$ is the mean DN value of the training set obtained for class c in Band k, and the covariance matrix C_c also derived from the training set. Given M_c and C_c, the statistical probability of a pixel to be a member of class c is computed as:

$$p_c = 2 \times \pi^{-0.5 \times n} |C_c|^{-0.5} \exp\left(\frac{-0.5 \times (X - M_c)^T \times}{C_c^{-1} \times (X - M_c)} \right) (6.21)$$

where X is the measurement vector of the pixel, n is the number of bands, $|C_c|$ is the determinant of the covariance matrix C_c and T is the matrix operation of transpose. For more information about the determinant and transpose of a matrix, please refer to a linear algebra textbook. For a given unknown pixel, the probability for each

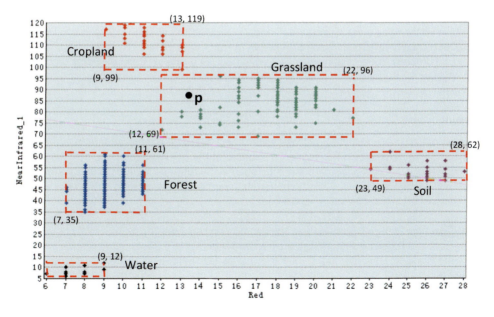

Figure 6.24 Decision regions used by the parallelepiped classifier.

class is calculated using Equation 6.21. The pixel gets k probability values for k classes. It is assigned to the class that has the highest probability (that is, the maximum likelihood). If the highest probability is smaller than a threshold probability value the analyst specifies, the pixel remains unclassified.

In order to improve computing efficiency, Equation 6.21 can be modified as follows:

$$-\ln(p_c) = \ln(|C_c|) + (X - M_c)^T \times C_c^{-1} \times (X - M_c) \tag{6.22}$$

The above equation does not affect the rank order of the values of P_c, $c = 1, 2, \ldots, k$. Both Equation 6.21 and Equation 6.22 assume that each of the k classes is equally likely to occur in the area covered by the image. However, in many applications, some classes may occur more often than others. The anticipated likelihood of occurrence or a priori probability of each class can be calculated based on the proportion of the area covered by each class estimated from scientific reports, land cover maps or aerial photographs. These prior probabilities can be incorporated into the maximum likelihood classifier. One way to do this is to subtract twice the logarithm of the prior probability of class c from the log likelihood of the class derived from Equation 6.22 (Strahler 1980). This extension is often termed the Bayesian classifier. With this method, a class with a higher prior probability is given a higher weight so that an unknown pixel is more likely assigned to this class than to other classes with lower prior probabilities.

The maximum likelihood classifier makes use of the statistics from the class signatures. Sufficient training data need to be sampled to allow estimation of the mean vector and the covariance matrix. This method cannot be applied if the DN values of the training sets do not follow the normal distribution. Box 6.11 shows how to perform a supervised classification using this method in ArcGIS.

Box 6.11   Supervised classification in ArcGIS   PRACTICAL

To follow this example, start ArcMap, and load the radiometrically corrected Landsat ETM+ image **wy00ar.tif** created in Box 6.4. This example shows how to classify the six-band ETM+ image of the Woori Yallock catchment into six types of land cover – forest, cropland, grassland, soil, water and residential – using the maximum likelihood classifier available in ArcGIS.

1. Display **wy00ar.tif** as a colour infrared image.
2. Click **Customize** in the main menu of ArcMap.
3. Point to **Toolbars**, then click **Image Classification**. The **Image Classification** toolbar appears with **wy00ar.tif** as the source image. All six bands associated with the image are to be used in the classification.

Select training sets

Training sets are normally selected and delineated manually on a computer screen using a drawing tool. The number of training sets depends on the nature of the classes and the complexity of the study area. In many applications, 100 or more training sets may be required in order to adequately represent the spectral variability in an image. For demonstration purposes, the training sets in this example are to be selected at the sites labelled in Figure 6.25.

4. Click ▦ on the **Image Classification** toolbar to open the **Training Sample Manager**.
5. Zoom in to the **Forest 1** site (indicated by an arrow in Figure 6.25).
6. Click the drawing polygons tool ⌵ on the **Image Classification** toolbar. Notice that there are three drawing tools available for drawing polygons, circles and rectangles.
7. Draw a polygon around the site, as shown in Figure 6.26. Note that the polygon should be placed carefully to avoid pixels located along the edge between land cover types. Once you have finished

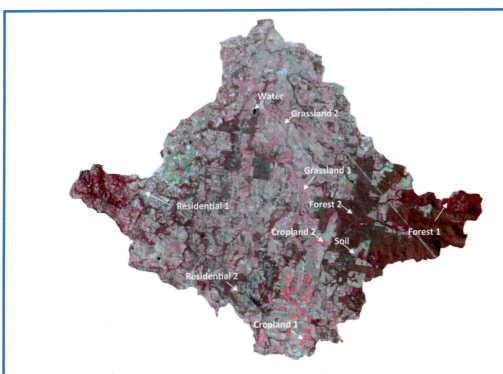

Figure 6.25 Woori Yallock Catchment and training sites.

drawing, a training set is created and shown in **Training Sample Manager** as a new class with a default name, value and colour. Change its name to **Forest 1** and its colour to **Green**.

8. Repeat Steps 5–7 above to create training sets at all other sites labelled in Figure 6.25. In **Training Sample Manager**, name them respectively **Forest 2**, **Grassland 1**, **Grassland 2**, **Cropland 1**, **Cropland 2**, **Residential 1**, **Residential 2**, **Water** and **Soil**, and change the colour of each class according to your preference.

9. In **Training Sample Manager**, select **Forest 1**. Hold down the **Ctrl** key, and select **Forest 2**. Click the **Merge training samples** ⊞⋅ button. The two training sets are merged into one. Change the name of the merged set to **Forest**.

10. Repeat the above step to merge **Grassland 1** and **Grassland 2** to form a new set **Grassland**; merge **Cropland 1** and **Cropland 2** to **Cropland**, and **Residential 1** and **Residential 2** to **Residential**. Six training sets for the six land cover classes are listed in **Training Sample Manager**.

Evaluate the training sets

11. In **Training Sample Manager**:

1) Select the training set for **Forest**.
2) Click the **Histograms** button 📊 . The image histograms for the **Forest** class on all the six bands available are displayed in the **Histograms** window. Histograms provide visual checks on the normality of the DN distributions.

(continued)

(continued)

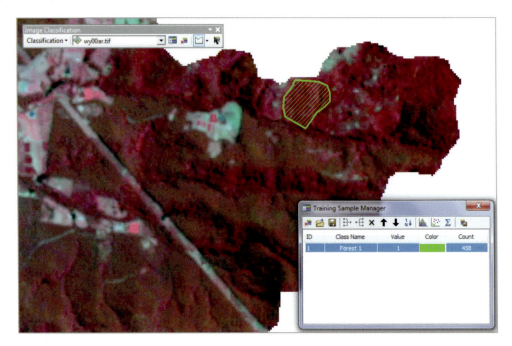

Figure 6.26 Selecting a training set.

   3)   Examine the histograms. If there is any histogram that is not normally distributed or has a multi-modal distribution, the training set needs to be removed by clicking the **Delete** button ✖ .

12. Repeat Step 11 to examine the histograms for all other training sets.
13. In **Training Sample Manager**:
   1)   Select the training sets for **Forest** and **Grassland**.
   2)   Click the **Histograms** button to show the histograms of the two training sets on each band.
   3)   Examine the histograms of the two training sets, and check whether the spectral response patterns of the two training sets overlap. If they do overlap, the two training sets need to be reviewed. One may need to be removed, or the two can be merged.

14. Repeat Step 13 to examine the histograms of other pairs of training sets.
15. Follow the same procedure in Steps 11–14, but instead click on the **Scatterplots** button 〰 and the **Statistics** button Σ to examine the scattergrams and statistics for different training sets. Their spectral response patterns should not overlap with each other on all the band combinations.
16. If some old training sets are removed and new ones are created, repeat Steps 11–15 to evaluate the new training sets. This process is repeated until all training sets are normally distributed and their spectral response patterns do not overlap.
17. Click the **Save training sample** button 💾 in **Training Sample Manager**, and save the training sets into a shapefile named **wy00ar_tif_training.shp**.

Create the class signatures

18. In **Training Sample Manager**:

    1) Click the **Create a signature file** button .

    2) Save the class signatures to a file in **C:\Databases\GIS4EnvSci\WooriYallock**, and name it **wy00ar.gsg**.

Perform the maximum likelihood classification

19. On the **Image Classification** toolbar, click **Classification > Maximum Likelihood Classification**. In the **Maximum Likelihood Classification** window:

    1) Select **wy00ar.tif** as the input raster bands.
    2) Enter **C:\Databases\GIS4EnvSci\WooriYallock\wy00ar.gsg** as the input signature file.
    3) Enter **C:\Databases\GIS4EnvSci\WooriYallock\wylc** as the output classified raster.
    4) Use the default values for other input parameters.
    5) Click **OK**. The land cover map **wylc** is created and displayed in the data view.

20. Save the current map document as **chapter6_mlc.mxd**.

Unsupervised classification

In contrast to supervised classification, unsupervised classification does not require the image analyst to specify land cover classes and provide training sets prior to classification. Rather, it automatically groups spectral clusters of pixels based on the DN values in the image using a particular clustering algorithm. The image analyst then matches them to land cover classes.

There are many clustering algorithms available for unsupervised classification. The techniques called k-means clustering (Hartigan and Wong 1979) and ISODATA (Iterative Self-Organising Data Analysis Techniques; Tou and Gonzalez 1974) are two of the widely used methods. Both involve iterative procedures and require the image analyst to specify the number of classes to be expected in the image. The k-means algorithm seeks to find the centres of the clusters in the multidimensional measurement space of the input bands that minimise the distance from the pixels to the cluster they belong to. It involves the following steps:

1. Randomly locate the user-specified number of cluster centres in the multidimensional measurement space, and calculate their mean vectors.
2. Compare the measurement vector of each pixel to the mean vector of each cluster, and assign each pixel to the cluster whose mean vector is closest.

3. Calculate new cluster mean vectors for each of the clusters with all the pixels assigned to the cluster.
4. Repeat Steps 2 and 3 until there is no significant change in the location of cluster mean vectors between successive iterations. The 'change' can be measured by the distance from the cluster mean vector to its member pixel's measurement vector that has changed from one iteration to the next, or by the percentage of pixels within the cluster that has changed between iterations.

ISODATA is a variant of the k-means method. It refines the k-means method by splitting and merging clusters. In each iteration, cluster mean vectors are recalculated and pixels are reclassified with respect to the new mean vectors. All pixels are classified to the nearest cluster unless a standard deviation or distance threshold is specified by the image analyst, in which case some pixels may be unclassified if they do not meet the selected criteria. Clusters are merged if either the number of member pixels in a cluster is smaller than an analyst-specified threshold or the centres of the clusters are closer than an analyst-specified distance threshold. Clusters are split into two different clusters if the cluster standard deviation exceeds an analyst-specified value and the number of member pixels is twice the threshold for the minimum number of members. Therefore, iterative cluster splitting, merging and deleting are carried out based on analyst-specified input threshold parameters. This process continues until the number of pixels in each class changes by less than the selected pixel change threshold or the maximum number of iterations is reached.

ISODATA allows for different numbers of clusters, while *k*-means assumes that the number of clusters is known a priori. Box 6.12 shows how to perform an unsupervised classification with the ISODATA method in ArcGIS.

Box 6.12   Unsupervised classification in ArcGIS   **PRACTICAL**

To follow this example, start ArcMap, and open **chapter6_mlc.mxd** saved in Box 6.11.

Classify the image

1. On the **Image Classification** toolbar, click **Classification > Iso Cluster Unsupervised Classification**. In the **Iso Cluster Unsupervised Classification** dialog:

    1) Select **wy00ar.tif** as the input raster bands.
    2) Enter **6** as the number of classes. (Note that the number of spectral classes depends on the complexity of the study area. As each land cover class may contain a variety of spectral response patterns, the number of spectral classes set for unsupervised classification usually exceeds the expected number of land use classes.)
    3) Enter **C:\Databases\GIS4EnvSci\WooriYallock\wylc_iso** as the output classified raster.
    4) Accept default values for other parameters.
    5) Click **OK**. **wylc_iso** is created and displayed in the data view, which is similar to Figure 6.27.

2. Compare **wylc_iso** with **wylc**, and identify the land use types for each of the six clusters or spectral bands in **wylc_iso**. This can be facilitated by using the **Swipe** tool as follows:

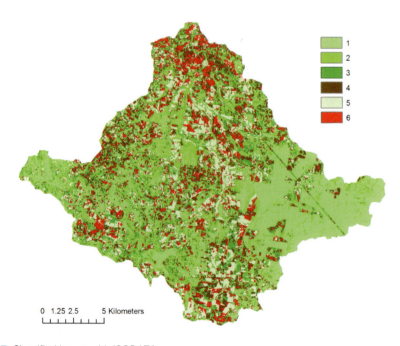

Figure 6.27 Classified image with ISODATA.

1) In the table of contents, check both **wylc_iso** and **wylc**, and uncheck other layers. Place **wylc** under **wylc_iso**.
2) Click **Customize** > **Toolbars** > **Effects** on the main menu of ArcMap.
3) On the **Effects** toolbar, select **wylc_iso** in the layer list.
4) On the **Effects** toolbar, click the **Swipe** tool to interactively peel the selected layer **wylc_iso** back and see **wylc** underneath it.

The six classes should be: 1 – forest, 2 – forest (with species different from class 1), 3 – grassland, 4 – soil/residential, 5 – cropland/grassland and 6 – residential/soil. Water bodies in the catchment are generally small. They are grouped into Class 1. The results indicate that several land cover classes are spectrally similar and cannot be differentiated in the multispectral image dataset with the unsupervised classification algorithm. This is one of the possible undesirable outcomes from unsupervised classification.

3. Repeat Steps 1 and 2 to run the **Iso Cluster Unsupervised Classification** tool multiple times with different numbers of spectral classes, and compare the results with **wylc** and **wy00ar.tif** to see whether the number of spectral classes needs to be set higher or lower.

Smooth the classified image

There are many isolated misclassified pixels or small patches of pixels in the classified image. These isolated pixels can be removed through a smoothing operation. A majority filter, which replaces pixel values with the majority value in their contiguous neighbourhoods, can be used to achieve this purpose.

4. Open **ArcToolBox**. In the **ArcToolBox** window, navigate to **Spatial Analyst Tools > Generalization**, and double-click **Majority Filter**.
5. In the **Majority Filter** dialog:

1) Select **wylc_iso** as the input raster.
2) Enter **C:\Databases\GIS4EnvSci\WooriYallock\wylc_iso_s** as the output raster.
3) Change the number of neighbourhoods to **EIGHT**.
4) Set **HALF** as the replacement threshold (that is, half of the pixels within the neighbourhood must have the same value and be contiguous).
5) Click **OK**. **wylc_iso_s** is created and displayed in the data view. Figure 6.28 shows the central part of the smoothed image compared with the classified image before smoothing.

 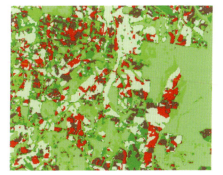

Before smoothing   After smoothing

Figure 6.28 Classified images before and after smoothing: (a) before smoothing and (b) after smoothing.

The clusters resulting from unsupervised classification are spectral classes identified based on the natural groupings of the DN values in an image. These classes may or may not correspond well to land cover classes of interest. The image analyst needs to determine and assign land cover types to each spectral class by using personal knowledge and some form of reference data, such as large-scale maps and aerial photos. Unsupervised classification often produces too many land cover classes, which may need to be combined to create a meaningful land cover map. In other cases, the classification may result in spectral classes that may be meaningless as they represent a mix of different cover types, as shown in the results produced in Box 6.12.

Unsupervised classification is often used when there is insufficient observational evidence and knowledge about the nature of the land cover types for the area covered by the image, or where the analyst cannot accurately and confidently select training sets of known cover types. It can also be used as an initial step prior to supervised classification. In this type of hybrid classification, unsupervised classification is used to determine the spectral class composition of the image to help the image analyst select training sets that adequately represent the spectral variability of each class for supervised classification. Some unsupervised classification programs output class signatures for all the identified clusters. After the analyst uses the reference data to associate the spectral classes with the land covers of interest, the class signatures can be used as input to a supervised classification tool to classify the image.

Both the supervised and unsupervised classifications discussed in this section are on a pixel-by-pixel basis. They assume that the measurement vectors associated with different pixels are independent, while they are indeed spatially correlated. For example, a pixel classified as pasture is likely to be surrounded by the same class of pixels. Due to this assumption of 'independence', image classification often creates many misclassified isolated pixels inside class interiors, commonly referred to as speckles. This may result in a salt-and-pepper appearance, as shown in Figures 6.27 and 6.28a. In practice, image classification is usually followed by post-classification smoothing in order to remove the speckles and improve the quality of the classified image, as demonstrated in Box 6.12.

Land cover information is fundamental for understanding landscape patterns and characteristics, estimating carbon, water and energy exchange between land and atmosphere, and for environmental modelling, monitoring and decision making. Remote sensing provides a major means of obtaining past and current land cover information. Box 6.1 and Case Studies 1, 5 and 12 in Chapter 10 provide examples of using remote sensing data to extract land cover information for impact assessment of an earthquake on giant panda habitats, rainfall runoff modelling, land evaluation and economic evaluation of ecosystem services.

Accuracy assessment

Although there are many image classification methods available, no single one is inherently superior to any other. Selection of a suitable method depends on the nature of the classification problem, the biophysical characteristics of the study area, the distribution of the remote sensing data and a priori knowledge. In addition, all image classification results are subjected to a thorough accuracy assessment before they can be used in scientific investigations and policy making. Assessment of image classification accuracy generally involves the collection of ground reference information at a number of sample locations, often referred to as test sites, in the study area that are representative of each land cover class. It normally requires the image analyst to:

1. randomly select a number of test sites for each class, ideally fifty or more per class, depending on the size and complexity of the study area;
2. visit these randomly selected sites on the ground, determine their true classes, and record their locations (for example, using a GPS);
3. locate these test sites in the classified image, and determine their cover classes derived by the classifier;
4. construct an error matrix using the observed true classes of the test sites (for example, ground reference data) and their derived classes (that is, classification data);
5. compute accuracy measures based on the error matrix.

An error matrix shows whether and to what extent a classification confuses classes by comparing classification data and ground reference data on a class-by-class basis. It is an $m \times m$ matrix, where m is the number of classes whose classification accuracy is being assessed. Each column of the matrix represents known ground reference data – the number of test sites assigned to each class on the ground and verified in the field. Each row represents classification data – the number of test sites actually assigned to each class by the classifier. Section 3.5 presented an example of an error matrix (see Table 3.3), which was constructed based on 1,056 test sites in a land cover classification study in the Jiuzhaigou Nature Reserve, China conducted by Zhu (2013) using TM images. Four measures of classification accuracy that can be derived from an error matrix are also discussed in that section. They include overall accuracy, producer's accuracy, user's accuracy and the kappa statistic.

6.6 SUMMARY

- Remote sensing data used in environmental applications mainly include aerial photographs, multispectral and hyperspectral images, thermal infrared images, radar images and LiDAR point clouds. They record the magnitude of electromagnetic radiation that is reflected or emitted from Earth surface features.

- Image preprocessing is normally required prior to image analysis and information extraction. It may involve geometric and radiometric calibration. Geometric calibration includes removing geometric distortions in an image and georeferencing an image. Radiometric calibration includes eliminating sensor irregularities and unwanted sensor or atmospheric noise in an image.

- Image enhancement aims to improve the interpretability or perception of information in images to facilitate visual interpretation and analysis. It involves techniques for increasing the visual distinctions between features in an image, including band combinations, pan-sharpening, contrast stretching and spatial filtering.

- Band ratioing is an image transformation operation which divides the pixels in one band by the corresponding pixels in any other band. It removes much of the effect of illumination in the analysis of spectral differences – that is, the ratio between an illuminated and unilluminated area of the same feature type will be the same. In addition, band ratioing can enhance differences between the spectral reflectance curves of feature types

- Principal components analysis is a statistical technique used to transform multispectral images that consist of possibly correlated bands into a few uncorrelated bands. The removal of inter-band correlation reduces data redundancy, improves interpretability and decreases processing time.

- Image classification is used to assign all pixels in an image to particular land cover classes based on the spectral information contained in one or more spectral bands. The two most commonly used methods for image classification are supervised and unsupervised classification.

- In supervised classification, the image analyst selects representative training sets for each land cover class in the image. The image classification software uses the training sets to identify the land cover classes in the entire image. The common supervised classification algorithms are parallelepiped and maximum likelihood classifier.

- In unsupervised classification, the image classification software generates clusters of pixels based on their spectral properties. The image analyst identifies each cluster with land cover classes. There are different image clustering algorithms, such as k-means and ISODATA. Unsupervised classification is generally used when no training sets exist.

- All image classification results are subjected to accuracy assessment. It usually involves the selection of test sites, construction of an error matrix and calculation of accuracy measures based on the error matrix.

REVIEW QUESTIONS

1. What is remote sensing? Why does remote sensing provide an important source of spatial data for environmental applications?
2. Describe the characteristics of multispectral, hyperspectral, thermal infrared, radar and LiDAR remote sensing data.
3. Define the spatial, radiometric and spectral resolutions of remote sensing data.
4. Describe three image resampling methods: nearest neighbour, bilinear interpolation and cubic convolution.
5. What is apparent radiance? What is apparent reflectance, or albedo?
6. Why is Sun elevation correction necessary when comparing images taken at different times?
7. What is a colour infrared image? What is it good for?
8. Describe the IHS pan-sharpening process.
9. Describe the following contrast stretching methods: min-max, standard-deviation, percent-clip and histogram equalisation.
10. What are low pass, high pass and edge detection filters used for?
11. How useful is band ratioing?

12. What is the purpose of principal components analysis?

13. Explain the processes of supervised and unsupervised classification.

14. Outline the parallelepiped and maximum likelihood classification algorithms.

15. Outline the *k*-means and ISODATA algorithms.

16. Describe a general procedure for accuracy assessment of image classification.

REFERENCES

Amro, I., Mateos, J., Vega, M., Molina, R. and Katsaggelos, A.K. (2011) 'A survey of classical methods and new trends in pansharpening of multispectral images', *EURASIP Journal on Advances in Signal Processing*, 2011: 79, accessed 29 February 2016 at http://link.springer.com/article/10.1186/1687-6180-2011-79/fulltext.html.

ASPRS (2013) *LAS Specification, Version 1.4 – R13*, Bethesda, MD: ASPRS.

Cocks, T., Jenssen, R., Stewart, A., Wilson, I. and Shields, T. (1998) 'The HyMap airborne hyperspectral sensor: the system, calibration and performance', in M. Schaepman, D. Schläpfer and K.I. Itten (eds) *Proceedings of the 1st EARSeL Workshop on Imaging Spectroscopy*, 6–8 October 1998, Zurich: EARSeL, 37–43.

Farebrother, R.W. (1974) 'Gram-Schmidt regression', *Applied Statistics*, 23(3): 470–476.

Hartigan, J.A. and Wong, M.A. (1979) 'A *k*-means clustering algorithm', *Applied Statistics*, 28: 100–108.

Jensen, J.R. (1996) *Introductory Digital Image Processing: A Remote Sensing Perspective*, 2nd edn, Upper Saddle River, NJ: Pearson Prentice Hall.

Jensen, J.R. (2007) *Remote Sensing of the Environment: An Earth Resource Perspective*, 2nd edn, Upper Saddle River, NJ: Pearson Prentice Hall.

Laben, C.A. and Brower, B.V. (2000) 'Process for enhancing the spatial resolution of multispectral imagery using pansharpening', US Patent 6,011,875, filed 29 April 1998 and issued 4 January 2000, Rochester, NY: Eastman Kodak Company.

Lillesand, T.M., Kiefer, R.W. and Chipman, J.W. (2008) *Remote Sensing and Image Interpretation*, 6th edn, New York: John Wiley & Sons.

Liu, X. (2008) 'Airborne LiDAR for DEM generation: some critical issues', *Progress in Physical Geography*, 32: 31–49.

Nikolakopoulos, K.G. (2008) 'Comparison of nine fusion techniques for very high resolution data', *Photogrammetric Engineering and Remote Sensing*, 74(5): 647–659.

Strahler, A.H. (1980) 'The use of prior probabilities in maximum likelihood classification of remotely sensed data', *Remote Sensing of Environment*, 10: 135–163.

Tou, J.T. and Gonzalez, R.C. (1974) *Pattern Recognition Principles*, Reading, MA: Addison-Wesley.

Yu, H., Zhao, Y., Ma, Y., Sun, Y., Zhang, H., Yang, S. and Luo, Y. (2011) 'A remote sensing-based analysis on the impact of Wenchuan Earthquake on the core value of World Nature Heritage Sichuan Giant Panda Sanctuary', *Journal of Mountain Science*, 8(3): 458–465.

Zhu, X. (2013) 'Land cover classification using moderate resolution satellite imagery and random forests with post-hoc smoothing', *Journal of Spatial Science*, 58(2): 323–337.

Terrain analysis

This chapter discusses the basic concepts, principles and applications of digital terrain analysis in GIS, introduces sources and structures of digital terrain data, and discusses different types of terrain visualisation based on digital terrain models, the main terrain features which can be extracted through digital terrain modelling, and the principles of terrain profiling, visibility analysis and solar radiation mapping.

LEARNING OBJECTIVES

After studying this chapter, you should be able to:

- understand digital terrain data;
- identify sources of terrain data;
- outline ways to visualise terrain data;
- know how to calculate topographic attributes;
- produce terrain profiles;
- explain the principle of visibility analysis;
- comprehend the principles of solar radiation mapping;
- understand applications of terrain analysis in environmental studies.

7.1 DIGITAL TERRAIN ANALYSIS AND TERRAIN MODELS

Terrain plays an important role in modulating Earth surface and environmental processes. For example, terrain influences the lateral transport and accumulation of water, sediments (such as sand, mud and clay) and other substances on the Earth's surface, affects climatic and meteorological characteristics and soil properties, and has a strong effect on the spatial and temporal distributions of light, heat, water and nutrients required by plants and animals (Wilson and Gallant 2000). Both the terrain or topographic attributes (such as elevation, slope and aspect) and biophysical processes or patterns tend to vary continuously over space in a correlated manner (Guisan and Zimmermann 2000). In addition, terrain attributes are much easier to measure than biophysical variables. Therefore, terrain attributes are often used as dependent variables or predictors to model and predict biophysical processes and patterns. Terrain analysis is the process of using topographic data to derive terrain attributes, which describe the morphology of terrain surface and the influence of terrain on environmental processes.

Traditionally, topographic data were mainly extracted from topographic maps. Terrain attributes were measured and calculated manually based on contours. Digital elevation models emerged in photogrammetry in the 1950s. Since then, digital point elevation data in the form of DEMs have gradually replaced contours on topographic maps, and become the main source of topographic data. Supported by DEMs and other forms of digital topographic data, digital terrain analysis takes the place of manual terrain analysis. It is now one of the most powerful tools in GIS for environmental modelling and landscape visualisation.

Digital terrain analysis is enabled by GIS and related computer technologies. It seeks to digitally construct terrain models, quantitatively measure terrain attributes and mathematically establish relationships between topography and environmental processes or patterns. Digital terrain analysis has a wide range of applications in environmental science. For example, it has been used to analyse soil–landscape relationships and infer soil properties from topographic conditions (Park et al. 2001; MacMillan 2008), derive hydrologically, geomorphologically and biologically significant terrain attributes and model hydrological and geomorphologic functions (Moore et al. 1991), establish species–environment relationships and predict species distributions (Franklin 1998), characterise the structural and erosive elements of fault morphology (Brunori et al. 2013), map active tectonic processes (Miliaresis 2013) and assess landslide susceptibility (Manzo et al. 2013).

GIS uses digital terrain models to represent and store terrain data. A digital terrain model provides a way of digitally recording the elevation values and other attributes of the terrain. DEM, TIN and contour vectors are three main types of digital terrain models. Contour vectors are digitised contour lines. They are only used in GIS for data capture and visualisation, not for analysis.

DEMs are the most widely used digital terrain models due to their simplicity and ease of computer implementation. A DEM is essentially a grid of point locations with terrain elevations (see Figure 2.2). Therefore, it is a raster data model stored in a raster data structure. Each point (the centre point in a grid cell) contains an elevation value. The smaller the distance between points, the more detail the model can represent. Elevation data in a DEM are relatively abundant and inexpensive. However,

its rigid grid structure does not adapt to the variability of terrain, and thus may lose important details of the terrain surface.

TINs are vector representations of terrain. They represent terrain using contiguous, non-overlapping triangle facets that consist of planes linking the three adjacent elevation points in the network (see Figure 2.23). The density of the triangles varies to match the roughness of the terrain. As discussed in Section 2.3, a TIN is created using a set of regularly or irregularly spaced height values. It can incorporate other information about the terrain surface, such as ridgelines, spot heights, troughs, coastlines, drainage lines, lakes, faults, peaks, pits and passes. Therefore, a TIN can incorporate discontinuities on the terrain surface. These discontinuities are represented by breaklines. Breaklines are line features or boundaries of area features with or without height measurements. They become sequences of one or more triangle edges in the network (Figure 7.1). There are two kinds of breaklines: hard and soft. Hard breaklines represent a discontinuity in the slope of the surface, capturing abrupt changes in the surface. For example, streams and road cuts could be included in a TIN as hard breaklines. Soft breaklines do not change the local slope of a terrain surface. For example, coastlines or study area boundaries could be included in a TIN as soft breaklines, to capture their position without affecting the shape of the surface.

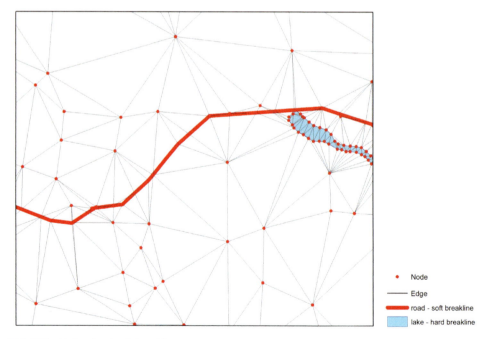

Figure 7.1 TIN with hard and soft breaklines.

7.2 HYDROLOGICALLY CORRECT DEMS

DEMs are used predominantly in environmental analysis and modelling, while TINs are employed mostly for visualisation and engineering applications (Hutchinson and Gallant 2005). DEMs can be readily integrated with remotely sensed environmental data. Well-established algorithms developed for digital image processing can be used to process DEMs. However, a DEM may contain erroneous sinks or depressions generated during the DEM's production and generalisation processes (for example, by incorrect or insufficient input data, interpolation defects during DEM generation, rounding elevation values to the nearest integer numbers, or by averaging elevation values within grid cells).

A sink is a cell or a set of spatially connected cells in a DEM that are surrounded by higher elevation values, representing an area of internal drainage. Some sinks are natural and real, such as quarries, potholes and sinkholes in Karst topography. But sinks are usually very rare in nature (Mark 1988), and they are most commonly caused by deficiencies in DEMs. Spurious sinks are common in coarse-resolution DEMs, particularly in low-relief areas (Martz and Garbrecht 1998). Sinks

affect the accurate determination of flow paths from the terrain data. If the sinks are not removed, a delineated drainage network from the DEM may be discontinuous. Figure 7.2 shows two hill-shaded DEMs for the same area with the same resolution: one with sinks, and the other with sinks removed. It also shows two different drainage lines derived from them. A DEM that has had all sinks removed is referred to as a hydrologically correct DEM, or depressionless DEM.

Sinks can be removed or 'filled' by raising the values of cells in depressions to the value of the depression's spill point – the cell with the lowest value surrounding the sink – as shown in Figure 7.3. A number of methods have been developed for sink filling (Wang and Liu 2006). The simplest method is to use a smoothing filter (see Section 4.4) to remove as many sinks as possible (Mark 1983). The Fill tool in ArcGIS is this type of method (Box 7.1). When a sink is filled, the boundaries of the filled area may create new sinks that then need to be removed. Thus, the identification and removal of sinks with a smoothing operation is an iterative process. Such a method is effective, especially for removing shallow depressions. But it may remove bona fide sinks, raise the average elevation of the terrain and create artificial flat areas. Hutchinson (1989) developed a drainage enforcement algorithm to control and fill spurious sinks. This algorithm infers drainage

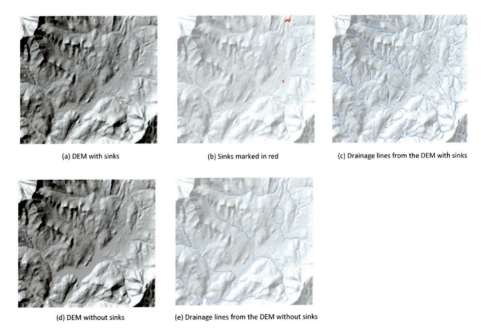

(a) DEM with sinks

(b) Sinks marked in red

(c) Drainage lines from the DEM with sinks

(d) DEM without sinks

(e) Drainage lines from the DEM without sinks

Figure 7.2 Hill-shaded DEMs with and without sinks and derived drainage lines: (a) DEM with sinks, (b) sinks marked in red, (c) drainage lines from the DEM with sinks, (d) DEM without sinks and (e) drainage lines from the DEM without sinks.

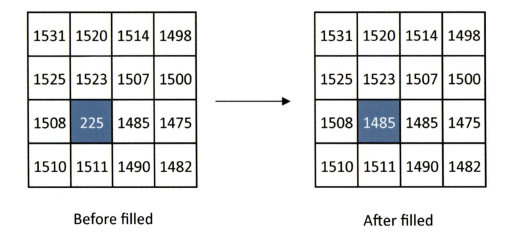

Before filled   **After filled**

Figure 7.3 A sink (the shaded cell) before and after being filled.

lines via the lowest saddle point in the drainage area surrounding each spurious sink, and attempts to clear spurious sinks to create a connected drainage structure and a correct representation of ridges and streams. It is essentially a spatial interpolation method used to create a hydrologically correct DEM from elevation point data or contours in conjunction with water feature data (streams and lakes), by following abrupt changes in terrain, such as streams and ridges. The algorithm is implemented in the Topo to Raster tool in ArcGIS (see Box 7.1). Soille (2004) developed a similar algorithm to fill sinks, which minimises overall modification of the DEM by optimising between raising the elevation of terrain within sinks and lowering the elevation of terrain along sink outflow paths. Wang and Liu (2006) provided a review of other sink filling methods.

Box 7.1 Creating hydrologically correct DEMs in ArcGIS   PRACTICAL

To follow this example, start ArcMap, and load the **boundary**, **dem**, **contours**, **spot_heights**, **lakes** and **rivers** feature classes from **C:\Databases\GIS4EnvSci\VirtualCatchment\Geodata.gdb**. **dem** is a DEM with a few sinks. Enable the **Spatial Analyst** extension.

Remove sinks from a DEM

1. Open **ArcToolBox**. In the **ArcToolBox** window, navigate to **Spatial Analyst Tools > Hydrology**, and double-click **Fill**.
2. In the **Fill** dialog:

   1) Select **dem** as the input surface raster.
   2) Name the output surface raster **C:\Databases\GIS4EnvSci\VirtualCatchment\Results\ filled_dem**.
   3) Click **OK**. The **Fill** tool finds and fills all the sinks. The resultant **filled_dem** is a depression-less DEM.

Create a hydrologically correct DEM from vector features

3. In the **ArcToolBox** window, navigate to **Spatial Analyst Tools > Interpolation**, then double-click **Topo to Raster**.

4. In the **Topo to Raster** dialog:

   1) Click the input feature data drop-down arrow, and select **contours**. Select **Contour** as both the field and the type.
   2) As in the previous step, add **spot_heights** as an input feature dataset, then select **elevation** as the field and **PointElevation** as the type.
   3) Add **lakes** as an input feature dataset, and select **Lake** as the type.
   4) Add **rivers** as an input feature dataset, then select **Stream** as the type.
   5) Add **boundary** as an input feature dataset, and select **Boundary** as the type.
   6) Name the output surface raster **C:\Databases\GIS4EnvSci\VirtualCatchment\Results\topo_dem**.
   7) Set the output cell size as **40**.
   8) Set Tolerance 1 as **50** (half the contour interval), and Tolerance 2 as **1000** (Tolerance 2 is at least six times greater than Tolerance 1).
   9) Click **OK**. After the process is complete, the interpolated depressionless DEM is created and added to the view.

The drainage enforcement algorithm used by the **Topo to Raster** tool identifies areas of local maximum curvature in each contour, and uses this information to detect the areas of steepest slope and create a network of streams and ridges (Hutchinson 1989). It ensures that the output DEM has proper hydrogeomorphic properties, and removes the need for post-processing to fill spurious sinks.

Most available DEMs were not generated by a hydrologically sound interpolation method, and contain spurious sinks. Removing sinks is an important first step in the use of these DEMs to derive terrain attributes for hydrological applications. But care should be taken not to remove real sinks or change the DEM so much as to introduce further errors into hydrological analyses.

7.3 DIGITAL TERRAIN DATA SOURCES

Digital terrain data can be generated by various laboratory, field and remote sensing techniques, such as ground surveys, photogrammetry, radar altimetry, interferometric synthetic aperture radar (InSAR) and LiDAR.

Most of the digital terrain data currently available are the product of photogrammetric data capture. They are produced from stereo pairs of remotely sensed images, including aerial photographs and satellite images. Photogrammetric data capture uses manual (analogue) or analytic (digital) stereoplotters to derive DEMs and extract spot elevations and contours (Jensen 2007). High-quality stereo aerial photographs are required to generate accurate, high-resolution DEMs. Considerable amounts of aerial photography and derived DEMs have been acquired by commercial photogrammetric companies or subcontractors for government agencies (for example, the US Geological Survey National Aerial Photography Program and the Victorian Coordinated Imagery Program in Australia). In Australia, the state governments are the major sources of medium- to large-scale aerial photography and derived DEM products covering their own states.

Some satellite remote sensing systems, such as ASTER (Advanced Spaceborne Thermal Emission and Reflection Radiometer), SPOT and QuickBird, are capable of obtaining stereoscopic panchromatic satellite images, which have been used to produce medium-resolution DEMs. ASTER GDEM and SPOT DEM are two examples of DEMs captured using satellite imagery through digital photogrammetry. ASTER GDEM is a global DEM generated using stereo-pair images collected by the ASTER instrument onboard the Terra satellite, covering all the Earth's land surface, and having a horizontal resolution of about 30m and a vertical resolution of about 15m (Hirt et al. 2010). It is freely available from the Land Processes Distributed Active Archive Centre (LP DAAC) of the US Geological Survey (http://gdex.cr.usgs.gov/gdex/). SPOT DEM is a global DEM with vegetation and man-made structures produced by automatic correlation of stereo-pairs acquired by the HRS (High Resolution Stereoscopy) instrument on the SPOT 5 satellite, which covers portions of Eurasia, Africa and Central America, and has a horizontal resolution of 15–30m and a vertical resolution of 10–20m (Spot Image 2005). SPOT DEM is a commercial product.

Radar techniques, particularly satellite radar altimetry and InSAR, have been employed to obtain digital elevation data. Satellite radar altimetry is a vertical-looking radar technique, which measures the ocean surface topography with an accuracy in the order of a few centimetres. It has been used to obtain DEMs of ocean floors (Sandwell and Smith 2001) and of the ice sheets of Antarctica and Greenland (Liu et al. 1999). InSAR acquires multiple radar images of the terrain to extract 3D information. It can provide digital terrain data that are as accurate as DEMs derived using photogrammetric techniques (Jensen 2007). InSAR technology has been implemented in the SRTM (Space Shuttle Radar Topography Mission) to obtain the global medium-resolution DEM (Rabus et al. 2003; Hirt et al. 2010). SRTM-derived DEMs are now available for a large proportion of the Earth's land surface at a spatial resolution of 90m. A considerable amount of data has been processed to 30m. SRTM-derived DEMs are freely available from LP DAAC. Those covering Australia are available from Geoscience Australia (http://www.ga.gov.au) as well. Also available from LP DAAC is GTOPO30, a global DEM with a spatial resolution of about 1km produced from a variety of map and remote sensing image sources.

As discussed in Section 6.1, airborne LiDAR creates masspoints. Each masspoint has a unique (x, y) location and an elevation value. They can be interpolated using one of the interpolation methods introduced in Section 4.5 to produce a digital surface model (DSM) that contains the elevation characteristics of trees, shrubs, buildings and other man-made structures. Figure 7.4 shows a perspective view of Glen Waverley Activity Centre and its surrounding area in Melbourne based on a DSM derived from LiDAR masspoints. A DEM of the bare ground can be derived from LiDAR data by removing masspoints that come from trees, shrubs, buildings and any other surface extrusions (Raber et al. 2002; Ma 2005). A LiDAR-derived DEM may have a horizontal resolution of less than 2m and a vertical accuracy within 2cm. Airborne LiDAR surveys have been applied to create large-scale and detailed DEMs as an alternative to ground surveying and photogrammetry. The horizontal and vertical accuracies of LiDAR-derived DEMs are comparable to that of photogrammetry. However, LiDAR offers no guarantee of collecting elevation points on breaklines such as ridgelines and steep changes of slope, while photogrammetry allows the analyst to selectively obtain the elevation points on breaklines.

Ground surveying carried out by means of conventional surveying instruments (such as total station) and GPS produces accurate elevation point data, which can then be interpolated into a DEM or TIN. It is used to acquire large-scale and detailed digital terrain data. In particular, the widespread use of GPS provides many new and affordable opportunities for the collection of large amounts of special-purpose, one-of-a-kind digital

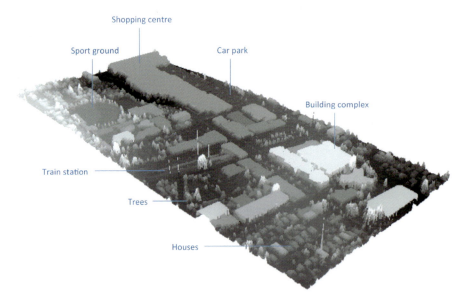

Figure 7.4 Perspective view of Glen Waverley Activity Centre in Melbourne based on LiDAR data.

terrain datasets for detailed, large-scale environmental studies (Allwardt 2007; Clark and Lee 1998; Florinsky and Eilers 2002; Wilson and Gallant 2000).

In addition, digital terrain data can also be acquired by digitising the contour lines on topographic maps. The digitised contours are tagged with elevations. During the digitising process, ancillary cartographic information can also be captured, such as elevation values for mountain summits, depression bottoms and structural lines. Algorithms have been developed to interpolate contour lines to DEMs or TINs (Hutchinson and Gallant 2005). This approach has been applied to produce national DEMs of the USA, Australia, Canada and other countries (USGS 1993; ANU Fenner School of Environment and Society and Geoscience Australia 2008; Natural

Resources Canada 1997). It is often used when the topographic maps are the only existing source of terrain data.

Digital terrain data are largely available in the form of DEMs. DEMs with spatial resolutions of 500m, 100m, 30m, 10m, 8m, 4m and 1m are increasingly available for different parts of the world from commercial data providers (for example, GEO Elevation from Airbus Defence and Space, http://www.astrium-geo.com/) and government agencies of many countries (for example Geoscience Australia, UK Ordnance Survey and US Geological Survey). TINs are mainly constructed from DEMs or contour lines, masspoints and/or breaklines. Box 7.2 shows how to use ArcGIS to build a TIN from contour lines, masspoints and breaklines. Box 8.8 illustrates how a TIN is constructed from a DEM in ArcGIS.

Box 7.2 Building a TIN from vector features in ArcGIS — PRACTICAL

To follow this example, start ArcMap, and load the **boundary**, **contours**, **spot_heights**, **lakes** and **rivers** feature classes from **C:\Databases\GIS4EnvSci\VirtualCatchment\Geodata.gdb**. Enable the **3D Analyst** extension.

1. Open **ArcToolBox**. In the **ArcToolBox** window, navigate to **3D Analyst Tools > Data Management > TIN**, and double-click **Create TIN**.
2. In the **Create TIN** dialog:
   1) Name the output TIN **C:\Databases\GIS4EnvSci\VirtualCatchment\Results\tin2**.
   2) Set the coordinate system of the output TIN as **boundary**.
   3) Click the input feature class drop-down arrow, then select **contours**. Set the height field as **Contour**, the SF type (surface feature type) as **Mass_Points**, and the tag field as **<None>**. Here, the height field contains elevation values for each contour line, and the vertices in the contour lines are used as masspoints for triangulation. The tag field is used when boundaries of area or polygon features are enforced in the triangulation as breaklines. Triangles inside these polygons are attributed with the tag values (for example, land use codes), which can be used to symbolise the triangles when rendering the TIN. No tag field is defined for **contours**.
   4) As in the previous step, add **spot_heights** as an input feature class, then set its height field as elevation, its SF type as **Mass_Points**, and its tag field as **<None>**.
   5) Add **lakes** as an input feature class, then set its height field as **height**, its SF type as **Hard_Replace**, and its tag field as **<None>**. Hard-replace or soft-replace sets the boundary and all interior heights of a polygon feature to the same elevation value. Hard and soft qualifiers indicate whether a distinct break in slope occurs on the surface at the location of a feature. The hard qualifier indicates a distinct break in slope, while the soft qualifier suggests a more gradual change in slope.
   6) Add **rivers** as an input feature class, then set its height field as **<None>** (there is no height field in the layer), its SF type as **Hard_Line**, and its tag field as **<None>**. Here, rivers are used as hard breaklines.

(continued)

(continued)

    7) Add **boundary** as an input feature class, then set its height field as **<None>**, its SF type as **Soft_Clip**, and its tag field as **<None>**. This input feature class is used as a clip polygon that defines the boundary for the TIN surface.

    8) Click **OK**. The TIN is created and added to the view. It is displayed with a default symbology.

3. Examine the TIN by zooming in and navigating through the surface. Notice that the lake surfaces are flat, defining areas of constant height.

4. Double-click **tin2** in the table of contents, and click **Properties**. In the **Properties** dialog:

    1) Click the **Symbology** tab.
    2) In the **Show** box, uncheck **Elevation**, and click **Edge types**.
    3) Click the **Add All Values** button.
    4) Click **OK**. Only the edges of the TIN are shown.

5. Turn on the contour lines and spot heights. Zoom in and navigate through the surface to understand how the TIN was made from the elevation values at the vertices of each contour line and the spot heights, and how breaklines were used to build the triangular network.

7.4 TERRAIN ATTRIBUTES

Digital terrain models provide a basis for extracting terrain-related attributes or features. Moore et al. (1991) provided a comprehensive list of terrain attributes that can be derived through digital terrain analysis. A distinction is generally made between primary and secondary (compound) terrain attributes. Primary attributes are computed directly from a digital terrain model, and secondary attributes are derived from two or more primary attributes.

Primary terrain attributes

Primary attributes are mainly used to describe the size, shape and other surface attributes of hillslopes, catchment boundaries and stream channels (Wilson and Gallant 2000). Typical primary attributes include elevation, slope, aspect, profile and plan curvature, catchment area, flow direction, flow accumulation, drainage network, flow length and upslope area. All can be obtained or calculated from a DEM or TIN. Elevation is a primary terrain attribute that has a significant impact on climate, vegetation, soil and potential energy. Obtaining an elevation value at a particular location is straightforward, as the digital terrain model represents elevation data. We can analyse the range of elevation by simply visualising or mapping the terrain based on a digital terrain model.

Slope

Slope, also called gradient or angle of terrain, measures the rate of change of elevation at a location. It is one of the most widely used terrain attributes in environmental analysis and modelling. For example, slope determines overland and subsurface flow velocity and runoff rate, and affects distributions of vegetation types, precipitation and soil water content. It has also been used as the primary means of classifying land capability, and as one of the significant parameters in soil erosion and landslide hazard modelling and wildlife habitat suitability assessment.

Slope is calculated as the amount of rise of a surface divided by the run over which this rise is measured along a straight line transect in a specific direction (Figure 7.5). The rise is defined as the vertical distance, and the run as the horizontal planar distance. Slope can be expressed as percent slope or degree slope. Percent slope is 100 times the ratio of rise over run, whereas degree slope is the arc tangent of the ratio of rise over run.

In practice, slope is calculated by measuring the amount of rise over a certain horizontal distance. For example, if the rise over a 10m distance in the horizontal plane was measured as 5m, the slope would be 5/10 = 50 per cent, or arctan(5/10) = 26.56°. In GIS, slope is computed for each grid cell of a DEM or every triangle of a TIN based on the concept of the normal vector. The

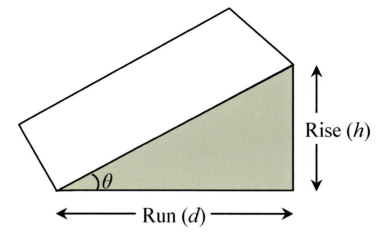

$$\text{Percent slope:} \quad \theta = 100 \times (h / d) \quad (\%)$$
$$\text{Degree slope:} \quad \theta = \arctan (h / d) \quad (\text{degree})$$

Figure 7.5 Slope measurement.

normal vector to a plane is a vector (directed line) perpendicular to it (Figure 7.6). Suppose a plane is given by:

$$f(x, y, z) = ax + by + cz + d = 0 \qquad (7.1)$$

Its normal vector will be n = (a, b, c). Then the slope of the plane will be:

$$\theta = \frac{\sqrt{a^2 + b^2}}{c} \qquad (7.2)$$

A triangle in a TIN made of three nodes with known coordinates (x_1, y_1, z_1), (x_2, y_2, z_2) and (x_3, y_3, z_3) determines a plane. Its normal vector can be computed as follows (Vince 2006):

$$a = (y_2 - y_1)(z_3 - z_1) - (y_3 - y_1)(z_2 - z_1) \qquad (7.3)$$
$$b = (z_2 - z_1)(x_3 - x_1) - (z_3 - z_1)(x_2 - x_1) \qquad (7.4)$$
$$c = (x_2 - x_1)(y_3 - y_1) - (x_3 - x_1)(y_2 - y_1) \qquad (7.5)$$

A slope map can be derived from a TIN by applying Equations 7.2–7.5 to every triangle facet in the TIN.

The normal vector of a grid cell in a DEM is estimated based on its adjacent neighbours. Jones (1998)

listed and compared eight algorithms to estimate slope from a DEM which used different neighbourhoods and methods to calculate the normal vector of a cell in the DEM, and found that the algorithms developed by Ritter (1987) and Horn (1981) are the most accurate ones. Both of the algorithms use a 3×3 neighbourhood centred on the grid cell whose slope is to be estimated. The algorithm by Ritter uses four immediate neighbours of the centre cell (Figure 7.7a), and calculates its slope as:

$$\frac{\sqrt{(z_1 - z_3)^2 + (z_4 - z_2)^2}}{2d} \qquad (7.6)$$

where d is the cell size and z_i (i = 1, 2, 3 and 4) is the cell value (elevation value). The a and b components of the normal vector to the centre cell are respectively $(z_1 - z_3)$ and $(z_4 - z_2)$.

Horn's algorithm uses eight neighbours of the centre cell (Figure 7.7b), and estimates the slope as:

$$\frac{\sqrt{[(z_1 + 2z_4 + z_6) - (z_3 + 2z_5 + z_8)]^2 + [(z_6 + 2z_7 + z_8) - (z_1 + 2z_2 + z_3)]^2}}{8d} \qquad (7.7)$$

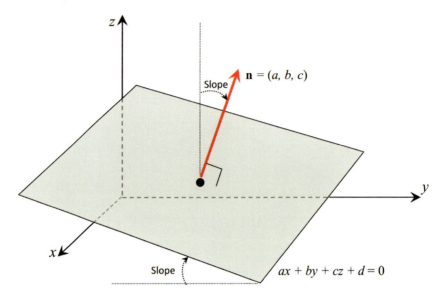

Figure 7.6 Normal vector and slope.

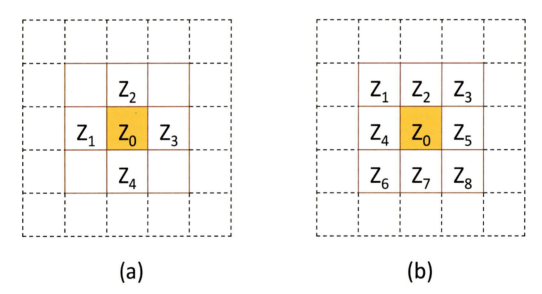

Figure 7.7 Neighbours for slope calculation used: (a) by Ritter (1987) and (b) by Horn (1981).

The a and b components of the normal vector to the centre cell are respectively $[(z_1 + 2z_4 + z_6) - (z_3 + 2z_5 + z_8)]$ and $[(z_6 + 2z_7 + z_8) - (z_1 + 2z_2 + z_3)]$. This algorithm is used in ArcGIS. Figure 7.8 shows the slope map derived with the algorithm based on the depressionless DEM created in Box 7.1.

Aspect

Aspect is the direction of maximum gradient of the surface at a particular point. It is an important factor that controls the spatial distribution of solar irradiance, thus affecting evapotranspiration, and flora and fauna

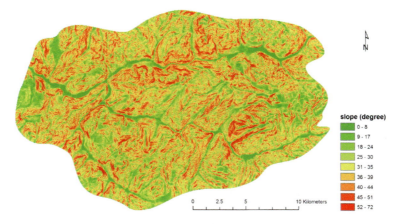

Figure 7.8 Slope map.

distribution and abundance. As with slope, it is calculated for each grid cell in a DEM or every triangle in a TIN based on the normal vector. The general formula for calculating the aspect of a plane with the normal vector n = (a, b, c) is written as follows (Hodgson and Galle 1999):

$$A = \frac{360}{2\pi} \times \arctan\left(\frac{b}{a}\right) \tag{7.8}$$

A is measured in degrees. Therefore, according to Equations 7.3–7.5, the aspect of a triangle in a TIN with three nodes (x_1, y_1, z_1), (x_2, y_2, z_2) and (x_3, y_3, z_3) can be computed as:

$$A = \frac{360}{2\pi} \times \arctan\left|\frac{\begin{aligned}&(z_2 - z_1)(x_3 - x_1) - \\ &(z_3 - z_1)(x_2 - x_1)\end{aligned}}{\begin{aligned}&(y_2 - y_1)(z_3 - z_1) - \\ &(y_3 - y_1)(z_2 - z_1)\end{aligned}}\right| \tag{7.9}$$

To derive aspect from a DEM, Ritter's algorithm uses the following equation:

$$A = \frac{360}{2\pi} \times \arctan\left(\frac{z_4 - z_2}{z_1 - z_3}\right) \tag{7.10}$$

while Horn's algorithm applies the formula:

$$A = \frac{360}{2\pi} \times \arctan\left|\frac{\begin{aligned}&(z_6 + 2z_7 + z_8) - \\ &(z_1 + 2z_2 + z_3)\end{aligned}}{\begin{aligned}&(z_1 + 2z_4 + z_6) - \\ &(z_3 + 2z_5 + z_8)\end{aligned}}\right| \tag{7.11}$$

Aspect is usually specified according to compass directions, which are measured clockwise in degrees from 0 (due north) to 360 (also due north) (Figure 7.9). The following rule is used to convert aspect values calculated using Equation 7.8 to compass direction values (0–360° from due north):

IF $A > 90.0$
    Aspect = $360.0 - (A - 90.0)$
ELSE
    Aspect = $90.0 - A$

Figure 7.10 is the aspect map derived with Horn's algorithm based on the depressionless DEM created in Box 7.1. Aspect in the flat areas is undefined, usually assigned the value −1.

Profiles

A terrain profile is a 2D diagram that depicts the landscape in vertical cross-section. It shows changes in elevation, slope and curvature of the terrain surface along a specific line called a profile line or transect, such as a hiking trail or a road. The profile line can be a straight line or a polyline. The x-axis of the diagram represents horizontal distance from the starting point of the profile line, and the y-axis represents the elevation. With a GIS, when a profile line is drawn on a DEM or TIN, a terrain profile will be plotted with a certain vertical exaggeration factor. Figure 7.11 shows an example. In this figure, the map shows a TIN with a profile line AB drawn on it. The diagram beside the map is the terrain profile along the line created by the GIS, whose vertical extent is exaggerated threefold.

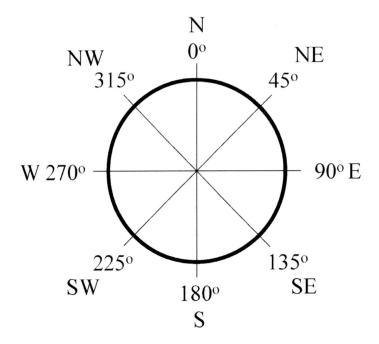

Figure 7.9 Aspect directions.

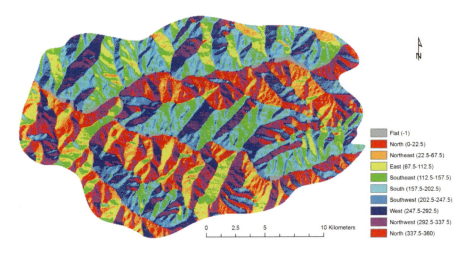

Flat (-1)
North (0-22.5)
Northeast (22.5-67.5)
East (67.5-112.5)
Southeast (112.5-157.5)
South (157.5-202.5)
Southwest (202.5-247.5)
West (247.5-292.5)
Northwest (292.5-337.5)
North (337.5-360)

0   2.5   5   10 Kilometers

Figure 7.10 Aspect map.

GIS methods for producing a terrain profile based on a digital terrain model generally involve:

1. finding all intersections of a specified profile line with sides of the triangles in a TIN or with grid cells in a DEM;

2. calculating the horizontal position along the profile line of every intersection and its elevation through interpolation;

3. drawing the x-axis with its length equal to the horizontal distance of the profile line at the scale of the digital terrain model;

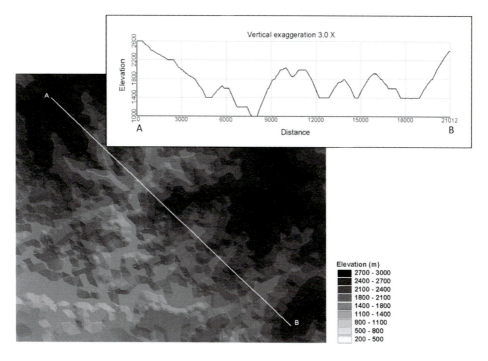

Figure 7.11 Terrain profile.

4. determining the height of the vertical axis according to the specified vertical exaggeration factor, and drawing the y-axis;

5. plotting the intersections in the diagram using their horizontal distances to the starting point of the profile line as x coordinates and their elevation times the vertical exaggeration factor as y coordinates;

6. connecting all the intersections into a curve.

It should be noted that a terrain profile shows slope and curvature of a terrain surface along the direction of a specific transect, which may not represent the direction of maximum gradient. Therefore, the slope measured at a particular point on the profile line may be different from the value calculated by using the slope function in GIS discussed above. The same applies to curvature, which is explained below.

Surface curvature

Surface curvature is the amount by which a surface deviates from being flat. It is an intrinsic property of a surface, whose application in hydrology and geomorphology has long been recognised. Curvature of a surface

at a point generally changes with orientation. There are many alternative measures of surface curvature (Schmidt et al. 2003). Among them, profile curvature and plan curvature are the ones most frequently used in GIS.

Profile curvature is the rate of change of slope in the direction of the maximum slope, and plan curvature is the curvature perpendicular to the direction of the maximum slope. A common method for computing surface curvature at a particular point based on a DEM is to use a quadratic polynomial function to approximate a continuous surface representing the surface covered by a 3×3 window centred at that point (Schmidt et al. 2003). Using the neighbourhood definition in Figure 7.7b and assuming the cell size is d, the polynomial function is expressed as follows (Zevenbergen and Thorne 1987):

$$Z = Ax^2y^2 + Bx^2y + Cxy^2 + Dx^2 + Ey^2 + Fxy + Gx + Hy + I \tag{7.12}$$

where

$$A = [(z_1 + z_3 + z_6 + z_8) / 4 - (z_2 + z_4 + z_5 + z_7) / 2 + z_0] / d^4$$

$$B = [(z_1 + z_3 - z_6 - z_8) / 4 - (z_2 - z_7) / 2] / d^3$$

$C = [(-z_1 + z_3 - z_6 + z_8)/4 + (z_4 - z_5)]/2]/d^3$
$D = [(z_4 + z_5)/2 - z_0]/d^2$
$E = [(z_2 + z_7)/2 - z_0]/d^2$
$F = (-z_1 + z_3 + z_6 - z_8)/4d^2$
$G = (-z_4 + z_5)/2d$
$H = (z_2 - z_7)/2d$
$I = z_0$

The profile curvature is calculated as

$$\varphi = -2 \times \frac{(D \times G)^2 + (E \times H)^2 + F \times G \times H}{G^2 + H^2} \quad (7.13)$$

and the plan curvature is computed as:

$$\omega = 2 \times \frac{(D \times H)^2 + (E \times G)^2 - F \times G \times H}{G^2 + H^2} \quad (7.14)$$

The overall curvature is measured as:

$$\chi = \varphi - \omega = -2 \times (E + D) \quad (7.15)$$

The units of φ, ω and χ are one-hundredth (1/100) of a z unit. A positive curvature indicates the surface is upwardly convex (for example, hills and ridges), a negative curvature indicates the surface is upwardly concave (for example, depressions and valleys) and flat surfaces have zero curvature (Figure 7.12). Figure 7.13 shows a profile curvature map created from the depressionless DEM created in Box 7.1.

Curvature has important implications for surface processes, such as erosion, deposition and runoff processes (Schmidt et al. 2003). While slope influences the overall rate of movement downslope and aspect determines the

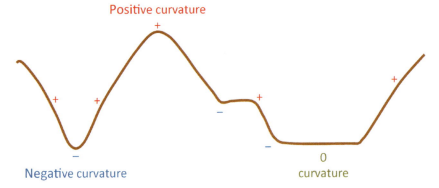

Figure 7.12 Positive, negative and zero curvature along a terrain profile.

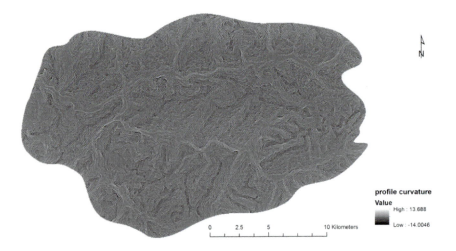

Figure 7.13 Profile curvature map.

direction of flow, profile curvature affects the flow acceleration and erosion/deposition rate, and plan curvature has an effect on convergence and divergence of flow. Box 7.3 shows how these surface attributes are derived in ArcGIS.

Box 7.3 Deriving primary surface attributes from a DEM in ArcGIS   **PRACTICAL**

To follow this example, start ArcMap, and load **filled_dem** created in Box 7.1. Enable the **3D Analyst** extension.

Slope

1. Open **ArcToolBox**. In the **ArcToolBox** window, navigate to **3D Analyst Tools > Raster Surface**, and double-click **Slope**.
2. In the **Slope** dialog:

    1) Click the input raster drop-down arrow, and select **filled_dem**.
    2) Name the output raster **C:\Databases\GIS4EnvSci\VirtualCatchment\Results\slope**.
    3) Select **DEGREE** as the output measurement, and enter **1** as the Z factor.
    4) Click **OK**. The slope map is created and added to the view, as shown in Figure 7.8.

Aspect

3. In the **ArcToolBox** window, navigate to **3D Analyst Tools > Raster Surface**, and double-click **Aspect**.
4. In the **Aspect** dialog, click the input raster drop-down arrow, then select **filled_dem**. Name the output raster **C:\Databases\GIS4EnvSci\VirtualCatchment\Results\aspect**.
5. Click **OK**. The aspect map is created, as shown in Figure 7.10.

Curvature

6. In the **ArcToolBox** window, navigate to **3D Analyst Tools > Raster Surface**, and double-click **Curvature**.
7. In the **Curvature** dialog:

    1) Click the input raster drop-down arrow, then select **filled_dem**.
    2) Name the output raster **C:\Databases\GIS4EnvSci\VirtualCatchment\Results\curvature**.
    3) Use the default value of 1 as the Z factor.
    4) Name the output profile curve raster **C:\Databases\GIS4EnvSci\VirtualCatchment\Results\curvature_pro**.
    5) Name the output plan curve raster **C:\Databases\GIS4EnvSci\VirtualCatchment\Results\curvature_pla**.
    6) Click **OK**. The three curvature maps are created. The profile curvature map is shown in Figure 7.13.

Profile

8. In the table of contents, turn on **filled_dem**, and turn off all other layers.
9. Click **Customize** in the main menu, point to **Toolbars**, and click **3D Analyst**. The **3D Analyst** toolbar appears.
10. Click the layer's drop-down arrow on the **3D Analyst** toolbar, and select **filled_dem**.
11. Click the **Interpolate Line** button on the **3D Analyst** toolbar.

(continued)

(continued)

12. Click the upper-left corner of the DEM, drag the line to the lower-right corner, and double-click to stop the line (you may draw a polyline by clicking to add a vertex each time and double-clicking to finish drawing).
13. Click the **Profile Graph Tool** button 📈 on the **3D Analyst** toolbar. A graph pops up which shows the terrain profile along the line.
14. Right-click the graph. From the pop-up menu, you may click **Properties** or **Advanced Properties** to edit the title, subtitle and other properties of the graph, or click **Add to Layout** to switch to the layout view and add the graph to the map layout.

Flow direction

Flow direction, flow accumulation, flow length, flow order, drainage network and catchment area are hydrological characteristics of a terrain surface. In GIS, they are mainly derived from depressionless DEMs and represented in raster layers.

Flow direction is fundamental in hydrological analysis because in order to determine where a landscape drains, it is necessary to determine the direction of flow for each location in the landscape. It is the direction of the steepest descent from a particular location on a surface. On a terrain surface represented in a DEM, flow direction is the direction in which water will flow out from each cell to its steepest downslope neighbour. Most GIS software packages use an eight-direction flow model known as the D8 algorithm (O'Callaghan and Mark 1984; Jenson and Domingue 1988) to calculate flow direction. The algorithm finds the locally steepest direction (rise over run) from a cell by calculating the distance-weighted drop in elevation to each of the cell's eight neighbours (as defined in Figure 7.7b):

$$D_{0j} = \frac{z_0 - z_j}{d_{0j}} \times 100 \qquad (7.16)$$

where D_{0j} is the distance-weighted drop in elevation from the centre cell to the neighbouring cell j (=1, 2, . . ., 8), and d_{0j} is the distance between the centres of the centre cell to the neighbouring cell j. For the four orthogonal neighbours, d_{0j} is the cell size. For the four diagonal neighbours, d_{0j} is the cell size × 1.414 (the square root of 2). Let D_{max} be the largest distance-weighted drop in elevation for a cell. The following rules are used to determine flow direction at the cell (Jenson and Domingue 1988):

IF $D_{max} < 0$

The cell is a single-cell sink, which does not occur for a depressionless DEM. The cell is filled to the lowest value of its neighbours, and has a flow direction to that neighbour.

ELSE IF $D_{max} \geq 0$ at only one neighbour
The cell is assigned the flow direction to that neighbour.

ELSE IF $D_{max} \geq 0$ at more than one neighbour (the cell has D_{max} in multiple directions)
IF the cell is part of a sink (or a flat area)
The flow direction is undefined.

ELSE
The cell is assigned the most likely flow direction according to a lookup table, as in Greenlee (1987).

Figure 7.14 illustrates how flow direction is derived using a simple example. The flow direction tool in ArcGIS implements the D8 algorithm to generate a flow direction raster. The tool uses the neighbour location codes shown in Figure 7.15 to encode the flow direction of a cell. For example, if a cell flows to the upper-right neighbour, its flow direction value would be 128. If a cell flows down, its flow direction value would be 4. If a cell has the largest drop in multiple directions and that cell is part of a sink, its flow direction value would be the sum of the neighbour location codes in those directions. For example, if a cell's largest weighted drop in elevation is both to the left and up, its flow direction value would be 16 + 64 = 80. If all neighbour elevations were equal to the centre cell, the centre cell would receive a flow direction value of 255. Thus, flow direction values range from 1 to 255. Figure 7.16 shows a flow direction raster derived from the same depressionless DEM created in Box 7.1.

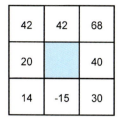

(a) DEM

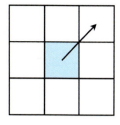

(b) Weighted drops

(c) Flow direction

Figure 7.14 Identification of flow direction: (a) DEM, (b) weighted drops and (c) flow direction.

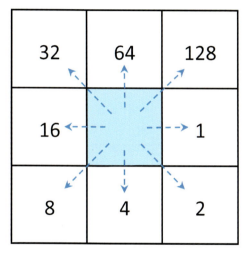

Figure 7.15 Neighbour location codes used by the flow direction tool in ArcGIS.

Flow accumulation

Flow accumulation refers to how much water accumulates from upstream areas and contributes to overland flow. It is basically calculated as the number of upstream cells that flow into each cell based on their flow directions. Figure 7.17 illustrates how flow accumulation is computed based on flow direction. The value of each cell in the flow accumulation raster is the number of cells that flow into it (the cell itself is not considered), and the arrows indicate flow directions. Locations having high flow accumulation values are the areas of concentrated flow, generally corresponding to stream channels, while locations having an accumulation value of zero correspond to the pattern of ridges. Therefore, flow accumulation provides a basis for delineating drainage networks and identifying ridgelines.

Weights can also be applied to each cell when calculating flow accumulation. For example, when multiplied by the area size of a cell, the flow accumulation value is equal to the size of the drainage area or the contributing area. When multiplied by average rainfall in each cell during a given storm, the flow accumulation value represents the amount of rain that flowed through each cell or the amount of rain that fell in the upslope area of each cell, on the assumption that all rain became runoff and there was no interception, infiltration and evaporation. Figure 7.18 is the flow accumulation map derived from the depressionless DEM created in Box 7.1.

Drainage network

Flow accumulation is used to create a drainage network based on the flow direction of each cell. By selecting cells with a flow accumulation value greater than some threshold value, a network of high-flow cells can be generated. These high-flow cells should lie on stream channels and at valley bottoms. However, by applying a threshold value, drainage networks can be extracted from a DEM with arbitrary drainage density. In general, the density of the drainage network decreases as the threshold value increases. The two drainage networks in Figure 7.19 were both derived from the flow accumulation map in Figure 7.18. Figure 7.19a was produced with a threshold value of 1,000 cells, while Figure 7.19b was based on a threshold value of 7,000 cells. Different threshold values need to be tried to select the most appropriate one so that the extracted drainage network corresponds to or is close to the one obtained by more traditional methods, such as from high-resolution topographic maps or fieldwork. Several other criteria have also been proposed for extracting drainage networks from DEMs, such as a threshold on local slope, a combination of drainage area

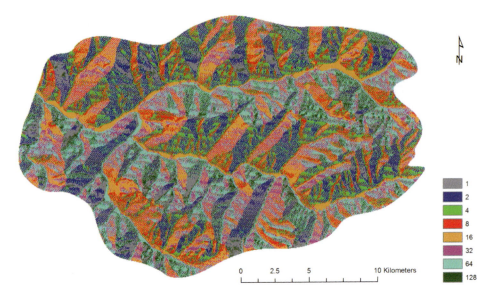

			1
			2
			4
			8
			16
			32
			64
			128

Figure 7.16 Flow direction raster.

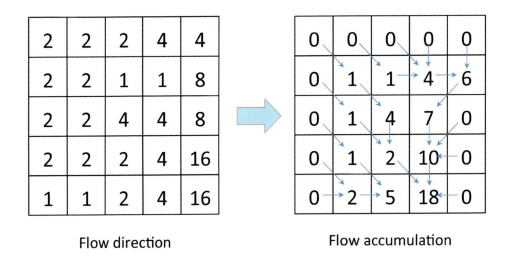

Figure 7.17 Flow accumulation calculated as the number of cells that flow into each cell.

and slope, a threshold on local curvature and a threshold slope-direction-change (Heine et al. 2004; Lashermes et al. 2007).

Stream order

As shown in Figure 7.20, a drainage network through which water travels to the outlet or pour point can be seen as a tree, with the root of the tree being the outlet downstream and the branches being stream channels. A point at which two upstream channels join to form one downstream channel is referred to as a node or junction. The sections of a stream channel connecting two successive junctions or a junction and the outlet are called stream links. Stream links are often ordered for the purposes of identifying and classifying types of

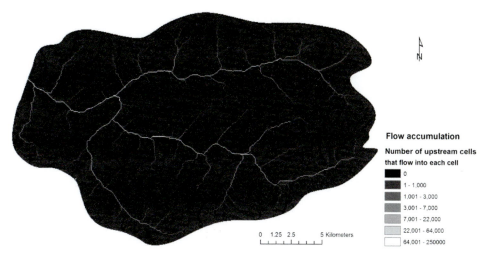

Flow accumulation

Number of upstream cells
that flow into each cell

■ 0
▨ 1 - 1,000
▨ 1,001 - 3,000
▨ 3,001 - 7,000
▨ 7,001 - 22,000
▨ 22,001 - 64,000
□ 64,001 - 250000

Figure 7.18 Flow accumulation map.

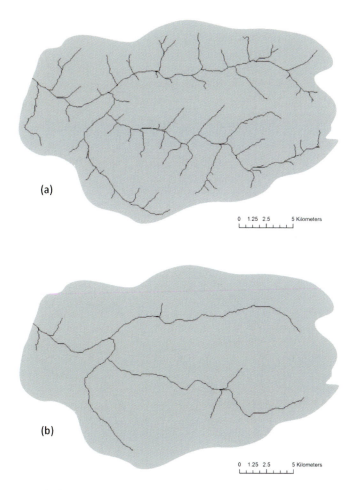

(a)

(b)

Figure 7.19 Drainage networks based on threshold values of: (a) 1,000 cells and (b) 7,000 cells.

streams based on their numbers of tributaries in order to infer their characteristics. There are two main methods for stream ordering: the Strahler method and the Shreve method.

With the Strahler method (Strahler 1952), stream links with no tributaries (that is, headwater flow paths) are assigned the number 1, referred to as first-order. Where two first-order links converge, the downslope link is assigned an order of 2. Where two second-order links meet, the downslope link is assigned an order of 3, and so on (Figure 7.21a). Where a lower-order stream (for example, first-order) joins a higher-order stream (for example, third-order), the downslope link will retain the higher order (that is, third-order). Only where two links of the same order join will the order increase. This is the most common method of stream ordering.

The Shreve method orders streams by adding order numbers at the confluence of each stream link (Shreve 1967). It also starts at the headwaters – that is, all stream links with no tributaries are assigned an order of 1. When two links join, their order numbers are added and assigned to the downslope link (Figure 7.21b). For example, where a first-order joins a second-order stream link, the downslope link is assigned an order of 3.

Stream ordering is an effective way to classify waterways, and is a crucial step in understanding and managing the many differences between streams of different sizes. It can be of assistance in studying the amount of sediment in an area, determining the number and types of organisms present in a stream of a given size, designing water quality monitoring and management plans, and using waterways as natural resources in an effective and sustainable manner. For example, many more plants can live in sediment-filled slower-flowing high-order streams than in a fast-flowing low-order tributary of the same river. As another example ordered using the Strahler method, third- or higher-order streams are likely to be valuable fish habitats, and hence could support viable fish populations. Therefore, fish passage barriers located in third- or higher-order streams should be considered for remediation.

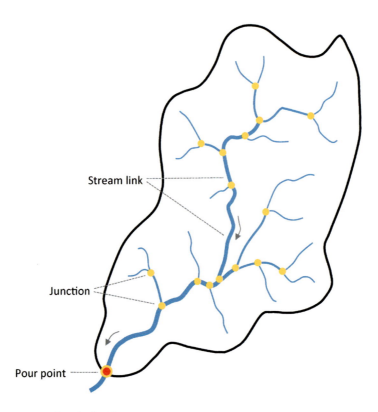

Figure 7.20 Drainage network and related concepts.

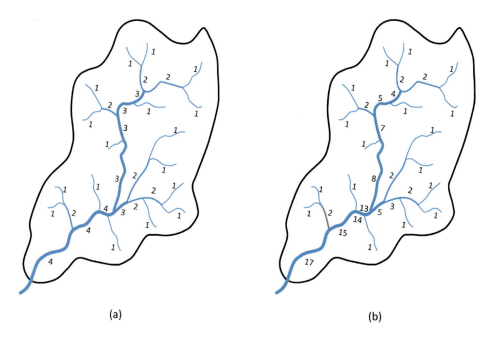

Figure 7.21 Stream ordering: (a) the Strahler method and (b) the Shreve method.

Flow length

Flow length is the distance or weighted distance along a flow path, either upstream or downstream. It can be used to calculate the length of the longest flow path. Upstream flow length is the longest distance along the flow path from each cell to the top of the drainage divide. Downstream flow length is the distance along the flow path from each cell to a sink or outlet on the edge of the mapped area. Upstream flow length is often used to calculate the time of flow concentration within a catchment, which is very useful for surface runoff modelling. Figures 7.22a and 7.22b show an upstream and a downstream flow length map respectively, derived from the depressionless DEM created in Box 7.1.

Catchment boundary

A catchment is an area that drains water and other substances to a common outlet. It is also called a drainage basin, watershed, basin or contributing area. All land is in one catchment or another. Catchment boundaries are dependent on topography, but can be drawn to show smaller parts of larger catchments based on the locations of selected pour points. Traditionally, a catchment boundary is traced on a topographic map by

starting at an outlet, determining flow directions, locating ridgelines and saddles, marking high points and connecting the high points with smooth lines that cross contours at right-angles. In GIS, catchment boundaries are delineated from DEMs by automatically identifying ridgelines between catchments based on a specific set of pour points. Figure 7.23 shows an example of catchment boundaries delineated from the depressionless DEM created in Box 7.1.

Delineation of catchment boundaries is important to any study of the water budget of an area, the impact of human activities on water quality or quantity or the development of water resources. Box 7.4 illustrates how the hydrology tools in ArcGIS are used to delineate catchments and extract other hydrology related terrain attributes from a DEM.

Secondary terrain attributes

Secondary attributes are used to characterise the spatial variability of specific environmental processes occurring in the landscape (Wilson and Gallant 2000). They can be applied to quantify the influence of topography on the spatial distribution and abundance of water, soil, fauna and flora. Examples of secondary attributes include the

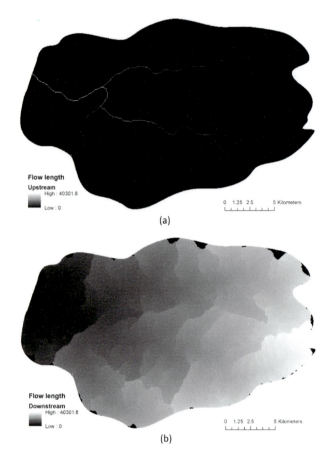

Figure 7.22 Flow length maps: (a) upstream and (b) downstream.

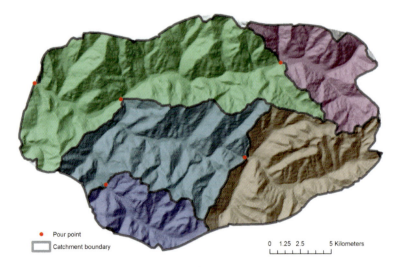

Figure 7.23 Five individual catchments delineated from a DEM.

Box 7.4 **Analysing hydrological characteristics of a terrain surface from a DEM in ArcGIS** **PRACTICAL**

To follow this example, start ArcMap, and load **filled_dem** generated in Box 7.1. Enable the **Spatial Analyst** extension.

Flow direction

1. Open **ArcToolBox**. In the **ArcToolBox** window, navigate to **Spatial Analyst Tools > Hydrology**, and double-click **Flow Direction**.
2. In the **Flow Direction** dialog:

    1) Click the input surface raster drop-down arrow, and select **filled_dem**.
    2) Name the output flow direction raster **C:\Databases\GIS4EnvSci\VirtualCatchment\ Results\flowdir**.
    3) Click **OK**. The flow direction map is created, as shown in Figure 7.16.

Flow accumulation

3. In the **ArcToolBox** window, navigate to **Spatial Analyst Tools > Hydrology**, then double-click **Flow Accumulation**. In the **Flow Accumulation** dialog:

    1) Select **flowdir** as the input flow direction raster.
    2) Name the output accumulation raster **C:\Databases\GIS4EnvSci\VirtualCatchment\ Results\flowacc**.
    3) Click **OK**. The flow accumulation map is created, as shown in Figure 7.18.

Drainage network

4. In the **ArcToolBox** window, navigate to **Spatial Analyst Tools > Conditional**, and double-click **Con**.
5. In the **Con** dialog:

    1) Select **flowacc** as the input conditional raster.
    2) Enter the expression: `Value > 1000`.
    3) Set the input true raster or constant as **1**.
    4) Name the output raster **C:\Databases\GIS4EnvSci\VirtualCatchment\Results\stream1k**.
    5) Click **OK**. **stream1k** is created, as shown as Figure 7.19a. It is a raster layer where the value 1 represents the drainage network on a background of NoData.

6. Repeat Step 5 to create a drainage network with a threshold value of **7000**. The result should be similar to Figure 7.19b.

Flow length

7. In the **ArcToolBox** window, navigate to **Spatial Analyst Tools > Hydrology**, and double-click **Flow Length**. In the **Flow Length** dialog:

    1) Select **flowdir** as the input flow direction raster.
    2) Set the output raster as **C:\Databases\GIS4EnvSci\VirtualCatchment\Results\flowlen_ up**.

(continued)

(continued)

3) Set the direction of measurement along the flow path as **UPSTREAM**.
4) Click **OK**. **flowlen_up** is created, as shown in Figure 7.22a.

8. Repeat the previous step with **DOWNSTREAM** as the direction of measurement along the flow path to calculate the downstream flow length. The result should look like Figure 7.22b.

Stream order

9. In the **ArcToolBox** window, navigate to **Spatial Analyst Tools > Hydrology**, and double-click **Stream Order**. In the **Stream Order** dialog:

1) Select **stream1k** as the input stream raster.
2) Select **flowdir** as the input flow direction raster.
3) Enter **C:\Databases\GIS4EnvSci\VirtualCatchment\Results\streamorder** as the output raster.
4) Select **STRAHLER** as the method of stream ordering.
5) Click **OK**. The layer showing stream orders using the Strahler method is created.

10. Repeat the previous step to create stream orders using the Shreve method.

Catchment boundary

To delineate catchment boundaries, pour points need to be specified. In ArcGIS, pour points should be represented in a point feature class or shapefile, which may be sampling sites, hydrometric stations or from other data sources. In this example, you will create your own pour points.

Create pour points

11. Follow Steps 16–19 in Box 5.1 to create a new, empty point shapefile, and name it **outlets**.
12. Turn on **flowdir** in the table of contents so that the flow accumulation layer is displayed.
13. Zoom in to your area of interest so that you are able to see the individual flow accumulation cells. Use the **Identify** tool to examine flow accumulation values. A pour point should be a natural outlet for the streams flowing into it, and must be on the high flow accumulation path. It essentially determines the end of a catchment. Everything upstream from this point will define a single catchment.
14. To add a pour point, open the **Editor Toolbar**, and start editing (refer to Step 1 in Box 3.7). In the **Start Editing** dialog, highlight the empty pour point layer, and click **OK**.
15. Click the **Create Features** tool 📝 on the **Editor Toolbar**. In the **Create Features** window, click **outlets** in the upper panel, then **Point** in the construction tools panel.
16. Move the cursor onto the flow accumulation map in the data view. Add a pour point by clicking at the centre of the high flow accumulation cell you have chosen as the pour point. Try to place a point in the centre of the cell. If a pour point is selected where two streams join, place the point one or two cells away from the stream confluence.
17. Click the **Attributes** tool 🔲 on the **Editor Toolbar**. In the **Attributes** window, enter **1** in the text box beside **Id**, which is one of the fields in the layer's attribute table and is used to store the unique ID number for a pour point.
18. One pour point defines one catchment. More pour points can be added as in Steps 14–17 to create several catchments. After adding a pour point for each catchment, click the **Editor** menu, and then **Stop Editing**. Click **Yes** when prompted to save the edits.

Snap pour points

19. In the **ArcToolBox** window, navigate to **Spatial Analyst Tools > Hydrology**, and double-click **Snap Pour Point**. This tool is used to snap the pour point(s) created above to the closest area of high flow accumulation. It converts the pour points to the raster format. In the **Snap Pour Point** dialog:

    1) Select **outlets** as the input raster or feature pour point data.
    2) Select **Id** as the pour point field.
    3) Select **flowacc** as the input accumulation raster.
    4) Name the output raster **C:\Databases\GIS4EnvSci\VirtualCatchment\Results\snapoutlets**.
    5) Leave the snap distance as the default. The snap distance is the distance (in map units) that the tool uses to search around the pour points for the cell of the highest accumulated flow. It should be set based on the resolution of the data, and it may take some trial and error to determine the best value.
    6) Click **OK**.

Delineate catchment boundaries

20. In the **ArcToolBox** window, navigate to **Spatial Analyst Tools > Hydrology**, and double-click **Watershed**. In the **Watershed** dialog:

    1) Select **flowdir** as the input flow direction raster.
    2) Select **snapoutlets** as the input raster or feature pour point data. The pour point field can be left as the default.
    3) Name the output raster **C:\Databases\GIS4EnvSci\VirtualCatchment\Results\catchments**.
    4) Click **OK** to run the tool. The catchment boundaries are delineated and displayed in the view. Figure 7.23 gives an example of the output.

21. Use the **Raster To Polygon** tool to convert the catchment raster to a vector layer (refer to Box 3.6).

---

topographic wetness index (TWI), stream power index (SPI), sediment transport index (STI) and radiation index. The radiation index is discussed in Section 7.6.

Topographic wetness index

TWI quantifies the effect of topography on the location and size of areas of water accumulation in soils (Beven and Kirkby 1979). It is defined as:

$$TWI = ln\left(\frac{A_s}{\tan\theta}\right) \tag{7.17}$$

where A_s is the upstream catchment area per unit contour length, and θ is the local slope angle. A_s is referred to as the specific catchment area. When a DEM is used, A_s can be estimated as the ratio of the upstream catchment area to the cell size (d) of the DEM unless the DEM has

a very high resolution (Chirico et al. 2005). The following procedure is generally used to estimate TWI from a depressionless DEM:

1. Derive the flow direction from the DEM.
2. Derive the flow accumulation (f_{acc}) from the flow direction.
3. Calculate the upstream catchment area (A) as:

$$A = \left(f_{acc} + 1\right) \times d^2 \tag{7.18}$$

4. Calculate the specific catchment area as:

$$A_s = \frac{A}{d} = \left(f_{acc} + 1\right) \times d \tag{7.19}$$

5. Calculate the slope (θ) from the DEM.
6. Calculate the TWI using Equation 7.17.

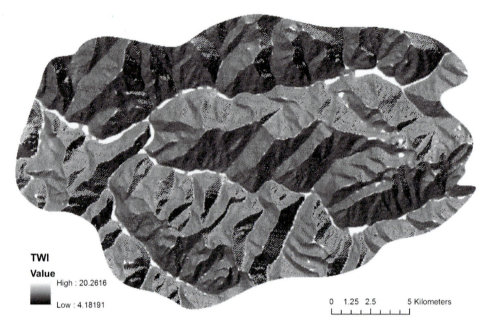

TWI

Value

High : 20.2616

Low : 4.18191

0 1.25 2.5   5 Kilometers

Figure 7.24 Topographic wetness index.

Because a ridge has a flow accumulation value of 0, we have to add 1 to all flow accumulation values in Equations 7.18 and 7.19, otherwise the size of the catchment area would be one grid cell size smaller. Figure 7.24 is a TWI map derived from the depressionless DEM created in Box 7.1.

TWI can provide insight into the spatial distribution and proportions of areas of relative wetness and areas of relative dryness. It serves as a physically based index quantifying the effect of topography on runoff generation and estimating the location of zones of surface saturation (Beven and Kirkby 1979). Areas generating saturation overland flow can be identified using a threshold TWI value. In a catchment, areas having similar TWI values are considered to have a similar hydrological response to rainfall when their other environmental conditions (such as land cover and soil) are the same (Qin et al. 2011). It is also found to correlate well with depth to groundwater, and soil pH (Sørensen et al. 2005). TWI has been used as an indicator to the pattern of potential soil moisture, an input for digital soil mapping, a parameter in modelling the spatial variability of crop yields and a criterion for delineating nitrogen management zones (Qin et al. 2011). In addition, it has been used to predict the spatial distribution of vascular plant species richness in the Swedish boreal forest (Zinko 2004). However, TWI assumes that the

soil hydraulic conductivity decreases exponentially with depth so that subsurface flow is confined to a shallow layer. If this is not the case, it may not be a good predictor of the spatial distribution of soil water (Wilson and Gallant 2000).

Box 7.5 demonstrates the calculation of TWI in ArcGIS.

Stream power index

SPI is a measure of the erosive power of overland flow, defined as follows (Moore et al. 1991):

$$SPI = A_s \times \tan\theta \qquad (7.20)$$

As the specific catchment area and slope gradient increase, the amount of water contributed by upslope areas and the velocity of water flow increase, hence stream power index and erosion risk increase.

SPI can be used to describe potential flow erosion at a given point of the terrain surface and related landscape processes. When used together, TWI and SPI provide good predictors of the location of ephemeral gullies (Moore et al. 1988). For example, ephemeral gullies formed where TWI > 6.8 and SPI > 18 for a small semiarid catchment in Australia (Moore et al. 1988), and they formed where TWI > 8.3 and SPI > 18 on a small

To follow this example, start ArcMap, and load **slope** and **flowacc** generated in Box 7.4. Enable the **Spatial Analyst** extension.

Calculate the specific catchment area

1. Open **ArcToolBox**. In the **ArcToolBox** window, navigate to **Spatial Analyst Tools > Map Algebra**, and double-click **Raster Calculator**.
2. In the **Raster Calculator** dialog:

    1) Enter the following expression (refer to Box 4.11): (`"flowacc"` + 1) * 40. Here, 40 (m) is the cell size of the flow accumulation raster.
    2) Name the output raster **C:\Databases\GIS4EnvSci\VirtualCatchment\Results\sca**.
    3) Click **OK**. The specific catchment area raster is created and added to the view.

Calculate the local slope *tan*(θ) in radians

3. Open **Raster Calculator**. In the **Raster Calculator** dialog:

    1) Enter the following expression (refer to Step 2 in Box 4.11): `Tan("slope"/57.296)`.
    2) Enter **C:\Databases\GIS4EnvSci\VirtualCatchment\Results\slope_radians** as the output raster name.
    3) Click **OK**.

Calculate TWI

4. Open **Raster Calculator**. In the **Raster Calculator** dialog:

    1) Enter the following expression: `Ln("sca" / "slope_radians")`.
    2) Name the output raster **C:\Databases\GIS4EnvSci\VirtualCatchment\Results\twi**.
    3) Click **OK**. **twi** is created, as shown as Figure 7.24. Note that the flat areas with no slope are assigned NoData.

catchment in Antigua in the Caribbean (Srivastava and Moore 1989). The threshold values of these indices vary from place to place due to differences in soil properties (Moore et al. 1988). Moore and Nieber (1989) also used SPI to identify areas where soil conservation measures should be taken to reduce the erosive effects of concentrated flow. Figure 7.25 shows an SPI map derived from the depressionless DEM created in Box 7.1. The calculation of SPI in ArcGIS is similar to TWI calculation described in Box 7.5.

Sediment transport index

STI accounts for the effect of topography on soil erosion. It was derived from the unit steam power theory (Moore and Wilson 1992). The index is calculated as:

$$STI = \left(\frac{A_s}{22.13}\right)^m \left(\frac{\sin\theta}{0.0896}\right)^n \tag{7.21}$$

where m and n are constants depending on slope length. Usually m is set to 0.4 and n to 1.3 (Moore et al. 1991). The value of STI may vary along the length of a stream.

STI can be used to characterise the processes of erosion and deposition, and to present the effect of topography on soil loss. For example, Pallaris (2000) applied this index to model erosion risk in the Cabuyal River catchment in Colombia, a typical hillside agro-ecosystem environment under the threat of soil degradation. The results were validated using local farmers' knowledge, and proved that STI is useful for initial erosion assessment at

Figure 7.25 Stream power index.

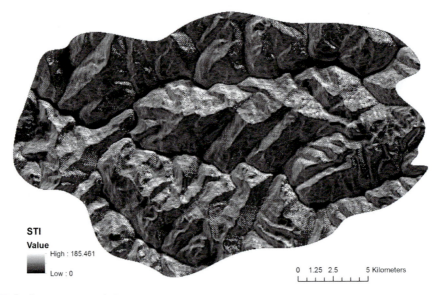

Figure 7.26 Sediment transport index.

the regional level. This index is sometimes also used in place of the length-slope factor in the revised universal soil loss equation for slope lengths < 100m and slope < 14° (Moore and Wilson 1992). Figure 7.26 is an STI map derived from the depressionless DEM created in Box 7.1. It was created in ArcGIS using a similar procedure to that used in Box 7.5.

7.5 VISIBILITY ANALYSIS

Visibility analysis involves determining what is visible from a location or a set of locations along given lines of sight or across the entire area. It is a useful tool in landscape planning for assessing scenic quality and the impact of new developments on the landscape. For example, it can be used to answer questions like 'Which areas can be seen from a forest fire observation station that is 20 metres high?', 'Which land areas are visible from a planned scenic road, and how often can those areas be seen from the road?' and 'How many wind turbines are visible from a village?'

Visibility from location to location is mainly affected by the shape of the terrain surface. Over large distances, it may need to take the curvature of the Earth into consideration. Two types of visibility analysis are commonly supported in GIS: line-of-sight analysis and viewshed analysis. They are based on digital terrain models.

Line-of-sight analysis

A line of sight is a line connecting two points that shows the visibility of different parts of the terrain surface along the line. It determines whether a given target is visible from a particular point of observation. As shown in Figure 7.27, it involves defining a viewpoint or observation point (and sometimes its height), a target (and its height) and a direct line of sight between the viewpoint and the target. If the line of sight is blocked by the terrain, the target is deemed invisible, otherwise it is visible.

In GIS, the user is allowed to decide the locations of a viewpoint and a target, set the offsets (heights) of the viewpoint and the target above the terrain surface, and to interactively create a line of sight between the

two locations on a DEM or TIN. The line of sight will show the parts of the terrain surface along the line that are visible or invisible from the viewpoint. If the target is blocked by the terrain, the line of sight will also show where the obstruction point is. Figure 7.28 is a typical output of a line-of-sight analysis over a DEM.

Viewshed analysis

Viewshed analysis expands line-of-sight analysis to cover every possible point in the study area as the target. It is used to define the areas of the terrain surface that are visible from a viewpoint or a set of viewpoints, thus allowing users to calculate what they can see from a given set of vantage points. The visible areas from a viewpoint are called the viewshed of that point. Through viewshed analysis, it is possible to determine what would be visible from planned developments, how visible a particular spot on the landscape is or what vegetative barriers might be required to reduce the visibility of some structure.

Viewsheds can be delineated from a DEM or TIN. Various algorithms have been developed for generating viewsheds (De Floriani and Magillo 2003). The basic algorithm for delineating a viewshed from a DEM is based on the estimation of the elevation difference of intermediate points between the viewpoint cell and target cells (Kim et al. 2004). The intermediate points between the viewpoint cell and a target cell are usually selected as

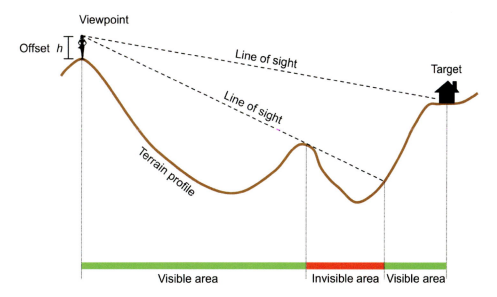

Figure 7.27 Line-of-sight analysis.

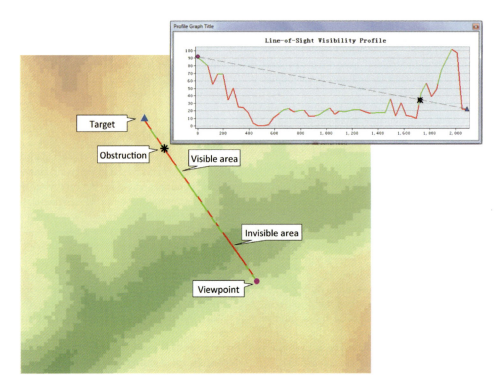

Figure 7.28 Typical GIS output of line-of-sight analysis.

the intersections of its line of sight and the grid lines of the DEM. The elevation of an intermediate point is estimated by interpolation from its neighbouring cells' elevation values. A line-of-sight analysis is performed from the viewpoint cell to each target cell by examining each of the intermediate points between the two cells. If the terrain surface rises above the line of sight, the target cell is invisible from the viewpoint. Otherwise, it is visible. The line-of-sight analysis is repeated for all cells from the viewpoint, and the set of cells that are visible from the viewpoint form the viewshed.

A viewshed on a TIN is basically computed by determining what part of each triangle in the TIN is visible from a given viewpoint. Ben-Moshe et al. (2008) described several algorithms for generating viewsheds on a TIN, among which the generic radar-like algorithm performs best. This algorithm divides the area around the viewpoint into a number of wedges. Each wedge is bounded by two rays emanating from the viewpoint. The algorithm then computes the visible segments along the two rays (that is, visible portions of the terrain profiles

in the two directions). If the two sets of visible segments (one set per ray) are similar enough, it extrapolates the visible region in the wedge defined by the two rays; otherwise, the wedge is subdivided into smaller wedges, and the algorithm will now consider each of the smaller wedges.

TIN-based viewshed analysis functions are not commonly available in GIS software packages. Viewshed analysis is mostly implemented based on DEMs. Typically, the inputs for DEM-based viewshed analysis include a DEM (or surface raster), a set of viewpoints (represented as point features or vertices for a line feature such as a road) and offset values for the viewpoints. Two types of output are typically produced from viewshed analysis in GIS. One is a viewshed raster, whose cell values represent the number of times that each cell location in the input surface raster can be seen by the input viewpoints (Figure 7.29b). The other type of output identifies which viewpoints are visible from each cell location (Figure 7.29c). Box 7.6 demonstrates how such viewsheds are generated from a DEM in ArcGIS.

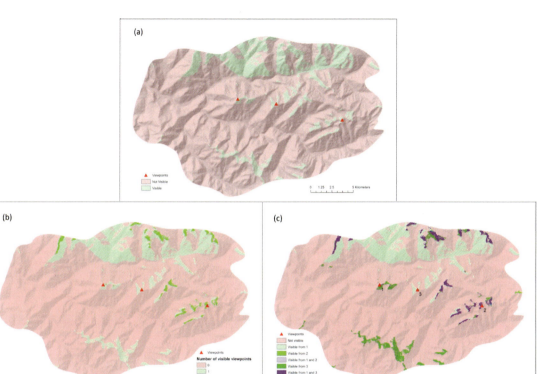

Figure 7.29 Typical GIS outputs of viewshed analysis.

Box 7.6   Visibility analysis in ArcGIS   **PRACTICAL**

To follow this example, start ArcMap, and load **dem** from **C:\Databases\GIS4EnvSci\ VirtualCatchment\Geodata.gdb** and **outlook** from **C:\Databases\GIS4EnvSci\VirtualCatchment\ Shapefiles**. Enable **3D Analyst**.

Create a line of sight

1. Click **Customize** in the main menu, point to **Toolbars**, then click **3D Analyst**.
2. On the **3D Analyst** toolbar, click the **Create Line of Sight** button ☞.
3. In the **Line of Sight** pop-up window, type **20** as the observer (viewpoint) offset (here the height of the lookout tower). The observer offset is the eye level of the observer used in determining what is visible from the observer's location. The units are the same as the DEM's z units (metres in this case). The target offset is the height of the target point above the surface. Set the target offset as **1**. Leave this window open.

(continued)

(continued)

4. Click the outlook point (the observer location) on the DEM surface, then click any target location. A line of sight is drawn between the two points, which is a graphic line. The green colour of the line indicates the locations where the terrain surface is visible, while the red colour indicates the locations where the surface is hidden or invisible. Along the line, the black dot represents the location of the viewpoint, the blue dot is the point of obstruction from the viewer to the target, and the red dot represents the location of the target.

5. Click the **Profile Graph Tool** button on the **3D Analyst** toolbar. A graph pops up showing the visibility profile along the line of sight. The outputs of the line-of-sight analysis are similar to Figure 7.28.

6. Repeat Steps 4 and 5 several times to create lines of sight for other targets.

7. After these experiments, click **Edit** on the main menu bar, then click **Select All Elements**. All the lines of sight created are selected. Click the **Cut** button on the main toolbar to delete them from the view.

Create viewsheds for one viewpoint

8. Right-click **outlook** in the table of contents, and click **Open Attribute Table**. The field **OFFSETA** in the attribute table is the default field name used by ArcGIS for storing offset heights of viewpoints. In this example, it stores the height of the outlook tower, which is 20m.

9. Open **ArcToolBox**. In the **ArcToolBox** window, navigate to **3D Analyst Tools > Visibility**, and double-click **Viewshed**. In the **Viewshed** dialog:

   1) Select **dem** as the input raster.
   2) Select **outlook** as the input point observer features.
   3) Enter **C:\Databases\GIS4EnvSci\VirtualCatchment\Results\viewshed1** as the output raster.
   4) Click **OK**. The viewshed raster is generated and displayed in the data view.

Calculate the number of viewpoints visible to each cell location

10. Display **dem** only, and turn off other layers.

11. Create a new empty point shapefile to store viewpoints by following Steps 16–19 in Box 5.1, and name it **viewpoints**.

12. Create three viewpoints of your choice based on the DEM by referring to the method and procedure for creating pour points described in Box 7.4, and save them into **viewpoints**.

13. Follow Step 9 above to conduct viewshed analysis using **viewpoints** as input point observer features, and name the output viewshed **C:\Databases\GIS4EnvSci\VirtualCatchment\Results\viewshed2**. Once **viewshed2** is created, it is displayed in the data view and shows NOT VISIBLE and VISIBLE areas, which is similar to Figure 7.29a.

14. In the table of contents, double-click **viewshed2**. In the **Layer Properties** dialog, change the symbology to map the cell values as unique values. The viewshed map should look similar to Figure 7.29b.

Identify which viewpoints are visible from each cell location

15. Turn off **viewshed2**.

16. In the **ArcToolBox** window, navigate to **3D Analyst Tools > Visibility**, and double-click **Observer Points**. In the **Observer Points** dialog, select **dem** as the input raster, **viewpoints** as the input point observer features, enter **C:\Databases\GIS4EnvSci\VirtualCatchment\Results\viewshed3** as the output raster, and click **OK**. **viewshed3** is a viewshed raster which identifies which viewpoints are visible from each cell. It should be similar to Figure 7.29c.

17. Open the attribute table associated with **viewshed3**. Every row lists the unique cell value and corresponding visible viewpoints.

7.6 SOLAR RADIATION MAPPING

Solar radiation mapping aims to estimate the spatial and temporal distribution of solar radiation on the Earth's surface. At a landscape scale, topography exerts a great impact on the amount of solar energy incident at a location on the Earth's surface (Moore et al. 1991). Variability in elevation, slope, aspect and shadows cast by terrain features creates substantial differences in solar radiation, which leads to significant spatial and temporal variations in local energy and water balance, thus affecting local environments (such as air and soil temperature, soil moisture and light availability) and biophysical processes (such as photosynthesis, evapotranspiration and primary production). These factors and processes in turn affect the distribution and abundance of flora and fauna. Therefore, solar radiation mapping at a landscape scale has many environmental applications. Meteorological stations generally provide solar radiation measurements, but they are point-specific and affected strongly by local topographic variations. The limited coverage of meteorological stations measuring solar radiation cannot describe the necessary variability. Simply interpolating the point solar radiation data usually produces inaccurate solar radiation maps. Accurate solar radiation mapping is more often based on solar radiation models integrated into a GIS.

Solar radiation is radiant energy emitted by the Sun. The solar radiation (insolation) reaching the Earth's surface through the atmosphere can be classified into three components: direct radiation, diffuse radiation and reflected radiation. Direct radiation propagates unimpeded in a direct line from the Sun to the receiving surface. Diffuse radiation is scattered by atmospheric constituents, such as aerosols, water vapour and dust. Reflected radiation is the direct and diffuse radiation reflected by nearby terrain towards the location of interest. The sum of the direct, diffuse and reflected radiation is referred to as the total or global solar radiation. Many solar radiation models have been developed to estimate insolation and its components at a given location (Ahmad and Tiwari 2011; Besharat et al. 2013). However, most of them do not incorporate topographic effects. GIS-based topographic solar radiation models emerged in the early 1990s (Dubayah and Rich 1995). They were developed to provide accurate estimates of solar radiation at a given location or over a large area, considering surface inclination, orientation and shadowing effects. Examples of the first GIS-based solar radiation models include SolarFlux (Hetrick et al. 1993a, 1993b), developed for ARC/INFO GIS, and Solei

(Miklánek 1993), which can be linked to IDRISI GIS. These models incorporate the local effect of topography by empirical relations. More advanced solar radiation models are implemented in GRASS GIS and the Solar Radiation toolbox of ArcGIS.

The *r.sun* model in GRASS estimates the three components of solar radiation for a given day, latitude, surface and atmospheric (clear-sky or overcast) conditions for large areas with complex topography. It accounts for sky obstruction (shadowing) by local terrain features calculated from the DEM. The model was applied to estimate the solar potential for photovoltaic systems in Europe (Šúri et al. 2005). Details of the model equations and applied approach can be found in Šúri and Hofierka (2004).

The model in the ArcGIS Solar Radiation toolbox calculates solar radiation across a landscape or for specific locations based on the hemispherical viewshed algorithm developed by Rich et al. (1994) and further developed by Fu and Rich (1999). It involves first generating an upward-looking hemispherical viewshed for every location. Such a hemispherical viewshed shows the angular distribution of sky visibility and obstruction, which is specified in a hemispherical coordinate system (Figure 7.30). A hemispherical viewshed is constructed based on a DEM by searching in a specified set of directions around the location of interest (Figure 7.31a) and determining the maximum angle of sky obstruction or horizon angle in each direction (Figure 7.31b). The horizon angles in other directions are interpolated. The resultant viewshed is a raster, in which each cell is assigned a value that corresponds to whether the sky direction is visible or obstructed (Figure 7.31c). The cell location (row and column) corresponds to a zenith angle θ and an azimuth angle α on the hemisphere of directions. The hemispherical viewsheds are then used together with sun position and sky direction information, represented by a sunmap and skymap respectively, to calculate solar radiation for each location.

A sunmap is a raster, in the same hemispherical coordinate system as for the hemispherical viewshed, that shows the suntrack, or apparent position of the Sun as it moves across the sky over a period of time. It divides the hemisphere of sky directions into sky sectors, each defining a suntrack (Figure 7.32a). The position of the Sun is calculated based on the latitude according to the time of day and year – for example, by half-hour intervals through the day and month intervals through the year. The sunmap is used to calculate the direct insolation originating from directions along the suntrack. A skymap is also a raster in the same hemispherical coordinate system as for the hemispherical viewshed.

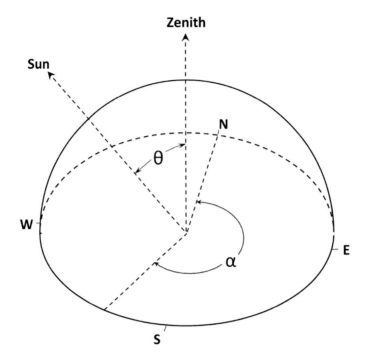

Figure 7.30 Hemispherical coordinate system.

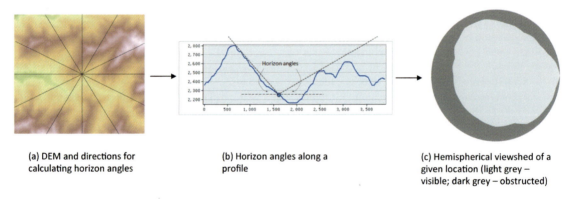

(a) DEM and directions for
calculating horizon angles

(b) Horizon angles along a
profile

(c) Hemispherical viewshed of a
given location (light grey –
visible; dark grey – obstructed)

Figure 7.31 Construction of a hemispherical viewshed: (a) DEM and directions for calculating horizon angles, (b) horizon angles along a profile and (c) hemispherical viewshed of a given location (light grey – visible; dark grey – obstructed).

It symmetrically divides the entire sky into a series of ranges of zenith and azimuth angles (Figure 7.32b). The skymap is used to compute diffuse radiation originating from all sky directions.

Direct solar radiation for a location is calculated as the sum of the direct insolation from all unobstructed sky sectors in the sunmap, which are identified by overlaying its hemispherical viewshed on the sunmap (Figure 7.33a).

The direct radiation for a sky sector with a centroid at (θ, α), I_{dir}, is calculated as follows (Fu and Rich 1999):

$$I_{dir} = SC \times \tau^{m(\theta)} \times SD_{\theta,\alpha} \times SG_{\theta,\alpha} \times \cos(A_{\theta,\alpha}) \qquad (7.22)$$

$$m(\theta) = \frac{e^{-0.000118 \times h - 1.638 \times 10^{-9} \times h^2}}{\cos\theta} \qquad (7.23)$$

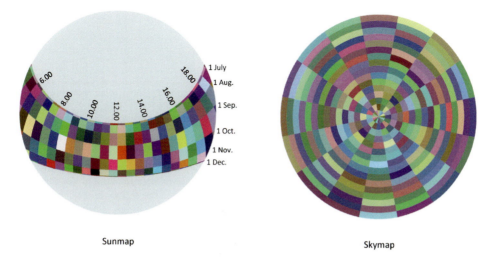

Sunmap

Skymap

Figure 7.32 Examples of sunmap and skymap: (a) sunmap and (b) skymap.

$$A_{\theta,\alpha} = \cos^{-1} \begin{bmatrix} \cos(\theta) \times \cos(S_z) + \sin(\theta) \times \\ \sin(S_z) \times \cos(\alpha - S_a) \end{bmatrix} \quad (7.24)$$

where SC is the solar constant (1,367 W/m²), τ is the transmissivity of the atmosphere (averaged over all wavelengths) in the direction of the zenith, $m(\theta)$ is the relative optical path length measured as a proportion relative to the zenith path length, h is the elevation above the sea level (in metres), $SD_{\theta,\alpha}$ is the time duration represented by the sky sector, $SG_{\theta,\alpha}$ is the gap fraction (proportion of visible sky) for the sky sector, $A_{\theta,\alpha}$ is the angle of incidence between the centroid of the sky sector and the axis normal to the surface, S_z is the surface zenith angle and S_a is the surface azimuth angle.

Diffuse solar radiation for a location is estimated as the sum of the diffuse insolation from all unobstructed sky sectors in the skymap. The unobstructed skymap sectors are found by overlaying the hemispherical viewshed on the skymap (Figure 7.33b). The diffuse

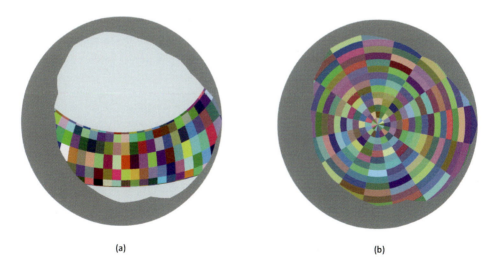

(a)

(b)

Figure 7.33 Overlaying a hemispherical viewshed on: (a) a sunmap and (b) a skymap.

radiation for a skymap sector with a centroid at (θ, α), I_{diff}, is calculated using the following equations (Fu and Rich 1999):

$$I_{diff} = R_g \times P_{diff} \times T_{int} \times KG_{\theta,\alpha} \times W_{\theta,\alpha} \times \cos(A_{\theta,\alpha}) \tag{7.25}$$

$$R_g = \frac{SC \times \Sigma \tau^{m(\theta)}}{1 - P_{diff}} \tag{7.26}$$

where R_g is the global normal radiation, P_{diff} is the proportion of global normal radiation flux that is diffused (typically about 0.2 for very clear sky conditions, and 0.7 for very cloudy sky conditions), T_{int} is the time interval for modelling, $KG_{\theta,\alpha}$ is the gap fraction for the sky sector and $W_{\theta,\alpha}$ is the proportion of diffuse radiation originating in a given sky sector relative to all sectors. When incoming diffuse radiation is assumed to be the same from all sky directions (for the uniform sky diffuse model):

$$W_{\theta,\alpha} = \frac{\cos(\theta_2) - \cos(\theta_1)}{d_a} \tag{7.27}$$

where θ_1 and θ_2 are the bounding zenith angles of the sky sector, and d_a is the number of azimuth divisions in the keymap. When diffuse radiation flux is considered to vary with zenith angle (for the standard overcast sky diffuse model):

$$W_{\theta,\alpha} = \frac{2\cos(\theta_2) + \cos(2\theta_2) - 2\cos(\theta_1) - \cos(2\theta_1)}{4d_a} \tag{7.28}$$

With this model, global solar radiation is calculated as the sum of direct and diffuse irradiance (that is, $I_{dir} + I_{diff}$) from all sky directions over a given time interval, assuming that the contribution of reflected radiation is negligible.

The above process of the construction of a hemispherical viewshed, sunmap and skymap, identification of unobstructed sky sectors in the sunmap and skymap, and calculation of direct, diffuse and global radiation can be repeated for every location on a topographic surface, resulting in a solar radiation or insolation map for a region. The ArcGIS Solar Radiation toolbox allows incoming solar radiation to be derived for a set of specific locations (point-specific) or mapped for every location over a landscape (area-based). While the *r.sun* model in GRASS is a useful tool for solar radiation mapping over a large region, the model in ArcGIS is suitable for detailed-scale studies (Šúri and Hofierka 2004). It requires a DEM, the atmospheric transmissivity and diffuse proportion as model input. The latter two input parameters can be estimated from nearby meteorological stations or using typical values. Box 7.7 illustrates how to map solar radiation in ArcGIS.

Box 7.7   Solar radiation mapping in ArcGIS   PRACTICAL

To follow this example, start ArcMap and load **dem** and **villages** from **C:\Databases\GIS4EnvSci\ VirtualCatchment\Geodata.gdb**. Enable **Spatial Analyst**.

Area solar radiation

To calculate global, direct and diffuse solar radiation for the virtual catchment within a day:

1. Open **ArcToolBox**. In the **ArcToolBox** window, navigate to **Spatial Analyst Tools > Solar Radiation**, and double-click **Area Solar Radiation**. In the **Area Solar Radiation** dialog:

    1) Select **dem** as the input raster. The latitude is filled out automatically after **dem** is selected.
    2) Enter **C:\Databases\GIS4EnvSci\VirtualCatchment\Results\globalrad** as the output global radiation raster.
    3) Set the sky size/resolution to **200**. (The resolution or sky size for the hemispherical viewshed, sunmap and skymap, and the units are cells. You may increase the sky size in order to get a more accurate result, but it will increase computation time significantly.)
    4) Select **Within a day** for the time configuration.

5) Set the day number of the year (the Julian date) to **173** (22 June 2014), the start time to **5.00** a.m. in Local Standard Time and the end time to **19.00** p.m. in Local Standard Time.

6) Uncheck **Create outputs for each interval** to calculate insolation integrated for the entire day (you may set an hour interval and tick to create outputs for hourly insolation values in a time series).

7) Click **Topographic parameters** to open the section. Select **FROM_DEM** as the slope and aspect input type so that the slope and aspect are calculated from the input DEM. Keep the default values for the Z factor and calculation directions.

8) Scroll down to the **Radiation parameters** section, and click it to open. Change the uniform sky diffuse model to the standard overcast sky diffuse model. Use the default values for transmittivity (0.5) and diffuse proportion (0.3) (although they should be set according to atmospheric conditions). Typical values for diffuse proportion are 0.2 for very clear sky conditions and 0.3 for generally clear sky conditions. Typical values for transmittivity are 0.6 or 0.7 for very clear sky conditions and 0.5 for only a generally clear sky. Also keep the default values for the zenith and azimuth divisions (all set to 8).

9) Scroll down to the **Optional outputs** section, and click it to open. Enter **C:\Databases\ GIS4EnvSci\VirtualCatchment\Results\dirrad** as the output direct radiation raster and **C:\ Databases\GIS4EnvSci\VirtualCatchment\Results\diffrad** as the name of the output diffuse radiation raster.

10) Click **OK**. Three rasters are created and added to the view. **globalrad**, as shown in Figure 7.34, represents the global radiation or total amount of incoming solar insolation (direct + diffuse) received during the day calculated for each location on the topographic surface, which is in the unit of watt hours per square metre (Wh/m^2). **dirrad** and **diffrad** represent direct and diffuse solar radiation received on the day respectively.

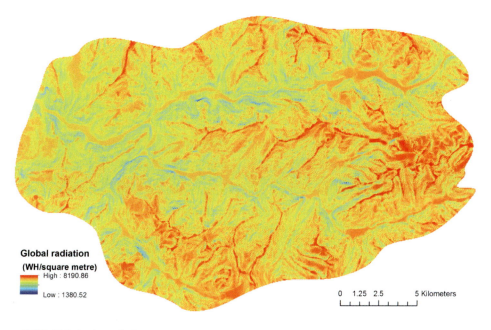

Global radiation
(WH/square metre)
High : 8190.86

Low : 1380.52

0 1.25 2.5 5 Kilometers

Figure 7.34 Global solar radiation map.

(continued)

(continued)

Points solar radiation

To calculate global solar radiation for seven villages in the virtual catchment within a year:

2. In the **ArcToolBox** window, navigate to **Spatial Analyst Tools > Solar Radiation**, and double-click **Points Solar Radiation**. This tool is similar to the **Area Solar Radiation** tool except that it requires an input point feature class or table specifying the locations for solar radiation modelling.
3. In the **Points Solar Radiation** dialog:

    1) Select **dem** as the input raster and **villages** as the input points feature. The latitude is filled out automatically.
    2) Enter **C:\Databases\GIS4EnvSci\VirtualCatchment\Results\globalpt.shp** as the name of the output global radiation features layer.
    3) Select **Whole year with monthly interval** for the time configuration, and set the year as **2013**.
    4) Check **Create outputs for each interval**.
    5) As in Step 1, select **FROM_DEM** as the slope and aspect input type, change the uniform sky diffuse model to the standard overcast sky diffuse model, and keep the default values for all other parameters.
    6) Click **OK**. **globalpt.shp** is created.

4. In the table of contents, right-click **globalpt.shp** and open its attribute table. The table contains twelve fields, **T0, T1, . . ., T11**, representing twelve months in sequence. It lists twelve output global radiation values for each location for each month. A time series of graduated symbol maps can be made to represent the monthly global radiation at the seven villages.

Solar radiation mapping provides a key tool to evaluate the solar energy potential of a particular region. It enables accurate prediction of the potential energy generation of all types of solar energy technology, such as solar photovoltaic (PV) power and thermal power generation and solar hot water, and provides information to support decision making concerning the location and feasibility of installation of solar power generation facilities. Box 7.8 presents a study on locating sites for solar PV panels based on DEM derived from LiDAR with solar radiation mapping.

Box 7.8 Locating sites for photovoltaic solar panels based on solar radiation mapping   CASE STUDY

Solar PV panels on the roofs of homes and businesses use solar radiation to generate electricity cleanly and quietly. The conversion of sunlight into electricity takes place in cells of specially fabricated semiconductor crystals. Small-scale solar PV systems generate electricity at the point of demand (that is, where people live and work) without the need to transfer energy over long distances using expensive electrical infrastructure. However, suitable sites for solar PV panels have specific characteristics and requirements. For example, solar panels should be located on top of buildings rather than at ground level, on south-facing slopes if in the Northern Hemisphere or north-facing slopes if in the Southern Hemisphere, which have a higher solar power output; on slopes of less than 35°, and at locations with high solar radiation. Chaves and Bahill (2010) designed a GIS approach to identify suitable locations for solar panels on rooftops based on the above criteria using a small area surrounding the University of Arizona in Tucson, Arizona as a case study. Figure 7.35 shows their approach.

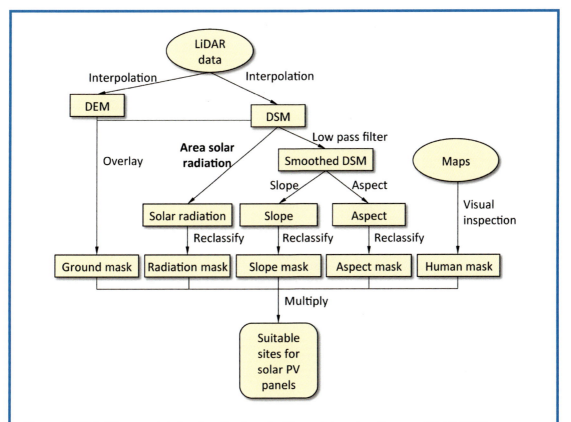

Figure 7.35 A GIS approach to locating sites for solar PV panels based on Chaves and Bahill (2010).

A DSM and a bare-earth DEM were derived from the LiDAR data, which were then georeferenced and saved in ArcGIS as floating-point rasters (see Box 8.3). Based on the DSM, a solar radiation map was created using the **Area Solar Radiation** tool in ArcGIS by specifying Tucson's latitude of 32° and a yearly interval. The solar radiation map takes into account the changes in the azimuth and position of the Sun as well as shadows cast by buildings and other surface features. The aspect and slope maps were derived from a smoothed DSM which was produced by applying a low pass filter on the DSM. Five binary masks were generated to select locations that meet the siting criteria:

- Ground mask – a binary raster in which a value of 1 was assigned for any cell having a surface elevation of greater than or equal to its bare-earth elevation plus 5 feet and a value of 0 was given to ground cells. Those cells with a value of 1 represented buildings and other objects with a certain height above ground. This mask was generated by comparing the DSM and DEM.
- Aspect mask – a binary map created by reclassifying the aspect map, in which a value of 1 represented south, southeast, southwest or flat aspects, and 0 represented other aspects.
- Slope mask – generated by reclassifying the slope map with 1 representing slopes of less than or equal to 35° and 0 for other slopes.
- Radiation mask – produced by specifying a minimum radiation threshold based on the desired power output and reclassifying the solar radiation map so that any cell having a radiation value

(continued)

(continued)

      equalling or exceeding the threshold value was assigned a value of 1, and other cells were assigned a value of 0.

- Human mask – a binary raster in which any cell that was deemed an undesirable site (such as athletic stadium seats, or roofs with pipes and air conditioning units) was assigned a value of 0, and others were assigned a value of 1.

The five masks were multiplied via map algebra into a final binary raster, in which a value of 1 represented a suitable site that met all the criteria and 0 indicated an unsuitable site. This binary raster was then converted to a vector layer using the **Raster To Polygon** tool. Each polygon represented a suitable area for solar PV panels. The number of panels for each suitable area was also calculated based on the size of the solar panels most commonly used in the study area with a 15 per cent buffer

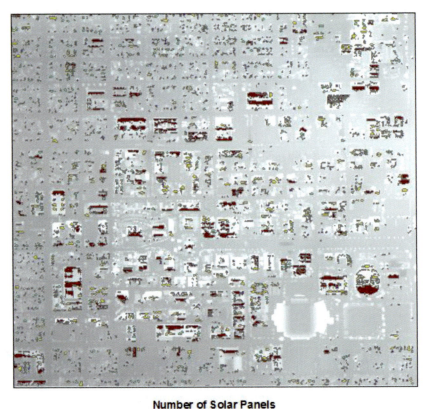

Number of Solar Panels

1- 10   11- 25   26- 50   51 - 150   151 - 1648

Figure 7.36 Suitable areas and theoretical number of solar PV panels (reproduced with permission of Andrea Chaves and A. Terry Bahill).

to accommodate the mounting frame and additional spacing. Figure 7.36 shows the final output, which identified the suitable areas (polygons) and the theoretical number of solar panels that could fit in each area.

The GIS approach did not incorporate other factors such as PV panel and inverter costs, return on investment or engineering concerns.

Source: Chaves and Bahill (2010).

7.7 SUMMARY

- Digital terrain models form an integral part of GIS, and are essential components of environmental analysis and modelling. DEM and TIN are two major digital terrain models used in digital terrain analysis.
- Hydrologically correct DEMs are free of sinks, and are normally required for hydrological applications.
- Digital terrain models can be used to extract terrain attributes that describe the morphology of the landscape, such as elevation at any point, slope, aspect, terrain profile and surface curvature, to derive hydrological parameters, such as flow direction, flow accumulation and stream order, to find features on the terrain, such as catchments and drainage networks, and to calculate measures for quantifying the influence of topography on environmental processes, such as topographic wetness, stream power, sediment transport and radiation indices.
- Visibility analysis is a method of modelling what can be seen from one or several locations. Two types of visibility analysis are available in GIS: line-of-sight analysis and viewshed analysis. Both are based on digital terrain models.
- Solar radiation mapping is another application of digital terrain models. It provides a useful tool to evaluate solar energy potential at a particular location or across a particular region.

REVIEW QUESTIONS

1. What is a digital terrain model? Discuss the differences between DEM and TIN.
2. What is a hydrologically correct DEM? Why do we need hydrologically correct DEMs for hydrological applications?
3. Explain how slope and aspect are derived from DEM and TIN.
4. Describe how a terrain profile is produced based on a digital terrain model.
5. What is the difference between profile and plan curvature? How are they computed from a DEM?
6. Describe the D8 algorithm for calculating flow direction from a DEM.
7. How is the drainage network extracted based on flow accumulation?
8. How is the catchment boundary delineated from a DEM?
9. Describe the Strahler method and the Shreve method for stream ordering.
10. What is the topographic wetness index? How is it calculated?
11. What is the stream power index? How is it calculated?
12. What is the sediment transport index? How is it calculated?
13. What is the difference between line-of-sight analysis and viewshed analysis?
14. What is a hemispherical viewshed used in solar radiation mapping?
15. What is the sunmap used in solar radiation mapping? What is the skymap?
16. Briefly describe the process of solar radiation mapping with the ArcGIS Solar Radiation tools.

REFERENCES

Ahmad, M.J. and Tiwari, G.N. (2011) 'Solar radiation models: a review', *International Journal of Energy Research*, 35:271–290.

Allwardt, P.F., Bellahsen, N. and Pollard, D.D. (2007) 'Curvature and fracturing based on global positioning system data collected at Sheep Mountain anticline, Wyoming', *Geosphere*, 3(6): 408–421.

ANU Fenner School of Environment and Society and Geoscience Australia (2008) *GEODATA 9 Second DEM and D8 Digital Elevation Model and Flow Direction Grid, User Guide*, Canberra: Geoscience Australia, accessed 29

February 2016 at http://www.ga.gov.au/metadata-gateway/metadata/record/gcat_66006.

Ben-Moshe, B., Carmi, P. and Katz, M.J. (2008) 'Approximating the visible region of a point on a terrain', *GeoInformatica*, 12(1): 21–36.

Besharat, F., Dehghan, A.A. and Faghih, A.R. (2013) 'Empirical models for estimating global solar radiation: a review and case study', *Renewable and Sustainable Energy Reviews*, 21: 798–821.

Beven K.J. and Kirkby M.J. (1979) 'A physically based, variable contributing area model of basin hydrology', *Hydrological Sciences Bulletin*, 24: 43–69.

Brunori, C.A., Civico, R. and Cinti, F.R. (2013) 'Characterization of active fault scarps from LiDAR data: a case study from Central Apennines (Italy)', *International Journal of Geographical Information Science*, 27(7): 1,405–1,416.

Chaves, A. and Bahill, A.T. (2010) 'Locating sites for photovoltaic solar panels', *ArcUser*, Fall: 24–27.

Chirico, G.B., Western, A.W., Grayson, R.B. and Blöschl, G. (2005) 'On the definition of the flow width for calculating specific catchment area patterns from gridded elevation data', *Hydrological Processes*, 19(13): 2,539–2,556.

Clark, R.L. and Lee, R. (1998) 'Development of topographic maps for precision farming with kinematic GPS', *Transactions of the American Society of Agricultural Engineers*, 41: 909–916.

De Floriani, L. and Magillo, P. (2003) 'Algorithms for visibility computation on terrains: a survey', *Environment and Planning B: Planning and Design*, 30(5): 709–728.

Dubayah, R. and Rich, P.M. (1995) 'Topographic solar radiation models for GIS', *International Journal of Geographical Information Systems*, 9: 405–419.

Florinsky, I. V. and Eilers, R. G. (2002) 'Prediction of the soil carbon content at micro-, meso- and macroscales by digital terrain modelling', in *Transactions of the 17th World Congress of Soil Science*, 14–21 August 2002, Bangkok, Thailand: ISSS (CD-ROM).

Franklin, J. (1998) 'Predicting the distribution of shrub species in southern California from climatic and terrain-derived variables', *Journal of Vegetation Science*, 9: 733–748.

Fu, P. and Rich, P.M. (1999) 'Design and implementation of the Solar Analyst: an ArcView extension for modeling solar radiation at landscape scales', in *Proceedings of the 19th Annual ESRI User Conference, San Diego, USA*, Redlands, CA: ESRI.

Greenlee, D.D. (1987) 'Raster and vector processing for scanned linework', *Photogrammetric Engineering and Remote Sensing*, 53(10): 1,383–1,387.

Guisan, A. and Zimmermann, N.E. (2000) 'Predictive habitat distribution models in ecology', *Ecological Modelling*, 135: 147–186.

Heine, R.A., Lant, C.L. and Sengupta, R.R. (2004) 'Development and comparison of approaches for automated mapping of stream channel networks', *Annals of the Association of American Geographers*, 94(3): 477–490.

Hetrick, W.A, Rich, P.M. and Weiss, S.B. (1993a) 'Modeling insolation on complex surfaces', in *Proceedings of the Thirteenth Annual ESRI User Conference*, Vol. 2, Redlands: ESRI, 447–458.

Hetrick, W.A, Rich, P.M., Barnes, F.J. and Weiss, S.B. (1993b) 'GIS-based solar radiation flux models', in *Proceedings of the ASPRS-ACSM Annual Convention*, Vol. 3, Bethesda, MD: ASPRS, 132–143.

Hirt, C., Filmer, M.S. and Featherstone, W.E. (2010) 'Comparison and validation of recent freely-available ASTER-GDEM ver1, SRTM ver4.1 and GEODATA DEM-9S ver3 digital elevation models over Australia', *Australian Journal of Earth Sciences*, 57(3): 337–347.

Hodgson, M.E. and Galle, G.L. (1999) 'A cartographic modeling approach for surface orientation-related applications', *Photogrammetric Engineering and Remote Sensing*, 65(1): 85–95.

Horn, B.K.P. (1981) 'Hill shading and the reflectance map', *Proceedings of the IEEE*, 69(1): 14–47.

Hutchinson, M.F. (1989) 'A new procedure for gridding elevation and stream line data with automatic removal of spurious pits', *Journal of Hydrology*, 106: 211–232.

Hutchinson, M.F. and Gallant, J.C. (2005) 'Representation of terrain', in P.A. Longley, M.F. Goodchild, D.J. Maguire and D.W. Rhind (eds) *Geographic Information Systems and Science*, 2nd edn, Chichester, England: John Wiley, 105–124.

Jensen, J.R. (2007) *Remote Sensing of the Environment: An Earth Resource Perspective*, 2nd edn, Upper Saddle River, NJ: Pearson Prentice Hall.

Jenson, S.K. and Domingue, J.O. (1988) 'Extracting topographic structure from digital elevation data for geographic information system analysis', *Photogrammetric Engineering and Remote Sensing*, 54(11): 1,593–1,600.

Jones, K.H. (1998) 'A comparison of algorithms used to compute hill slope as a property of the DEM', *Computers and Geosciences*, 24(4): 315–323.

Kim, Y., Rana, S. and Wise, S. (2004) 'Exploring multiple viewshed analysis using terrain features and optimisation techniques', *Computers and Geosciences*, 30(9): 1,019–1,032.

Lashermes, B., Foufoula-Georgiou, E. and Dietrich, W.E. (2007) 'Channel network extraction from high resolution topography using wavelets', *Geophysical Research Letters*, 34: L23S04.

Liu, H., Jezek, K.C. and Li, B. (1999) 'Development of an Antarctic digital elevation model by integrating cartographic and remotely sensed data: a GIS based approach', *Journal of Geophysical Research*, 104(B): 23,199–23,213.

Ma, R. (2005) 'DEM generation and building detection from LiDAR data', *Photogrammetric Engineering and Remote Sensing*, 71(7): 847–854.

MacMillan, R.A. (2008) 'Experiences with applied DSM: protocol, availability, quality and capacity building', in A.E. Hartemink, A. McBratney and M.L. Mendonca-Santos (eds) *Digital Soil Mapping with Limited Data*, Dordrecht, The Netherlands: Springer, 113–135.

Manzo, G., Tofani, V., Segoni, S., Battistini, A. and Catani, F. (2013) 'GIS techniques for regional-scale landslide susceptibility assessment: the Sicily (Italy) case study', *International Journal of Geographical Information Science*, 27(7): 1,433–1,452.

Mark, D.M. (1983) 'Automatic detection of drainage networks from digital elevation models', *Cartographica*, 21: 168–178.

Mark, D.M. (1988) 'Network models in geomorphology', in M.G. Anderson (ed.) *Modelling Geomorphological Systems*, New York: John Wiley, 73–97.

Martz, L.W. and Garbrecht, J. (1998) 'The treatment of flat areas and depressions in automated drainage analysis of raster digital elevation models', *Hydrologic Processes*, 12: 843–855.

Miklánek, P. (1993) 'The estimation of energy income in grid points over the basin using simple digital elevation model', *Annales Geophysicae, Suppl. II*, 11: 296.

Miliaresis, G.C. (2013) 'Terrain analysis for active tectonic zone characterization: a new application for MODIS night LST (MYD11C3) data set', *International Journal of Geographical Information Science*, 27(7): 1,417–1,432.

Moore, I.D. and Nieber, J.L. (1989) 'Landscape assessment of soil erosion and nonpoint source pollution', *Journal of the Minnesota Academy of Science*, 55: 18–25.

Moore, I.D. and Wilson, J.P. (1992) 'Length-slope factors for the Revised Universal Soil Loss Equation: simplified method of estimation', *Journal of Soil and Water Conservation*, 47(5): 423–428.

Moore, I.D., Burch, G.J. and Mackenzie, D.H. (1988) 'Topographic effects on the distribution of surface soil water and the location of ephemeral gullies', *Journal of American Society of Agricultural Engineers*, 31(4): 1,098–1,107.

Moore, I.D., Grayson, R.B. and Ladson, A.R (1991) 'Digital terrain modelling: a review of hydrological, geomorphological, and biological applications', *Hydrological Processes*, 5(1): 3–30.

Natural Resources Canada (1997) *Canadian Digital Elevation Data: Standards and Specifications*, Sherbrooke, Canada: Centre for Topographic Information, Natural Resources Canada.

O'Callaghan, J.F. and Mark, D.M. (1984) 'The extraction of drainage networks from digital elevation data', *Computer Vision, Graphics and Image Processing*, 28(3): 323–344.

Pallaris, K. (2000) 'Terrain modelling for erosion risk assessment in the Cabuyal river catchment: comparison of results with farmer perceptions', *Advances in Environmental Monitoring and Modelling*, 1(1): 149–177.

Park, S. J., McSweeney, K. and Lowery, B. (2001) 'Identification of the spatial distribution of soils using process-based terrain characterization', *Geoderma*, 103: 249–272.

Qin, C., Zhu, A., Pei, T., Li, B., Scholten, T., Behrens, T. and Zhou, C. (2011) 'An approach to computing topographic wetness index based on maximum downslope gradient', *Precision Agriculture*, 12(1): 32–43.

Raber, G.T., Jensen, J.R., Schill, S.R. and Schuckman, K. (2002) 'Creation of digital terrain models using an adaptive LiDAR vegetation point removal process', *Photogrammetric Engineering and Remote Sensing*, 68(12): 1,307–1,315.

Rabus, B., Eineder, M., Roth, A. and Bamler, R. (2003) 'The Shuttle Radar Topography Mission: a new class of digital elevation models acquired by spaceborne radar', *ISPRS Journal of Photogrammetry and Remote Sensing*, 57: 241–262.

Rich, P.M., Dubayah, R., Hetrick, W.A. and Saving, S.C. (1994) 'Using viewshed models to calculate intercepted solar radiation: applications in ecology', *American Society for Photogrammetry and Remote Sensing Technical Papers*, 524–529.

Ritter, P. (1987) 'A vector-based slope and aspect generation algorithm', *Photogrammetric Engineering and Remote Sensing*, 53(8): 1,109–1,111.

Sandwell, D.T. and Smith, W.H.F. (2001) 'Bathymetric estimation', in L. Fu and A. Cazenave (eds) *Satellite Altimetry and Earth Sciences*, San Diego, CA: Academic Press, 441–458.

Schmidt, J., Evans, I.S. and Brinkmann, J. (2003) 'Comparison of polynomial models for land surface curvature calculation', *International Journal of Geographical Information Science*, 17(8): 797–814.

Shreve, R.L. (1967) 'Infinite topologically random channel networks', *Journal of Geology*, 75: 178–186.

Soille, P. (2004) 'Optimal removal of spurious pits in grid digital elevation models', *Water Resources Research*, 40(12): W12509.

Sørensen, R., Zinko, U. and Seibert, J. (2005) 'On the calculation of the topographic wetness index: evaluation of different methods based on field observations', *Hydrology and Earth System Sciences Discussions*, 2: 1,807–1,834.

Spot Image (2005) *SPOT DEM Product Description*, accessed 29 February 2016 at http://www2.geo-airbusds.com/files/pmedia/public/r467_9_spot_dem_product_description.pdf.

Srivastava, K.P. and Moore, I.D. (1989) 'Application of terrain analysis to land resource investigations of small catchments in the Caribbean', in *Proceedings of the 20th International Conference of the Erosion Control Association, 15–18 February 1989, Vancouver, BC, Canada*, Steamboat Springs, CO: International Erosion Control Association, 229–249.

Strahler, A.N. (1952) 'Dynamic basis of geomorphology', *Geological Society of America Bulletin*, 63: 923–938.

Šúri, M. and Hofierka, J. (2004) 'A new GIS-based solar radiation model and its application to photovoltaic assessments', *Transactions in GIS*, 8: 175–190.

Šúri, M., Huld, T.A. and Dunlop, E.D. (2005) 'PV-GIS: a web-based solar radiation database for the calculation of PV potential in Europe', *International Journal of Sustainable Energy*, 24(2): 55–67.

USGS (1993) *Digital Elevation Models: Data Users Guide 5*, Reston, VA: US Geological Survey.

Vince, J. (2006) *Mathematics for Computer Graphics*, 2nd edn, London: Springer.

Wang, L. and Liu, H. (2006) 'An efficient method for identifying and filling surface depressions in digital elevation models for hydrologic analysis and modelling', *International Journal of Geographical Information Science*, 20(2): 193–213.

Wilson, J.P. and Gallant, J.C. (2000) 'Digital terrain analysis', in J.P. Wilson and J.C. Gallant (eds) *Terrain Analysis: Principles and Applications*, New York: John Wiley & Sons, 1–27.

Zevenbergen, L.W. and Thorne, C.R. (1987) 'Quantitative analysis of land surface topography', *Earth Surface Processes and Landforms*, 12: 47–56.

Zinko, U. (2004) *Plants Go with the Flow: Predicting Spatial Distribution of Plant Species in the Boreal Forest* (PhD thesis), Umeå, Sweden: Swedish University of Agricultural Sciences.

Spatial visualisation

This chapter discusses the basic principles and methods for using and designing effective visual and cartographic representations to explore and visualise environmental data, including cartographic representations, multivariate mapping, 3D visualisation and cartographic animation.

LEARNING OBJECTIVES

After studying this chapter, you should be able to:

- describe the basic elements of a map;
- understand different types of cartographical representations;
- explain the process of map design;
- comprehend the principles of map design;
- understand and apply the concept of visual hierarchy in map design;
- critically evaluate the designs of maps;
- design and make publication-quality maps;
- understand the nature of multivariate analysis;
- outline ways to approach multivariate mapping;
- produce interactive and animated maps.

8.1 CARTOGRAPHIC REPRESENTATIONS

In a GIS environment, environmental data analysis often starts with maps and also ends with maps. Maps are a major means of exploring data, examining intermediate analysis results and presenting final results or findings. GIS provides easy-to-use and flexible mapping tools that allow users to produce their own maps without applying cartographic rules. However, such a map may be understandable by the analyst who produced the map, but may not be effective in communicating spatial information and research findings to a wide audience. Generally speaking, the design of maps for data exploration and analysis does not necessarily have to follow cartographic principles, but maps for data presentation require cartographic design so that they can be used to communicate the spatial information and disseminate research outcomes effectively. Maps that symbolise the spatial distributions and properties of geographical features according to cartographic conventions and rules are called cartographic representations.

Cartographic elements

A cartographic representation is made up of a number of elements. These include the title, map body, legend, scale, orientation and other supplementary elements (Figure 8.1).

The map body contains the geographical features mapped using symbols, annotated using text labels and referenced in geographical coordinates or projected map coordinates. The legend explains the meaning of the symbols used to represent geographical features on the map body. Every symbol that appears on the map body should appear in the legend. The scale shows the relationship between distances on the map and on the ground. A graphic scale is preferred in a GIS environment because representative fractions change when the map is drafted at different scales. The orientation of the map is indicated using a north arrow, which may be supplemented by graticules of latitude and longitude. The above elements are found on virtually all maps. The title describes what the map is about. It is often useful to include the location of the mapped area.

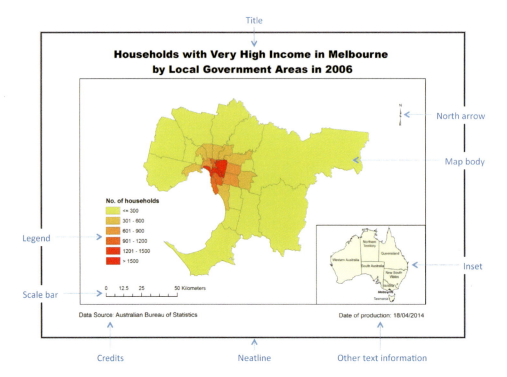

Figure 8.1 Cartographic elements.

Other supplementary elements include insets, neatlines, credits, data sources, short content descriptions, information on the map projection, supplementary tables, graphs and photos. An inset is either an enlarged or close-up map that shows a specific portion of the map body with detailed information, or a reduced map designed to place the map body in context by showing its geographical location. An inset may have its own set of cartographic elements. Neatlines are used to frame a map. Supplementary elements are used selectively to assist effective communication.

Map types

Maps are broadly categorised into two types in terms of function: general-purpose maps and thematic maps. General-purpose maps are designed to provide geographical information for reference purposes. The most common type of general-purpose map is a topographic map. Topographic maps provide comprehensive reference information about a region, including elevation (represented using contours), surface hydrology (for example, rivers, streams, lakes and coastlines), transport networks, settlements, political boundaries, and other physical, socio-economic and cultural features. Topographic maps are usually made by government agencies and issued in series of individual sheets. Figure 8.2 shows a portion of a topographic map issued by Geoscience Australia.

Another type of general-purpose map is the outline map, also called a base map. A base map provides a foundation for superimposing additional layers of information. It typically contains some physical features such as rivers and coastlines, and some man-made features such as roads and political boundaries. Figure 8.3 shows a base map of the world.

Thematic maps are used to map the spatial distribution of a single attribute or variable, or the relationship among several of them. They usually build on top of a base map in order to convey a specific theme, such as precipitation, land use, population by state or air pollution by country. Figure 8.1 is an example of a thematic map. Thematic maps symbolise geographical features according to their dimensionality (point, line, area or surface) and the level of measurement (nominal, ordinal, interval or ratio) of their attributes. As data layers in GIS are separated into point, line, area feature layers and raster surface layers, a data layer or a combination of layers can be used to make thematic maps on top of a base map. The following types

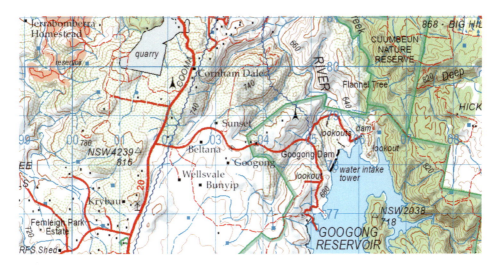

Figure 8.2 Topographic map (source: Geoscience Australia. © Commonwealth of Australia. This material is released under the Creative Commons Attribution 4.0 International Licence).

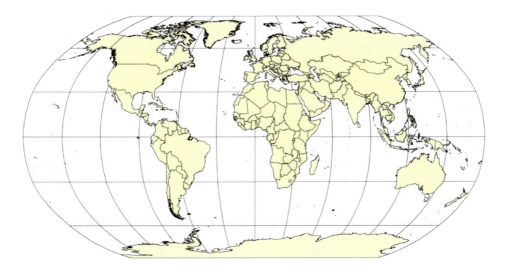

Figure 8.3 Base map of the world.

of thematic maps represent typical methods for producing cartographic representations with data layers in GIS.

Dot maps

A dot map uses dots to represent the spatial distribution of point features. Figure 5.10 is an example, showing the distribution of the mainland serow. Dots can also represent quantities. For example, Figure 8.4 shows a dot map representing population distribution against the stream network in a mountainous area. One dot in the map represents 100 persons. Therefore, areas of high dot density indicate regions of greater population, while low dot density areas indicate sparsely populated regions.

Proportional and graduated symbol maps

A proportional symbol map uses proportionally scaled symbols or icons to represent quantitative values of

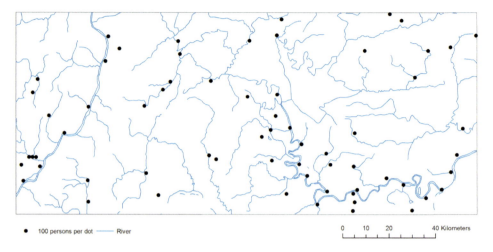

Figure 8.4 Dot map.

geographical features. Symbols are scaled in propor-tion to the magnitude of data so that the sizes (areas or widths) of the symbols are in the same ratio as the attrib-ute values they represent. A graduated symbol map uses scaled symbols or icons to represent ranges of quanti-tative values of geographical features. Different from proportional symbol maps, a graduated symbol map is constructed by first grouping the attribute values into a number of ordered ranges or classes, then assigning each class a graduated symbol from smallest to largest.

Figures 8.5 and 8.6 show the differences between the two types of symbol maps, both representing village populations.

Graduated and proportional symbol maps are mainly used to map point and line features. Symbols can be any shape, such as a circle, triangle or rectangle, with any col-our or texture (pattern). Symbols can also be pictures or icons. GIS software often provides a rich set of symbols for making maps. Box 8.1 illustrates how to make gradu-ated and proportional symbol maps in ArcGIS.

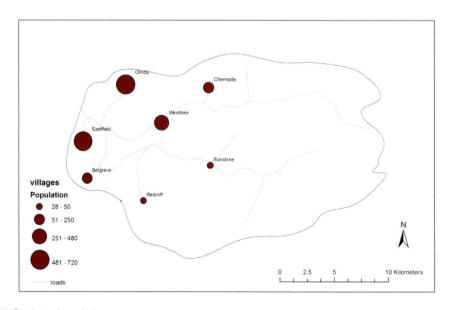

Figure 8.5 Graduated symbol map.

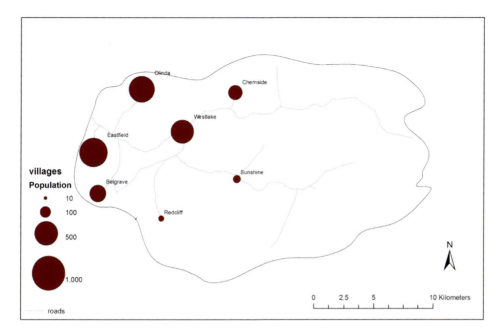

Figure 8.6 Proportional symbol map.

Box 8.1   Graduated and proportional symbol mapping in ArcGIS   PRACTICAL

To follow this example, start ArcMap, and load the **villages**, **roads** and **boundary** feature classes from **C:\Databases\GIS4EnvSci\VirtualCatchment\Geodata.gdb**. This example involves producing symbol maps using graduated and proportional symbols to show the distribution of population in the villages on a base map of a road network.

Make the base map

1. In the table of contents, click the **List By Drawing Order** button. Use the mouse to drag **villages** to the top and **boundary** to the bottom.
2. Click the symbol under **boundary**. The **Symbol Selector** dialog opens. In this dialog, a symbol palette is on the left, providing a range of symbols available in ArcGIS from which the user may choose for symbolising the features in the layer. The **Current Symbol** box is on the right, which allows the user to modify basic symbol properties.
3. In the **Current Symbol** box, change the fill colour to **No Color**. Click **OK** to close **Symbol Selector**. The **boundary** layer is displayed in the data view with no fill colour.
4. Click the symbol under **roads**. The **Symbol Selector** dialog opens:

    1) In the symbol palette, scroll down the list of line symbols to find the symbol for **Roads, Undefined**. Click the symbol, which becomes the current symbol for displaying roads.
    2) Click **OK**. Now the base map is created for mapping the populations of villages.

(continued)

(continued)

Make a graduated symbol map

5. Double-click **villages** in the table of contents. In the **Layer Properties** dialog:

   1) Click the **Symbology** tab.
   2) Click **Quantities** in the **Show** box on the left side of the dialog box, then click **Graduated symbols**.
   3) In the **Fields** box, select **Population** to map.
   4) In the **Classification** box, change the number of classes to **4**. The population values are classified into 28–50, 41–250, 251–480 and 481–720 using the default natural breaks classification method (to be discussed in Section 8.2).
   5) Change the minimum and maximum symbol sizes to **10** and **30** respectively.
   6) Click **OK**. Graduated symbols are displayed representing population classes and placed at the locations of each village. The map in the data view should look similar to Figure 8.5.

6. In the table of contents, right-click **villages**, and select **Label Features**. The village names appear besides each graduated symbol.
7. Click **View** on the main menu, and select **Layout View**. If the **Layout** toolbar is not shown, go to the main menu, click **Customize**, point to **Toolbars**, and select **Layout**.
8. In the **Layout** toolbar, click the **Change Layout** button 🖻 . Click the **ISO (A) Page Sizes** tab, and select **ISO A4 Landscape.mxd**. Click **Finish**. The map is displayed in the A4 landscape page layout.
9. Click **Insert** on the main menu, and select **Legend**. In the **Legend Wizard** dialog:

   1) In the **Legend Items** box, click **boundary**, then the **To left** button ⬅ . Click **Next**.
   2) Delete the legend title. Click **Next**. Click **Next** again, then again.
   3) Click **Finish**. The legend appears in the layout view.
   4) Drag the legend to the position shown in Figure 8.5.

10. Click **Insert** on the main menu, and select **Scale Bar**. In the **Scale Bar Selector** dialog:

    1) Click on the first scale bar in the left-side box.
    2) Click the **Properties** button. In the **Scale Bar** dialog, select **Kilometers** as the division units, then click **OK**.

11. Click **OK**. Drag the scale bar to the position shown in Figure 8.5.
12. Click **Insert** on the main menu, and select **North Arrow**. In the **North Arrow Selector** dialog, select **ESRI North 3**, and click **OK**. Drag the north arrow to the position shown in Figure 8.5. The graduated symbol map is made.
13. Click **File** on the main menu, and select **Export Map**. You can save the map as an image with the required resolution.

Make a proportional symbol map

14. Click **View** on the main menu, and select **Data View**. Now return to the data view.
15. Double-click **villages** in the table of contents. In the **Layer Properties** dialog:

    1) Click the **Symbology** tab.
    2) Click **Quantities** in the **Show** box, then click **Proportional symbols**.
    3) In the **Fields** box, select **Population** to map.

4) Click **Min Value** button in the **Symbol** box. In the **Symbol Selector** dialog, change the size of the symbol in the **Current Symbol** box to **8**, then click **OK**.
5) Change the number of symbols to display in the legend to **4**.
6) Click **OK**.

16. Change **Data View** to **Layout View**. The proportional symbol map appears as shown in Figure 8.6.

Choropleth and area qualitative maps

A choropleth map uses range-graded colours to symbolise unit areas (such as local government areas, states, countries or other kinds of enumeration districts) according to their quantitative values of a statistical variable (for example, population by state or agriculture production by country). It is made by first classifying the statistical data into ranges or classes, then shading or colouring each unit area using graduated colours according to the class they fall in. Figure 5.12 is an example of a choropleth map in which the unit areas are postcode areas, the aged population density data

are classified into five classes and each class is symbolised using a shade of a particular colour (changing gradually from light to dark as values increase).

While a choropleth map depicts a quantitative attribute of area features, an area qualitative map represents categorical or qualitative data associated with area features. On an area qualitative map, each area is symbolised using a distinctive colour or pattern representing its class. The maps in Figure 4.2 are examples of area qualitative maps. Box 8.2 shows how to make choropleth and area qualitative maps in ArcGIS.

Box 8.2   Choropleth and area qualitative mapping in ArcGIS   **PRACTICAL**

To follow this example, download the shapefile of the digital chart of the world from **http://world map.harvard.edu/data/geonode:Digital_Chart_of_the_World**. This is an ESRI product originally developed for the US Defence Mapping Agency, and is now freely available. The shapefile contains the boundaries of all countries in the world as of 1992, captured at a scale of 1:1,000,000. For the specifications of the dataset, refer to **http://earth-info.nga.mil/publications/specs/ printed/89009/89009_DCW.pdf**.

Now start ArcMap, and load the downloaded shapefile and the spreadsheet **Annual mean PM10 (WHO).xlsx** from **C:\Databases\GIS4EnvSci\Others**. The spreadsheet contains the annual mean PM10 (particulate matter with diameter of 10μm or less) data for each country (some countries have no data) sourced from the World Health Organization (WHO). When adding the spreadsheet to ArcMap, select **Sheet1$**.

Make an area qualitative map

1. In the table of contents, click **Digital_Chart_of_the_World**. A box will appear around the name, and the cursor will blink inside. Type in the new name **PM10**.
2. In the table of contents, right-click **PM10**, and select **Properties**. In the **Layer Properties** dialog:

    1) Click the **Symbology** tab.
    2) Click **Categories** in the **Show** box, then click **Unique values**.
    3) In the **Value Field** box, select **COUNTRY** to map. This field lists the country names.

(continued)

(continued)

    4)   Click the **Add All Values** button. Every area object (country or territory) is assigned a colour.

    5)   In the **Color Ramp** box, select your preferred colour series to re-assign colours to each area.

    6)   Click **OK**. An area qualitative map is generated, which is a world map. Land use, soil and vegetation maps can be produced in a similar way.

Make a choropleth map

3.   In the table of contents, right-click **PM10**, and select **Open Attribute Table**. In the **Table** window, click the **Table Options** button ▤ , point to **Joins and Relates**, then click **Join** (for more information about the **Join** operation, see Section 2.4). In the **Join Data** dialog:

    1)   Click the **What do you want to join to this layer?** arrow, then click **Join attributes from a table**.

    2)   Select **ISO_A2** as the field on which the join will be based.

    3)   Choose the table **Sheet1$** to join to the layer.

    4)   Select the field **ISO_codes** in **Sheet1$** on which the join will be based. Both **ISO_A2** and **ISO_codes** fields store country codes published by the International Organization for Standardization. The two fields are used as the key to join the table to the layer.

    5)   In **Join Options** box, check **Keep all records**.

    6)   Click **OK**. Now the data on PM10 in **Sheet1$** are joined to the map layer on a country basis. As the table **Sheet1$** does not include all the countries represented in the map data layer, many records in the joined table have no data.

    7)   Close the table window.

4.   In the table of contents, right-click **PM10**, and select **Properties**. In the **Layer Properties** dialog:

    1)   Click the **Symbology** tab.

    2)   Click **Quantities** in the **Show** box, then click **Graduated colors**.

    3)   In the **Fields** box, select **PM10_ug_m3** to map. The PM10 data (in µg/m³) are classified into five classes or ranges using the default natural breaks method (see Section 8.2).

    4)   Select a colour gradient or colour ramp for symbolisation.

    5)   Click **Label**, then click **Format Labels**. In the **Number Format** dialog, click **Numeric** in the **Category** box, change the number of decimal places to **0**, and click **OK**.

    6)   Click **OK**. A choropleth map is created, but those countries or areas with no data do not appear on the map. To show them, add another copy of **Digital_Chart_of_the_World** to ArcMap, and symbolise this new layer as a single, unique colour or pattern as described below.

5.   Add **Digital_Chart_of_the_World** by clicking the **Add Data** button.

6.   Change the name of the newly added layer to **No data**.

7.   In the table of contents, click the symbol under **No data**. In the **Symbol Selector** dialog, select **Hollow** from the symbol palette to symbolise the countries and areas with no data with their outlines, but no fill colour. Click **OK**.

8.   Follow the instructions in Box 8.1 to make a layout view of the map, and add a legend and scale bar. The choropleth map of annual mean PM3 in the world should look similar to Figure 8.7.

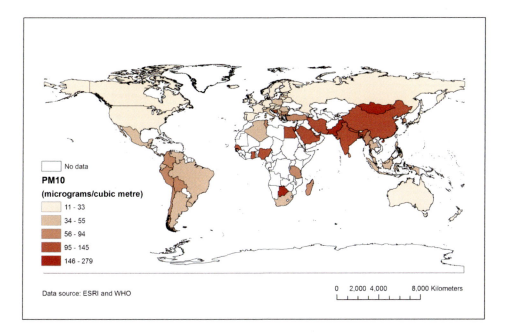

Figure 8.7 Annual mean PM3 in the world.

Isoline maps

Isoline maps consist of lines joining points of equal value of a continuous variable, and are used to map continuous surface features such as temperature and air pressure. An isoline is also called an isogram or isarithm. The isoline interval is the difference in value between two adjacent isolines. Isolines never cross or divide. An isoline map representing terrain is usually called a contour map. Figure 2.2b is an example of a contour map. In order to improve visual effects, a hypsometric map is often created by means of shading, tinting or batching the space between contour lines, which represents a terrain surface effectively as a series of elevation zones.

In most environmental applications of GIS, surface features are often represented using rasters. Although isolines can be interpolated using raster data, surfaces are often mapped as unclassed surface maps, which display continuous raster cell values across a gradual ramp of colours, or classified surface maps, which group cell values into ranges and shade or colour each range using a gradual ramp of colours. In an unclassed map, each data value can theoretically be displayed as a different symbol (for example, a raster with 200 different cell values might be depicted using 200 different shades of grey). For example, Figure 2.2c is an unclassed terrain map. A classified surface map is similar to a hypsometric map, in which different zones are shaded through hypsometric tinting (also called layer tinting). Figures 8.8a, 4.33 and 4.34 are examples of classified surface maps.

Hill-shaded maps

A hill-shaded (or shaded relief) map is a 2D representation of 3D terrain surfaces produced by simulating the appearance of the effects of sunlight illuminating the surface of the land. A mountain slope directly facing incoming light will be very bright, while a slope opposite to the light source appears dark. It provides a bird's eye view of the terrain, as shown in Figure 2.2a.

By tradition, hill-shaded maps are shaded as though they are illuminated from the northwest. In GIS, hill-shaded maps can be created by setting different Sun altitudes and azimuths, simulating the Sun's relative position at different times during a day or different seasons. The Sun's altitude is the angle of the incoming light measured above the horizon. It ranges from 0 to 90°, with 0° indicating that the Sun is on the horizon, and 90° indicating that the Sun is directly overhead. The Sun's azimuth is the direction of the incoming light relative to

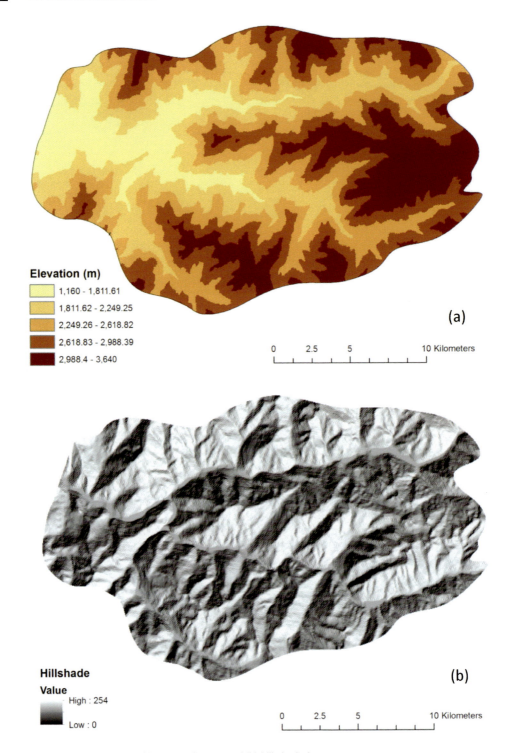

Elevation (m)

- ☐ 1,160 - 1,811.61
- ☐ 1,811.62 - 2,249.25
- ☐ 2,249.26 - 2,618.82
- ☐ 2,618.83 - 2,988.39
- ☐ 2,988.4 - 3,640

(a)

0   2.5   5   10 Kilometers

Hillshade

Value
High : 254

Low : 0

(b)

0   2.5   5   10 Kilometers

Figure 8.8 2D surface maps: (a) Hypsometric map and (b) hill-shaded map.

the north, ranging from 0 to 360° in a clockwise direction. An azimuth of 0° indicates north, east is 90°, south is 180° and west is 270°. Figure 8.8b shows a hill-shaded map produced with the Sun's altitude as 45° and the Sun's azimuth as 135°.

Hill shading adds to the visual quality of the map and makes the shape of terrain features more discernible. But by itself it does not really add to the information provided by the map. It is particularly useful for depicting mountainous areas. Some maps combine hill shading and contour lines or hypsometric tinting to produce maps which are easy to interpret due to the hill shading and provide actual elevation information from the contour lines or elevation zoning. Hill shading also provides a useful background or base map for thematic mapping.

Box 8.3 shows how to make 2D surface maps in ArcGIS.

Box 8.3   2D surface mapping in ArcGIS   **PRACTICAL**

To follow this example, start ArcMap, and load the **boundary** feature class and the **dem** raster layer from **C:\Databases\GIS4EnvSci\VirtualCatchment\Geodatabase.gdb**. The **boundary** layer is used as a base map. The **dem** raster is a digital terrain model of the hypothetic virtual catchment. There are two types of rasters supported by ArcGIS: integer raster and floating-point raster. An integer raster uses whole numbers to record the cell values. In a floating-point raster, cell values are represented as numbers with a decimal portion. Integer rasters commonly represent categorical data, and floating-point rasters commonly represent continuous surfaces. **dem** is a floating-point raster. When a floating-point raster is loaded into ArcMap, it is displayed as an unclassed map by default.

Make a classified surface map

1. Map **boundary** with no fill colour as in Step 7 in Box 8.2.
2. In the table of contents, right-click **dem**, and select **Properties**. In the **Layer Properties** dialog:

    1) Click the **Symbology** tab.
    2) Click **Classified** in the **Show** box. The elevation is classified into five ranges using the default natural breaks classification method.
    3) Choose a colour ramp.
    4) Click **Label**, then click **Format Labels**. In the **Number Format** dialog, click **Numeric** in the **Category** box, change the number of decimal places to **2**, and click **OK**.
    5) Click **OK**. A classified elevation or hypsometric map is created, as shown in Figure 8.8a.

Make an unclassed surface map

3. In the table of contents, right-click **dem**, and select **Properties**. In the **Layer Properties** dialog:

    1) Click the **Symbology** tab.
    2) Click **Stretched** in the **Show** box.
    3) Choose a colour ramp. Click **OK**. The DEM is displayed as an unclassed map, as in Figure 2.2c.

Make a contour map

4. Set both the ArcGIS current workspace and scratch workspace as **C:\Databases\GIS4EnvSci\ VirtualCatchment\Results** by clicking **Geoprocessing > Environments** on the ArcMap main menu bar.

(continued)

(continued)

5. Open **ArcToolBox**. Navigate to **Spatial Analyst Tools > Surface**, and double-click **Contour**. In the **Contour** dialog:

    1) Select **dem** as the input raster.
    2) Enter **contour200** as the output polyline feature.
    3) Enter **200** as the contour interval.
    4) Click **OK**. The contour map is created and saved in the workspace. It appears similar to the map in Figure 2.2b.

Make a hill-shaded map

6. In **ArcToolBox**, navigate to **Spatial Analyst Tools > Surface**, and double-click **Hillshade**. In the **Hillshade** dialog:

    1) Select **dem** as the input raster.
    2) Name the output **hillshade**.
    3) Enter **135** as the azimuth and **45** as the altitude.
    4) Click **OK**. The hill-shaded map is created, as shown in Figure 8.8b. Compare this hill-shaded map with that in Figure 2.2a to see the difference.

8.2 MAP DESIGN

As discussed in the previous section, maps symbolise geographical features according to their dimensionality (point, line, area or surface) and the level of measurement (qualitative or quantitative) of their attributes. What geographical features are to be presented, how they are to be classified, what symbols are to be used, how texts are to be placed and how cartographic elements are to be laid out are the major issues in map design. In general, map design involves selection, classification, symbolisation and layout design.

Selection

Map design depends on the purpose of the map. Once the purpose has been decided, the geographical area to be included on the map must be determined. This may be a city, a catchment, a region or the entire world. The larger the area covered, the greater the Earth's curvature involved in the map. Therefore, selection may involve making a choice of map projections, particularly for small-scale mapping over a large region or when detailed measurements need to be made from the map.

All projected maps are distorted in some way (see Section 2.2). A map projection should be chosen so that it does not distort the spatial pattern, or to relegate the distortions away from the area of main interest on the map.

Equal area or equivalent projections are generally best for thematic mapping of data distributions. For example, for correct interpretation of point patterns and densities, an equivalent projection is required. Equidistant map projections should be used when measuring distances from a point. Conformal projections are often used for large-scale reference maps and for maps showing routes or shapes, as they preserve angles rather than areas. Shape, size and location of the mapped area are also important determinants in making decisions about map projections. Table 2.2 provides some useful recommendations.

The most important task in the selection stage is cartographic abstraction and generalisation. Not all geographical features and their properties can be represented in a map. The selection process aims to identify what is relevant given the purpose of the map, and select and organise information to help users interpret spatial patterns and processes. Section 2.1 discussed the principles and methods of cartographic abstraction and generalisation.

Classification

For most maps, it is necessary to classify data before mapping them. It is not usually practical to have a unique symbol for each data record for every feature because the human eye only has a limited ability to discriminate a large number of different symbols or colour

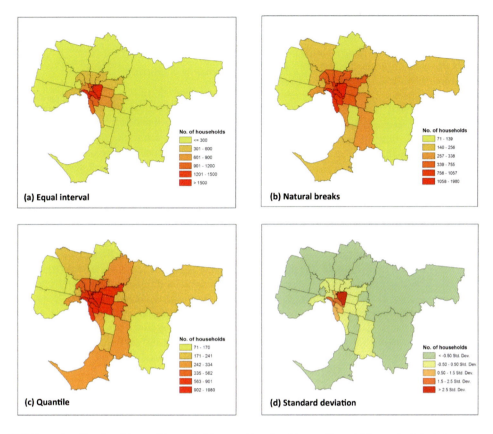

Figure 8.9 Data classification: (a) equal interval, (b) natural breaks, (c) quantile and (d) standard deviation.

shades. Data classification is a means of generalising the-matic maps. Generalisation helps make sense of complex data. Classification involves grouping similar features into classes to reduce complexity and organise data. It is mainly based on the attribute values associated with geographical features. Qualitative attribute data are often organised into classes (categories) and subclasses. For example, production from relatively natural environ-ments is a class of land use. It can be divided into two subclasses: grazing natural vegetation and production forestry. Production forestry can be further divided into two subclasses: wood production and other forest pro-duction. When mapping qualitative attribute values of geographical features, they are grouped into classes or subclasses. Each class or subclass is symbolised using a distinct colour or pattern.

Quantitative attribute data, describing counts, amounts, ratios or ranked values, are generally classified into a number of classes, each representing a range of numerical values. Many data classification methods exist. The classification methods most commonly available in

GIS include equal interval, natural breaks, quantile and standard deviation.

Equal interval

This classification method divides the data into equal-sized ranges. For example, if features have values rang-ing from 0 to 1,000 and they are to be grouped into five classes, each class will have an interval of 200 and the five classes will be 0–200, 201–400, 401–600, 601–800 and 801–1,000. In general, the class interval is determined by dividing the whole range of all the data (the highest data value minus the lowest data value) by the number of classes. Figure 8.1 is a wealthy household map clas-sified into six classes using the equal interval method, with a constant class interval of 300. It is reproduced in Figure 8.9a for comparison.

The equal interval method is easy to understand and most suitable for familiar data ranges such as percent-ages. It is also useful for comparing the amount of a value with other values. For example, we can show that an agricultural production region is part of the group of agricultural production areas that made up the top one-fifth of the gross annual revenues of agricultural production. However, this classification method may create classes with unequal distribution of features, and in some cases may produce classes containing no features. In the example shown in Figure 8.9a, the fifth class has no data (that is, no local government areas have a number of wealthy households within the range 1,201–1,500), therefore it is not shown in the map.

Natural breaks

The natural breaks method is a variance-minimised (within classes) classification method. It is based on natural groupings of data values associated with geographical features. This method identifies natural breaks by identifying patterns inherent in the data. The data are divided into classes whose boundaries are set where there are relatively large changes in the data values, so that the differences in data values between classes are maximised while the differences in data values in the same class are minimised. This method is carried out in GIS by using a specific algorithm, such as the Jenks algorithm (Jenks and Caspall 1971). The Jenks algorithm is based on minimising the sum of squared deviations of class members from the mean values. It starts with a random set of classes, and adjusts the class boundaries to reduce the sum of squared deviations in each class in an iterative process (Slocum 2005).

This method is used when natural breaks occur, but often data are too scattered to provide clear breaks. It is generally good at revealing the patterns inherent in the data. However, class breaks are typically uneven, unusual and data-specific. Therefore, it is unsuitable for comparing maps based on different underlying information. Figure 8.9b shows the wealthy household map classified using the natural breaks method.

Quantile

This method groups data so that each class contains an equal number of features. Each class is approximately equally represented on the map. It is well suited to linearly distributed data and very useful for ordinal data, as the class assignment of quantiles is based on ranked data. Class ranges are usually irregular as features are grouped by the number in each class, as shown in Figure 8.9c. The method may produce a visually attractive, but misleading map. This is because the classification may result in a situation where similar features are placed in adjacent classes, while features with widely different values are put in the same class. This distortion could be minimised by increasing the number of classes.

Standard deviation

This method classifies data according to their deviation from the mean of the dataset. It first calculates the mean value, then generates class breaks by successively adding or subtracting the standard deviation from the mean. Class intervals can be one standard deviation, ½ standard deviation, ⅓ standard deviation or ¼ standard deviation. This often results in a central class defined as the mean value +/− half a class interval. For example, if the class interval is one standard deviation, the central class will range from the mean minus 0.5 standard deviations to the mean plus 0.5 standard deviations, with additional classes at +/− 1 standard deviation intervals beyond this central class. Figure 8.9d provides an example. This method is particularly useful when the purpose is to show deviation from the mean.

The maps shown in Figure 8.9 are classified using different data classification methods based on the same dataset. Obviously, different methods can lead to different representations of the same data, from which substantially different conclusions may be drawn. There is no single unique method for data classification. Experimentation and judgement are critical. An appropriate classification method should help reveal patterns and anomalies that otherwise might be obscured. The general principle is that the chosen classification method should produce meaningful patterns, convey clear messages and be easily interpretable by users. For ease of interpretation, an important consideration in any method of classification is selecting an appropriate number of classes. Due to users' inability to distinguish between differing symbols, it is generally recommended to have five to eight classes. More classes may not add much to the maps, and will make interpreting the main features less straightforward. In addition, the entire range of data should be included with no overlapping classes and no gaps within the data range.

Box 8.4 demonstrates how to classify quantitative data for thematic mapping in ArcGIS.

Symbolisation

Symbolisation is the process of choosing symbols and using visual variables to represent the geographical features in some logical manner. It is critical to a map's success, as everything on a map is a symbol. GIS provides a large collection of symbols, and allows users to set visual variables to create new ones. However, map symbols

---

Box 8.4 **Quantitative data classification for thematic mapping in ArcGIS**

PRACTICAL

To follow this example, start ArcMap, and load the **Digital_Chart_of_the_World** shapefile and the spreadsheet **Annual mean PM10 (WHO).xlsx**.

1. Rename **Digital_Chart_of_the_World** to **PM10** in the table of contents.
2. Refer to Step 3 in Box 8.2 to join the spreadsheet to the shapefile.
3. In the table of contents, right-click **PM10**, and select **Properties**. In the **Layer Properties** dialog:
   1) Click the **Symbology** tab.
   2) Click **Quantities** in the **Show** box, then click **Graduated colors**.
   3) In the **Fields** box, select **PM10_ug_m3**.
   4) Click the **Classify** button.
   5) In the **Classification** dialog, click the **Method** arrow, and select **Equal Interval**. Click the up/down arrow of the **Classes** input box, then select **6** as the desired number of classes. Click **OK**. The system will determine the class interval (or interval size) according to the range of the data and the specified number of classes, and calculate the class boundaries.
   6) Click **Label**, then click **Format Labels**. In the **Number Format** dialog, click **Numeric** in the **Category** box, change the number of decimal places to **0**, and click **OK**.
   7) Click **OK**. The annual mean PM10 world map is created, which is classified using the equal interval method.
4. Refer to Steps 5–8 in Box 8.2 to add another copy of **Digital_Chart_of_the_World** in order to show the countries with no data, and to make a layout view of the map.
5. Click **File** on the main menu, and select **Export Map** to save the map as an image.
6. Return to the **Data View**, and repeat Steps 3–5 above to classify **PM10_ug_m3** data using the **Classification** dialog with the natural breaks, quantile and standard deviation methods respectively and produce individual maps.
7. Compare the maps produced and Figure 8.7, and assess the differences among these maps.

ArcGIS also supports manual and defined interval methods for data classification. Manual classification allows the user to specify the class boundaries (or break values). The defined interval method is a variant of equal interval in which the user defines a class interval, and the system will automatically determine the number of classes based on the interval size and calculate the break values of each class. The two methods are useful when mapping data according to certain standards and guidelines, or when grouping features with particular threshold values – for example, producing a temperature map classified with a 10° temperature band. All the classification methods discussed in this section can be applied to classifying quantitative data for graduated symbology (including graduated colours and symbols), and can be used to make choropleth maps, graduated (point or line) symbol maps and classified surface maps.

---

must be chosen consciously and appropriately. As a general principle, map symbols should be simple, intuitive, systematic and consistent.

Map symbols are graphic marks that relate to the actual geographical features in the real world. They need to be simple so that the map is easy to read and conveys clear information while representing a sufficient level of detail about geographical features. Wherever possible, map symbols should be self-explanatory and resemble the features they represent, so that users can quickly relate the symbols to the real-world features. For example, an airplane symbol can be used to show the location

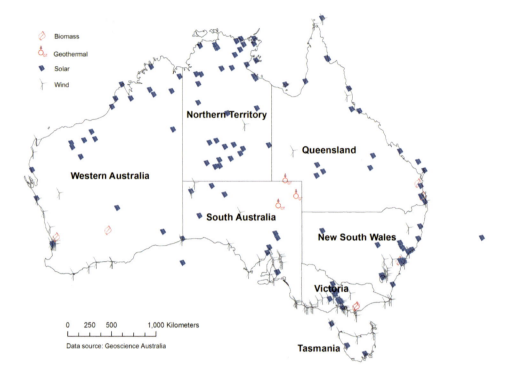

Figure 8.10 Use of shape to symbolise different types of renewable energy power stations in Australia.

of airports, and a group of tree symbols regularly placed in an area may be used to represent a forest. In Figure 8.10, the turbine symbol represents wind farms, and the solar panel symbol suggests the location of solar power stations. When map symbols do not resemble what they symbolise, they should be symbols by convention. For example, most maps use black for cultural features, blue for water features, brown for contours representing relief, green for vegetation and red or black for roads. When less familiar or unconventional symbols are used, they are often labelled to prevent misunderstanding. More importantly, map symbols should be systematic, so that readers can easily distinguish between classes and sub-classes or relative importance of features on the map. For instance, when showing road networks using line symbols, we may use a double solid line symbol for dual carriageways, a single solid line symbol for roads and a single dashed line symbol for tracks. In this case, the size and form of the symbols correspond to the road classification system. In addition, map symbols should be consistent with standards. National agencies in most countries have established cartographic standards for topographic map series and for soil and geologic mapping. These standards should be adopted where applicable.

Visual variables

A map symbol is a visual mark connected to the data at a point, along a line, or in an area. According to feature types they represent, map symbols may be point, line or area symbols. Point symbols can take the form of dots, icons, pictures, letters, circles, squares or other geometric shapes. Line symbols can be single lines, double lines, dotted lines, dashed lines or hatched lines. Area symbols take the form of colouring, line-shading, crosshatching and dot patterns. Every map symbol is defined by shape, size, colour, pattern and orientation, which are called visual variables (Figure 8.11). Visual variables are used to represent qualitative or quantitative differences of geographical features.

Shape is a visual variable of point, line and area symbols. However, with regard to line and area symbols, it refers to the individual graphic elements that are used to construct the symbol, rather than the overall form of the symbol. Shape suggests a qualitative difference, and is therefore used for symbolising qualitative data. Figure 8.10 is an example of the use of shape to differentiate different types of renewable energy power stations in Australia, including biomass, geothermal, solar and wind power stations.

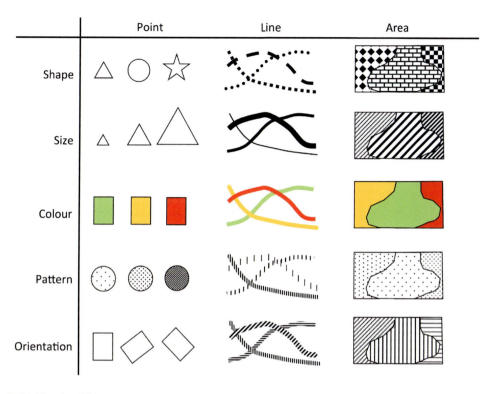

Figure 8.11 Visual variables.

Size refers to the dimensions of point and line symbols, or the dimensions of individual graphic elements of area symbols. This variable is used to represent quantitative differences, and is thus used to map quantitative data. The graduated and proportional symbols used in Figures 8.5 and 8.6 are examples of using the symbol size to depict differences in quantity.

Colour is a complex visual variable, but the most powerful and most frequently used one. In GIS, colour is specified by RGB (red, green and blue triplets) or HSI (hue, saturation and intensity) values. The RGB system uses integer numbers (0–255) to specify the intensities of the red, green and blue light in the colour. For example, (0, 0, 0) specifies black, (255, 255, 255) is white, (255, 0, 0) represents red and (128, 128, 128) denotes mid-grey.

The HSI model uses hue, saturation and intensity to specify a colour. The hue is the basic colour we perceive, determined by the dominant wavelength of light, going from red at the long-wave end of the visible spectrum to blue at the other extreme. The intensity is the illumination effect or brightness of the colour, determined by the actual amount of light, with more light corresponding to more intense colours. The saturation is the amount of the colour per unit display area, determined by the excitation purity, and depending on the amount of white light mixed with the hue. A pure hue is fully saturated – that is, no white light is mixed in. Less-saturated colours, if available, are more suitable for mapping. Hue and saturation together determine the chromaticity for a given colour. By cartographic convention, hue is used to show qualitative differences, while the variation in saturation or intensity of a colour implies quantitative difference (usually, light means less and dark means more).

Pattern is an arrangement of graphic elements (such as dots, lines, markers or pictures) used to fill in a symbol in a recognisable and repeatable structure. Variations in the density, shape and size of the graphic elements in a symbol can be used to show both qualitative and quantitative differences. When showing quantitative differences, patterns should appear gradually and smoothly from light to dark, with dark usually being high, and light low.

Orientation is the direction in which a point symbol or the individual graphical elements in a line or area symbol are placed. Orientation is mainly used to show qualitative differences. When orientation and density are used together, the symbol may show both qualitative and quantitative differences.

Visual variables selected for map symbolisation need to match the data. If the data are qualitative, a visual variable that implies qualitative differences such as shape, colour hue, pattern or orientation should be chosen. If the data are quantitative, a visual variable that suggests quantitative differences such as size, colour intensity or saturation, or a gradual shading pattern can be applied. Several visual variables can be used in combination to represent both qualitative and quantitative characteristics of geographical features.

Visual hierarchy

In addition to the selection of appropriate visual variables to symbolise the data, map design must consider the overall visual effect of the map, and build a clear, effective visual hierarchy in order to aid comprehension and reinforce the message the map intends to convey. A visual hierarchy is the perceptual organisation of mapped features in such a way that they appear visually to lie in a set of layers of increasing importance as they approach the user. It structures some features as foreground (at the top of the hierarchy) and some as background (at the lowest level of the hierarchy). The foreground features are most visually prominent, and should be those directly related to the map theme. They are the primary features and the most important aspects of the map, and so should be noticed first. The secondary and supporting features that supply additional information should have decreasing visual importance, placed in the lower levels of the visual hierarchy or in the background.

A visual hierarchy on a map is established by discernible visual contrasts or visual differences in colour intensity, colour hue, shading pattern and size of map symbols. Features with the highest contrast to their surrounding features are recognised first by the human mind. Generally, dark colours stand before light colours. Vivid or intense colours are more visually prominent than dull colours. Isolated coarser and complex textures of patterns tend to stand out and move to higher visual levels. Larger symbols grab our attention first, and so come across as more important. Therefore, we tend to use vivid, dark, prominent or large symbols for primary features, and dull, light, less distinct or small symbols for secondary and supporting features. In this way, the layers and order of the features on the map can be clearly presented.

There are three primary methods for building a visual hierarchy on a map: stereogrammic, extensional and subdivisional (Robinson et al. 1995). The stereogrammic method uses a number of graphical mechanisms (such as shading) to enhance the separation of mapped features from their surrounding background so that the mapped features stand out prominently. For example, Figure 8.12a is a portion of the map of urban population in Europe without visual differences between cities (represented as circles), land and ocean. All features are located in the same visual plane. If you are not very familiar with Europe, you may not know whether Locations A and B labelled on the map are land or water. Urban population as the map theme is not visually prominent. The graduated circle symbols intersect with country boundaries and coastlines, creating an unintelligible mess. In Figure 8.12b, the water is shaded in blue, which makes the land stand out. The red graduated symbols representing urban population rise above the visual plane of the land. A three-level visual hierarchy is effectively established in Figure 8.12b. Graticules or grids of latitude and longitude can also be added so that the user can easily identify the important mapping area as well as the background or non-analysed portions of the map, as shown in Figure 8.12c. The stereogrammic method mainly uses progression in colour intensity and differences in hue and shading pattern to enhance visual depth. The depth perception is known as stereogrammic hierarchy.

The extensional method is mainly used to present the order or ranking of line or point features, such as stream orders in a drainage network, road classes in a road network and a hierarchy of settlements. It utilises the visual variables of size, colour intensity, colour hue and pattern to show the visual differences separating different orders and rankings and to make the top-ranking or high-order features stand out. For example, Figure 8.13 uses the size (thickness) and pattern of line symbols to portray different classes of roads in a road network, in which dual carriageways are most prominent, principal roads are wider than secondary roads, secondary roads are wider than minor roads, and tracks are shown as dashed lines. While the stereogrammic method involves creating a stereogrammic hierarchy that makes the important features appear above the background base data layers, or forms layers of information in accordance to importance, the extensional method aims to show relative importance among features.

The subdivisional method uses the visual variables of colour and pattern to distinguish between classes and subclasses of area features. Each class has distinctive visual homogeneity. Distinctions between subclasses of each class are made less prominent than those between classes, so that the user can instantly recognise the differences

(a) (b) (c)

Figure 8.12 Examples of the stereogrammic method for establishing a visual hierarchy.

Dual Carriageway
Principal Road
Secondary Road
Minor Road
Track

Figure 8.13 An example of the extensional method for establishing a visual hierarchy.

between classes. For example, we can use various line hatch patterns to depict secondary classes of the primary land use class 'dryland agriculture and plantations' that is coloured in golden yellow, different crosshatch patterns to describe secondary classes of the primary land use class 'irrigated and agriculture and plantations' coloured in blue, and variations of stipple patterns to portray secondary classes of the primary land use class 'conservation and natural environment' coloured in green, as shown in Figure 8.14.

Map lettering

Text is an integral part of a map. In addition to text in the title, legend, scale and credit on a map, text is often used to label or annotate features in the map body.

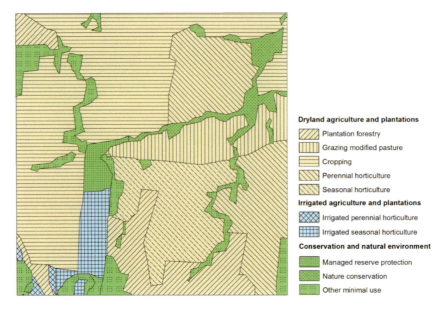

Dryland agriculture and plantations

- Plantation forestry
- Grazing modified pasture
- Cropping
- Perennial horticulture
- Seasonal horticulture

Irrigated agriculture and plantations

- Irrigated perennial horticulture
- Irrigated seasonal horticulture

Conservation and natural environment

- Managed reserve protection
- Nature conservation
- Other minimal use

Figure 8.14 An example of the subdivisional method for establishing a visual hierarchy.

Annotation can be used to name geographical features (such as the name of a city, a lake or a highway), describe qualitative attributes (such as the dominant tree species of a forest stand or water quality of a well) and to attach numerical values (such as the elevation value of a contour line, the depth of a river or the width of a road). Text information can enrich the map contents and provide information about geographical features that cannot be represented using graphic symbols. However, map lettering should not reduce the legibility of the map. The use of text must adhere to the following principles:

1. Text should be easily recognisable.
2. Text should show a hierarchy among geographical features, and help distinguish between classes and subclasses.
3. Annotation should clearly identify the feature to which it refers without ambiguity, and should not obscure the adjacent important features.
4. Annotation should not be used excessively, and should only be used where necessary.
5. The density of annotation distributed across the map should reflect the relative density of the distribution of the features.

These principles can be applied through the choice of font, colour, size and position of text for map lettering.

Different fonts of characters have different styles (for example, Arial, Times New Roman and Calibri), various weights (the thickness relative to its height, ranging from thin through light, regular, bold to extra-bold or black) and different widths (from narrow to wide). They may be upright or slanted, regular or italic and uppercase or lowercase. Generally, font style and colour suggest differences in categories; and size and boldness imply relative importance, order, ranking or quantitative differences. The weight of a font determines readability. The thinner the letter, the harder it is to read. However, bold and large lettering may mask other important features. Careful selection of font, colour and size is required so that map lettering will not change the effective visual hierarchy established by map symbols.

Text placement should reflect characteristics of the location being labelled. There are a number of rules for text placement. For a point feature, the most preferred position for annotation is above and to the right of the feature. Figure 8.15 shows the possible positions of annotation for a point feature ranked in order of decreasing preference. In addition, the names of places on the shoreline of large water bodies should be placed on the water, land features should be labelled on the land and water features on the water, and places on one side of a river should be annotated on the same side.

Text for line features should follow line directions, curved or slanted if they are rivers, and is better placed

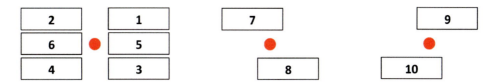

Figure 8.15 Preferred positions for labelling a point feature.

Figure 8.16 Map lettering applied with text placement rules.

above than below the features. Horizontal text is easiest to locate and read. Text should never be upside down. If vertical, text should read up on the left of the map and down on the right. Names along long features such as rivers and highways need to be repeated rather than spaced out. Area features should be labelled along the longest direction or a gently curved line following the shape of the features. Area labels should be kept as horizontal as possible, be away from borders and avoid hyphenation. Figure 8.16 shows text placement for point, line and area features following some of the rules discussed above, and Box 8.5 illustrates how ArcGIS is used to symbolise and label those features.

Box 8.5   Map symbolisation and lettering in ArcGIS   **PRACTICAL**

To follow this example, start ArcMap, and load the **villages**, **roads**, **parks** and **boundary** feature classes from **C:\Databases\GIS4EnvSci\VirtualCatchment\Geodata.gdb**. The stereogrammic method is used to build a visual hierarchy.

Select visual variables to symbolise features

1.   In the table of contents, move the layer **villages** to the top, the layer **roads** immediately below, the layer **parks** under **roads**, and the layer **boundary** to the bottom. Generally, point and line feature layers should be placed above area feature layers when making a map.

(continued)

(continued)

2. In the table of contents, click the symbol under **boundary**. The **Symbol Selector** dialog appears. In this dialog:

    1) In the **Current Symbol** box on the right, select **Gray 10%** as both the fill and outline colours (set the colour hue and intensity values).
    2) Click **OK**. The mapped area is effectively separated from the background.

3. Click the symbol under **parks** in the table of contents. In the **Symbol Selector** dialog, set the fill and outline colours of the parks as **Leaf Green**, then click **OK**.
4. Click the symbol under **roads** in the table of contents. In the **Symbol Selector** dialog, change its colour to **Gray 40%**, then click **OK**.
5. Click the symbol under **villages** in the table of contents. In the **Symbol Selector** dialog, select the **Square 1** symbol from the marker palette on the right to change the shape of the village symbol to a square. Set its colour to **Mars Red** and size to **9**, then click **OK**. A three-level visual hierarchy is established. Parks stand out most prominently, and villages are above roads.

Label features

6. Double-click **parks** in the table of contents. In the **Layer Properties** dialog, click the **Labels** tab, then:

    1) Tick **Label features in this layer**.
    2) Set the method as **Label all the features the same way**.
    3) Select **name** as the label field.
    4) In the **Text Symbol** box, select the font **New Times Roman**, set its size as **10**, and change the text colour to **white** (as parks are shaded in leaf green).
    5) Click the **Placement Properties** button. In the **Placement Properties** dialog:

        a. Click the **Placement** tab.
        b. Select **Always straight** (each label will be placed by following the longest direction of the area feature being labelled).
        c. Tick **Only place label inside polygon**.
        d. Select **Place one label per feature**.
        e. Click the **Conflict Detection** tab, and set both the label weight and feature weight as **High**. (This step assigns relative importance to labels and features to be used when there is an overlap between a label and a feature. As a general rule, a feature cannot be overlapped by a label with an equal or lesser weight. More important labels are assigned higher label weights. Setting a feature weight of high for parks ensures that no labels will be placed on the outline of the parks.)
        f. Click **OK**.

    6) Click **OK**. The names of the parks are placed on the map.

7. Double-click **roads** in the table of contents. In the **Layer Properties** dialog, click the **Labels** tab, then:

    1) Repeat Steps 1), 2) and 3) in Step 6 above.
    2) In the **Text Symbol** box, select the font **Arial**, set its size as **8**, and the text colour as **black**.
    3) Click the **Placement Properties** button. In the **Placement Properties** dialog:

        a. Click the **Placement** tab.
        b. Select **Curved** in the **Orientation** box, **Above** in the **Position** box, and **Remove duplicate labels** in the **Duplicate Labels** box (the text will be placed on a curved line above the line feature following its direction).

c. Click the **Conflict Detection** tab, and set both the label weight and feature weight as **Medium** (so that road names cannot overlap parks). Click **OK**.

4) Click **OK**. The names of the roads are placed on the map.

8. Double-click **villages** in the table of contents. In the **Layer Properties** dialog, click the **Labels** tab, then:

1) Repeat Steps 1), 2) and 3) in Step 6 above.
2) In the **Text Symbol** box, select the font **Arial Black**, set its size as **8**, and the text colour as **black**.
3) Click the **Placement Properties** button. In the **Placement Properties** dialog:

a. Click the **Placement** tab.
b. Select **Offset label horizontally around the point**.
c. Click the **Conflict Detection** tab, and set both the label weight and feature weight as **Low** (so that village names cannot overlap parks and roads). Click **OK**.

4) Click **OK**. The names of the villages are placed on the map. The map should look similar to that in Figure 8.16.

Layout design

A map layout is a set of cartographic elements laid out and organised on a page. Layout design involves bringing together disparate elements like those shown in Figure 8.1 into an effective visual arrangement that is aesthetic, coherent and balanced. As a general principle, the map body should be placed at the visual centre and other elements arranged harmoniously around the map body. The visual centre of a map is slightly above the actual centre of the page. As the visual centre is an area where the eye naturally tends to land on its own, the most important cartographic elements should be located in that area. Figure 8.17 shows a map layout with a poor design, in which the scale bar is inappropriately placed in the area of the visual centre.

Second, reading a map is like reading a page in a book, mostly following a path from the top to the bottom and from the left to the right. Therefore, the cartographic elements that should be read first need to be arranged in the upper or upper left part of the map, such as the map title. Those elements that are not important and provide supplementary information should be located in the lower or bottom part of the map, such as the scale bar, credits and projection information. In Figure 8.17, the legend, scale bar, north arrow and inset are all above the map body, and are arranged in an incorrect reading order. Figure 8.1 shows a better-designed version of Figure 8.17.

In practice, map layout design also involves optimisation of space around the often irregular-shaped map body, as shown in Figure 8.1. Figure 8.18 shows several other map layouts which effectively utilise the space around the map body to arrange cartographic elements according to the shape of the mapped area.

In ArcGIS, a map layout is designed using the layout view (see Box 8.1). The layout view provides a page space for users to arrange cartographic elements. The elements can be resized and repositioned within the layout.

8.3 MULTIVARIATE MAPPING

Multivariate mapping involves showing two or more attributes or variables of geographical features simultaneously. For example, we may want to show temperature, rainfall and evaporation all together on a map. The simultaneous display of multivariate attributes allows for comparison of multiple attributes or analysis of cross-correlation between them.

Traditionally, multivariate maps are made by combining and representing several map layers with different types of symbols, comparing multiple attributes using charts or graphs, or representing multiple attributes using visual variables. For example, Figure 8.4 illustrates how the population distribution is related to the river system in a mountainous area by overlaying the population layer represented by dots and the stream layer represented using line symbols. Figure 8.19 shows the estimated above-ground in-use copper stocks in the greater Sydney areas in Australia. Patterns of line symbols represent the estimated total amount of above-ground in-use copper stocks in

Very High Income Households in Melbourne
Data source: Australian Bureau of Statistics

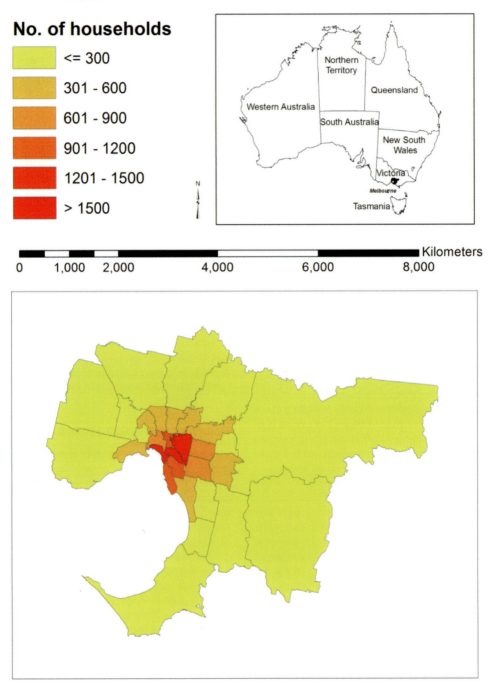

Figure 8.17 A poorly designed map layout.

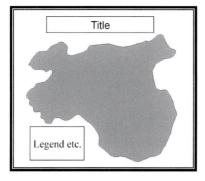

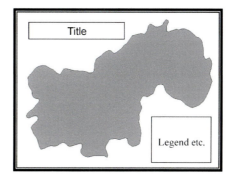

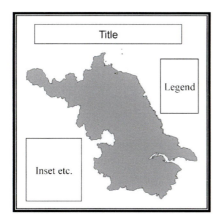

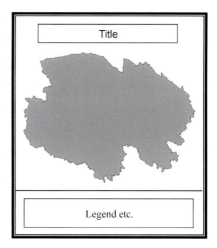

Figure 8.18 Examples of effective map layout design.

each area. Column charts represent in-use copper stocks in private dwellings, private motor vehicles, consumer products, and telecommunication and power distribution infrastructure respectively. Figure 8.20 is a bivariate or two-variable choropleth map, which uses colour variations to show arable land (percentage of land area) and rural population density simultaneously. Box 8.6 illustrates how a bivariate choropleth map is made in ArcGIS.

Because the number of classes the human eye can distinguish is limited, traditional multivariate mapping is generally restricted to combinations of either two or three variables. Appropriate selection of symbols (colours) is important for map readability. For higher-dimensional multivariate mapping, two visualisation techniques have been used and integrated with GIS mapping techniques: parallel coordinate plots (PCPs) and self-organising maps (SOMs).

A parallel coordinate plot is a plot of multivariate data, in which a set of equally spaced parallel vertical axes are drawn for each variable or attribute, and then a given row of attributes in an attribute table is represented by drawing a polyline whose vertices on the vertical axes represent the value of that row on each corresponding axis (Inselberg 1985). The vertical axes corresponding to each variable are usually scaled linearly from the minimum to the maximum value of that variable. A row represents the profile of a data item (a row of attributes). Consider, for example, an attribute table that lists four soil properties for twenty-three mapping units: soil organic matter (OM), total nitrogen (TN), total phosphorus (TP) and total potassium (TK) (Table 8.1). For each mapping unit (a row or data item), the PCP plots a profile of the four soil properties. In the PCP in Figure 8.21, four vertical axes

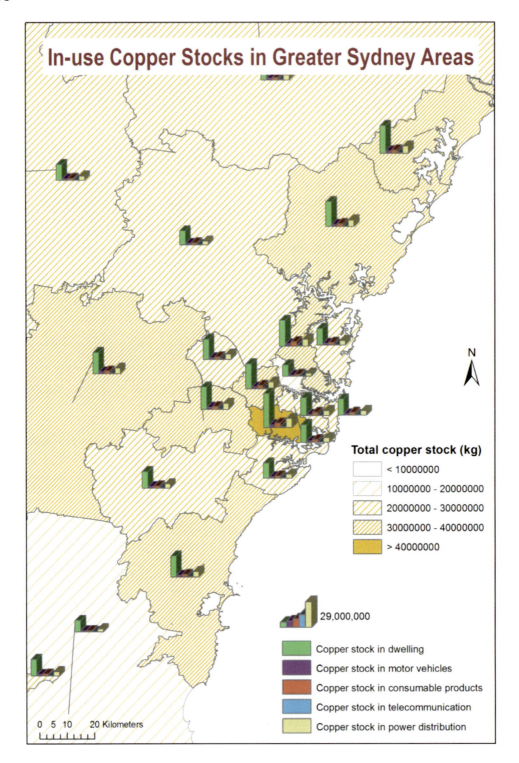

Figure 8.19 Multivariate data map of above-ground in-use copper stocks.

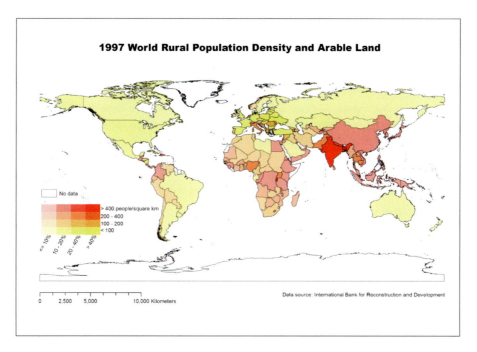

Figure 8.20 Bivariate choropleth map of 1997 world rural population density and arable land.

Box 8.6   Bivariate choropleth mapping in ArcGIS   PRACTICAL

To follow this example, start ArcMap, and load the **Digital_Chart_of_the_World** shapefile down-loaded in Box 8.2 and the spreadsheet **World arable land and rural pop 1997 (IBRD).xlsx** from **C:\Databases\GIS4EnvSci\Others**. The spreadsheet contains the rural population density (**RuralPopDen**) and per cent arable land (**PercentArable**) for each country, sourced from the International Bank for Reconstruction and Development. When adding the spreadsheet to ArcMap, select **Sheet1$**.

Make a base choropleth map classified based on rural population density

1. Join the table in **Sheet1$** to **Digital_Chart_of_the_World** (see Step 3 in Box 8.2).
2. In the table of contents, right-click **Digital_Chart_of_the_World**, and select **Properties**. In the **Layer Properties** dialog:

   1) Click the **Symbology** tab.
   2) Click **Quantities** in the **Show** box, then click **Graduated colors**.
   3) In the **Fields** box, select **RuralPopDen** as the value to map.
   4) Click the **Classify** button. In the **Classification** dialog box, change the classes to **4**, change the break values to **100**, **200**, **400** and **6260** (the largest density value in the dataset), and click **OK**.
   5) Select the yellow-to-red colour ramp, and click **OK**. A choropleth map of world rural population density is made with four classes: <100, 100–200, 200–400 and >400 people/km^2.

(continued)

(continued)

3. In the table of contents, right-click **Digital_Chart_of_the_World**, and click **Copy**.
4. In the table of contents, right-click **Layers**, and click **Paste Layer(s)**. A copy of **Digital_Chart_of_the_World** is added.
5. Repeat Steps 3 and 4 three times to make three other copies of **Digital_Chart_of_the_World**.
6. Rename the four copied layers **<=10**, **10–20**, **20–40** and **>40** respectively, corresponding to four classes of percentage of arable land: <= 10, 10–20, 20–40 and >40 per cent.

Represent classes of per cent arable land using different levels of layer transparency

7. In the table of contents, right-click **<=10**, and click **Properties**. In the **Layer Properties** dialog:

   1) Click the **Definition Query** tab. Click **Query Builder**. In **Query Builder**, enter the following query expression: `"Sheet1$.PercentArable"` <=10. Click **OK**. Only those countries with less than or equal to 10 per cent of arable land are selected to display.
   2) Click the **Display** tab. Set the transparency to **75%**. Click **OK**.

8. In the table of contents, right-click **10–20**, and click **Properties**. In the **Layer Properties** dialog:

   1) Click the **Definition Query** tab. Click **Query Builder**. In **Query Builder**, enter the following query expression: `"Sheet1$.PercentArable"` <=20 AND `"Sheet1$.PercentArable"` > 10. Click **OK**. Only those countries with 10–20 per cent of arable land are selected for display.
   2) Click the **Display** tab. Set the transparency to **50%**. Click **OK**.

9. In the table of contents, right click **20–40**, and click **Properties**. In the **Layer Properties** dialog:

   1) Click the **Definition Query** tab. Click **Query Builder**. In **Query Builder**, enter the following query expression: `"Sheet1$.PercentArable"` <=40 AND `"Sheet1$.PercentArable"` > 20. Click **OK**. Only those countries with 20–40 per cent of arable land are selected for display.
   2) Click the **Display** tab. Set the transparency to **25%**. Click **OK**.

10. In the table of contents, right-click **>40**, and click **Properties**. In the **Layer Properties** dialog:

   1) Click the **Definition Query** tab. Click **Query Builder**. In **Query Builder**, enter the following query expression: `"Sheet1$.PercentArable"` > 40. Click **OK**. Only those countries with > 40 per cent of arable land are selected for display.
   2) Click the **Display** tab. Set the transparency to **0%**. Click **OK**.

11. In the table of contents, move **Digital_Chart_of_the_World** to the top. Right-click **Digital_Chart_of_the_World**, and click **Properties**. In the **Layer Properties** dialog:

   1) Click the **Symbology** tab.
   2) Click **Features** in the **Show** box on the left side of the dialog box, then click **Single symbols**.
   3) Click the **Symbol** button in the **Symbol** panel. Select the **Hollow** symbol from the **Symbol Selector**.
   4) In the **Legend** panel, enter **No Data** as the label appearing next to the symbol in the table of contents. This layer is used to draw boundaries of the countries or areas with no data.

Make the legend

12. Click **View** on the main menu bar, and select **Layout View**. If the **Layout** toolbar is not shown, go to the main menu bar, click **Customize**, point to **Toolbars**, and select **Layout**.

13. In the **Layout** toolbar, click the **Change Layout** button 🖼️ . Click the **ISO (A) Page Sizes** tab, and select **ISO A4 Landscape.mxd**. Click **Finish**.
14. Click **Insert** on the main menu bar, and select **Legend**. In the **Legend Wizard** dialog:

    1) In the **Legend Items** box, select **Digital_Chart_of_the_World**, **<=10**, **10–20**, **20–40** and **>40** as the layers to be included in the legend. Click **Next**.
    2) Delete the legend title. Click **Next**. Click **Next** again, then again.
    3) Change the spacing between patches (vertically) to **0** points. Keep the other default values for spacing. Click **Finish**. The legend appears in the layout view.

15. Right-click the legend, and select **Convert to Graphics**, which converts the legend to graphics.
16. Right-click the legend again, and select **Ungroup**.
17. Repeat Step 16. Now all the elements in the legend become individual graphics. Each can be individually manipulated.
18. Click the **Select elements** button ▶ on the main standard toolbar.
19. Use the mouse to select all symbols in the legend for **<=10**, right-click the group, and select **Group**. Right-click the group, point to **Rotate or Flip**, then click **Flip vertically**. Drag the group to the position indicated in Figure 8.20.
20. Repeat Step 19 to move the legends for other layers to the corresponding positions, as shown in Figure 8.20. Select all other irrelevant graphics in the layout, and delete them by pressing the **Delete** key on the keyboard.
21. Click **Insert** on the main menu, and select **Text** to add the labels to the legend, as shown in Figure 8.20. Double-click a label. In the **Text properties** dialog, change the font, size and angle of the label, as shown in Figure 8.20.
22. Click **Insert** on the main menu bar, and select **Title** to add the map title. Change the font and size of the text as necessary.
23. Add other map elements as you wish.

represent OM, TN, TP and TK respectively, and each polyline (in a different colour) represents a particular mapping unit. A vertex of a polyline for a mapping unit on a vertical axis indicates the value of the soil property the axis represents for that mapping unit. With the PCP, we can now see which mapping units are similar to each other in the composition of the four soil properties, by comparing the profiles to each other. This is where the PCP is most useful – comparing profiles in order to find similarities.

Table 8.1 An attribute table of soil properties

Mapping unit	OM (%)	TN (g/kg soil)	TP (g/kg soli)	TK (g/kg soil)
1	2.18	0.15	0.07	2.99
2	1.79	0.12	0.07	3.02
3	0.96	0.08	0.04	3.24
4	5.28	0.36	0.07	3.32
5	18.3	0.64	0.12	2.14
6	6.65	0.46	0.15	1.34

(continued)

Table 8.1 (continued)

Mapping unit	OM (%)	TN (g/kg soil)	TP (g/kg soli)	TK (g/kg soil)
7	25.62	1.01	0.13	1.52
8	18.11	0.63	0.07	2.81
9	3.23	0.09	0.02	2.96
10	0.16	0.03	0.03	2.03
11	25.62	1.01	0.13	1.52
12	25.62	1.01	0.13	1.52
13	25.62	1.01	0.13	1.52
14	0.16	0.03	0.03	2.03
15	0.26	0.03	0.04	2.1
16	25.62	1.01	0.13	1.52
17	25.62	1.01	0.13	1.52
18	3.52	0.19	0.06	1.51
19	2.15	0.14	0.06	1.42
20	2.11	0.12	0.08	1.48
21	2.04	0.09	0.14	1.52
22	25.62	1.01	0.13	1.52
23	4.66	0.23	0.16	1.58

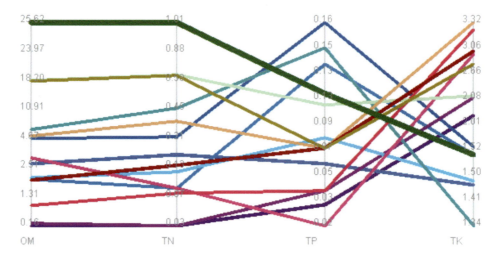

Figure 8.21 PCP from Table 8.1.

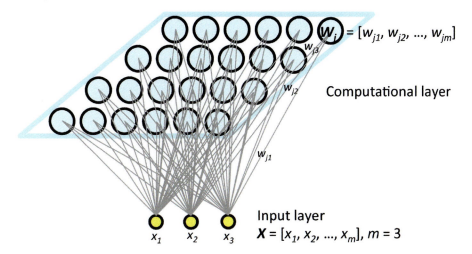

Figure 8.22 Structure of a SOM.

A self-organising map is a kind of artificial neural network (ANN) (see Section 9.5), which is trained to produce a low-dimensional, discretised representation of high-dimensional data (Kohonen 2001). It projects a set of given data items (as the input layer) onto a regular, usually two-dimensional, grid of nodes or neurons (as a single computational layer), as shown in Figure 8.22. Each node is fully connected to the input layer. Associated with each node is a weight vector of the same dimension as the input vectors and a specific position in the grid. The procedure for mapping an input vector onto the grid is to find the node with the closest or most similar weight vector to the input vector and to assign the grid coordinates of this node to the input vector. Where the node's weights match the input vector, that area of the grid is selectively optimised to more closely resemble the data for the class the input vector belongs to. From an initial distribution of random weights, and over many iterations, the SOM eventually settles into a map of stable zones formed by similar neighbouring nodes. Each zone is effectively a feature classifier. Any new, previously unseen input vectors presented to the SOM will stimulate nodes in the zone with similar weight vectors. Therefore, the SOM automatically classifies a new input vector. The set of the nodes can be regarded as constituting a similarity graph and SOM of the distribution of the given data items.

The SOM process involves several steps over many iterations:

1. Initialise weight vectors for each node.
2. Select an input vector at random from the set of training data.
3. Examine every node to determine how similar its weight vector is to the input vector. The node whose weights are most like the input vector is declared the winner, or best matching node.
4. Calculate the radius of the neighbourhood of the best matching node, and find any nodes within this radius that are deemed to be inside the winner's neighbourhood.
5. Adjust each neighbouring node's weights to make them more like the input vector. The closer a node is to the best matching node, the more its weights get altered.
6. Repeat Steps 2–5 for N iterations.

Suppose $X = [x_1, x_2, \ldots, x_m]$, which is the m-dimensional input vector, and $W_j = [w_{j1}, w_{j2}, \ldots, w_{jm} \mid j = 1, 2, \ldots, n]$, which is the corresponding weight vector of m dimensions for node j, where n is the number of nodes in the computational layer. Initialisation is to initialise weight vectors for each node with small standardised random values (usually $0 < w_{ji} < 1$, $i = 1, 2, \ldots, m$; $j = 1, 2, \ldots, n$). Similarity between an input vector and a node is determined by a discriminant function. This function is defined as the Euclidean distance between the input vector X and the weight vector W_j for each node j, calculated as:

$$d_j = \sqrt{\sum_{i=1}^{m}\left(x_i - w_{ji}\right)^2} \tag{8.1}$$

For each input vector, the nodes compute their respective values of the discriminant function. The node with the smallest value is declared the best matching node – that is, its weight vector is most similar to the input vector.

After the best matching node is determined, the next step is to calculate which of the other nodes are within its neighbourhood. A neighbourhood is often defined in practice by the Gaussian function expressed as:

$$\eta\left(r_w, r_j\right) = \exp\left(-\frac{\left\|r_w - r_j\right\|^2}{2\sigma^2}\right) \quad (8.2)$$

where r_w and r_j are the positions of the best matching node and node j in the neighbourhood on the grid, $\|r_w - r_j\|$ is the Euclidean distance between the two nodes and σ represents the effective range or the size of the neighbourhood (Yin 2008). This neighbourhood function has several important properties: it is maximal at the best matching node, it is symmetrical about that node, it decreases monotonically to zero as the distance goes to infinity and it is independent of the location of the best matching node. In addition, a unique feature of SOMs is that the area of the neighbourhood decreases with time. A popular time dependence is an exponential decay:

$$\sigma(t) = \sigma_0 \exp\left(-\frac{t}{\lambda}\right) \quad (8.3)$$

where σ_0 denotes the width of the neighbourhood at the beginning of the process (typically set to the 'radius' of the grid), t is the current time-step (iteration of the loop) and λ is a time constant. In other words, the area of the neighbourhood shrinks over time.

The point of the neighbourhood is that not only does the winning node get its weights updated, but its neighbours will have their weights updated as well, although by not as much as the winner itself. In practice, if a node is found to be within the neighbourhood, its weight vector is adjusted according to the following equation:

$$w_{ji}(t+1) = w_{ji}(t) + \alpha(t)\eta\left(r_w, r_j\right)$$
$$\left[x_i - w_{ji}(t)\right] \quad (8.4)$$

where t represents the time-step, and $\alpha(t)$ is called the learning rate, which decreases with time. Basically, this equation suggests that the new adjusted weight of node j for input i is equal to the old weight plus a fraction of the difference between the old weight and the input. The decay of the learning rate is calculated each iteration using the following equation:

$$\alpha(t) = \alpha_0 \exp\left(-\frac{t}{\lambda}\right) \quad (8.5)$$

α_0 is generally initialised with values between 0.4 and 0.1. It then gradually decays over time so that during the last few iterations it is close to 0. The effect of each weight update is to move the weight vectors of the winning node and its neighbours towards the input vector. Repeated presentations of the training data thus lead to topological ordering, in which weights of the whole neighbourhood are moved in the same direction and similar data items tend to excite adjacent nodes. Therefore, the SOM forms a semantic map where similar data items are mapped close together and dissimilar ones spaced out. This may be visualised by a hexagonal grid shaded in different shades of grey or colours, as shown in Figure 8.23. Each hexagon represents a node in the SOM. The greyscale or colour or intensity of the colour of a hexagon represents the Euclidean distance between the weight vector of the hexagon and the weight vectors of its immediate neighbouring hexagons. Conventionally, light colours depict closely spaced node weight vectors (that is, similar multivariate data vectors) and darker colours indicate more widely separated node weight vectors (that is, dissimilar multivariate data vectors). Thus, areas of light colours are considered as clusters, and the dark areas as the boundaries between the clusters. This kind of graphic display is called a unified distance matrix, commonly known as a U-matrix (Ultsch and Siemon 1990). It provides a similarity graph of the input data, and can help visualise the clusters in the high-dimensional, multivariate datasets.

Guo et al. (2005) proposed an approach to integrate PCPs and SOMs in multivariate mapping. With their

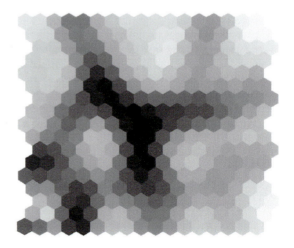

Figure 8.23 A U-matrix.

approach, the SOM assigns data items (features such as mapping units and statistical area units) to nodes and colours the nodes according to a systematically designed colour scheme so that nearby or similar nodes have similar colours. A node may be assigned several data items or may have no data items. The SOM itself does not show the original data values. But a node with data items contains four pieces of information: the data items assigned to the node, the number of data items in the node, the mean vector (the mean values of all the data items) and the colour of the node. These pieces of information are utilised by the PCP to draw profile lines of each data item. Every data item (not each node) is then mapped, geographically, with the colour assigned to the node that contains the data item. This map is a multivariate map linked to the non-empty nodes in the SOM's U-matrix and profiles in the PCP. From a thematic mapping perspective, the SOM serves as a multivariate classification method – providing colours for mapping, and the PCP serves as the legend – giving the meanings of those colours. This makes it easy to understand the multivariate

data and discover the spatial distribution of various multivariate patterns.

For example, Figure 8.24 shows a multivariate map of four soil properties (that is, four variables) and its linked SOM and PCP created from Table 8.1 using multivariate mapping software SOMVIS (Guo 2008). In the U-matrix, the scaled circle in a node hexagon depicts the number of data items in the node. Each circle is filled with a colour. The colour is used to draw the profile lines of the data items (soil mapping units) contained in the node in the PCP, and to map the data items onto the soil property map. Similar colours in the PCP and map represent similarities in terms of the four soil properties (input variables). The thickness of a profile line in the PCP is also scaled to the total number of data items contained in a node. By interpreting the PCP, the distribution of the four soil properties in each mapping unit can be understood. The map is effectively a multivariate map with four variables. Box 8.7 shows how SOMVIS is used to produce the three displays in Figure 8.24.

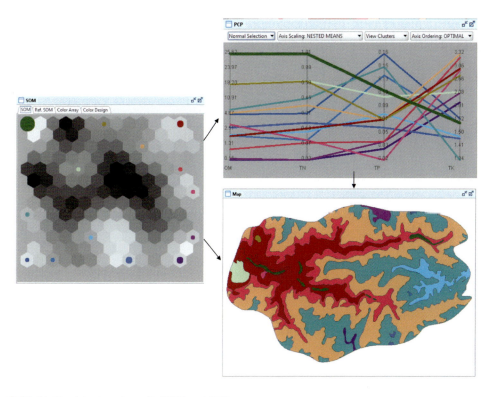

Figure 8.24 Multivariate mapping with SOM and PCP.

Box 8.7 Multivariate mapping with PCP and SOM **PRACTICAL**

To follow this example, download SOMVIS from **http://www.spatialdatamining.org/software/ somvis** after creating an account by following the instructions on the Web page, and save **somvis2.0_0.jar** to a directory of your choice. If not already installed on your system, download Java from **http://java.com/en/download/index.jsp**, and install Java (SOMVIS is developed in the Java programming language and requires the Java Virtual Machine to run). Start ArcMap, and load the **soils** feature class and **soilProperties** table from **C:\Databases\GIS4EnvSci\VirtualCatchment\ Geodata.gdb**.

Prepare data in ArcGIS and Excel

1. Follow Step 3 in Box 8.2 to append the **soilProperties** table to the **soils** attribute table using the **PatchID** field for the join.
2. In the table of contents, right-click **soils**, point to **Data**, and click **Export Data**.
3. In the **Export Data** dialog, change the **Output feature class** to **C:\Databases\GIS4EnvSci\ VirtualCatchment\Results\soils2.shp**, then click **OK**. **soils2.shp** contains all joined soil properties from the **soilProperties** table.
4. After **soils2.shp** is created, start **Microsoft Office Excel**.
5. In **Excel**, click the **Microsoft Office Button** 🔘 , then click **Open**. In the **Open** dialog:

   1) Navigate to **C:\Databases\GIS4EnvSci\VirtualCatchment\Results**.
   2) Click **All Files** in the **Files of type** box.
   3) Click the file **soils2.dbf**, then click **Open**. **soils2.dbf** is opened in **Excel**.

6. In **Excel**, click the **Microsoft Office Button**, then click **Save as**. In the **Save as** dialog:

   1) Click **CSV (Comma delimited) (*.csv)** in the **Save as type** list.
   2) Click **Save**. The file **soils2.dbf** is saved as **soils2.csv** (SOMVIS requires the attribute table to be saved as a CSV file).

Multivariate mapping with SOMVIS

7. In the directory where you downloaded SOMVIS, double-click the file **somvis2.0_0.jar**.
8. In SOMVIS, click **Load Data** on the **File** menu.
9. In the **Open Shape File** dialog, navigate to **C:\Databases\GIS4EnvSci\VirtualCatchment\ Results**, and select **soils2.shp**. Click **Open**. The **soils2** layer is displayed in the **Map** window, and the attributes in **soils2.csv** associated with each mapping unit (data item) are listed in the **Control** window.
10. Hold down the **Ctrl** key, and click **OM**, **TN**, **TP** and **TK** in the **Control** window to select the four variables.
11. Click **Submit Variables** in the **Control** window, then click **Multivariate/Univariate**. The SOM and PCP windows appear, as shown in Figure 8.24. The Map, SOM and PCP windows interact with each other. You may drag the mouse to run across the profile lines to select them, while their corresponding mapping units and SOM nodes are highlighted respectively in the map and SOM. Or you may drag the mouse to draw a rectangle in the SOM window to select nodes, while highlighting the corresponding profile lines for the PCP and mapping units in the map. For more information about how to interact with the map, SOM and PCP, and options for PCP axis scaling in SOMVIS, see Guo (2008).

With modern visualisation technology such as SOMVIS, multivariate mapping provides a visual approach to exploring and understanding spatiotemporal and multivariate patterns in spatial data facilitated by interactive visual interfaces.

8.4 3D MAPPING

Modern GIS provides not only traditional 2D mapping tools, but also 3D mapping capabilities. 3D mapping makes spatial data visualisation more effectively by scaling each geographical feature along the z-axis to represent values of a particular attribute. It can be used to map 2.5D and 3D phenomena.

A 2.5D phenomenon can be considered as a two-dimensional phenomenon with the third dimension representing a single attribute value. Elevation above sea level, air temperature, precipitation and population density are examples of continuous 2.5D phenomena. In such a surface, every point is characterised by its location on the surface of the Earth and a single value of a particular attribute associated with it. Therefore, 2.5D data consist of a pair of (x, y) coordinates (longitude and latitude, or easting and northing) and exactly one z value for each point. Figure 8.25 shows a 3D perspective view of an elevation surface, but it is not a true 3D phenomenon. Discontinuous 2.5D surface data, such as

Figure 8.25 3D perspective view of an elevation surface.

areal aggregated population data, can be displayed in 3D perspective as a stepped statistical surface (Figure 8.26). In essence, 3D mapping of 2.5D phenomena simply involves extruding 2D points, lines, polygons or surfaces along the z-axis by a set of specified attribute values to show them as 3D features.

In contrast, a 3D phenomenon is a true three-dimensional feature, which can be described by its location in a 3D Euclidean space and the associated values of an attribute or variable. The attribute values vary not only at different horizontal locations, but also at different vertical positions in the space. Generally, a point in 3D data is defined by a pair of (x, y) coordinates (horizontal location), the height above (or depth below) a zero point

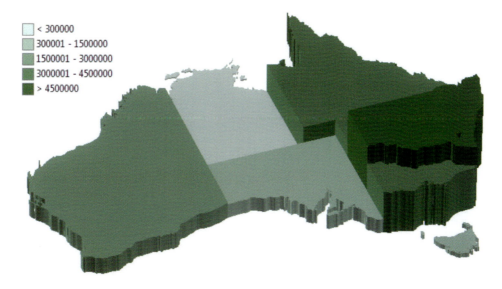

☐	< 300000
☐	300001 - 1500000
☐	1500001 - 3000000
☐	3000001 - 4500000
■	> 4500000

Figure 8.26 Stepped statistical surface of Australian population (by state).

(vertical location) and the value of an attribute or variable associated with the point. In other words, true 3D features have multiple z values associated with every pair of (x, y) points. For example, carbon dioxide concentrations in the atmosphere are a 3D phenomenon. To describe this phenomenon, every point can be defined by its horizontal location (for example, longitude and latitude), height above sea level and an associated level of carbon dioxide. Water temperature in the ocean and contaminant plumes in the air are also 3D phenomena. Figure 8.27 shows a 3D model of layers of hydrostratigraphical units of an aquifer in the Barwon Downs region, southwest to Melbourne. 3D phenomena can be symbolised by using opaque colours to shade the surface of the 3D structure, as in Figure 8.27, or by cutting and slicing through the 3D structure to 'peer into' the 3D data, as in Figure 8.28. More advanced visualisation techniques such as isosurface and volume rendering have also been developed and applied to visualise true 3D data (Slocum 2005; Copsey 2002). However, these techniques have not been incorporated into general-purpose GIS software systems.

3D visualisation in ArcGIS mainly supports the presentation of 2.5D data. But it can be used to represent true 3D geometric structures by multipatch features, which are made of planar 3D rings and triangles stitched together to model 3D objects like trees and buildings with overhanging features (Kennedy 2009). Multipatches represent the outer surface or isosurfaces of 3D features. They store texture, colour, transparency and geometric information representing different parts of the features. They are particularly useful for representing urban features to create 3D urban models to provide contexts for studying environmental and economic development issues in cities. Figure 8.29 shows such an example, in which the annual mean PM2.5 distribution is mapped in a 3D virtual model of Clayton Campus, Monash University, created in ArcGIS's ArcScene module. The buildings in the 3D view are all multipatch features. ArcGIS allows multipatch features to be created from 3D point and line symbols and extruded polygons, but it does not provide tools to create complex photo-realistic 3D multipatch features. Other software packages such as Google SketchUp, AutoCAD, OpenFlight and VRML are often used to create multipatch features to be imported into ArcGIS. Box 8.8 illustrates how to make 3D visualisations in ArcGIS.

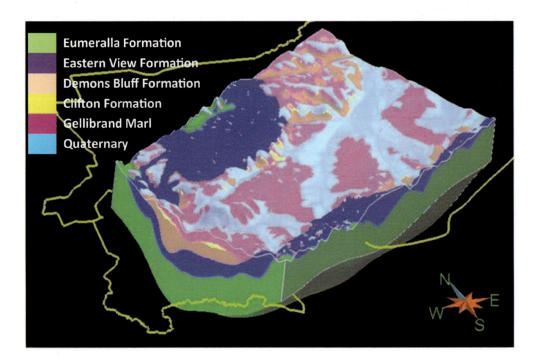

Figure 8.27 A 3D model of the Barwon Downs Graben aquifer (created using GOCAD; courtesy of Sultana Nasrin Nury).

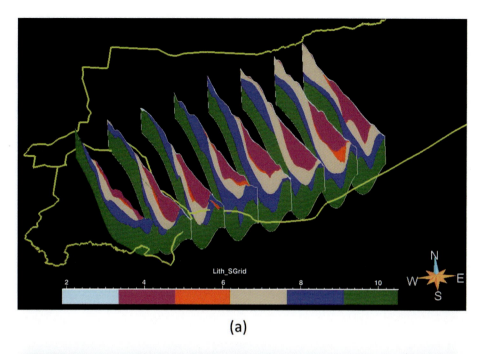

(a)

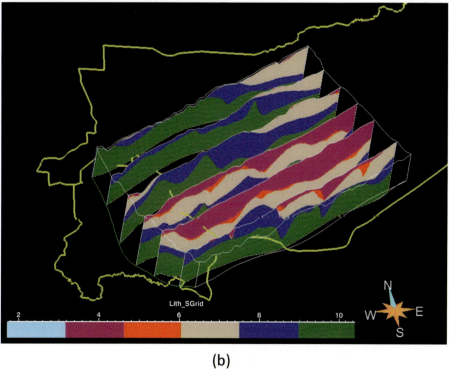

(b)

Figure 8.28 Vertical cross-sections of the 3D model of the Barwon Downs Graben aquifer: (a) in the northwest–southeast direction and (b) in the southwest–northeast direction (created using GOCAD; courtesy of Sultana Nasrin Nury).

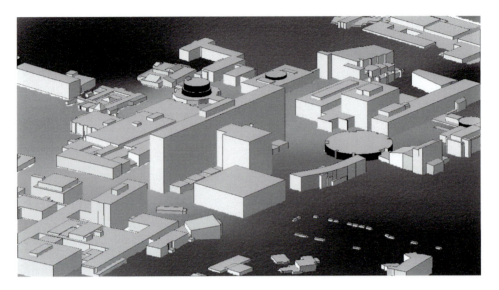

Figure 8.29 Visualisation of the distribution of PM2.5 with a 3D virtual model of Clayton Campus, Monash University. Darker colours on the ground indicate higher levels of PM2.5 (data source: Monash University).

Box 8.8   3D visualisation with ArcGIS   **PRACTICAL**

To follow this example, **ArcScene** (an ArcGIS software module) and the **3D Analyst** extension are required.

Visualise an elevation raster and 2D features in 3D with ArcScene

Raster data and 2D vector features do not store 3D geometry. They are required to be draped onto an elevation surface (DEM or TIN) to obtain base heights (z-values) at which cells or features are drawn in 3D.

1. Click the **Start** button on the **Windows** taskbar, point to **All Programs**, point to **ArcGIS**, and click **ArcScene**. Open a blank scene.
2. Click the **Add Data** button ✛ . Navigate to **C:\Databases\GIS4EnvSci\VirtualCatchment\Geodata.gdb**, and double-click **dem**. The digital elevation model (elevation raster) of the virtual catchment is added to the scene.
3. Click ✛ to add the **roads** feature class from **Geodata.gdb**. Both the **dem** and **roads** layers are drawn on a plane, with a base height value of zero.
4. Right-click **dem** in the table of contents, and select **Properties**. In the **Layer Properties** dialog:

   1) Click the **Base Heights** tab.
   2) Click the option **Floating on a custom surface**. Because the **dem** layer is the only surface data in the scene, it appears in the surface drop-down list.
   3) Click **OK**. As the **dem**'s base heights are set to its cell values (that is, the raster uses its own values as a source of base heights), it is now displayed in 3D, as shown in Figure 8.25. However, base heights for roads have not been set; therefore they lie underneath the elevation raster.

5. Right-click **roads** in the table of contents, and select **Properties**. In the **Layer Properties** dialog, click the **Base Heights** tab. Click the option **Floating on a custom surface**, and select **dem**, from which the base heights for the 2D road features are obtained. Click **OK** to close the **Layer Properties** dialog. Now the roads drape over the elevation surface. However, you may notice that the roads are not displayed continuously.

6. Right-click **roads** in the table of contents, and select **Properties**. In the **Layer Properties** dialog, click the **Base Heights** tab. Type **100** in the **Add a constant elevation offset in meters** text box. Click **OK** to close the **Layer Properties** dialog. This offsets the base height of the elevation surface by a small amount and makes the roads appear completely above the terrain surface.

Create and visualise a TIN in 3D with ArcScene

A TIN is a vector representation of an elevation surface (see Section 2.3). Every node in a TIN stores a *z*-value. TINs are automatically drawn in 3D when loaded into ArcScene. A TIN can be created from contour vectors and spot heights, or from an elevation raster.

7. Click **Customize** in ArcScene, and click **Extensions**.
8. Check **3D Analyst,** and click **Close**. The 3D Analyst extension is enabled.
9. Click the **ArcToolBox** button 🗺 on the toolbar.
10. In the **ArcToolBox** window, expand **3D Analyst Tools > Conversion > From Raster**. Double-click **Raster to TIN**. In the **Raster to TIN** dialog:

    1) Select **dem** as the input raster.
    2) Set **C:\Databases\GIS4EnvSci\VirtualCatchment\Results\tin** as the output TIN.
    3) Click **OK**.

11. Uncheck **dem** in the table of contents. The TIN is shown in 3D in the scene, which may resemble Figure 8.30.

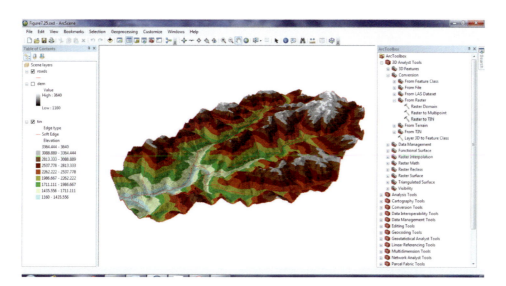

Figure 8.30 TIN in 3D.

(continued)

(continued)

12. Right-click **Scene layers** in the table of contents, and click **Scene Properties**. In the **Scene Properties** dialog, click the **General** tab, type **2** in the **Vertical Exaggeration** combo box, then click **OK**. The apparent height of the terrain is now doubled. Vertical exaggeration allows subtle features in the surface to be brought out and enhances the sense of depth in the scene. You may also change other scene properties shared by all the layers in the scene using the **Scene Properties** dialog, including the background colour, the coordinate system and extent of the data, and the way in which the scene is illuminated (the position of the light source relative to the surface).

13. Click the **Navigate** button ⊕ on the toolbar. Click and hold the scene with the mouse pointer, and slowly drag up and down, from right to left. The scene is tipped, turned and tilted. In this way, you can navigate around the 3D scene.

14. Click the **Save** button 💾 on the toolbar to save the scene to **C:\Databases\GIS4EnvSci\ VirtualCatchment\results\3dscene.sxd**.

Create a stepped statistical surface with ArcScene

15. Click the **New** button 🗋 on the toolbar to open a new blank scene.

16. Add the **rivers**, **dem** and **animalDensity** layers from **Geodata.gdb**. **animalDensity** is a grid map. Each grid is a 2D polygon feature with pheasant and mainland serow densities, and will be mapped as a 3D feature extruded from the 2D polygon.

17. Right-click **animalDensity** in the table of contents, and select **Properties**. In the **Layer Properties** dialog:

    1) Click the **Extrusion** tab.
    2) Check to enable **Extrude features in layer**, and click the **Calculate Extrusion Expression** button 🖩 . This extrudes a grid to the height equal to its pheasant density. This information is stored in the **pheasant** field.
    3) In the **Expression Builder** dialog, click **pheasant** in the **Fields** list box. **[pheasant]** appears in the **Expression** box. Click **OK**.
    4) Click the drop-down arrow to apply the extrusion expression by **adding it to each feature's base height**.
    5) Click **OK**. As the density values are too small relative to the horizontal extent of the mapped area, the 3D polygon (grid) features still seem to be lying on a flat plane.

18. Right-click **Scene layers** in the table of contents, and click **Scene Properties**. In the **Scene Properties** dialog, click the **General** tab, type **200** in the **Vertical Exaggeration** combo box, then click **OK**. Now you can see the new 3D grids in the scene. Their heights are proportional to their pheasant densities.

19. Right-click **dem** in the table of contents, and select **Properties**. In the **Layer Properties** dialog, click the **Base Heights** tab, type **2** as the layer offset, and click **OK**. This makes the **dem** layer fully visible.

20. Right-click **rivers** in the table of contents, and select **Properties**. In the **Layer Properties** dialog, click the **Base Heights** tab, type **3** as the layer offset, and click **OK**. Both **dem** and **rivers** are displayed as 2D features on a plane at the base height. They are used as the base map for geographical reference. The scene is now as shown in Figure 8.31.

8.5 ANIMATED MAPPING

Animated mapping, or cartographic animation, provides one of the most powerful means for visualising and communicating spatial or environmental data. It uses animation to show changes over time, space or attributes. Generally, it involves processing or playing back a sequence of individual maps, referred to as a track. A map layer in a track is called a frame. Each frame is

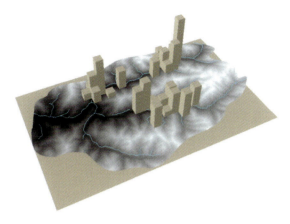

Figure 8.31 3D scene of pheasant density.

a snapshot of mapped features at a certain point in the animation. Animation is based on a principle of human sight known as persistence of vision. When we view a sequence of related frames in rapid succession (24–30 frames per second), we perceive them as continuous motion. The most elementary animation method is to create every single frame separately, then combine all the frames into an animation. One minute of animation might require 1,440–1,800 separate frames, depending on the quality of the animation. In cartographic animation using computers, the keyframing technique is often

used. With this technique, the most characteristic frames that form an outline of the most important movements, called key frames, are created first, then the computer uses mathematical algorithms to interpolate the frames in between. Changes in all the components of spatial data (location, attribute and time) presented in a map can be animated. A distinction is made between temporal and non-temporal cartographic animations (Lobben 2003; Slocum 2005).

Temporal cartographic animations

Temporal cartographic animations show change over time, such as changing patterns of wind speed and direction. They produce time-series animations to show changes in the locational or attribute components of spatial data through data layers or maps as a time-lapse. For example, Figure 9.9 includes four temporal data layers showing states of forest fires at four different time steps. When animated, the locations of forest fires and forest cover change as time passes. For another example, Figure 8.32 shows the 1997 time stamp of the change in the percentage of arable land in each country from 1961 to 2011. When animated, the percentage of arable land increases in some countries and decreases in others with each successive year.

Temporal cartographic animations can show the movement of geographical features (such as storms,

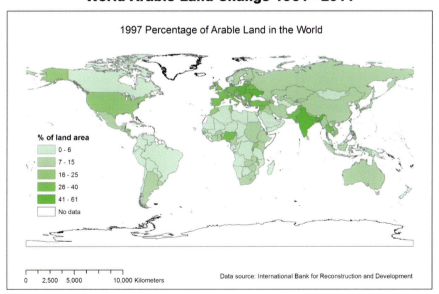

Figure 8.32 The percentage of arable land worldwide in 1997.

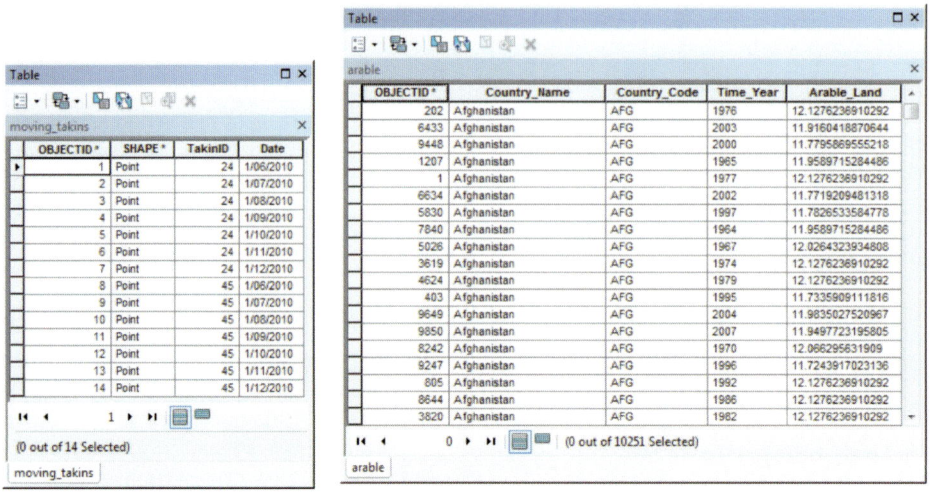

Table

moving_takins

OBJECTID *	SHAPE *	TakinID	Date
1	Point	24	1/06/2010
2	Point	24	1/07/2010
3	Point	24	1/08/2010
4	Point	24	1/09/2010
5	Point	24	1/10/2010
6	Point	24	1/11/2010
7	Point	24	1/12/2010
8	Point	45	1/06/2010
9	Point	45	1/07/2010
10	Point	45	1/08/2010
11	Point	45	1/09/2010
12	Point	45	1/10/2010
13	Point	45	1/11/2010
14	Point	45	1/12/2010

1

(0 out of 14 Selected)

moving_takins

(a)

Table

arable

OBJECTID *	Country_Name	Country_Code	Time_Year	Arable_Land
202	Afghanistan	AFG	1976	12.1276236910292
6433	Afghanistan	AFG	2003	11.9160418870644
9448	Afghanistan	AFG	2000	11.7795869555218
1207	Afghanistan	AFG	1965	11.9589715284486
1	Afghanistan	AFG	1977	12.1276236910292
6634	Afghanistan	AFG	2002	11.7719209481318
5830	Afghanistan	AFG	1997	11.7826533584778
7840	Afghanistan	AFG	1964	11.9589715284486
5026	Afghanistan	AFG	1967	12.0264323934808
3619	Afghanistan	AFG	1974	12.1276236910292
4624	Afghanistan	AFG	1979	12.1276236910292
403	Afghanistan	AFG	1995	11.7335909111816
9649	Afghanistan	AFG	2004	11.9835027520967
9850	Afghanistan	AFG	2007	11.9497723195805
8242	Afghanistan	AFG	1970	12.066295631909
9247	Afghanistan	AFG	1996	11.7243917023136
805	Afghanistan	AFG	1992	12.1276236910292
8644	Afghanistan	AFG	1986	12.1276236910292
3820	Afghanistan	AFG	1982	12.1276236910292

0

(0 out of 10251 Selected)

arable

(b)

Figure 8.33 Time values in attribute tables.

animals and vehicles), changes in attribute values (such as land use/land cover, sea surface temperature and wind speed and direction) or both (such as wildfires). Spatial data used for temporal cartographic animation must have a time component, referred to as temporal data. Temporal data in GIS can be stored and managed in both vector and raster structures. Usually, temporal data associate a time value or values with every vector feature or every raster layer. Time values in the data can represent a point in time sampled at a regular or irregular interval (for example, second, minute, hour, day, week, month or year), or a duration of time. Time values representing a point in time are stored in a single field (called the time field) in the attribute table associated with a vector feature layer or a collection of raster layers, and those representing a duration of time are stored in two fields (called the start and end time field) in the attribute table (Figure 8.33). A raster layer with a time value represents a state at that particular moment of time, or time step. With feature layers, features that change in location or shape over time are stored as separate features with different time values. For example, as shown in Figure 8.33a, multiple point features (rows or records) represent the locations of the same takin at different times. A feature that changes in attribute values but not in location or shape is also represented by multiple features, each representing the same feature at the same location with a different attribute value for each time moment or time span. Figure 8.33b shows such an example, in which each country is represented by multiple features for the same country with different percentage values of arable land for different years.

Temporal data may come from many sources, ranging from field observations to data capture using satellite remote sensors or produced from simulation models. Time values in temporal data enable temporal cartographic animations. When animated, they provide useful insights into patterns, trends and processes of many environmental variables over time. Box 8.9 illustrates temporal data and their animations in ArcGIS.

Non-temporal cartographic animations

Non-temporal cartographic animations show changes caused by variables other than time. These may include changes along a path, through the landscape and with varying data presentation methods or view properties. There are many types of non-temporal cartographic animations (Peterson 1995; Sidiropoulos et al. 2005). The most frequently used non-temporal cartographic animations in environmental applications include fly-through animation, spatial trend animation and scene animation.

Fly-through animation provides the appearance of flying through a 3D landscape by showing a series of oblique views of the landscape. It is usually constructed by draping an aerial photo or satellite image over a DEM. Spatial trend animation sequentially displays a series of related data without a time stamp. For example, an animation through a set of data layers representing the population density in age groups from younger to older

Box 8.9 Temporal cartographic animation in ArcGIS PRACTICAL

To follow this example, start ArcMap.

Animate the movement of features

Workflow: (1) Symbolise the features (Step 3); (2) enable time on the feature layer (Step 4); (3) animate the features (Steps 6–8).

1. Load the **moving_takins** feature class from **C:\Databases\GIS4EnvSci\VirtualCatchment\ Geodata.gdb**. This feature layer contains point features representing locations of two takins' movement from 1 June to 1 December 2010.

2. Open and examine the attribute table associated with the feature class, which should be the same as Figure 8.33a. The **TakinID** field corresponds to each takin's unique ID number, and **Date** is the time field, denoting the date when the takin's location was recorded. In general, ArcGIS requires temporal data to contain records in the attribute table, each of which is specific to an individual feature and to a single point or duration in time.

3. Right-click **moving_takins** in the table of contents, and click **Properties**. In the **Layer Properties** dialog:

   1) Click the **Symbology** tab.
   2) Click **Categories** in the **Show** box, then click **Unique values**.
   3) In the **Value Field** box, select **TakinID**, and click the **Add All Values** button. Takin 24 and Takin 45 are assigned a dot symbol with two different colours.
   4) Double-click the dot symbols to show the **Symbol Selector** dialog, through which you can change the symbols' shape, size and colour. After finishing changing the symbols, click **Apply** in the **Layer Properties** dialog. The two takins' movement locations are shown in the data view with your selected symbols.

4. the **Layer Properties** dialog:

   1) Click the **Time** tab.
   2) Check **Enable time in this layer**. The **moving_takins** layer is now time-enabled.
   3) Set **Date** as the time field. As the **Date** field is already a date-formatted field, the field format is set by default.
   4) Set **Time Step Interval** as **1 month** (the duration after which the map will be refreshed to display the data valid at that time).
   5) Click **OK**.

5. Add the **dem** raster from **Geodata.gdb** to the view as the background for animation. Zoom out the data view to the full extent by clicking 🌐.

6. Click the **Time Slider** button 🕐 on the **Tools** toolbar. In the **Time Slider** window, click the **Options** button ▦ . In the **Time Slider Options** dialog:

   1) Click the **Time Display** tab.
   2) Keep the **Time step interval** as **1 month**
   3) Set the time window as **0.0** month so that moving the **Time Slider Control** by one month (time-step interval) will display data for that time stamp. (If you set the time window as **3** months, the data will be displayed for the past four months, although the time slider will move by only one month (time-step interval).)

(continued)

(continued)

4) Click the **Playback** tab, then move the speed slider closer to the slower end.
5) Click **OK**.

7. Click the **Play** button ▶ on the **Time Slider** window. The animation of the takin movement is shown in the data view.
8. Click the **Export To Video** button ⅠⅠ to export the temporal animation to a video.

Animate changes in attribute values of features

Workflow: (1) Re-format temporal data (Step 12); (2) join tables (Steps 13–14); (3) symbolise the layer (Step 16); (4) enable time on the feature layer (Step 16); (5) animate the features (Steps 17–18).

9. Click the **New** button 🗋 on the toolbar to open a new blank scene without saving the changes to the current map document.
10. Load the **Digital_Chart_of_the_World** shapefile downloaded in Box 8.2 and the spreadsheet **world arable land 1961 - 2011.xlsx** from **C:\Databases\GIS4EnvSci\Others**. The spreadsheet contains the statistics of arable land (percentage of land area) for each country in the world from 1961 to 2011, sourced from the International Bank for Reconstruction and Development. When adding the spreadsheet to ArcMap, select **Data$**.
11. Open the **Data$** table. In the table, time values (years) are stored as column headings. In other words, temporal data are in columns or fields. To visualise the data through time, the table must be reformatted to transpose data in columns into rows.
12. Click the **ArcToolBox** button 🗃 on the toolbar. In the **ArcToolBox** window, expand **Data Management Tools > Fields**. Double-click **Transpose Fields**. In the **Transpose Fields** dialog:

    1) Set **Data$** as the input table.
    2) Check the fields from **1961** to **2011** to transpose.
    3) Set the output table as **C:\Databases\GIS4EnvSci\VirtualCatchment\Results\MyGeo database.mdb\arable**.
    4) Define the **transposed field** as **Time_Year** and the **value field** as **Arable_Land**. The transposed field will store field name values of the fields that are selected to be transposed (years in this case), and the value field will store the values from the input table (here percentages of arable land corresponding to each country in each year).
    5) In the **Attribute Fields** box, check **Country_Name** and **Country_Code** as additional attributes to be included in the output table.
    6) Click **OK**. The **arable** table is created and added to the table of contents. Each row in the output table has a time value, as shown in Figure 8.33b.
    7) Open the **arable** table. In the table window, click the **Table Options** button ⊟ . Select **Add Field**. In the **Add Field** dialog, type the name **Percent_**, select **Double** as the field type, and click **OK**.
    8) Right-click the field heading **Percent_**. Click **Field Calculator**.
    9) In the **Field Calculator** window, double-click **Arable_Land** in the **Fields** list box. [Arable_ Land] appears under **Percent_ =**. Click **OK**. This converts values in **Arable_Land** from text to numbers in **Percent_**.

13. In the **ArcToolBox** window, expand **Conversion Tools > To Geodatabase**. Double-click **Feature Class to Feature Class**. In the **Feature Class to Feature Class** dialog:

    1) Choose **Digital_Chart_of_the_World** as the input feature.
    2) Set **C:\Databases\GIS4EnvSci\VirtualCatchment\Results\MyGeodatabase.mdb** as the output location.

3) Enter **world** as the name of the output feature class.
4) Click **OK**. This converts the **Digital_Chart_of_the_World** shapefile to a feature class and saves it to the geodatabase where the **arable** table is stored.

14. In the **ArcToolBox** window, expand **Data Management Tools > Layers and Table Views**. Double-click **Make Query Table** for joining a table (**arable**) to a feature class (**world**) in the same geodatabase with one-to-many relationships. In the **Make Query Table** dialog:

1) Click the **Input Tables** drop-down arrow, and add **world** and then **arable** to join.
2) Choose the following fields to be included in the output layer: **world.Shape**, **world. COUNTRY**, **arable.Time_Year** and **arable.Percent_**.
3) In the **Expression** box, enter the join statement `world.ISO_A3 = arable.Country_ Code`. Here, the **ISO_A3** field in **world** and the **Country_Code** field in **arable** are the common fields to be used to append the **arable** table to the **world** layer.
4) Type **WorldArable** as the table name (actually the name of the output layer).
5) Click the **Key Field Options** drop-down arrow, then click **USE_KEY_FIELDS** and select **arable.OBJECTID** as the key field (used to define the dynamic objectID column).
6) Click **OK**. After the tool is run, the layer **WorldArable** is created and added to the view.

15. Open and examine the attribute table associated with **WorldArable**. Each feature or row represents a country repeated for each year.
16. Right-click **WorldArable** in the table of contents, and click **Properties**. In the **Layer Properties** dialog:

1) Click the **Symbology** tab. Make a choropleth map using **arable.Percent_** as the value field, as in Step 4 in Box 8.2.
2) Click the **Time** tab. Check **Enable time in this layer**. The **WorldArable** layer is now time-enabled. Make sure the time field is set to **arable.Time_Year**, the field format to **YYYY**, and the time step interval to **1 Year**.
3) Click **OK**.

17. Click the **Time Slider** button 🕓 on the **Tools** toolbar. In the **Time Slider** window, click the **Enable time on map** button 🔄 , then click the **Options** button ⊟ . In the **Time Slider Options** dialog:

1) Click the **Time Display** tab. Change the format of the displayed date **yyyy**, as we are using annual data.
2) Click the **Playback** tab, tick **Play in a specified duration (seconds)**, and enter **51**. As we have 51 years' data, each year will be visible for 1 second.
3) Click **OK**.

18. Click the **Play** button ▶ on the **Time Slider** window to start the animation.

Animate raster layers

Workflow: (1) Create an empty raster catalogue (Step 20); (2) add rasters to the raster catalogue (Step 21); (3) add a time field to the attribute table associated with the raster catalogue; (4) enable time on the raster catalogue (Step 22).

19. Click the **New** button 🗋 on the toolbar to open a new blank scene without saving.
20. Click the **Catalog** button 🗐 on the main toolbar. In the **Catalog** window, navigate to **C:\Databases\ GIS4EnvSci\VirtualCatchment\Results\MyGeodatabase.mdb**. Right-click **MyGeodatabase. mdb**, point to **New**, and select **Raster Catalog**. In the **Create Raster Catalog** dialog, enter

(continued)

(continued)

FireDynamics as the raster catalogue name, and click **OK**. An empty raster catalogue is created. (A raster catalogue is an ArcGIS raster data format, which is a collection of rasters.)

21. Under **MyGeodatabase.mdb** in the **Catalog** window, right-click **FireDynamics**. Point to **Load**, and click **Load Raster Datasets**. In the **Raster To Geodatabase** dialog:

    1) Click the **Open file** button 📂 for input rasters, and navigate to **C:\Databases\GIS4EnvSci\ Others\fires**.
    2) In the **Input Rasters** dialog, click **fires0**, then hold down the **Shift** key and click **fire9**. Click **Add**. **fires0**, **fires1**, **fires2**, . . ., **fire9** raster layers are added together to the **FireDynamics** raster catalogue. The ten raster layers were produced by the forest fire model in Box 9.2. In these layers, a cell value of 0 stands for an empty site, 1 for a tree and 2 for fire.
    3) Click **OK**.

22. Right-click **FireDynamics** in the table of contents, and click **Properties**. In the **Layer Properties** dialog, click the **Time** tab, check **Enable time in this layer**, use **OBJECTID** as the time field, set the field format as **YYYY**, and specify the time step interval as **1 Year** (one step). (You may add a time field to the attribute table with **FireDynamics** and assign a time value to every raster in the raster catalogue.) Click **OK**.

23. Click the **Time Slider** button 🕰 to open the **Time Slider** window. Click the **Play** button ▶ in the **Time Slider** window to start the animation. You may change the animation speed and other options through the **Time Slider Options** dialog.

may show a spatial trend that the older populations tend to concentrate in the inner suburbs and younger populations in the outer suburbs. Scene animation animates a 3D scene by changing the Sun's position in the scene, altering the background colour of the scene or modifying other scene properties.

In ArcGIS, non-temporal cartographic animations are built by capturing three types of key frames: camera key frames, layer key frames and scene key frames. A camera key frame is a snapshot of the view in a scene, which is mainly used to create fly-through animations. A layer key frame is a snapshot of a data layer's properties, which can be used to build spatial trend animations or animate an object's movement along a path. A scene key frame stores properties of a scene for scene animations. They are used as snapshots at a particular time to interpolate in between a track. Time in the context of non-temporal cartographic animation is presentation time (time to show the frames), not real time as in temporal cartographic animation. Box 8.10 shows how to use the three types of key frames for non-temporal cartographic animations in ArcGIS.

Box 8.10 Non-temporal cartographic animation in ArcGIS   **PRACTICAL**

Create a fly-through animation

Workflow: *(1) Create a 3D view of a landscape (Steps 2–4); (2) select view points of the virtual camera (Figure 8.34); (3) capture a series of camera key frames (Steps 5–9); (4) play back the animation (Steps 10–11).*

1. Start **ArcScene**.
2. Add the **dem** raster from **C:\Databases\GIS4EnvSci\VirtualCatchment\Geodata.gdb**, and follow Steps 2–4 in Box 8.8 to show a 3D perspective of the DEM.

3. Click the **Add Data** button ✛ . In the **Add Data** dialog, Navigate to **C:\Databases\GIS4EnvSci\ VirtualCatchment\images**, click **2002.tif**, then click **Add**. A satellite image of the area is added to the scene.

4. Right-click **2002.tif** in the table of contents, and select **Properties**. In the **Layer Properties** dialog:
    1) Click the **Base Heights** tab.
    2) Click the option **Floating on a custom surface**. **dem** is listed.
    3) Type **100** in the **Add a constant elevation offset in meters** text box. Click **OK** to close the **Layer Properties** dialog. Now the satellite image is draped onto the DEM, giving a realistic 3D view of the catchment, as shown in Figure 8.34. The figure also shows several view points of the virtual camera at which camera key frames will be captured.

Figure 8.34 3D landscape and view points of the virtual camera.

5. Click **Customize**, point to **Toolbars**, and click **Animation**. The **Animation** toolbar appears.
6. On the **Animation** toolbar, click the **Capture View** button 📷 to create a camera key frame showing the entire scene.
7. Click **Zoom In** on the **Tools** toolbar, and zoom to View Point 1. Click the **Capture View** button to create a camera key frame showing the area around the first view point.
8. Click **Zoom In** on the **Tools** toolbar, and zoom to View Point 2. Click the **Capture View** button to create a camera key frame showing the area around that view point.
9. Repeat Step 7 for other four view points to capture four new camera key frames. Click the **Full Extent** button. The captured views are stored as a set of camera key frames in a camera track. When the track is played, it shows a smooth animation between the key frames.
10. Click the **Open Animation Controls** button ▣ on the **Animation** toolbar.

(continued)

(continued)

11. Click the **Play** button ▶ on the **Animation Controls** toolbar. The fly-through animation is played back by interpolating the view points between the key frames in the track.

Animate scene properties

Workflow: *(1) Create scene key frames (Steps 13–17); (2) animate the scene track (Step 18).*

12. Click **Animation** on the **Animation** toolbar, and click **Clear Animation**. The camera track created above is removed from the scene.
13. Click the **Animation** drop-down menu, then click **Create Keyframe**. In the **Create Animation Keyframe** dialog:

    1) Click the **Type** drop-down arrow, and select **Scene**.
    2) Click the **New** button to create a new track with the default name **Scene track 1**. Keep the default name **Scene keyframe 1** for the key frame.
    3) Click **Create** once. The first scene key frame shows the current scene.

14. In the table of contents, right-click **Scene layers**, and click **Scene Properties**. In the **Scene Properties** dialog, change the vertical exaggeration, background colour, Sun azimuth, Sun altitude or contrast, then click **OK**.
15. Click **Create** once in the **Create Animation Keyframe** dialog. The second scene key frame is created and named **Scene keyframe 2**.
16. Repeat Steps 13 and 14 to create four more key scene frames for the scene animation track by changing the scene properties.
17. Close the **Create Animation Keyframe** dialog.
18. Open **Animation controls** to play the animation as in Steps 10 and 11. The scene animation is played back by interpolating the scene key frames in the track.

Move an object along a path

Workflow: *(1) Load a point feature layer representing the object to move and a line feature layer containing the path (Step 20); (2) select the path to move the object (Step 22); (3) create a layer animation track (Step 23); (4) animate the movement of the object (Step 24).*

19. Click **Animation** on the **Animation** toolbar, then click **Clear Animation**. The scene track created above is removed from the scene.
20. Add the **car** and **roads** feature classes from **Geodata.gdb**. The **car** layer contains a point feature representing the current position of a car (the object to move).
21. Drape the two layers over the DEM with an **elevation offset** of **150** metres for **roads**, and **200** metres for **car**.
22. Click the **Select Features** button 🔖 on the **Tools** toolbar, and select a particular road segment as a path.
23. Click the **Animation** drop-down menu on the **Animation** toolbar, then click **Move Layer along Path**. In the **Move Layer along Path** dialog, select **car** as the layer to move and the selected line feature (a road segment) as the path source, type **1500** in the **Vertical offset** text box as the height of the **car** layer so that the car appears to fly above the surface, then click **Import** to import the selected road section as a moving path. The car movement is stored as a set of layer key frames in a layer animation track.
24. Open **Animation controls** to play the animation as in Steps 10 and 11.
25. Close **ArcScene** without saving the scene.

Animate a group of layers without a time stamp

Workflow: (1) Add a group of layers to be animated (Step 27); (2) create a layer animation track (Step 29); (3) animate the group of layers (Steps 30–31).

26. Start **ArcMap**.
27. Add the **fires9**, **fires8**, **fires7**, . . ., **fire1** raster layers sequentially from **C:\Databases\GIS4EnvSci\ Others\fires**. Use the same colour to show the same value for all ten raster layers.
28. Click **Customize**, point to **Toolbars**, and click **Animation**.
29. On the **Animation** toolbar, click the **Animation** drop-down menu, then click **Create Group Animation**. In the **Create Group Animation** dialog, tick **One layer at a time**, then click **OK**. A layer animation track is created.
30. Open **Animation controls**. Click the **Options** button on the **Animation controls** toolbar. Select **By number of frames**, and enter **10** in its text box. Set **Frame duration** as **1.0** second.
31. Click the **Play** button. The layer animation track is played back, where each frame is a map layer and is displayed sequentially every second. Creating a group animation is the major means of building spatial trend animations in ArcGIS.

Animation can be used as an exploratory tool to understand changes, trends and processes, detect similarities or differences in distribution within a series of maps, and to gain insights into spatial relations. For example, it has been successfully used in visualisation of hydrological temporal data and processes (Fuhrmann 2002), simulation of storm surge flooding (Zhang et al. 2006) and animation of climate change (Weber and Buttenfield 1993).

More recently, cartographic animation, 3D visualisation and virtual globes have been combined to support the presentation, exploration and dissemination of environmental data and communication of scientific findings from environmental research. Box 8.11 provides a case study of the use of the three types of spatial visualisation to communicate projected climate change data and their implications for dairy production in Victoria, Australia.

Box 8.11   Application of spatial visualisation to communicate climate change information to local farming communities in Victoria, Australia   CASE STUDY

Climate change is the greatest global challenge the human race has ever faced. Protecting people and ecosystems from unavoidable changes in climate to whatever degree possible requires significant public engagement in the issue so that difficult decisions can be made by members of the public and policy makers. However, climate change has predominantly been a highly complex and abstract discussion of science, policy and the economic issues. Making climate change information relevant, interesting and engaging for the public is an ongoing challenge.

The former Victorian Department of Primary Industries in Australia investigated the use of spatial visualisation to communicate climate change information to the community under the Victorian Climate Change Adaptation Program (VCCAP). As part of a VCCAP visualisation project, they developed Virtual DemoDairy. DemoDairy, located near Terang, Victoria, is a research demonstration dairy farm with more than 300 members from the southwest region of Victoria (http://www.demodairy.org.au/). It supports dairy-focused research, and provides applied demonstration and educational activities to members. Virtual DemoDairy aims to communicate some information on the long-term impacts of climate

(continued)

(continued)

change at both the regional and local scale, in the southwestern region of Victoria, and to make it accessible to DemoDairy managers and stakeholders. It uses Google Earth, a virtual globe, as a development platform, which provides an interactive environment to visually explore and interact with scientific data on climate change. The multi-level resolution remote sensing imagery and terrain data contained in Google Earth allow users to contextualise data within their own catchment or farming area.

At the regional scale, Virtual DemoDairy uses cartographic animation to show the projected changes in temperature and rainfall for the whole of Victoria from a baseline year 2000 to year 2050. Figure 8.35 shows two frames from the animation representing the mean annual temperatures in Victoria in 2020 and 2050. The time step slide bar at the upper-left corner of the view allows users to animate the temperature surfaces. The temperature and rainfall surfaces (rasters) of each year were created in ArcGIS based on monthly climate records and climate change projections (using the CSIRO Mark 3.5 climate change model). They were then converted to KML for display in Google Earth. The KML time-span function was used to automatically specify the dates and duration for each temperature or rainfall surface to display.

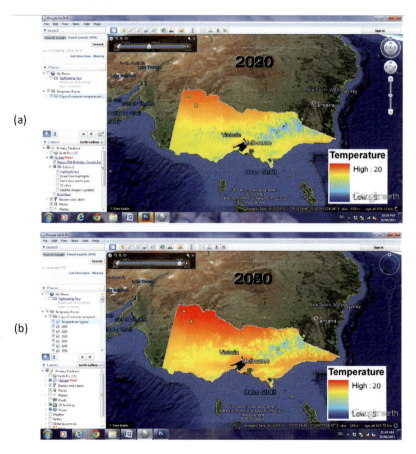

Figure 8.35 Two frames from Victorian temperature animation: mean annual temperature of: (a) year 2020 and (b) year 2050 (created using 2000-2050_temp.kmz freely available at http://vro.depi.vic.gov.au/dpi/vro/vrosite.nsf/pages/geovis_virtual_demoDAIRY, © the State of Victoria).

At the local scale, Virtual DemoDairy combined climate data and model outputs with 3D visualisation in the Google Earth environment to provide general information about the farm and data on pasture growth under different climate change scenarios. It also used 3D photo-realistic images to describe local infrastructure on the ground and possible futures. Figure 8.36a shows the current environment, local infrastructure and conditions on the farm. The 3D objects in the view were created using 3D Studio Max from photographs, which were then converted into KML and placed on the aerial photos using Google SketchUp. The cow icons shown in Figure 8.36a are clickable point features (called Google Earth placemarks), to which the outputs from a locally calibrated pasture growth (SGS) model are linked. For example, by clicking the 'Demo Dairy 2070' placemark on the right in Figure 8.36a, you can get a report on the likely impact on pasture yield and growth pattern under different climate change scenarios on DemoDairy farm in 2070, as shown in Figure 8.36b.

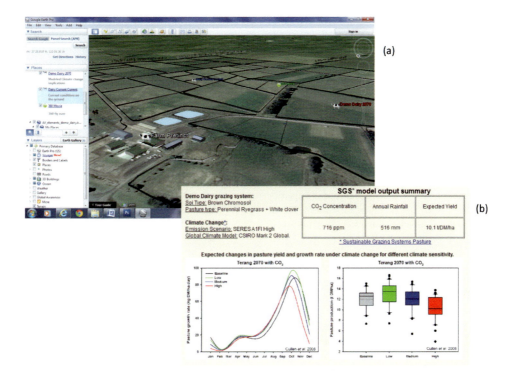

Figure 8.36 Visualisation of the current conditions of the DemoDairy farm and the likely climate change impact in 2070 (created using Demo_dairy_application_Points_of_interest.kmz freely available at http://vro.depi.vic.gov.au/dpi/vro/vrosite.nsf/pages/geovis_virtual_demoDAIRY, © the State of Victoria).

Virtual DemoDairy was evaluated by twenty local stakeholders in a workshop. The end-user evaluation indicated that a combination of animation and 3D visualisation in a virtual globe could enhance the capacity to communicate the result of complex climate change modelling. However, the evaluation also suggested that there was greater support for more conventional mapping and graphic representations than for interactive 3D visualisation and realistic renderings.

Source: Aurambout et al. (2013).

8.6 SUMMARY

- Maps are the major means of spatial visualisation. Modern GIS supports traditional 2D mapping as well as multivariate, 3D and animated mapping.
- The cartographic elements on a map include the title, map body, legend, scale, orientation and other supplementary elements such as neatline, inset and credit. All these elements should be brought together into an effective visual arrangement so that a hierarchy of information is established in accordance with their importance to understanding the mapped information.
- There are two broad categories of maps: general-purpose and thematic maps. According to the types of geographical features mapped and the methods of cartographic representation, thematic maps are further classified into dot, proportional symbol, graduated symbol, choropleth, area qualitative, isoline, hypsometric and hill-shaded maps.
- Map design is a process of applying cartographic knowledge and experience to make maps for some particular purpose or set of purposes. It mainly involves selection of the area, geographical features and their properties to map, classification of data, and design of map symbols and map layout.
- Several methods have been developed to classify quantitative data for mapping, including equal interval, natural breaks, standard deviation and quantile. Different methods may result in different cartographic representations of the same dataset. A suitable classification method should produce meaningful patterns and allow for easy interpretation.
- Geographical features and their properties are represented using map symbols. The shape, size, colour and other visual variables of a map symbol must be selected appropriately according to the data, and cartographic convention and standards.
- An effective visual hierarchy should be built on a map so that the most important features are visually prominent. Three methods are usually used to build a visual hierarchy: stereogrammic, extensional and subdivisional.
- Text is often used to label features on a map. However, its use should not reduce legibility and change the effective visual hierarchy of the map.
- Multivariate mapping uses visual variables of map symbols, multiple map layers, charts or more advanced visualisation techniques such as parallel coordinate plots and self-organising maps to display multiple variables simultaneously.
- 3D mapping provides a 3D view of geographical distribution, and is becoming a popular method of spatial visualization.
- Animated mapping involves displaying sequences of spatial data. It produces dynamic representations of changes over time, space or attributes. Temporal cartographic animations deal with temporal data and produce time-series maps to show changes over time. Non-temporal cartographic animations create a series of related map presentations sequentially to show non-temporal changes. Combined with 3D mapping, animation provides a powerful tool for discovering trends, patterns and processes.

REVIEW QUESTIONS

1. What are the key cartographic elements?
2. What are general-purpose maps? What are thematic maps?
3. What is the graduated symbol map? How is it constructed?
4. What is the choropleth map? How is it designed?
5. How does a hill-shaded map portray the terrain surface? Discuss the potential uses of a hill-shaded map.
6. Describe the process of map design.
7. The most commonly used methods for data classification include equal interval, natural breaks, standard deviation and quantile. Describe each of the methods, and discuss their advantages and disadvantages using examples.
8. What are the major considerations in data classification?
9. How should map symbols be designed?
10. What is visual hierarchy? How is it established on a map?
11. Discuss the principles of map lettering.
12. How should a map layout be designed?
13. How can multivariate mapping be accomplished?
14. Describe how to design a bivariate choropleth map.
15. What is a parallel coordinate plot? What is a self-organising map? How can they be used in multivariate mapping?
16. What is temporal cartographic animation? What is non-temporal cartographic animation?
17. What is the keyframing technique in animated mapping?

REFERENCES

Aurambout, J., Sheth, F., Bishop, I. and Pettit, C. (2013) 'Simplifying climate change communication: an application of data visualisation at the regional and local scale', in A. Moore and I. Drecki (eds) *Geospatial Visualisation*, Berlin, Germany: Springer, 119–136.

Copsey, R.D. (2002) 'Visualizing environmental data', in K.C. Clarke, B.O. Parks and M.P. Crane (eds) *Geographic Information Systems and Environmental Modelling*, Upper Saddle River, NJ: Prentice Hall, 252–286.

Fuhrmann, S. (2002) 'Designing a visualization system for hydrological data', *Computers and Geosciences*, 26(1): 11–19.

Guo, D. (2008) *SOMVIS: A Multivariate Mapping and Visualization Tool – User Manual, Version 2.0*, available at http://www.spatialdatamining.org/ for registered users.

Guo, D., Gahegan, M., MacEachren, A.M. and Zhou, B. (2005) 'Multivariate analysis and geovisualization with an integrated geographic knowledge discovery approach', *Cartography and Geographic Information Science*, 32(2): 113–132.

Inselberg, A. (1985) 'The plane with parallel coordinates', *The Visual Computer*, 1: 69–97.

Jenks, G.F. and Caspall, F.C. (1971) 'Error on choroplethic maps: definition, measurement, reduction', *Annals of American Geographers*, 61: 217–244.

Kennedy, K.H. (2009) *Introduction to 3D Data: Modeling with ArcGIS 3D Analyst and Google Earth*, Hoboken, NJ: Wiley.

Kohonen, T. (2001) *Self-Organizing Maps*, New York: Springer.

Lobben, A. (2003) 'Classification and application of cartographic animation', *The Professional Geographer*, 55(3): 318–328.

Peterson, M.P. (1995) *Interactive and Animated Cartography*, Englewood Cliffs, NJ: Prentice Hall.

Robinson, A.H., Morrison, J.L., Muehrcke, P.C., Kimerling, A.J. and Guptill, S.C. (1995) *Elements of Cartography*, 6th edn, New York: Wiley.

Sidiropoulos, G., Pappas, V. and Vasilakos, A. (2005) 'Virtual reality: the non-temporal cartographic animation and the urban (large scale) projects', *Soft Computing*, 9: 355–363.

Slocum, T.A. (2005) *Thematic Cartography and Visualization*, 2nd edn, Upper Saddle River, NJ: Pearson Prentice Hall.

Ultsch, A. and Siemon, H. (1990) 'Kohonen's self organizing feature maps for exploratory data analysis', in B. Widrow and B. Angeniol (eds) *Proceedings of the International Neural Network Conference (INNC-90), Paris, France, July 9–13, 1990*. Dordrecht, The Netherlands: Kluwer, 305–308.

Weber, C.R. and Buttenfield, B.P. (1993) 'A cartographic animation of average yearly surface temperatures for the 48 contiguous United States: 1897–1986', *Cartography and Geographic Information Systems*, 20(3): 141–150.

Yin, H. (2008) 'The self-organizing maps: background, theories, extensions and applications', in J. Fulcher and L.C. Jain (eds) *Computational Intelligence: A Compendium*, Berlin, Germany: Springer, 715–762.

Zhang, K., Chen, S.C., Singh, P., Saleem, K. and Zhao, N. (2006) 'A 3D visualization system for hurricane storm-surge flooding', *IEEE Computer Graphics and Applications*, 26(1): 18–25.

Spatial decision analysis and modelling

9

This chapter discusses the concepts, principles and techniques of multi-criteria decision analysis, and introduces several spatial modelling tools which have been implemented in the GIS environment for environmental modelling, including cellular automata, agent-based modelling, weights-of-evidence modelling and artificial neural networks.

LEARNING OBJECTIVES

After studying this chapter, you should be able to:

- describe the major components of multi-criteria decision analysis;
- be familiar with popular multi-criteria decision analysis techniques;
- understand how multi-criteria decision analysis can be used to integrate technical information, scientific knowledge and value judgements to support decision making;
- grasp the fundamental concepts of cellular automata and agent-based modelling and identify potential environmental applications;
- describe the principle of weights-of-evidence modelling and potential environmental applications;
- explain the principle of the neural network, its strength and weakness, and potential environmental applications;
- know how models and decision analysis techniques are implemented in GIS.

9.1 MULTI-CRITERIA DECISION ANALYSIS

Environmental management is a complex socio-technical process. The decision making process may involve clarifying, refining and resolving a problem situation through information exchange, discussion and negotiation among stakeholders. Therefore, decisions in environmental management do not rely on technical information derived from GIS spatial analysis and modelling alone. While a technique might offer insights regarding the information, the interpretation of the results depends on the values and attitudes of those who examine the technical information. The decisions may be influenced by hard facts, supported by scientific evidence, as well as by subjective considerations. There is a need for a systematic approach to decision analysis that can make explicit the necessary value judgements, and integrate and incorporate the values of the decision makers, public opinion and policy and management goals with technical information to examine the overall implications of each alternative management plan. Multi-criteria decision analysis (MCDA) techniques provide a framework for facilitating such an approach.

General principle of MCDA

MCDA refers to a set of procedures designed to help decision makers investigate a number of possible choices with regard to multiple criteria and derive rankings of alternative choices. It involves the utilisation of data and decision makers' preferences, and the manipulation of those data and preferences according to certain decision rules. MCDA has three key components:

- a number of decision alternatives or options;
- a set of criteria by which the alternatives are to be assessed;
- a numerical method for prioritising the alternatives according to how well they satisfy the criteria.

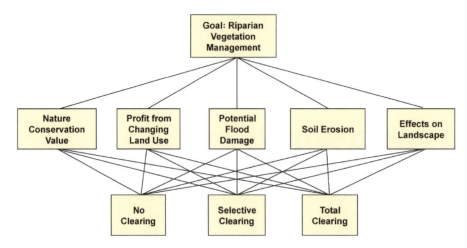

Figure 9.1 A decision hierarchy for riparian vegetation management.

Let us consider an example of riparian vegetation management. The objective is to identify feasible management options for riparian vegetation. There could be three possible management options: no clearing, selective clearing and total clearing. The criteria used to judge the management options include maintaining nature conservation value, increasing profit from changing land use, reducing potential flood damage, reducing soil erosion and minimising effects on landscape. The goal, criteria and alternatives can be structured into a hierarchy, as shown in Figure 9.1. The best alternative is not obvious. Alterative 1, 'no clearing', might provide the best level of preservation and recreation, but with undesirable social and economic consequences. While Alternative 3, 'total clearing', might provide the greatest income from changing land use, it might not satisfy preservation and recreation needs. Alternative 2, 'selective clearing', might be a compromise between Alternatives 1 and 3, but might have the highest level of environmental damage in terms of soil erosion.

For a particular riparian zone, we can create a performance matrix, or consequence table, in which each row describes an alternative and each column describes the performance or preference of the alternatives against each criterion. Table 9.1 shows the performance matrix for the example of riparian vegetation management. Each performance value in the matrix is a score or rating measured on a strength of preference scale. More preferred alternatives score higher on the scale, and less preferred alternatives score lower. A scale may extend from 0 to 1, where 0 represents the least preferred alternative and 1 is associated with the most preferred alternative. All alternatives considered in the MCDA would then fall between 0 and 1. Other scales may extend from 0 to 10 or from 0 to 100 and so on. Performance values are determined based on facts and scientific evidence (including modelling results). In Table 9.1, a scale of 0–1 is used. According to the performance matrix, 'no clearing' is the best in the light of the criterion 'maintaining nature conservation value', and the worst with respect to the criterion 'increasing profit from changing land use', while 'total clearing' is not a preferred option for 'maintaining nature conservation value', but has a very good outcome on 'increasing profit from changing land use'.

Table 9.1 Performance matrix

(Weights)	Conservation (0.21)	Profit (0.10)	Flood (0.12)	Erosion (0.18)	Landscape (0.39)
No clearing	1	0	0.8	0.9	0.3
Selective clearing	0.7	0.5	0.6	0.4	0.9
Total clearing	0.1	0.9	0.2	0.1	0.2

In a performance matrix, rankings or numerical weights are also assigned to each criterion to measure their relative importance. In the example shown in Table 9.1, the criterion 'effects on landscape' has the highest weight of 0.39, meaning it is the most important criterion among the five criteria. Weights are usually measured on a 0–1 scale. The sum of the weights for all the criteria should equal 1. The assignment of different rankings or sets of weights allows the effects of different points of view to be explored. In other words, weights reflect stakeholders' or decision makers' value judgements. The higher the importance, the higher the weighting applied. Weights represent social preferences.

The weights and scores are then combined with a linear additive model:

$$S_j = \Sigma_{i=1}^n W_i X_{ij} \qquad (9.1)$$

where S_j is the overall score of the jth alternative, W_i is the weight of the ith criterion, X_{ij} is the score of the jth alternative on the ith criterion, n is the number of criteria and $\Sigma W_i = 1$. For example, the overall score of 'no clearing' in the above example is:

$$0.21 \times 1 + 0.10 \times 0 + 0.12 \times 0.8 + 0.18 \times 0.9 + 0.39 \times 0.3 = 0.58$$

Using the same equation, we can calculate the overall scores for 'selective clearing' and 'total clearing', which are respectively 0.69 and 0.23. The overall scores are used to rank or prioritise the alternatives. In this example, 'selective clearing' gets the highest score, 0.69. Therefore, it is the most preferred option.

In general, MCDA involves six steps:

1. Establishing the decision context – this involves identifying the aims of the MCDA, decision makers and other stakeholders, and establishing a shared understanding of the problem situation.
2. Structuring the decision problem – this identifies objectives, alternatives and criteria. Here, objectives are desired outcomes, such as the level of economic growth and sustainable development. They should be specific, measurable, agreed, realistic and time-dependent. Alternatives are the options that may contribute to the achievement of these objectives. Criteria are uncontrolled variables that reflect performance in meeting the objectives. Each criterion must be measurable, in the sense that it must be possible to assess, at least in a qualitative sense, how well a particular alternative is expected to perform in relation to the criterion. Objectives, criteria and alternatives are often arranged into a hierarchy. A basic form of a hierarchical model of a decision problem is a pyramid with a broad overall objective (or goal) at the highest level. Lower levels list the criteria and respective sub-criteria used to choose among alternatives. At the lowest level are the alternatives to be evaluated. Figure 9.1 shows such a hierarchy.
3. Scoring the alternatives in relation to each criterion – this involves constructing scales measuring the performance or preference of the alternatives against each criterion. Scales can be ordinal, interval or ratio. An ordinal scale represents an order only, indicating that one alternative scores higher than another alternative, but not by how much. An interval scale measures the difference between two alternatives, but does not indicate actual magnitude. For instance, on an interval scale, the scores 5 and 10 are equivalent to the scores 15 and 20. They suggest that one alternative scores 5 units higher than another, but does not suggest that one alternative has a double score compared with another or that the difference in performance of two alternatives is moderate. A ratio scale provides a measure of both difference and magnitude. For example, on a ratio scale the scores 5 and 10 indicate a doubling of the scores, while the scores 15 and 20 indicate an increase of 33 per cent in performance. The type of measurement scale to be used depends on the objectives, the available information and the evaluation method for combining scores and weights. Scores can be obtained through subjective judgement, or through direct measurement of real attributes.
4. Assigning weights to the criteria to reflect their relative importance to the decision – weights can be assigned directly by the individuals carrying out the analysis, derived from the views or opinions of a group of people, or generated mathematically from limited information on rankings or comparative judgements (see the SMARTER and AHP methods discussed below). The weight on a criterion reflects both the range of difference of the alternatives and how much that difference matters. Any numbers on a ratio scale can be used for weights, but most commonly they range from 0 to 1.
5. Evaluating the alternatives – this involves applying a mathematical procedure to combine the weights and scores for each alternative to produce a ranking of alternatives. There are many different evaluation methods. Equation 9.1 is the simplest and most widely used method, which is also called weighted

summation. With this method, both the weights and scores must be quantitative. Some methods are designed for qualitative data, while others can deal with both quantitative and qualitative data. Some methods rank alternatives, some identify a single optimal alternative, some provide an incomplete ranking, and others differentiate between acceptable and unacceptable alternatives. Two methods, SMARTER and AHP, will be introduced below, and detailed descriptions of other popular methods used in environmental applications can be found in Linkov and Moberg (2012).

6. Conducting a sensitivity analysis – sensitivity analysis in MCDA is used to assess how different scoring and weighting systems affect the overall ranking of alternatives. Potentially, it can help resolve disagreements between interest groups. For example, a sensitivity analysis may reveal that two or three alternatives always appear best under different weighting systems, though their order may shift. If the differences between these best options are small, accepting a second-best alternative can be shown to be associated with little loss of overall benefit. More importantly, the insights gained from a sensitivity analysis may suggest that part of the decision analysis should be revisited, more information collected or additional criteria introduced. The steps above can be repeated until an MCDA model is good enough to resolve the issues at hand.

SMARTER

As described above, scoring and weighting are two important steps in MCDA. Weighting is perhaps the most difficult judgement step. In practice, it is not always easy for decision makers to use precise numerical values to express the relative importance of the different criteria. SMARTER (Simple Multi-Attribute Rating Technique – Exploiting Ranks) provides an approach to identify a single set of weights that is representative of all the possible weight combinations, which maintains the rank order of importance of the criteria.

SMARTER is an approximate method based on an elicitation procedure for weights, developed by Edwards and Barron (1994). This procedure includes nine steps:

1. identifying the purpose of decision making and decision makers whose values or judgements should be elicited and whose utilities are to be maximised;
2. eliciting a list of criteria which are relevant to the purpose of the value elicitation from the decision makers;

3. identifying alternatives or the outcomes of possible actions to be evaluated and building an MCDA hierarchy;
4. evaluating how well each alternative would perform on each criterion;
5. eliminating dominated alternatives (that is, those with anticipated high performance) in order to reduce the total number of alternatives;
6. converting the measures of attainment of each alternative on each criterion into a value score, with 0 representing the worst plausible score and 1 the best plausible score;
7. eliciting rank order of the criteria;
8. calculating the weights for each criterion based on the rank order;
9. calculating overall scores for alternatives by applying the linear additive model in Equation 9.1.

SMARTER uses the centroid method to estimate the set of weights based on the rank order of the criteria. The centroid method assigns weights as follows.

Assume W_1 is the weight of the most important criterion, W_2 is the weight of the second most important criterion and so on. For n criteria:

$$W_1 = (1 + 1/2 + 1/3 + \ldots + 1/n) / n$$

$$W_2 = (0 + 1/2 + 1/3 + \ldots + 1/n) / n$$

$$\ldots \quad \ldots$$

$$W_n = (0 + 0 + 0 + \ldots + 1/n) / n$$

Generally, if $W_1 \geq W_2 \geq \ldots \geq W_n$, then the weight of the ith criterion is:

$$W_i = \frac{1}{n} \Sigma_{j=i}^{n} \left(\frac{1}{j} \right) \quad (9.2)$$

According to the simulation studies by Edwards and Barron (1994), SMARTER locates the best available alternatives on 75–85 per cent of occasions. In particular, when the best alternative is not located, the average loss of the overall score is only 2 per cent. Given that decision makers' weight estimates are imprecise, this method can be very effective. Ranking order is a decision task that is easier than developing numerical weights. Using an ordinal approximation, SMARTER alleviates the discomfort many people feel when forced to allocate hard numbers (weights) to subjective judgements. However, order ranking does not provide decision makers with an opportunity to carefully weigh the relative importance of criteria during which the insights could emerge. Case Study 4 in Chapter 10 provides an example of the

Table 9.2 The pair-wise comparison scale (after Saaty 1990)

Intensity of importance	How important is *A* relative to *B*?
1	Equally important
3	Moderately more important
5	Strongly more important
7	Very strongly more important
9	Overwhelmingly more important
2, 4, 6, 8	Intermediate values between the two adjacent judgements
Reciprocal of above numbers	If *A* has one of the above numbers assigned to it when compared with *B*, then *B* has the reciprocal value when compared to *A*.

application of SMARTER to sugarcane land allocation in Northern Queensland, Australia.

AHP

AHP (Analytic Hierarchy Process) was developed as a methodology for MCDA which allows critical examination of the underlying assumptions, consistency of value judgements and facilitates the identification of trade-offs among social, economic and environmental considerations for a wide range of decision problems (Saaty 1980). It uses three underlying principles: decomposition, comparative judgements and synthesis.

Decomposition involves identifying key elements of a decision problem, including objectives, criteria and alternatives, and building a decision hierarchy. In other words, it is used to structure the decision problem, the second step in the general procedure of MCDA discussed above. The hierarchy structuring is a relatively subjective activity, based on decision makers' experience and knowledge about the decision problem. A hierarchy should focus on those criteria that are most important. Due to the limited capacity of the human mind to compare things simultaneously, the AHP calls for a maximum of five to nine branches in any one node in a decision hierarchy in order to have great efficiency and consistency in making assessments.

Based on the hierarchy, the principle of comparative judgements is applied to determine the relative importance of the criteria and the relative preference of the alternatives through pair-wise comparisons. Elements in a given level in a hierarchy are compared in pairs with respect to a common property or criterion in the level above. AHP uses a fundamental scale of absolute numbers to express individual preferences

or judgements. Table 9.2 lists these fundamental scale values. If there are many people participating, multiple judgements can be combined by taking the geometric means of individual judgements. Pair-wise comparison judgements should be made based on the best information available, and decision makers' intuition, knowledge and experience.

Pair-wise comparisons against a given criterion or property result in a matrix. Table 9.3 shows a matrix of pair-wise comparisons of the criteria with respect to the overall goal in Figure 9.1. The numbers in the matrix are the AHP scale values whose definitions are described in Table 9.2. They correspond to the judgements obtained by comparing the elements in the left-hand column with the elements in the top row. When an element is regarded less favourably than another, the judgement is a fraction. It can be noted that when comparing one element with itself, the comparison must result in 1. Therefore, the diagonal values of a pair-wise comparison matrix are always 1. A pair-wise comparison matrix is also reciprocal. The judgements are only needed for the upper triangular part of a matrix. The lower triangular part is their reciprocals. Once a pair-wise comparison matrix is generated, AHP derives the weights or priorities for the relevant elements by solving for the principal eigenvector of the matrix.

Suppose we have the elements $C_1, C_2, \ldots, C_n$ of some level in a hierarchy. We wish to find their priority weights, $w_1, w_2, \ldots, w_n$. The value of w_i represents the degree of influence or importance of C_i with respect to the criterion element in the level above. We use a_{ij} to denote the pair-wise comparison value of C_i over C_j. The pair-wise comparison matrix for these elements is denoted as:

$$A = \left(a_{ij} \right) \tag{9.3}$$

Table 9.3 Pair-wise comparison matrix

	Conservation	Profit	Flood	Erosion	Landscape
Conservation	1	2	2	1	1/2
Profit	1/2	1	1	1/2	1/4
Flood	1/2	1	1	1	1/3
Erosion	1	2	1	1	1/2
Landscape	2	4	3	2	1

Obtaining the weights $w = (w_1, w_2, \ldots, w_n)$ based on A is an eigenvalue problem:

$$Aw = \lambda_{max} w \qquad (9.4)$$

where λ_{max} is the largest or principal eigenvalue of A. The solution to w is obtained by raising the matrix to a sufficiently large power, then summing over the rows and normalising.

The weights derived from a pair-wise comparison matrix are called local weights with respect to a particular criterion. After they are weighted by the weight of their criterion element, they are called global weights. After all judgements have been made and all pair-wise comparison matrices have been obtained, the global weights for the elements will be calculated. Coming down the hierarchy from the second level, the global weight for each element in one level is computed by multiplying its local weight by the priorities of the criterion elements in the level immediately above, and adding them. This global weight is in turn used to weight the local weights of the relevant elements in the level below, and so on to the bottom level. This process in AHP is called synthesis, which results in a set of overall priorities for the alternatives in the bottom level. The details of the synthesis and overall results for the riparian vegetation management decision problem are shown in Figure 9.2.

AHP also measures the inconsistency of judgements. The pair-wise comparison matrix A is said to be perfectly consistent if

$$a_{ik} a_{kj} = a_{ij} \, i, j, k = 1, 2, \ldots, n \qquad (9.5)$$

and

$$\lambda_{max} = n \qquad (9.6)$$

The closer λ_{max} is to n, the more consistent the judgements are. The consistency index is defined in the AHP as:

$$\frac{\lambda_{max} - n}{n - 1} \qquad (9.7)$$

In order to derive a meaningful interpretation of the consistency index, Saaty (1980) defined the consistency ratio, also called the inconsistency ratio, as the ratio of the consistency index for a particular set of judgements to the average consistency index for random comparisons for a matrix of the same size. A set of perfectly consistent judgements produces an inconsistency ratio of 0, while a set of inconsistent judgements equivalent to what would be expected from random judgements produces an inconsistency ratio of 1.0. As a rule of thumb, when the inconsistency ratio exceeds 0.10, the judgements often need re-examination.

Case Study 5 in Chapter 10 involves the use of AHP in land evaluation in northern Laos along the border with China.

MCDA with GIS

MCDA with GIS is a process of transforming and combining spatial data and value judgements to derive information for decision making. Some GIS software packages integrate basic MCDA functions, for example the weighted overlay tool in ArcGIS and the Multi-Criteria Evaluation module in IDRISI. In GIS, alternatives can be locations defined by point, line or area features on vector data layers, or grid cells on raster data layers. Criterion scores or ratings are represented as attribute values associated with geographically defined decision alternatives (locations). They are stored as data layers called criterion maps.

There are two types of criterion maps: natural-scale and constructed-scale (Malczewski 1999). Natural-scale criterion maps record the quantitative or numerical attribute values measured on their natural scales – for example, distance to rivers in metres, temperature in Celsius degrees, land value in dollars and number of animal species per square metre. These measures are

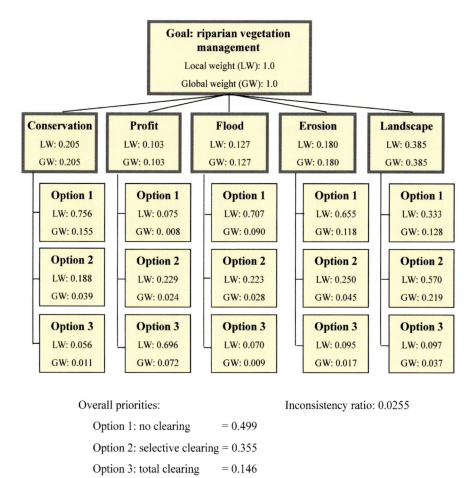

Figure 9.2 Local, global weights and overall results for the riparian vegetation management problem.

objective and not based on value judgement, but they can be used directly as ratings or scores of alternative locations regarding the attributes or criteria. Constructed-scale criterion maps represent scores or ratings defined subjectively using the value judgements of experts or decision makers – for example, ratings for different ranges of slope and suitability scores for different types of land use.

Given the variety of measurement scales on which attributes or criteria can be measured, MCDA requires that the values contained in the various criterion maps be transformed to comparable units. It is hard to compare different attributes (for example, temperature on an interval scale versus proximity to rivers on a ratio scale). If we want to combine the various criterion maps, the measurement scales must be commensurate. To achieve this, the values of the criteria need to be converted into standardised criterion scores. There are several techniques for

score standardisation, including linear scale transformation, value/utility functions, fuzzy membership functions and probabilistic reasoning (Malczewski 1999). Two linear scale transformation methods are often used: score range and maximum score.

The score range procedure calculates the standardised scores as follows:

1. If the value of a criterion is to be maximised:

$$Score_i = \frac{X_i - X_{min}}{X_{max} - X_{min}} \tag{9.8}$$

2. If the value of a criterion is to be minimised:

$$Score_i = \frac{X_{max} - X_i}{X_{max} - X_{min}} \tag{9.9}$$

where $Score_i$ is the standardised score for the criterion for the ith alternative location on the criterion map, X_i is the value of the criterion associated with the ith location, X_{max} is the maximum value of the criterion and X_{min} is the minimum value of the criterion. $Score_i$ ranges between 0 and 1. The higher the score, the more attractive the alternative is in relation to the criterion.

The maximum score approach standardises scores as follows:

1.   If the value of a criterion is to be maximised:

$$Score_i = \frac{X_i}{X_{max}} \qquad (9.10)$$

2.   If the value of a criterion is to be minimised:

$$Score_i = \frac{X_{min}}{X_i} \qquad (9.11)$$

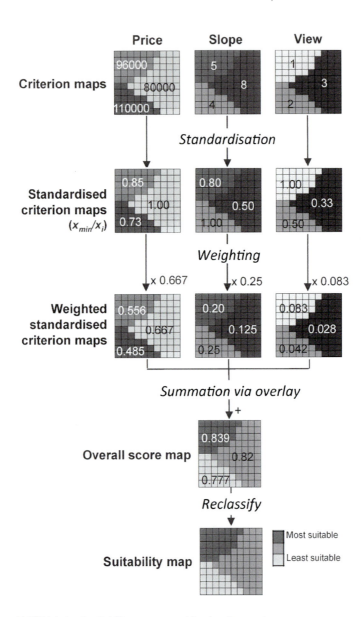

Figure 9.3 GIS-based MCDA in land suitability assessment for development.

Figure 9.3 shows a procedure of MCDA with GIS using land suitability assessment for development as an example. In this example, three criteria are considered: land price, slope and view (visibility). Each grid cell is an alternative site for development. The land price map is a natural-scale criterion map, and other two are constructed-scale criterion maps. The three criterion maps are then standardised using the maximum score method. The weights for price, slope and view are respectively 0.667, 0.25 and 0.083 (adding up to 1). It uses weighted summation via overlay to calculate the final scores for each cell. The final scores are then classified into land suitability for development.

Overlay or map algebra functions are readily available in GIS for implementing the linear additive model in MCDA. However, GIS in general does not provide weighting methods. There are a few third-party MCDA add-ins for GIS – for example, MCDA4ArcMap (an open source MCDA extension to ArcGIS available at http://mcda4arc map.codeplex.com/) and AHP Extension to ArcGIS (available at http://arcscripts.esri.com/details.asp?dbid=13764). Weights can also be directly specified by decision makers or calculated by using a stand-alone MCDA software tool, then input into GIS. Box 9.1 shows an example of MCDA with ArcGIS and a stand-alone MCDA package, Decision Analyst, which supports AHP and SMARTER.

Box 9.1   MCDA with ArcGIS and Decision Analyst for non-hazardous landfill site selection   **PRACTICAL**

To follow this example, download **Decision Analyst** from **http://wfw-atlas.monash.edu/Decision_ Analyst/default.html**. Unzip the downloaded file, and double-click **da_windows_1.exe** to install Decision Analyst following the on-screen instructions. Then start ArcMap, and load the **boundary**, **villages**, **rivers**, **faults** and **roads** feature classes as well as **degreeSlope** and **landcv_r** grids from **C:\ databases\GIS4EnvSci\VirtualCatchment\Geodata.gdb**. Set both the ArcGIS current workspace and scratch workspace as **C:\Databases\GIS4EnvSci\VirtualCatchment\Results** (refer to Step 4 in Box 8.3). In this example, six criteria are selected to assess site suitability for non-hazardous landfills in the hypothetical catchment, including slope, land cover, proximity to rivers, proximity to faults, proximity to roads and proximity to villages. There are several constraints for landfill siting. These constraints include that a landfill should not be located within 1km from villages and faults and 500m from rivers, not in crop fields, and not in water bodies or on slopes of more than 30°.

Generate criterion maps

1. Use the **Euclidean Distance** tool to create a map of proximity to rivers using **40** as the cell size, and save the output raster to the workspace as **dis_rivers**. Before implementing the tool, click the **Environments** button in the **Euclidean Distance** dialog. In the **Environment Settings** dialog, expand **Processing Extent**, change the extent to **Same as layer boundary**, then expand **Raster Analysis**, change the cell size to **As Specified Below** and enter **40**, then select **Boundary** layer as the mask. Click **OK** to close the **Environment Settings** dialog, then run the **Euclidean Distance** tool.
2. In the same way, use the **Euclidean Distance** tool to create raster maps of proximity to roads, faults and villages using **40** as the cell size, and save them to the workspace as **dis_roads**, **dis_faults** and **dis_villages** respectively.
3. Use the **Reclassify** tool to create six constructed-scale criterion maps according to Table 9.4. A scale of 1–5 is used, where 1 = least suitable, 2 = marginally suitable, 3 = suitable, 4 = moderately suitable and 5 = most suitable. A score of 0 indicates restricted. The six criterion maps are saved to the workspace as **slope_cr**, **landcover_cr**, **roads_cr**, **rivers_cr**, **faults_cr** and **villages_cr**.

(continued)

(continued)

Table 9.4 Suitability scores for criteria

	0	1	2	3	4	5
Dis_roads		> 2,000	1,500–2,000	1,000–1,500	500–1,000	< 500
Dis_rivers	< 500	500–1,000	1,000–1,500	1,500–2,000	2,000–2,500	> 2,500
Dis_faults	< 1,000	1,000–1,250	1,250–1,500	1,500–1,750	1,750–2,000	> 2,000
Dis_villages	< 1,000	1,000–2,000	2,000–3,000	3,000–4,000	4,000–5,000	> 5,000
Slope	> 30	20–30	15–20	10–15	5–10	< 5
Land cover	Water body, cropland	Forests	Shrubs		Meadow	Bare rock

As all constructed-scale criterion maps are on the same scale, no standardisation is required.

Use AHP to derive weights

4. Start **Decision Analyst** from the **Start** button on the Windows taskbar.
5. In the model area, double-click anywhere in the first column under Level 1. A dark blue rectangle appears. Type in **landfill** as the name of the element representing the top element of the decision hierarchy.
6. Move the cursor to somewhere in the second column under Level 2, and double-click the mouse. Another dark blue rectangle appears. Type in **slope** as the name of the element. Similarly, in Level 2, add the elements **landcover**, **dis_roads**, **dis_rivers**, **dis_faults** and **dis_villages**. These are the second-level elements of the decision hierarchy.
7. Click the **landfill** element in Level 1. The element is highlighted in dark blue. Go to the **Edit** menu, and select **Connect All**. The **landfill** element will be connected to all elements in Level 2, which forms a two-level decision hierarchy, as shown in Figure 9.4.
8. Go to the **Assessment** menu, and select the **AHP Pairwise** menu item. In the **Select an Expression** dialog, select **Importance**, then click **OK**. In the **Pairwise Assessment** dialog, enter the pair-wise comparison matrix, as shown in Figure 9.4. After the pair-wise comparison matrix is complete, the weights for the six criteria are listed in the **Weight** column. They are 0.097 for slope, 0.135 for land cover, 0.047 for distance to roads, 0.318 for distance to rivers, 0.126 for distance to faults and 0.277 for distance to villages. The inconsistency ratio is 0.047 (< 0.1).

Derive overall scores with weighted overlay

9. Return to ArcMap. In **ArcToolBox**, navigate to **Spatial Analyst Tools > Overlay**, and double-click **Weighted Overlay**. In the **Weighted Overlay** dialog:

   1) Click the **Add Raster** button ➕ . Click the **Input raster** arrow, then click **slope_cr**. Select **VALUE** as the input field. Click **OK**. **slope_cr** is added to the Weighted Overlay table.
   2) Repeat the previous step to add **landcover_cr**, **roads_cr**, **rivers_cr**, **faults_cr** and **villages_cr** one by one to the Weighted Overlay table.
   3) Set the **Evaluation scale** to **1 to 5 by 1**.
   4) In the weighted overlay table, change values in the **Scale Value** column to the corresponding values listed in the **Field** column. But when the field value equals to 0, set the corresponding scale value as **Restricted** (Figure 9.5).

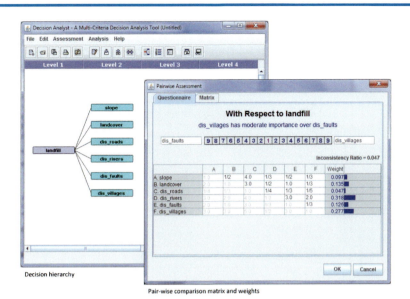

Decision hierarchy

Pair-wise comparison matrix and weights

Figure 9.4 Pair-wise comparison assessment for landfill siting.

5) In the **%Influence** column, enter the corresponding weights (×100) derived from AHP in Step 8 above (Figure 9.5). Note that the weight values must be integers and add up to 100.

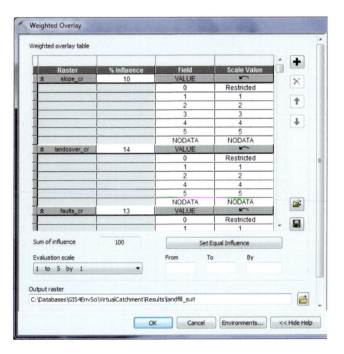

Figure 9.5 Weighted overlay.

(continued)

(continued)

6) Name the output raster **landfill_suit**, and save it to the workspace.
7) Click **OK**. After **landfill_suit** is created, it is added to the data view, as in Figure 9.6.

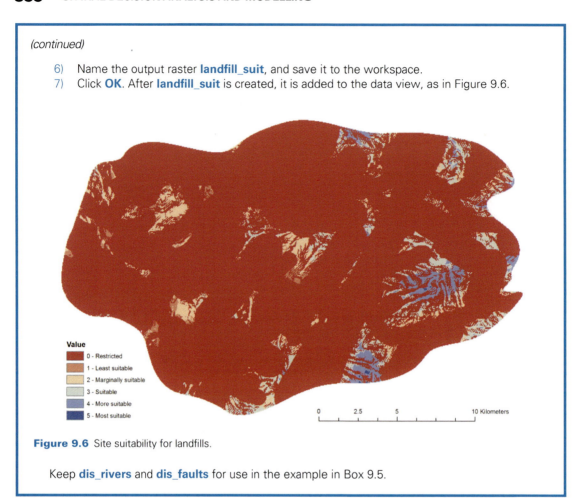

Figure 9.6 Site suitability for landfills.

Keep **dis_rivers** and **dis_faults** for use in the example in Box 9.5.

9.2 CELLULAR AUTOMATA

Simply stated, an automaton is a self-operating machine. The term automata is plural, often used to describe non-electronic moving machines that are made to resemble human or animal actions. Cellular automata (CA) are not machines, but computer models used to simulate the behaviour of self-organising and self-replication systems, such as forest fire propagation, urban development, land use/land cover change, species dispersal, snowflake formation and crystallisation.

A CA simulated system is made up of many discrete cells on a grid. Each cell may be in one of a finite number of states denoted by attribute values. In the simplest situation, a cell may have a binary state – for example, either presence/on/alive (denoted by 1) or absence/off/dead (denoted by 0). A cell may change the state (value) according to a set of rules, referred to as transition rules, based on the cell's own state and the states of its neighbouring cells. Basically, a CA model applies a specific set of transition rules to a regular grid representing a particular configuration of cells in specific states to simulate the evolution of a system through a number of discrete time steps (sometimes called ticks or generations). The grid can be in any finite number of dimensions. A one-dimensional grid is a line of cells; a two-dimensional grid resembles a raster, with each cell represented by a square. We will focus on two-dimensional CA only.

A simple and well-known two-dimensional CA example is John Conway's Game of Life, also known simply as Life (Gardner 1970). The CA uses a raster to represent the system, in which each cell is in one of two possible states, dead (= 0) or alive (= 1) (Figure 9.7).

Every cell interacts with the eight surrounding neighbours immediately adjoining it horizontally, vertically or diagonally. At each step in time, the state of every cell changes synchronously, according to the following transition rules based on its own state and the old states of its neighbours:

1. A cell that is alive will stay alive if the cell has exactly two or three live neighbours; otherwise, the cell will die of loneliness or overcrowding.
2. A cell that is dead will come to life if the cell has exactly three live neighbours, as if by reproduction; otherwise, the cell will remain dead.

The initial configuration makes up the seed of the system. The second generation is produced by applying the above rules simultaneously to all cells in the seed. The rules continue to be applied repeatedly to generate further generations. Each generation is a function of the previous one. As the process is repeated over and over again, a dynamic system is obtained that exhibits surprisingly complex

behaviour. Figure 9.7 shows four successive generations of Life on an 8×7 grid.

From the example described above, it can be seen that CA can be defined in terms of four elements:

1. An *n*-dimensional grid of cells – cells can be differently shaped, for example squares or hexagons. With GIS, we mainly deal with 2D grids of squares – that is, rasters.
2. States – at each discrete time step, each cell is in one and only one state.
3. Neighbourhood – the two most common types of neighbourhoods are the von Neumann neighbourhood and the Moore neighbourhood (Figure 9.8). John Conway's Game of Life uses the Moore neighbourhood.
4. Transition rules – the dynamics of the system are determined by the transition rules. Following the rules, at time step *t* each cell updates its state according to the states of the neighbouring cells at the immediately previous time step $t - 1$.

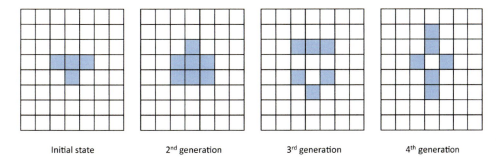

| Initial state | 2nd generation | 3rd generation | 4th generation |

Figure 9.7 Three successive generations of Life (dark cells are alive and white cells are dead).

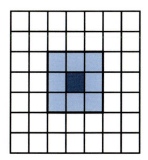

Moore neighbourhood

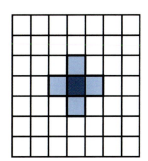

von Neumann neighbourhood

Figure 9.8 Neighbourhoods in CA.

The distinctive feature of CA is that they display complex global patterns and emergent behaviour, starting from simple cells deterministically following simple local rules. Thus, CA are particularly suitable for modelling any system that is composed of simple components, where the global behaviour of the system is dependent on the behaviour and local interactions of the individual components.

The neighbourhood (focal) functions and conditional operators in map algebra discussed in Section 4.4 are the core functions for implementing CA in GIS. Box 9.2 demonstrates how a simplified CA model simulating forest fire spread is executed in ArcGIS. This model incorporates probabilities of lightning strike and tree growth, thus it can be considered a probabilistic CA model.

Box 9.2 A cellular automata forest fire model in ArcGIS   PRACTICAL

To follow this example, start ArcMap, and load the **inittrees** grid from **C:\Databases\GIS4EnvSci\ VirtualCatchment\Grids**. This raster layer shows the existing trees in a forest. The cell value 1 represents a tree, and 0 represents an empty site. The example illustrates a CA model simulating how forest fires spread after trees ignite randomly through lightning strike. The Spatial Analyst extension is required.

This CA model is from Drossel and Schwabl (1992), which is defined by four transition rules that are applied simultaneously:

1. A burning tree becomes an empty cell.
2. A tree ignites if at least one neighbour is burning.
3. A tree becomes a burning tree with probability f of lightning strike even if no neighbour is burning.
4. A tree grows at an empty cell with probability p.

It should be noted that these rules are not a realistic representation of how real forests behave. This example seeks to demonstrate how a CA model is executed in ArcGIS, not how a forest fire model is created.

Spatial modelling in ArcGIS can be conducted with ModelBuilder or through scripting. ModelBuilder is a diagramming tool in ArcGIS for creating and executing spatial models using sequences of spatial analysis tools, feeding the output of one tool into another as input. It is also a useful tool for constructing and executing workflows represented using spatial influence diagrams discussed in Section 1.4. In principle, ModelBuilder can be used for dynamic modelling via an iterative process. However, it is far more efficient to use a scripting language for dynamic modelling. The major scripting language for model building and implementation in the current version of ArcGIS is Python, which is a free, cross-platform and open source programming language. In this example, the CA model was written in Python. Comments (starting with the hash character #) are provided to help understand the Python script. To learn Python programming in ArcGIS, consult Zandbergen (2013).

Examine the CA model

1. Click the **Start** button on the Windows taskbar, point to **All Programs**, click **ArcGIS**, click **Python 2.7** (used for ArcGIS 10.1; a different version of Python may be used for other versions of ArcGIS), and select **IDLE (Python GUI)**. The window **Python Shell** is launched.
2. In the **Python Shell** window, click **File** on the main menu bar, and select **Open**. Navigate to **C:\Databases\GIS4EnvSci\VirtualCatchment\Scripts**, and select **forest_fire_model. py**. .py is the suffix attached to a Python script file name. The CA model is contained in the Python script, which is opened in a new window, **forest_fire_model.py**. The script is listed in Table 9.5.

Table 9.5 Python script for the CA forest fire model in Box 9.2

```
1    import arcpy
2    from arcpy import env
3    from arcpy.sa import *
4
5    # Check out the ArcGIS Spatial Analyst extension license
6    arcpy.CheckOutExtension("Spatial")
7    # Allow output files to be overwritten
8    arcpy.env.overwriteOutput = True
9    # Set the workspace
10   env.workspace = "C:/Databases/GIS4EnvSci/VirtualCatchment/Results/fires"
11
12   # The input raster showing initial trees
13   # In the input raster, 0 - empty site, 1 - tree
14   inRaster = Raster("inittrees")
15   # The output rasters named as "fires" plus the step number (e.g. fires0,
16   # fires1, . . .). In the output rasters, 0 - empty site, 1 - tree,
17   # 2 - burning
18   outRaster = "fires"
19   # Set the map extent
20   arcpy.env.extent = "inittrees"
21
22   # Set probability with which a tree grows at an empty site (cell)
23   p = 0.01
24   # Set probability with which a tree ignites by lightning strike
25   f = 0.001
26   # Time steps
27   steps = 10
28   # Set seeds for generating random numbers
29   seedValue1 = 0.01
30   seedValue2 = 0.001
31   # Set the cell size of the rasters
32   cellSize = 40
33
34   # Define the Moore neighbourhood
35   neighborhood = NbrRectangle(3, 3, "CELL")
36
37   # Loop for running time steps
38   for i in range(steps):
39       # Apply a neighbourhood function with "MAXIMUM" operator on the input
40       # raster
41       outFocal = FocalStatistics(inRaster, neighborhood, "MAXIMUM", "")
42       # Create random rasters to represent probabilities of no tree growth
43       # and no lightning strike for each cell
44       randomRaster1 = CreateRandomRaster(seedValue1, cellSize, "")
45       seedValue1 = seedValue1 + 0.001
46       randomRaster2 = CreateRandomRaster(seedValue2, cellSize, "")
47       seedValue2 = seedValue2 + 0.00001
48
49       # If a cell is burning, it becomes empty (Rule a)
50       outCon1 = Con((inRaster == 2), 0, 0)
```

(continued)

(continued)

Table 9.5 *(continued)*

```
51      # If a cell is empty, a tree grows with the probability p (Rule d)
52      outCon2 = Con((inRaster == 0) & (randomRaster1 <= p), 1, 0)
53      # If a cell is a tree and at least one neighbour is burning, the tree
54      # ignites (Rule b)
55      outCon3 = Con((inRaster == 1) & (outFocal == 2), 2, 0)
56      # If a cell is a tree, it becomes a burning tree with probability f
57      # even if no neighbour is burning (Rule c)
58      outCon4 = Con((inRaster == 1) & (randomRaster2 <= f) &
59                     (outFocal != 2), 2, 0)
60      # If a cell is a tree, Rule b and Rule c do not apply, it remains a
61      # tree
62      outCon5 = Con((inRaster == 1) & (randomRaster2 > f) &
63                     (outFocal != 2), 1, 0)
64
65      # Combine the above rules to create a data layer
66      outCon = outCon1 + outCon2 + outCon3 + outCon4 + outCon5
67
68      # Save the combined layer with name "fires" + the time step number i
69      outCon.save(outRaster+str(i))
70      # Set the saved layer as the input raster for the next step
71      inRaster = Raster(outRaster+str(i))
72
73  print('The process is over')
```

In this script, ArcPy is a Python package, which provides Python access for all ArcGIS tools. A line beginning with # indicates that the remainder of the line is a comment explaining the following line of the program. Line numbers are inserted purely for the reader's convenience and are not part of the program. Line 10 specifies a directory as the workspace for ArcGIS geoprocessing tool inputs and outputs. You may change the workspace where you saved **inittrees**. The outputs from the CA model will also be written to the workspace.

Lines 23 and 25 set the probabilities of tree growth and lightning strike. Typically, the probability of a lightning strike is very low. p is larger than f so that large structures can be developed. Lines 44 and 46 randomly assign each cell a probability of growth and a probability of lightning strike respectively at a given time step. Line 41 identifies whether there are any fires in neighbouring cells. As the maximum statistic is used in the **FocalStatistics** function, if the function returns a value of 2 for a cell, that cell has at least a burning neighbour. Line 55 uses the result from the **FocalStatistics** function to determine whether a tree will become a burning tree.

Lines 50–63 implement the transition rules using the ArcGIS **Con** tool. This tool controls the output value for each cell based on whether the cell value is evaluated as true or false in a specified conditional statement. For example, Line 52 means if the value of a cell in inRaster is equal to 0 (empty cell) and at the same time the probability of no tree growth is less than or equal to p, 1 will be assigned to that cell location (true) (which is filled by a tree) on the output raster **outCon2**; otherwise, the cell will be assigned 0 (false) on **outCon2**.

Line 27 sets the number of time steps as 10. You may change it to 20, 30 or any other number. Line 38 indicates that the CA will run for the specified number of time steps or ticks. Line 71 means that the output or states of each cell at time step i will be used as the input for modelling at the next time step $i + 1$.

Run the CA model

3. In the **forest_fire_model.py** window, click **Run > Run Module** to execute the script in the window. Wait until the message 'The process is over' appears in the **Python Shell** window.
4. In ArcMap, add the raster layers **fires0**, **fires1**, **fires2**, **fires3**, . . . from the workspace set in the script. These layers represent fires and burnt areas at time steps 0, 1, 2, 3, ... Figure 9.9 shows the states of fires at time steps 3, 5, 7 and 9. Because of the stochastic nature of the modelling, your outputs may look different.

9.3 AGENT-BASED MODELLING

Agent-based models (ABMs) are the natural extension of CA, in which agents (such as people, animals, infectious diseases, insects, fires, pollutants and land use activities) are able to move around in space, rather than being restricted to the cells of a raster. Like CA, ABM simulates the behaviour of a complex system in which agents interact with each other and with their environment using simple local rules. However, in ABM agents typically do not simultaneously perform actions at constant time

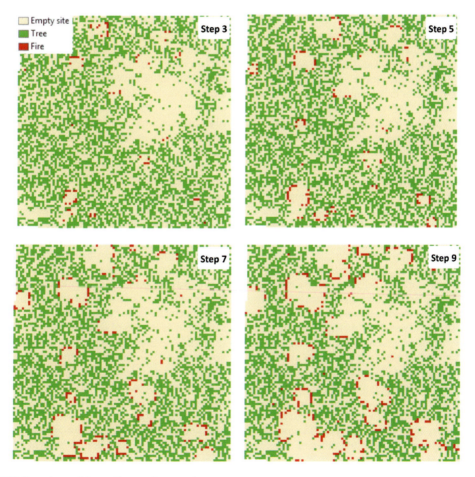

Figure 9.9 Snapshots of forest fires at discrete time steps.

steps as in CA. Rather, their actions follow a sequential schedule of interactions, which allows for the cohabitation of agents with different environmental experiences. In general, an ABM consists of agents that interact within an environment and the rules governing the interactions. Therefore, to build an ABM, four elements need to be defined: agents, rules, interactions and the environment.

Agents can be any representations of discrete, autonomous entities or individual system components with their own goals and behaviours and without the influence of centralised control. They vary in terms of their attributes or states. Agents' states are represented by discrete or continuous variables (for example, the land use for a land parcel agent and the concentration of a pollutant agent). Given the choice of an agent's state variable or variables, a state transition may occur when the agent interacts with another agent or with the environment. Agents can be designed to change their state depending on their previous state, as in CA (in which a cell is an agent). They can also engage with their environment, adapt to changes in their environment according to their expectations and goal, and evolve in a learning cycle of acting through evaluating the results of the actions dependent on the response of the environment and updating the goal or the actions. Agents are generally goal-driven. They may seek to satisfy a basic need, such as meeting basic energetic requirements, or a complex goal, such as maintaining sustainable urban growth (such an ABM may comprise both social and natural systems and respective agents).

Behaviours or actions of agents are determined by a collection of rules. These rules range from simple IF . . . THEN . . . rules to algorithms comprising thousands of lines of code. As in CA, rules are iterated over one or more discrete time steps. Rules can be derived from scientific evidence, empirical observations, heuristic knowledge or value judgements. For example, an ABM can be used to model the foraging behaviour of the mainland serow as a rule-guided walk over the landscape in which the animal agent assesses its environment and moves in a deliberate fashion to search for preferred food sources. According to empirical observations in the study area, the mainland serow grazes on grass, shoots and leaves from along beaten paths. This foraging habit could be incorporated into rules. When modelling human agents, rules are typically elicited through interviews and surveys, in which subjects may be asked to describe how they would respond in a certain situation. The responses from the human subjects regarding different scenarios can be used to generate rules.

Agents are free to interact with other agents. They typically only have access to information about their immediate neighbourhood, and do not have global knowledge about the whole system they inhabit. Local interactions of agent-to-agent or agent-to-environment lead to the emergence of global behaviour or patterns which are unknown to the individual agents. Interactions among agents may occur via direct or indirect communication. Direct communication often takes place through the message passing mechanism, which allows agents to compose and send messages that are then interpreted by receiving agents. The messages may represent spoken dialogue between human agents or tags displayed by agents that can be decoded by other agents. Indirect communication is accomplished via modifications to a shared environment. For example, in an ABM model of the behaviour of ants, ant agents initially roam randomly. Once they find food, they return to their colony while leaving pheromone trails. When other ants find such a trail, they are likely to follow it and reinforce it with their own pheromone if they also find food. However, the pheromone along the trail evaporates over time, reducing its attractive strength. The longer it takes for an ant to travel along the trail, the weaker and less attractive the pheromone it has laid down becomes. Over a short path, the strength of the pheromone remains high as it is reinforced before it can evaporate or decay. Ants tend to choose their trails with stronger pheromone concentrations. Therefore, when one ant finds a shorter path, other ants are more likely to follow that trail, and positive feedback eventually leads all the ants to choose the shortest trail. This example indicates that communication between agents reinforces and amplifies the effects of local interactions, giving rise to group-level or global-level dynamics.

Agents can roam the space – for example, animals. They can also be fixed – for example, land parcels. The environment where agents are situated may be represented in different ways, such as a raster, a network or a vector map representing a continuous space. Commonly, environments represent geographical spaces – for example, in models concerning urban growth, where the environment simulates some of the physical features of a city, and in models of foraging behaviour of animals, where the environment represents the spatial distribution of foods. They may also include the effects of other agents in the surrounding locality, and the influence of factors such as resource depletion and topography. In GIS, agents can be represented as points, polygons or grid cells in separate layers, and can be connected by networks or areas of various types, or move along a network or over a continuous surface represented by a raster or polygon map (such as an urban district map).

ABMs are often used to explore questions related to the evolution of cooperation and self-organisation in human and animal societies, to study dynamics of heterogeneous populations of individuals and to deduce group-level behaviours. Typical applications include modelling of urban growth (Batty 2005), simulation of ecological and evolutionary processes (DeAngelis and Mooij 2005), exploration of dynamics of land use and land cover change (Parker et al. 2003), and analysis of wildlife movements (Watkins et al. 2011). Box 9.3 introduces the Agent-based Rural Land Use New Zealand (ARLUNZ) model.

Box 9.3   Agent-based Rural Land Use New Zealand model   **CASE STUDY**

ARLUNZ is an agent-based model developed by Landcare Research, New Zealand to assess the impact of climate change policies on rural land use, farm net revenue and environmental indicators such as greenhouse gas (GHG) emissions, nutrient loadings and soil erosion. It can also be used to analyse the resulting land use effects caused by changes in farming demographics, social networks and decision making. The model implements a spatially and behaviourally heterogeneous population of farmers that operate in line with a real-world population, and provides information on how farmers will adapt both economically and socially to global change.

The model involves two types of agents: a landscape (the environment) on which the agents make decisions, and the economic information associated with both the landscape and the agents. The 'farm' agent is defined as the centroid of a farm parcel delineated by the cadastral boundary. Farm agents do not have any decision making ability. The farm agent generates a 'farmer' agent at the same location. The farmer agent is defined with a range of social and economic attributes, such as age, size of social networks, succession, potential revenues and net revenue from the last step of the model. The landscape is represented by a cadastral layer defining the boundaries of farm parcels that have different types of land use (dairy, sheep, beef, pine plantation and so on) and productivity.

The model uses the information from the farm and farmer agents to determine the optimal use of the farm parcel based on yields, input costs, output prices and environmental constraints to generate the expected net revenue for the possible land use enterprises available in the model. The land use that produces the highest net revenue for the farm is defined as the land use the farmer agent assesses for conversion. The landscape, farm and farmer agents interact through the development of dependencies and feedback loops between them. It is the farmer agents who make land use decisions by considering and integrating farm, farmer and economic information.

Once the model is run, the attribute values of farmer agents and market variables (commodity prices) are updated, which are then used to calculate the expected net revenue values for all the potential enterprises that could be undertaken on each farm. The land use with the best expected net revenue value is the enterprise of a farmer agent proposed by the model. For any given farm, if the proposed enterprise is the same as the farmer agent's current enterprise, the net revenue of the farmer agent is updated. If not, the farmer agent decides whether to change the current land use based on its likelihood of land use conversion. An agent's likelihood of land use conversion is a function of the information received from the agent's social networks as well as their current and proposed enterprises. It is initially set at 0.2 for all farmer agents. To simulate the decision to undertake a land use conversion, a random number generator is used to produce probabilities for evaluation against the farmer agent's likelihood of land use conversion. If the random number is greater than the farmer agent's likelihood of land use conversion, the current land use of the farm will be converted to the proposed enterprise. After all farmer agents have weighed up their potential for land use conversion, those which have reached the end of the farming life cycle without finding a successor sell the farms. This concludes a time step of the model. The model is run repeatedly until the specified time step is reached. Each time step in the model represents a five-year period. Figure 9.10 shows the process.

(continued)

(continued)

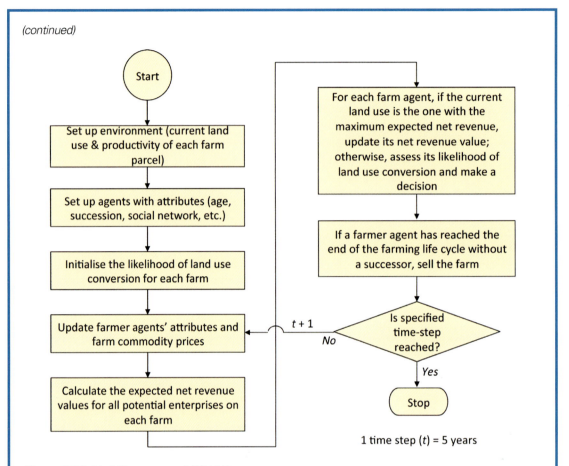

Figure 9.10 Modelling process of ARLUNZ.

This model was applied to assess the impact of a GHG reduction policy on farm-level land use in the Hurunui and Waiau catchments in the North Canterbury region of New Zealand's South Island. The case study focused on the effect of GHG prices on land use/cover change, and the impact of land use/cover changes through feedback effects from the farmers' social networks. The effect of GHG prices on land use/cover change was simulated by imposing a series of GHG prices on emissions (sequestration) from agricultural (forest) enterprises, ranging from NZ$0 to NZ$60 per tonne carbon dioxide equivalent (tCO_2e). The impact of feedback effects from the social networks was modelled by comparing the impacts of farmer agents utilising, and not utilising, information received through their social networks within their land use conversion decisions. The model assumed an annual increase in farm commodity prices (milk, meat and timber) of 2 per cent and a farm succession success rate of 75 per cent. It also assumed that the climate and available technology (hence farm productivity) were held constant over the entire model simulation. The model ran ten incremental time steps, covering a time horizon of fifty years.

Key results from the model showed that total farm net revenue for the catchments was estimated to increase over time regardless of the GHG price, due to increasing commodity prices and farmers switching to more profitable enterprises. However, imposing a GHG price policy reduced farm net revenue by about 1–2 per cent over fifty years compared with the no GHG price baseline. More immediate effects could see a reduction of 7–13 per cent in early periods, as farmers had yet to fully adjust to the policy change. Net GHG emissions were estimated to decline over time, even under a no GHG price baseline, due to an expansion of forestry on low-productivity land. Higher GHG prices provided a greater net reduction of emissions. While social network effects had minimal impact on net revenue and environmental outputs for the catchment, they had an effect on the land use pattern, and in particular the clustering of enterprises. Figure 9.11 shows the spatial distributions of three enterprises (dairy, sheep and beef, and forestry) under different scenarios of the GHG prices and social network effects resulting from the model.

Source: Morgan and Daigneault (2015).

GIS itself does not provide agent-based modelling capabilities, but it provides spatial data manipulation, analysis and visualisation tools for developing spatial representations of agents and the environment, and visualising the modelling results. Scripting languages in a GIS can be used to build and implement ABMs with GIS spatial analysis functions (Box 9.2 is an example). Several third-party ABM tools have also been developed for GIS to facilitate agent-based modelling. Among them, Agent Analyst is an ABM extension to ArcGIS, and Open Map,

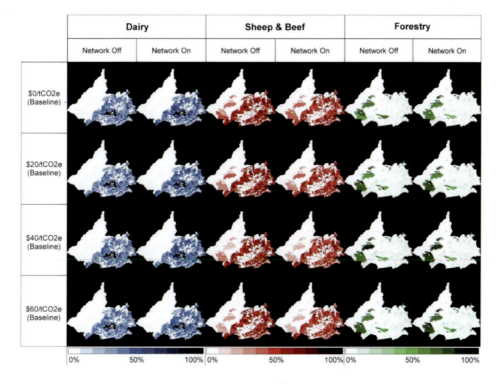

Figure 9.11 Simulated spatial patterns of dairy, sheep and beef, and forestry under different scenarios of the GHG prices and social network effects (darker colours represent a higher likelihood of the farm being used in the enterprise at the end of the model runs) (© Morgan and Daigneault, used under the terms of the Creative Commons Attribution License, which permits unrestricted use, distribution, and reproduction in any medium, provided the original author and source are credited).

which allows users to define agents, environments and actions, create, edit and interpret rules, and to simulate the actions by the agents according to the rules in a GIS environment. It is free software. Box 9.4 demonstrates how an ABM is created and implemented with Agent Analyst with ArcGIS.

Box 9.4   Urban growth modelling using Agent Analyst with ArcGIS   PRACTICAL

To follow this example, download Agent Analyst from **http://resources.arcgis.com/en/help/agent-analyst/**, run the downloaded executable file, and install Agent Analyst following the setup wizard. Then start ArcMap, and load the **mytown**, **ugmout** and **myroads** shapefiles from **C:\databases\GIS4EnvSci\VirtualCity**. The **mytown** layer represents land parcels with a town centre and a village close to a hypothetical large city. Its attributes include **PARCELID**, **STATUS** (development status, either developed or undeveloped) and **OWNERSHIP** (land ownership, either public or private). The **myroads** layer represents the road network in the same region. **ugmout** is used as the output layer from the model. This example involves simulating urban growth around the town and village centres with an expectation that the population of the town will grow over the years.

In general, the process of building and implementing an ABM involves seven stages:

1.   defining the research question;
2.   defining agents including their attributes, goals (desires), intents (actions they intend to carry out to meet its goals) and the environment in which the agents reside and act;
3.   defining agent-to-agent and agent-to-environment interactions;
4.   formulating rules governing agent's actions;
5.   specifying the number of time steps and the duration of a time step;
6.   running the model;
7.   verifying and validating the model.

Define the research question

The research question to be addressed is: 'What future urban growth patterns could evolve in the hypothetical study area?' The purpose of this example is to demonstrate the major elements of agent-based modelling using simple rules based on the environmental and neighbourhood conditions. It does not intend to produce realistic urban growth scenarios.

Start Agent Analyst

1.   Open the **Catalog** window by clicking [icon] in ArcMap's toolbar.
2.   In the **Catalog** window, navigate to the folder or geodatabase where you want the toolbox to be created.
3.   Right-click the folder or geodatabase, and click **New > Toolbox**. The default name of the toolbox (Toolbox.tbx or Toolbox) appears, and is highlighted in the folder or geodatabase. Change the toolbox name to **Urban growth**. If the toolbox name is not highlighted, right-click the toolbox, and click **Rename**.
4.   Right-click the **Urban growth** toolbox, point to **New**, and click **Agent Analyst**. Wait until the **Agent Analyst** window is open. If it does not open automatically, right-click the new tool and select **Edit**, and the **Agent Analyst** window will appear (Figure 9.12).

Define and create agents

Land parcels (usually from cadastral data) are modelled as agents. For simplicity's sake, the state of a parcel agent is either developed or undeveloped. In other words, the main attribute of the parcel

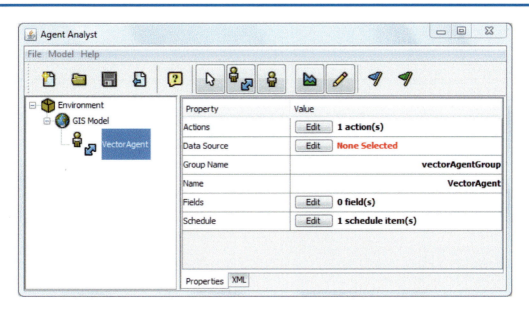

Figure 9.12 Agent Analyst.

agents is their development status. All parcel agents share the same attributes and follow the same rules for actions. The **mytown** layer provides the data on the location (geometry) and attributes of each parcel agent. Parcel agents are defined and created in Agent Analyst following the next two steps:

5. In the **Environment** panel on the left, click **GIS Model**. In the **Property-Value** panel on the right:

    1) Make sure **ArcGIS** is selected as the GIS package.
    2) Click the **Value** box of the **Model Name** property, and change it to **UrbanGrowthModel**.
    3) Similarly, change the value of the property of **Display Name** to **Urban Growth Model**.

6. In the **Environment** panel, click **VectorAgent**. In the **Property-Value** panel:

    1) Click the **Value** box of the **Name** property, and type **Parcel**.
    2) Click the **Value** box of the **Group Name** property, and type **Parcels**. This defines all the parcels as an agent group named Parcels, which will have the same attributes and behaviour.
    3) Click the **Edit** button to the right of the **Data Source** property. In the **Data Source Editor** dialog, click **Browse,** navigate to the folder where **mytown** is stored, and select **mytown.shp**. Click **Open**. The **Data Source Fields** are automatically populated with the attributes (fields) from the **mytown** shapefile, including **PARCELID**, **STATUS**, **OWNERSHIP** and **the_geom** (geometry of an agent). These are the attributes of parcel agents to be used in the model. Click **OK**. A collection of parcel agents is created.
    4) Click the **Edit** button to the right of the **Fields** property. In the **Fields Editor** dialog, define two additional attributes (fields) for parcel agents: **NeighbourList** (a list of neighbours) and **NumNeighbours** (the number of neighbours). First, key in **NumNeighbours** in the name box, **int** in the type box (it means the attribute data are integer), and **0** as the default value. Check **Accessible** so that this attribute can be accessed by other components of the model.

(continued)

(continued)

Click **Add**. Second, type **NeighbourList** in the name box, **java.util.ArrayList** in the type box (it means that **NeighbourList** is an array list of neighbouring parcels; as Agent Analyst was developed using the Java programming language, this array list is defined as a Java class of ArrayList), and check **Accessible**. Click **Add**. Click **OK**.

Define the environment and create environment layers or data

The urban environment is a complex system. The environment of parcel agents can be characterised by a number of physical and socio-economic factors, such as terrain, geology, natural hazards (for example, flooding and landslides) vulnerability, accessibility, land values, land use policies and planning regulations. A parcel agent decides whether to change its development status by assessing its current environmental and neighbourhood conditions. For the purpose of demonstrating the basic concepts of ABM, this example only considers parcel agents' accessibility to roads and the development status of their neighbours. It is assumed that there is no physical constraint on development in the area. Accessibility to roads is measured as the straight-line distance from parcels to roads using the **Near** tool in ArcGIS. The neighbouring parcels can be identified using the **Polygon Neighbors** tool in ArcGIS. The two environmental conditions are derived as follows.

Measure accessibility to roads

7. In **ArcMap**, open **ArcToolBox**. Navigate to **Analysis Tools > Proximity**, and double-click **Near**.
8. In the **Near** dialog:

   1) Select **mytown** as the input features and **myroads** as the near features.
   2) Click **OK**. After the operation is complete, the following two fields are added to the attribute table of **mytown**: **NEAR_FID**, which stores the feature ID of the nearest road segment, and **NEAR_DIST**, which stores the distance from each land parcel to the nearest road segment. **NEAR_DIST** represents the accessibility of parcel agents to roads.

9. Go back to the **Agent Analyst** window. Click the **Parcel** agent in the **Environment** panel, then click the **Edit** button next to the **Data Source** property to reload the **mytown** shapefile. **NEAR_FID** and **NEAR_DIST** are added as two new attributes to the list of **Data Source Fields**.

Create a table of parcel neighbours

10. In **ArcToolBox** in ArcMap, navigate to **Analysis Tools > Proximity**, and double-click **Polygon Neighbors**.
11. In the **Polygon Neighbors** dialog:

    1) Set **mytown** as the input features.
    2) Specify **parcel_nbrs** as the name of the output table, and save it to your ArcGIS as **Default.gdb**.
    3) Tick **PARCELID** as the field to report.
    4) Check **Include both sides of neighbour relationship**.
    5) Enter **50 Meters** as the XY tolerance (two parcels are considered neighbours if they are within 50m apart).
    6) Keep the output linear units as **METERS**.
    7) Click **OK**. Wait until the table **parcel_nbrs** is created.

12. Right-click **parcel_nbrs** in the table of contents, and open the table. It lists four attributes: **src_PARCELID**, the object ID of the source polygon; **nbr_PARCELID**, the object ID of the neighbouring

polygon of the source polygon; **LENGTH**, the total length of coincident edges between a source polygon and a neighbouring polygon; and **NODE_COUNT**, the number of times a source polygon and a neighbouring polygon cross or touch at a point. Delete the **LENGTH** and **NODE_COUNT** fields from the table as they are not useful for this analysis.

13. Right-click **parcel_nbrs** in the table of contents, point to **Data**, and click **Export**. In the **Export Data** dialog, click the **Browse** button, navigate to the directory of your choice, select **Text File** as the save-as type, and type the name **parcel_nbrs.txt**. Click **Save** to return to the **Export Data** dialog. Click **OK**. A table of parcel neighbours is created and saved as a text file.

Initialise the environmental conditions

14. Return to **Agent Analyst**. In the **Environment** panel in the **Agent Analyst** window, click **Urban Growth Model**. Click the **Edit** button right to the **Fields** property.
15. Click the **Edit** button right to the **Actions** property. The **Actions Editor** dialog appears.
16. In the **Actions Editor** dialog:

    1) In the drop-down list located above **Java Imports**, select **initAgents**. A single line of code `def initAgents():` appears in the **Source** box. It is used to define a function named **initAgents** to initialise the ABM model. It is often used to load environment data, and initialise the environmental conditions and states of the agents. This function is always executed by the model first at the beginning of the simulation. Here, we are using it to load the environment data and initialise the agents and environmental conditions. As the distance to roads has been added to **mytown** as two attributes of parcel agents, only parcel neighbours need to be initialised. Add the Python code shown in Table 9.6 under `def initAgents()`.

Table 9.6 Python script for initialising the ABM model in Box 9.4

```
def initAgents():
    # Construct a hash table (or hash map) to store parcel IDs and parcel
    # objects, which facilitates the search of a particular parcel in the
    # table using the parcel ID as a key.
    ParcelsByID = HashMap()
    # Initialise each parcel agent using a for-loop.
    for theParcel as Parcel in self.Parcels:
        # Create an empty list of neighbours for the parcel and set the
        # number of neighbours as 0
        theParcel.NeighbourList = ArrayList()
        theParcel.NumNeighbours = 0
        # Add the parcel ID and the parcel object to the hash table
        ParcelsByID.put(Integer(theParcel.PARCELID), theParcel)
        # Open the parcel_nbrs.txt file
        reader = BufferedReader(FileReader("C:\Databases\GIS4EnvSci\
          VirtualCity\parcel_nbrs.txt"))
        # Read the first line of the table of parcel neighbours, which contains
        # the field names.
        line = reader.readLine()
        # Read the second line of the table of parcel neighbours. Starting
        # from the second line, the table contains the IDs of
        # source and neighbour parcels
        line = reader.readLine()
```

(continued)

(continued)

Table 9.6 *(continued)*

```
# Extract the neighbour parcel IDs and add them to the neighbour list
# line by line with a while-loop until the end of the file.
while (line):
    # Break the line into several parts separated by ","
    tokenizer = StringTokenizer(line, ",")
    # Get the first part of the line, which is the object ID in
    # parcel_nbrs.txt (not the parcel ID).
    objID = Integer(tokenizer.nextToken().trim())
    # Get the ID of the source parcel
    parcelID = Double(tokenizer.nextToken().trim()).intValue()
    # Find the source parcel with parcelID in the hash table.
    theParcel = (Parcel)ParcelsByID.get(Integer(parcelID))

    # Get the ID of a neighbour parcel of the source parcel
    nghID = Double(tokenizer.nextToken().trim()).intValue()
    # Find the neighbour parcel from the hash table
    nbr = (Parcel)ParcelsByID.get(Integer(nghID))

    # Add the neighbour parcel to the neighbour list of the source parcel
    theParcel.NeighbourList.add(nbr)
    theParcel.NumNeighbours = theParcel.NumNeighbours + 1

    # Read the next line of the table of parcel neighbours
    line = reader.readLine()

# Close the parcel_nbrs.txt file
reader.close()
```

2) In the text box right below **Java Imports**, add the following four lines. They import four Java libraries to be used in the function **initAgents** for reading files, breaking a string into tokens and constructing hash tables:

```
java.io.BufferedReader
java.io.FileReader
java.util.StringTokenizer
java.util.HashMap
```

17. In the same **Actions Editor** dialog:

1) In the drop-down list located above **Java Imports**, select **writeAgents**. In the **Source** text box, under `def writeAgents():`, add `self.writeAgents(self.Parcels, "C:\Databases\ GIS4EnvSci\VirtualCity\ugmout.shp")`. This action writes the output to **ugmout.shp**.
2) The **updateDisplay** function is used to update the map display in ArcMap. Make no change to the code of the function.
3) Click **OK**.

Define actions and formulate rules for the parcel agents

In this overly simplified urban growth model, the goal of a parcel agent is to become developed if it is undeveloped, and the action a parcel agent can take is to change its development status from undeveloped to developed. The rules controlling this action are formulated based on the current state of the parcel agent and the current conditions of the environment (that is, the accessibility to roads and the states of neighbours). At each time step or tick, a parcel agent assesses its own state and the environment and acts according to the following rules:

- An undeveloped parcel remains undeveloped if it is public land.
- An undeveloped parcel remains undeveloped if its distance to roads is greater than 50m.
- An undeveloped, private parcel becomes developed if one of its neighbours is developed and its distance to roads is within 50m.

A parcel agent interacts with the environment by consulting the environmental conditions and with other parcel agents by consulting its neighbours regarding their states, then acts when suitable conditions are met. When an action is taken, it will affect the neighbourhood conditions, and hence the environment for the next time step. Actions and behaviour rules are defined and created in Agent Analyst using the Python scripting language as described below.

18. In the **Environment** panel, click **Parcel**.
19. In the **Property-Value** panel, click **Edit** right to **Actions**. In the **Actions Editor** dialog, def step(): appears in the **Source** text box. It is used to define actions and rules for each parcel agent. This function is called and implemented at each tick during modelling, and is applied to every parcel agent. Add the lines of Python code in Table 9.7 under def step(). Click **OK**.

Table 9.7 Python script for defining actions and rules for parcel agents in Box 9.4

```
# Agent actions and rules performed during each time step.
def step():
  # The parcel agent being processed is referred to as self. If the
  # parcel agent's ownership is public, do nothing.
  if (self.OWNERSHIP == "public"):
    return

  # If the parcel agent's status is developed, do nothing.
  if (self.STATUS == "developed"):
    return

  # If the parcel agent's distance to roads is > 50m, do nothing.
  if (self.NEAR_DIST > 50):
    return

  # Check neighbours.
  numDevelopedNeighbors = 0
  for i as Parcel in self.NeighbourList:
    #print "ID: ", i.OBJECTID,i.STATUS
    if (i.STATUS == "developed"):
      numDevelopedNeighbors = numDevelopedNeighbors + 1

  # If at least one of the parcel agent's neighbours is developed, change
  # its development status to developed.
  if (numDevelopedNeighbors > 0):
    self.STATUS = "developed"
```

(continued)

(continued)

Specifying the number of time steps

20. In the **Environment** panel, click **Parcel**.
21. In the **Property-Value** panel, click **Edit** right to **Schedule**. In the **Schedule Editor** dialog:

   1) Select step as **Actions**.
   2) Click the drop-down arrow right to **Execution**, and select **EVERY TICK**. The function `step()` will be executed at every time step.
   3) Click **Add**, then **OK**.

22. In the **Environment** panel, click **Urban Growth Model**.
23. In the **Property-Value** panel, click **Edit** right to **Schedule**. In the **Schedule Editor** dialog:

   1) Click the drop-down arrow right to **Actions**, and select **writeAgents**. Click the drop-down arrow right to **Execution**, and select **EVERY TICK**. Click **Add**. It will allow the output from the model to be saved at the end of every tick.
   2) Click the drop-down arrow right to **Actions**, and select **updateDisplay**. Click the drop-down arrow right to **Execution**, and select **EVERY TICK**. Click **Add**. This will allow the output from the model to be displayed in **ArcMap** at the end of every tick.
   3) Click **OK**. Depending on the rules and constraints placed on urban growth (for example, the maximum amount of land allowed to be developed each year), a time step could be one year, five years and so on.

Run the model

24. In **ArcMap**, move **ugmout** to the top in the table of contents. Use **STATUS** as the value field to display the layer with two different colours to show developed and undeveloped areas in the data view.
25. Switch back to **Agent Analyst**. On the **Agent Analyst** toolbar, click the **Run** button ✔ . The **Repast** toolbar appears, as in Figure 9.13.

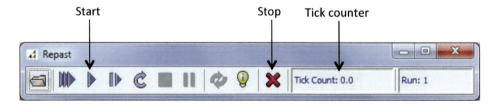

Figure 9.13 Repast toolbar.

26. Click the **Start** button on the **Repast** toolbar. As modelling progresses, the **ugmout** map in the **ArcMap** data view is updated to show the urban growth scenario simulated at every time step. The tick count changes from 0 to 1, 2, ... Click the **Stop** button on the **Repast** toolbar when the tick count changes to 4. If a time step represents one year, the map shown in **ArcMap** represents a scenario of urban growth after four years. (Ignore the warning message 'Field Length for NEAR_ DIST set to 19 preserving length, but should be set to max of 18 for dbase III specification.')
27. Close the **Repast** toolbar by clicking **Exit**. You may save the model by clicking the **Save** button on **Agent Analyst**'s toolbar.

Once a model is built and the results are produced, it has to be verified and validated. Model verification is about building the model correctly. It is a process of checking whether a program does what it was planned to do. Model verification is often carried out using a set of test cases. Model validation is about building the right model – that is, it concerns whether the simulation is a valid model of the real system. If the model reflects the behaviour of the system, it is a valid model. A common method of model validation is to compare the modelling results with real-world data or observations collected about the system. Santé et al. (2010) introduced several methods for validating urban growth models. The model built in this example is hypothetical and overly simplified.

Using **Agent Analyst** requires basic knowledge of the Java and Python programming languages. To learn this modelling tool, please read the book *Agent Analyst: Agent-Based Modeling in ArcGIS* (Johnston 2013). A more complete urban growth model can be found in the same book.

9.4 WEIGHTS-OF-EVIDENCE

Environmental analysis and decision making are often based on multiple lines of evidence. Line of evidence means a set of information that pertains to an important aspect of the environment. For example, studying species distributions requires the investigation of ecological evidence, which may include resource availability (for example, availability of water and light), direct environmental variables that have physiological importance but are not consumed by animals or plants (for example, temperature) and indirect environmental variables that have no direct physiological relevance (for example, aspect and elevation). Weights-of-evidence (WofE) modelling provides an approach to combining the information from these multiple lines of evidence into a single measure for decision making.

WofE is based on the concept of statistical weights of evidence. It measures the likelihood of a hypothesis being true conditioned on the evidence having been observed. For instance, in species distribution modelling, the evidence consists of a set of environmental conditions at a site, and the hypothesis is that the species is present at the site. For each environmental condition, a pair of weights is calculated: one for presence of the environmental condition in favour of the hypothesis, and the other for absence of the condition in favour of the hypothesis. The magnitude of the weights is determined based on the statistical association between the environmental condition and the occurrence of the species in a large group of sample sites in the study area. The weights are then used to calculate the probability that the species is present at unsampled sites based on the presence or absence of the environmental conditions. Environmental conditions are represented by the values of relevant environmental variables, such as temperature, precipitation, radiation, wind, evaporation, topography, soil moisture and vegetation. These variables are also called predictor variables. In GIS, they are typically represented in rasters, called evidence themes.

The integration of WofE with GIS started with mineral potential mapping (Bonham-Carter et al. 1988). Such integration allows the distribution of geographical entities to be predicted based on geographically referenced observation data. WofE modelling with GIS involves combining a set of evidence themes with the known distribution of the modelled entities (for example, tree species, landslides and mineral deposits) to generate a response theme (a raster) representing the probabilities of occurrence of the entities at every location.

In predicting the potential spatial distribution of geographical entities, each evidence theme is classified according to its characteristics. A pair of weights is calculated for each class: W^+ measuring the influence of the presence of the class on the occurrence of the entities, and W^- measuring the influence of the absence of the class on the occurrence of the entities. Suppose there is a set of evidence themes and the number of classes in the ith evidence theme is m_i. The weights for class j of the ith evidence theme, W_{ij}^+ and W_{ij}^-, are calculated as log ratios of conditional probabilities (Bonham-Carter et al. 1988; Agterberg et al. 1990):

$$W_{ij}^+ = \ln \frac{P(E_{ij} \mid S)}{P(E_{ij} \mid \overline{S})} \qquad (9.12)$$

$$W_{ij}^- = \ln \frac{P(\overline{E}_{ij} \mid S)}{P(\overline{E}_{ij} \mid \overline{S})} \qquad (9.13)$$

where $P(E_{ij} \mid S)$ is the conditional probability of the presence of E_{ij} (class j of evidence theme i) given S (the presence of the entities), $P(E_{ij}|\bar{S})$ is the conditional probability of the presence of E_{ij} given $\bar{S}$ (the absence of the entities), $P(\bar{E}_{ij}|S)$ is the conditional probability of the absence of E_{ij} (denoted as $\bar{E}_{ij}$) given S and $P(\bar{E}_{ij}|\bar{S})$ is the conditional probability of $\bar{E}_{ij}$ given $\bar{S}$. W_{ij}^{+} can be viewed as how much more likely it is we would see the presence of the class given that the entities are present, compared to the likelihood of observing the same class given the absence of the entities. Similarly, W_{ij}^{-} measures how much more likely we observe the absence of the class given that the entities are present, relative to the likelihood of the absence of the class given the absence of the entities. Both weights are measured on a log scale. The above conditional probabilities are calculated from:

$$P\left(E_{ij}|S\right)=\frac{NS_{ij}}{NS} \tag{9.14}$$

$$P\left(E_{ij}|\bar{S}\right)=\frac{N_{ij}-NS_{ij}}{N-NS} \tag{9.15}$$

$$P\left(\bar{E}_{ij}|S\right)=\frac{NS-NS_{ij}}{NS} \tag{9.16}$$

$$P\left(\bar{E}_{ij}|\bar{S}\right)=\frac{\left(N-N_{ij}\right)-\left(NS-NS_{ij}\right)}{N-NS} \tag{9.17}$$

where N is the total number of cells covering the study area, NS is the number of cells containing the entities in the study area, N_{ij} is the number of cells containing E_{ij} and NS_{ij} is the number of cells containing the entities in E_{ij} .

If more occurrences are found within E_{ij} than would be expected by chance, W_{ij}^{+} will be positive and W_{ij}^{-} will be negative. Conversely, W_{ij}^{+} will be negative and W_{ij}^{-} positive when fewer occurrences are found within E_{ij} than would be expected by chance. The difference between W_{ij}^{+} and W_{ij}^{-} is called contrast, expressed as:

$$C_{ij}=W_{ij}^{+}-W_{ij}^{-} \tag{9.18}$$

This measures the strength of spatial association or correlation between the jth class of evidence theme i and the known occurrences of the species. The larger the C_{ij} value, the stronger the spatial association. Under normal conditions, the maximum value of C_{ij} $(j = 1, 2, ..., m_i)$ for the ith evidence theme gives the cut-off at which the predictive accuracy of the evidence theme is maximised. Therefore, the maximum value of C_{ij} $(j = 1, 2, ..., m_i)$ is used to obtain the optimum cut-off (a class value) for

reclassifying a continuous evidence theme into a binary map with a presence/absence pattern. The statistical significance of C_{ij} can be tested by its studentised value CS_{ij} (Bonham-Carter et al. 1988), which is an approximate student's t-test:

$$CS_{ij}=\frac{C_{ij}}{\sigma\left(C_{ij}\right)} \tag{9.19}$$

where $\sigma\left(C_{ij}\right)$ is the standard deviation of C_{ij}, which can be estimated as follows (Agterberg et al. 1990):

$$\sigma\left(C_{ij}\right)=\sqrt{\frac{1}{NS_{ij}}+\frac{1}{N_{ij}-NS_{ij}}+\frac{1}{NS-NS_{ij}}+\frac{1}{\left(N-N_{ij}\right)-\left(NS-NS_{ij}\right)}} \tag{9.20}$$

When $|CS_{ij}|$ is ≥ 1.96, it is statistically significant at a significance level of 0.05. We call a class 'predictive' if the absolute value of its studentised contrast is ≥ 1.96. If an evidence theme has no class that is predictive, it is considered 'insignificant', as it is not significantly associated with the spatial distribution of the entities. Such an evidence theme has no prediction power, and thus will be discarded from modelling.

After weights and contrasts are calculated for all evidence themes, the significant evidence themes which are strongly associated with the spatial distribution of the entities are identified according to studentised contrast values. These significant evidence themes are then combined to create a response theme or a probability map representing posterior probabilities of the occurrence of the entities. The posterior probabilities are calculated as follows.

Suppose there are n significant evidence themes. Let P_{priori} be the priori probability of the occurrence of the entities, O_{priori} be the priori odds, P_{post} be the posterior probability of the occurrence of the entities and O_{post} be the posterior odds.

$$P_{priori}=\frac{NS}{N} \tag{9.21}$$

$$O_{priori}=\frac{P_{priori}}{1-P_{priori}} \tag{9.22}$$

$$O_{post}=exp\left[\ln O_{priori}+\Sigma_{i=1}^{n}\Sigma_{j=1}^{m_i}W_{ij}^{k}\right] \tag{9.23}$$

where

$$
W_{ij}^k = \begin{cases} W_{ij}^+ \text{ if } E_{ij} \text{ is a predictive class AND} \\ \text{it is present} \\\\ W_{ij}^- \text{ if } E_{ij} \text{ is a predictive class AND} \\ \text{it is absent} \\\\ 0 \text{ if } E_{ij} \text{ is not a predictive class OR it} \\ \text{is a predictive class, but its presence} \\ \text{or absence is unknown} \end{cases}
$$

and the posterior probability is:

$$
P_{post} = \frac{O_{post}}{1 - O_{post}} \qquad (9.24)
$$

Equations 9.23 and 9.24 are applied to every cell in the combined map. After the posterior probabilities for all cells are calculated, a response theme is generated.

WofE assumes that the evidence themes are conditionally independent from one another with respect to the observed occurrences of the species. Lack of conditional independence (CI) between two evidence themes may result in an inflated posterior probability where predictive classes from the two themes are present (Agterberg and Cheng 2002). Basically, there are two types of CI tests: pair-wise test and overall test. The pair-wise test of CI uses the χ^2 test statistic. Suppose there are two evidence themes A and B. The pair-wise test involves the construction of a contingency table, as shown in Table 9.8, and calculation of the χ^2 test statistic as follows:

$$
\chi^2 = \frac{N\left(\left|N_{AB}N_{\sim A\sim B} - N_{\sim AB}N_{A\sim B}\right| - \frac{N}{2}\right)^2}{\left(N_{AB} + N_{\sim AB}\right)\left(N_{A\sim B} + N_{\sim A\sim B}\right)} \\ \left(N_{AB} + N_{A\sim B}\right)\left(N_{\sim AB} + N_{\sim A\sim B}\right) \qquad (9.25)
$$

where N_{AB} is the number of occurrences where predictive classes from both A and B are present, $N_{\sim AB}$ is the number of occurrences where no predictive class of A is present while a predictive class of B is present, $N_{A\sim B}$ is the number of occurrences where a predictive class of A is present while no predictive class of B is present, $N_{\sim A\sim B}$ is the number of occurrences where predictive classes from both A and B are absent, and

$$
N = N_{AB} + N_{\sim AB} + N_{A\sim B} + N_{\sim A\sim B} \qquad (9.26)
$$

Table 9.8 2×2 contingency table for conditional independence test

	A	~A
B	N_{AB}	$N_{\sim AB}$
~B	$N_{A\sim B}$	$N_{\sim A\sim B}$

If the χ^2 test fails, the evidence themes A and B are conditionally dependent.

Theoretically, conditional independence of evidence themes implies that the sum of the posterior probabilities for all grid cells in the study area T is equal to NS (Agterberg and Cheng 2002). If $T > NS$, there is a lack of CI among the evidence themes. The overall test of conditional independence is based on this assumption. According to Agterberg and Cheng (2002), the hypothesis of overall conditional independence could be accepted with a probability of 95 per cent if

$$
\frac{T - NS}{\sigma(T)} < 1.645 \qquad (9.27)
$$

where $\sigma(T)$ is the standard deviation of T. Equation 9.27 is called the Agterberg-Cheng conditional independence test.

The validity of the independence assumption can also be assessed by applying a CI ratio, calculated by dividing the actual number of training points by the predicted number of points of occurrence – that is, NS/T. This value ranges from 0 to 1. A value of 1 (which never occurs in practice) indicates conditional independence among the evidence themes. Values much smaller than 1 indicate a violation of the conditional independence assumption. Generally, a CI ratio of <0.85 may indicate a lack of conditional independence (Bonham-Carter 1994, pp. 267–302).

If one or all of the above CI tests fail, one evidence theme should be discarded, or otherwise they should be combined so that there is approximate conditional independence verified by new conditional independence tests.

After the final response theme is produced, the likelihood of the occurrence of the modelled entity can be classified into four classes: high, moderate, low and uncertain, using the following method (Carranza and Hale 2000):

High: when $P_{post}/P_{priori} > 5$ and $P_{post}/\sigma(P_{post}) > 1.5$

Moderate: when $5 \geq P_{post}/P_{priori} \geq 1$ and $P_{post}/\sigma(P_{post}) > 1.5$

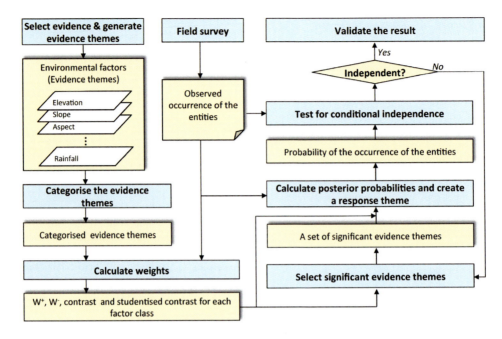

Figure 9.14 Procedure of WofE modelling with GIS.

Low: when $P_{post}/P_{priori} < 1$ and $P_{post}/\sigma(P_{post}) > 1.5$

Uncertain: when $P_{post}/\sigma(P_{post}) \leq 1.5$

where $\sigma(P_{post})$ is the standard deviation of the posterior probabilities. $P_{post}/\sigma(P_{post})$ is called studentised posterior probability.

WofE modelling with GIS generally involves eight steps (Figure 9.14): collecting evidence and generating evidence themes, collecting training sites (sample data), categorising the evidence themes, calculating weights for each class on all evidence themes, selecting significant evidence themes, calculating posterior probabilities and creating the response theme, conducting the test for conditional independence among the evidence themes and validating the result. Box 9.5 shows an example of WofE using ArcSDM (Spatial Data Modeller) for landslide susceptibility assessment. ArcSDM is an ArcGIS extension for spatial data modelling using WofE, logistic regression, fuzzy logic and neural networks developed by the US Geological Survey and the Geological Survey of Canada. Case Study 11 in Chapter 10 gives another example of habitat potential mapping with WofE in ArcSDM.

Box 9.5   WofE modelling with ArcGIS for landslide susceptibility assessment   PRACTICAL

To follow this example, download ArcSDM from **http://www.ige.unicamp.br/sdm/default_e.html**. Unzip the downloaded file to C:\. A directory named **C:\SDM** is created, which contains all ArcSDM files. Then start ArcMap, and load the **landslides** shapefile from **C:\Databases\GIS4EnvSci\VirtualCatchment\Shapefiles**, **dis_rivers** and **dis_faults** created in Box 9.1, the **boundary**, **rocks** and **soils** feature classes as well as **degreeSlope** and **landcv_r** grids from **C:\Databases\GIS4EnvSci\VirtualCatchment\Geodata.gdb**. **landslides** is a landslide inventory containing the locations of landslide occurrence over sixty years. The points in this layer are used as training sites. In this hypothetical landslide susceptibility assessment problem, the evidence to be used includes slope, distance to rivers, distance to faults, lithology, soils and land cover. Basically, the WofE modelling is used to examine the conditions under which landslides occurred in the past, establish associations between those conditions and the occurrence of landslides, and to predict the possible occurrence of landslides based on the associations.

Generate evidence themes in raster

1. Use the **Polygon to Raster** tool to convert the rocks feature class (representing lithology) to a raster using **NAME** as the value field and **40** as the cell size, and save the output raster as **C:\ Databases\GIS4EnvSci\VirtualCatchment\Results\rocks_r**.
2. Use **Polygon to Raster** to convert soils to a raster using **Soil_order** as the value field and **40** as the cell size, and save the output raster as **C:\Databases\GIS4EnvSci\VirtualCatchment\ Results\soils_r**.
3. Use **Polygon to Raster** to convert the boundary layer to a raster using **OBJECTID** as the value field and **40** as the cell size, and save the output raster as **C:\Databases\GIS4EnvSci\ VirtualCatchment\Results\bnd_r**.
4. **degreeSlope** and **landcv_r** are already in rasters.

Load ArcSDM, and set up the modelling environment

5. In the **ArcToolBox** window, right-click **ArcToolBox**. Click **Add Toolbox**. In the **Add Toolbox** dialog, navigate to **C:\SDM**, click **Spatial Data Modeller Tools.tbx**, then click **Open**. The **Spatial Data Modeller Tools** toolkit is added to the ArcToolBox window.
6. Click **Geoprocessing > Environments** from the ArcMap main menu bar. In the **Environment Settings** dialog:
   1) Expand **Workspace**, and enter **C:\Databases\GIS4EnvSci\VirtualCatchment\Results** as the current workspace and **C:\temp** as the scratch workspace.
   2) Expand **Output Coordinates**, click the drop-down arrow under **Output Coordinate System**, and select **Same as Layer "boundary"**.
   3) Scroll down the panel, and expand **Raster Analysis**. Set the cell size as **40**. Select **bnd_r** as the mask.
   4) Click **OK**.

Categorise the evidence themes

7. Use the **Reclassify** tool to reclassify **dis_rivers** to the following ten classes: **0–100**, **100–200**, **200–400**, **400–600**, **600–800**, **800–1000**, **1000–1500**, **1500–2000**, **2000–2500** and **>2500**. Save the output raster as C:\Databases\GIS4EnvSci\VirtualCatchment\Results\dist_rivers_c.
8. Use the **Reclassify** tool to reclassify **dis_faults** to the same ten classes as those for the distance to rivers in the previous step, and save the output raster as C:\Databases\GIS4EnvSci\ VirtualCatchment\Results\dist_faults_c.
9. Use the **Reclassify** tool to reclassify **degreeSlope** to the following nine classes: **0–5**, **5–10**, **10–15**, **15–20**, **20–25**, **25–30**, **30–40**, **40–50** and **>50**. Save the output raster as **C:\Databases\ GIS4EnvSci\VirtualCatchment\Results\slope_c**.
10. **rock_r**, **soils_r** and **landcv_r** are categorical data.

Calculate and interpret weights

11. In **ArcToolBox**, navigate to **Spatial Data Modeller Tools > Weights of Evidence**, and double-click **Calculate Weights**. In the **Calculate Weights** dialog:

(continued)

(continued)

1) Select **soils_r** as the evidence raster layer, and **Value** as the evidence raster code field.
2) Select **landslides** as the training points.
3) Select **Categorical** as the type of weights table.
4) Enter **C:\Databases\GIS4EnvSci\VirtualCatchment\Results\soils_CT.dbf** to save the output weights table.
5) Change 2 to **1.96** as the confidence level of studentised contrast (that is, set the significance level at 0.05).
6) Enter **0.0016** as the unit area (the area size of the cell).
7) Click **OK**. The weights table **soils_CT.dbf** is created and added to the table of contents. Open the weights table. It should appear as in Figure 9.15.

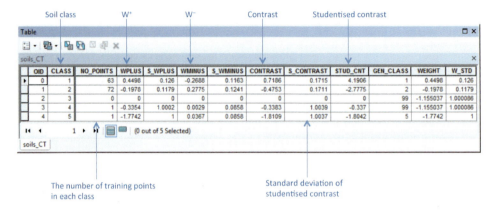

Figure 9.15 Weights table for categorical data.

As shown in the table, soil class 1 – semi luvisols – has a contrast of 0.7186, and its studentised contrast is 4.1906 > 1.96, which is significant. This soil class has a W^+ of 0.4498 and a low W^- of −0.2688, therefore it has a positive association with landslides. Soil class 2 (alpine soil) also has a significant contrast, but it has negative W^+ and positive W^- values. Negative W^+ and positive W^- indicate that landslides are less likely to occur at sites with these soil classes than would be expected by chance. Class 3 has no association with landslides as its W^+ and positive W^- are all 0. Class 4 and class 5 have a negative association with landslides, but statistically insignificant.

12. Repeat Step 11 to calculate weights for the evidence themes, **landcv_r**, **rocks_r** and **slope_c**, and generate weights tables and save them respectively as **landcv_CT.dbf**, **rocks_CT.dbf** and **slope_CT.dbf**. Here, land cover classes, rock types (lithology) and slope classes are considered as categorical data, therefore **landcv_r**, **rocks_r** and **slope_c** are categorical evidence themes. Interpret the weights tables as in the previous step. The results indicate that one land cover class (cold temperate coniferous forest), six rock types (mudstone and metamorphic siltstone, limestone and siliceous limestone, and so on) and one slope class (40–50°) have significant positive associations with the occurrence of landslides.
13. Repeat Step 11 to calculate weights for the evidence themes, **dist_rivers_c** and **dist_faults_c**. But select **Ascending** as the type of weights table, as the distance to rivers and faults is quantitative

data (not categorical) and the weights are calculated cumulatively from lowest to highest distance class (ascending). Specify the output weights tables as **rivers_CA.dbf** and **faults_CA.dbf** under the directory **C:\Databases\GIS4EnvSci\VirtualCatchment\Results**. Figure 9.16 gives the weights table of **faults_CA.dbf**.

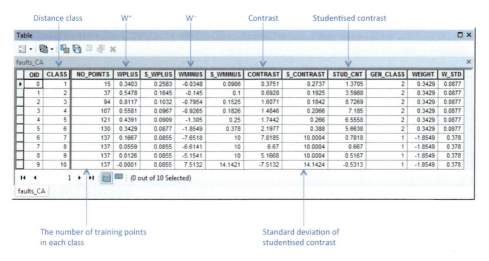

Distance class W⁺ W⁻ Contrast Studentised contrast

	OID	CLASS	NO_POINTS	WPLUS	S_WPLUS	WMINUS	S_WMINUS	CONTRAST	S_CONTRAST	STUD_CNT	GEN_CLASS	WEIGHT	W_STD
▶	0	1	15	0.3403	0.2583	-0.0348	0.0906	0.3751	0.2737	1.3705	2	0.3429	0.0877
	1	2	37	0.5478	0.1645	-0.145	0.1	0.6928	0.1925	3.5988	2	0.3429	0.0877
	2	3	94	0.8117	0.1032	-0.7954	0.1525	1.6071	0.1842	8.7269	2	0.3429	0.0877
	3	4	107	0.5581	0.0967	-0.9265	0.1826	1.4846	0.2066	7.185	2	0.3429	0.0877
	4	5	121	0.4391	0.0909	-1.305	0.25	1.7442	0.266	6.5558	2	0.3429	0.0877
	5	6	130	0.3429	0.0877	-1.8549	0.378	2.1977	0.388	5.6638	2	0.3429	0.0877
	6	7	137	0.1667	0.0855	-7.6518	10	7.8185	10.0004	0.7818	1	-1.8549	0.378
	7	8	137	0.0559	0.0855	-6.6141	10	6.67	10.0004	0.667	1	-1.8549	0.378
	8	9	137	0.0126	0.0855	-5.1541	10	5.1668	10.0004	0.5167	1	-1.8549	0.378
	9	10	137	-0.0001	0.0855	7.5132	14.1421	-7.5132	14.1424	-0.5313	1	-1.8549	0.378

The number of training points in each class

Standard deviation of studentised contrast

Figure 9.16 Weights table for quantitative data.

The weights table indicates that proximity to fault lines shows significant positive associations with landslides from 100m to 1,000m, as the studentised contrasts for the second to the sixth distance classes are all greater than 1.96. Therefore, landslides are more likely to occur within 100–1000m from fault lines. A similar conclusion can be drawn for proximity to rivers by interpreting the weights table in **rivers_CA.dbf**.

Based on the analysis of weights tables associated with the evidence themes, we may conclude that all six evidence themes have significant spatial associations with landslides, and thus have predictive power. They are all used to calculate the response theme below.

Calculate the response theme

14. In **ArcToolBox**, navigate to **Spatial Data Modeller Tools > Weights of Evidence**, and double-click **Calculate Response**. In the **Calculate Response** dialog:

1) Add the evidence themes in the following sequence: **landcv_r**, **soils_r**, **rocks_r**, **slope_c**, **dist_rivers_c** and **dist_faults_c**.
2) Add the weights tables in the same order as the evidence themes they are associated with – that is, add **landcv_CT.dbf**, **soils_CT.dbf**, **rocks_CT.dbf**, **slope_CT.dbf**, **rivers_CA.dbf** and **faults_CA.dbf** subsequently.
3) Select **landslides** as the input training sites feature class.
4) Tick **Ignore Missing Data**, and enter **0.0016** as the unit area.
5) Click **OK**. After the process is complete, three layers are created and added to the table of contents: **W_pprb**, **W_Conf** and **W_Std**. **W_pprb** is the response theme representing the

(continued)

(continued)

posterior probability of landslides at every location in the study area. **W_Conf** shows the confidence that the reported posterior probability is not zero, measured by an approximate student *t*-test. **W_Std** represents the standard deviation of the posterior probability due to the weights.

Test for conditional independence

15. In **ArcToolBox**, navigate to **Spatial Data Modeller Tools > Weights of Evidence**, and double-click **Agterberg-Cheng CI Test**. In the **Agterberg-Cheng CI Test** dialog:

    1) Select **W_pprb** as the post-probability raster.
    2) Select **W_Std** as the post-probability std raster.
    3) Select **landslides** as the training sites.
    4) Enter **0.0016** as the unit area.
    5) Click **OK**. After the process is complete, **AC_ModelName** is added to the table of contents.

16. Open **AC_ModelName**. The table reports a CI ratio and the Agterberg-Cheng test of conditional independence. A CI ratio of less than 0.85 indicates the there is conditional dependence. When the Agterberg-Cheng conditional independence test result is less than 1.645, some conditional dependence occurs. In this case, the CI ratio is 0.76 (< 0.85) and the Agterberg-Cheng test value is 4.155658 (> 1.645). Therefore, there is conditional dependence of the evidence. In order to meet the conditional independence, let us remove **landcv_r** and **dist_rivers_c** evidence themes from the model.

Re-calculate the response theme and test for conditional independence

17. Repeat Step 14 with four evidence themes: **soils_r**, **rocks_r**, **slope_c** and **dist_faults_c**. The results are saved as **W_pprb1**, **W_Conf1** and **W_Std1**.
18. Repeat Step 15 with **W_pprb1** as the post-probability raster and **W_Std1** as the post-probability std raster. Save the output CI test file as **AC_ModelName1**.
19. Open **AC_ModelName1**. Now the CI ratio is 0.95, which is close to 1; the Agterberg-Cheng test value is 0.68, much less than 1.645. Therefore, the evidence is conditionally independent at the significance level of 0.05. The posterior probability in **W_pprb1** is statistically reliable. The larger the probability values in this map, the higher the susceptibility for landslides.

Produce the landslide susceptibility map

20. In the main menu, click **Geoprocessing > Results**. In the **Results** window:

    1) Expand **Current Session**, then **CalculateResponse**.
    2) Right-click **Messages** in the **Results** window, and select **View**. In the **Messages** pop-up window, search for the prior probability, which should be 0.00055. Note down the prior probability value, then close the **Messages** window.

21. Open **ArcToolBox**. Navigate to **Spatial Analyst Tools > Map Algebra**, and double-click **Raster Calculator**.
22. In the **Raster Calculator** dialog, enter the following expression: `"W_pprb1" / 0.00055`.
23. In the **Raster Calculator** dialog, enter **C:\Databases\GIS4EnvSci\VirtualCatchment\Results\prob_ratio** as the output raster. Click **OK**. The layer prob_ratio is created and added to the table of contents.

24. Start the **Raster Calculator** dialog, and enter the following expression to calculate the studentised posterior probabilities: `"W_pprb1" / "W_Std1"`.
25. In the **Raster Calculator** dialog, enter **C:\Databases\GIS4EnvSci\VirtualCatchment\Results\ stud_post** as the output raster. Click **OK**. The layer stud_post is created and added to the table of contents.
26. Start the **Raster Calculator** dialog, and enter the following expression: `Con(("stud_post" <= 1.5), 4, Con (("prob_ratio" < 1), 3, Con(("prob_ratio" <= 5) & ("prob_ ratio" >= 1), 2, 1)))`.

Here, 1 represents high susceptibility, 2 moderate susceptibility, 3 low susceptibility and 4 uncertain.

27. In the **Raster Calculator** dialog, enter **C:\Databases\GIS4EnvSci\VirtualCatchment\Results\ likelihood** as the output raster. Click **OK**. The layer **likelihood** is created and added to the table of contents, as shown in Figure 9.17.

9.5 ARTIFICIAL NEURAL NETWORKS

Artificial neural networks (ANNs) are computer algorithms developed to emulate biological neural systems (Graupe 2013). Like human beings, ANNs learn by example. An ANN is configured for a particular application through a learning process. With its remarkable ability to derive meaning from complicated or imprecise data, an ANN can be used to extract patterns and detect trends that are too complex to be noticed by either humans or other computer techniques. A trained neural network can be thought of as an 'expert' in the type of information it has been given to analyse. This expert can then

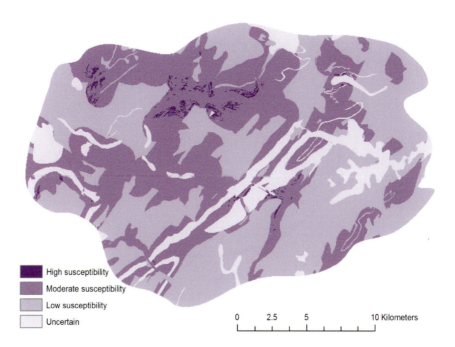

High susceptibility
Moderate susceptibility
Low susceptibility
Uncertain

0   2.5   5   10 Kilometers

Figure 9.17 Landslide susceptibility map derived by WofE.

be employed to provide projections given new situations and answer 'what if' questions.

An ANN processes information in a similar way to a biological neural system. The basic building blocks of biological neural systems are neurons. A neuron consists of a cell body, an axon and dendrites. A neuron is connected through its axon with a dendrite of another neuron. The connection point between neurons is called a synapse. Basically, a neuron receives signals from the environment. When the signals are transmitted to the axon of the neuron – that is, the neuron is fired – the cell sums up all the inputs, which may vary by strength of connection or frequency of input signals, processes the input sum, then produces an output signal, which is propagated to all connected neurons. An artificial neuron is a simplified computational model of a biological neuron, performing the above basic functions.

In an artificial neuron, the cell body is modelled by an activation or transfer function (Figure 9.18). The artificial neuron receives one or more inputs (representing signals received through dendrites that are excited or inhibited by positive or negative numerical weights associated with each dendrite), calculates the weighted sum of the inputs and passes the sum to the activation function. The activation function is a non-linear function, and its output represents an axon, which propagates as an input to another neuron through a synapse. An activation function can be binary, sigmoid, hyperbolic tangent or other forms (Graupe 2013).

Let $S = \Sigma_{i=1}^{n}(w_i x_i)$ be the net input signal and θ be a threshold value, also referred to as the bias. An activation function receives the net input signal and bias, and determines the output (or firing strength) of the neuron. A binary or step activation function is expressed as:

$$f(S-\theta)=\begin{cases} \gamma_1 \text{ if } S \geq \theta \\ \gamma_2 \text{ if } S < \theta \end{cases} \tag{9.28}$$

where γ_1 and γ_2 are constants. Usually, a binary output is produced for which $\gamma_1 = 1$ and $\gamma_2 = 0$; or a bipolar

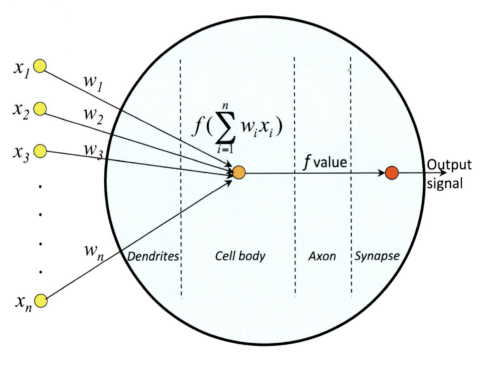

$x_1, x_2, \ldots x_n,$ — inputs
$w_1, w_2, \ldots w_n,$ — weights
f — non-linear activation function

Figure 9.18 An artificial neuron.

output is used where $\gamma_1 = 1$ and $\gamma_2 = -1$. A sigmoid function is written as:

$$f(S-\theta) = \frac{1}{1+e^{-\beta(S-\theta)}} \qquad (9.29)$$

where β is a constant. A hyperbolic tangent function is in the following form:

$$f(S-\theta) = \frac{e^{\beta(S-\theta)} - e^{-\beta(S-\theta)}}{e^{\beta(S-\theta)} + e^{-\beta(S-\theta)}} \qquad (9.30)$$

An ANN is a layered network, composed of a large number of highly interconnected artificial neurones (that is, processing elements) working in parallel to solve a specific problem. It typically consists of an input layer, one or more hidden layers and an output layer (Figure 9.19). Each layer is composed of a number of artificial neurons, also called nodes. The artificial neurons in one layer are connected by weights to the artificial neurons in the next layer. It is essentially a model representing a non-linear mapping between an input vector and output vector. As the output from an artificial neuron or node is a function of the sum of the inputs to the node modified by a non-linear activation function, an ANN superposes many simple non-linear activation functions used by the nodes constituting the network, which enables it to

approximate highly non-linear functions, thus introducing complex non-linear behaviour to the network. These functions can be trained to accurately generalise with new data. The adaptive property is embedded within the network by adjusting the weights that interconnect the nodes during the training phase. After the training phase, the ANN parameters are fixed and the system is deployed to solve the problem at hand. Therefore, an ANN is an adaptive, non-linear system that learns to perform a function from data.

There are several types of ANN, including multilayer feed-forward, recurrent, temporal, probabilistic, fuzzy and radial basis function ANNs (Haykin 1999). The most popular is the multilayer feed-forward neural network, as shown in Figure 9.19. It is also referred to as the multilayer perceptron network. In this type of ANN, the output of a node is scaled by the connection weights and fed forward as an input to the nodes in the next layer. That is, information flow starts from the nodes in the input layer, then moves along weighted links to the nodes in the hidden layers for processing. The input layer plays no computational role, but provides the inputs to the network. The connection weights are normally determined through training. Each node contains an activation function that combines information from all the nodes in the preceding layer. The output layer is a complex

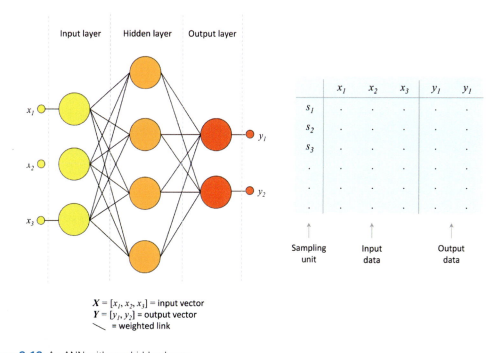

Figure 9.19 An ANN with one hidden layer.

function of the outcomes resulting from internal network transformations.

Multilayer perceptron networks are able to learn through training. Training involves the use of a set of training data with a systematic step-by-step procedure to optimize a performance criterion or to follow some implicit internal constraint, which is commonly referred to as the learning rule. Training data must be representative of the entire dataset. A neural network starts with a set of initial connection weights. During training, the network is repeatedly fed with the training data (a set of input–output pattern pairs obtained through sampling), and the connection weights in the network are modified until the learning rule is satisfied. One performance criterion could be a threshold value of an error signal, which is defined as the difference between the desired and actual output for a given input vector. Training uses the magnitude of the error signal to determine how much the connection weights need to be adjusted so that the overall error is reduced. The training process is driven by a learning algorithm, such as back-propagation (Rumelhart et al. 1986) and scaled conjugate gradient algorithms (Hagan et al. 1996). Once trained with representative training data, the multilayer perceptron network gains sufficient generalisation ability and can be applied to new data.

Radial basis function neural networks (RBFNN) are feed-forward neural networks consisting of an input layer, a hidden layer and an output layer, but hidden nodes implement a set of radial basis functions instead of activation functions, and the output nodes implement linear summation functions as in a multilayer perceptron network. A radial basis function for a data point x is a function whose value depends only on the distance from some other data point c, called a centre, so that:

$$\varnothing(x,c) = \varnothing(\|x-c\|) \qquad (9.31)$$

where $\| . \|$ denotes the Euclidean norm, usually Euclidean distance. The radial basis function $\varnothing$ is univariate, and usually continuous. The name 'radial' indicates that all data points equidistant from c produce the same value. Different types of radial basis functions could be used, but the most common one is the Gaussian function, expressed as:

$$\varnothing(x,c) = e^{-\frac{\|x-c\|^2}{2\sigma^2}} \qquad (9.32)$$

where σ is the radius around the centre (also called spread) to which each neuron responds. Each neuron has a different centre and may have a different radius.

Let X be the m-dimensional input vector and Y be the n-dimensional output vector. $X = [x_1, x_2, \ldots, x_m]$ and $Y = [y_1, y_2, \ldots, y_n]$, where m is the number of input nodes, and n is the number of output nodes. Set the number of hidden nodes as k. The output value of the jth output node produced by an RBFNN is given by:

$$y_j = \Sigma_{i=1}^k \lambda_{ij} e^{-\frac{\|X-c_i\|^2}{2\sigma_i^2}} \qquad (9.33)$$

where λ_{ij} is the weight on the connection between the ith hidden node to the jth output node (Figure 9.20a). Here X can be seen as a data point with m attributes or variables. It feeds the values to each of the neurons in the hidden layer. Each hidden node or neuron consists of a radial basis function centred on a point with as many dimensions as the input vector. When presented with the X vector of input values from the input layer, a hidden node computes the Euclidean distance of the test case from the node's centre, then applies the radial basis function to this distance using the radius. The resulting value is then multiplied by a weight assigned to the node and passed to the summation layer, which adds up the weighted values and presents this sum as the output of the network. The optimal number of hidden nodes, the centres and spreads of each radial basis function are determined by the training process. RBFNNs are able to model complex mappings, while perceptron networks can only model by means of multiple intermediary layers.

As an extension to RBFNNs, radial basis functional link networks (RBFLNs) learn more quickly and are more accurate in operation. An RBFNN is a non-linear model, while an RBFLN incorporates an additional linear model (Looney 2002). As shown in Figure 9.20b, an RBFNN adds links that directly connect the input nodes to the output nodes, and these links have another set of weights $\{w_{ij}\}$. With the new links and weights, Equation 9.33 is modified as:

$$y_j = \Sigma_{i=1}^k \lambda_{ij} e^{-\frac{\|X-c_i\|^2}{2\sigma_i^2}} + \Sigma_{i=1}^m w_{ij} x_i \qquad (9.34)$$

When $\{w_{ij}\}$ are set to zero, an RBFLN becomes an RBFNN. Therefore, the RBFLN is a generalisation of the RBFNN, a more complete model of a general non-linear mapping.

ANNs are applicable when a relationship between the independent variables (outputs) and dependent variables (inputs) exists. They are able to learn the relationship from a given dataset without any assumptions about the statistical distribution of the data. In addition, ANNs perform a non-linear transformation of input data to approximate output data, learning from training data

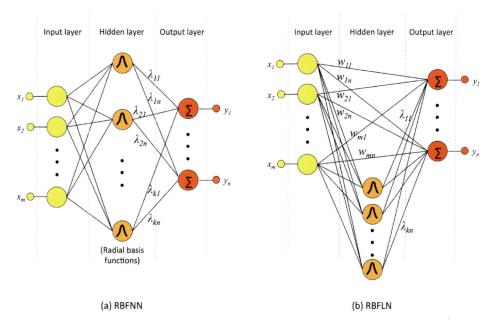

Figure 9.20 (a) RBFNN and (b) RBFLN.

and exhibiting the ability of generalisation beyond training data. The capability of learning from data and modelling non-linear relationships make ANN particularly suitable for pattern classification (that is, classifying data into the pre-determined discrete classes) and predictive

modelling. Zhu (2014) provided a review of ANN applications in Earth and environmental sciences. Box 9.6 shows an example of landslide susceptibility assessment using RBFLN in ArcGIS with the same data and some results from the example in Box 9.5.

Box 9.6   ANN modelling with ArcGIS and GeoXplore for landslide susceptibility assessment   **PRACTICAL**

To follow this example, load **ArcSDM**, and set up the modelling environment as in Box 9.5. Start **ArcMap**, and load the **boundary** feature class from **C:\Databases\GIS4EnvSci\VirtualCatchment\ Geodata.gdb**, the **landslides** shapefile from **C:\databases\GIS4EnvSci\VirtualCatchment\ Shapefiles**, the rasterised study area **bnd_r** and the categorised evidence themes created in Box 9.5, including **soils_r**, **rocks_r**, **slope_c** and **dist_faults_c**, as well as the response theme **W_pprb1**. GeoXplore is a program accompanying ArcSDM that consists of three methods of data or pattern classification: fuzzy clustering, RBFLN and probabilistic neural networks. The program is located at **C:\ SDM\Scripts\geoxplore061106.exe**.

In this example, an RBFLN neural network will be built to predict the presence and absence of a landslide occurrence based on four variables: soils, rock types, slope and distance to fault lines. Therefore, the neural network will consist of an input layer with four input nodes (representing the four variables) and an output layer with two output nodes (representing two classes: presence and absence). The number of hidden nodes or radial basis functions is to be determined through training.

(continued)

(continued)

The training points will be a combination of presence training points and non-presence or absence training points.

Generate a random set of non-presence training points

The RBFLN neural network in ArcSDM requires two sets of training points: the first set defines the presence of the entities or conditions to be predicted (that is, landslide occurrences), and the second defines the absence of these entities or conditions (that is, locations where landslide occurrences are known not to occur). The **landslides** shapefile contains the first set of training points. The data on the absence of the entities are often not readily available. In this example, we will approximate such a dataset by generating a set of random points in areas of very low probability (lower than the prior probability) of landslide occurrence based on **W_pprb1**.

1. Use the **Reclassify** tool (refer to Box 4.3) to reclassify **W_pprb1** to two classes: ≤ 0.0003 and > 0.0003, and assign **1** to the first class and **NoData** to the second class. Save the output as **lowProb** to the workspace **C:\Databases\GIS4EnvSci\VirtualCatchment\Results**. Note the prior probability of landslide occurrence is 0.00055 in this case.
2. Use the **Raster to Polygon** tool (refer to Box 3.6) to convert **lowProb** to a feature class, and save the output as **lowProbVector.shp** to the workspace.
3. Add a new field **Area** (field type: double) to the attribute table of **lowProbVector**, and calculate the area for each polygon in the layer by following the instructions in Box 4.4.
4. Add a new field **Sample** (field type: short integer) to the attribute table of **lowProbVector**, and conduct a proportionate stratified point sampling by following the instructions in Box 5.1 with a sample size of 100. Save the sample points as **nonPresencePts**, which contains about 100 non-presence training points.

Create a unique conditions raster

5. In **ArcToolBox**, navigate to **Spatial Analyst Tools > Local**, and double-click **Combine**. In **Combine** dialog:

   1) Add the categorised evidence themes as the input rasters one by one: **soils_r**, **rocks_r**, **slope_c** and **dist_faults_c**.
   2) Name the output raster **unique**, and save it to the workspace.
   3) Click **OK**. The unique conditions raster **unique** is created and added to the table of contents. In the unique conditions raster, each cell value represents a unique combination of evidence theme classes. The original evidence theme class values from each of the inputs are recorded in the attribute table of the unique conditions raster.

Generate neural network input files

6. In the **ArcToolBox** window, navigate to **Spatial Data Modeller Tools > Neural Network**, and double-click **Neural Network Input Files**. In the **Neural Network Input Files** dialog:

   1) Select **unique** as the input unique conditions raster.
   2) Select **landslides** as the training sites.
   3) Select **nonPresencePts** as the ND (non-deposit, that is non-presence) training sites.
   4) Tick **Classification file**.
   5) Click **OK**. Two files, named **unique_train.dta** and **unique_class.dta**, are generated and saved to the workspace.

These two files will be used as the input files to **GeoXplore**. Both files are text files with a similar format. For the file formats, see the GeoXplore documentation available at **C:\SDM\Help_ArcSDM\ Reprints**. **unique_train.dta** is the training data file, containing information from unique conditions in which training points are located. Each row in the file represents a unique condition – that is, an input feature vector for a neural network, with a class membership value of 1 or 0. 1 indicates the presence of landslides, and 0 the absence of landslides. This training data file is used for training the neural network algorithms. **unique_class.dta** is the classification data file, storing data of unknown class to be classified (a complete set of unique conditions). Each row represents a unique condition present in the study area (including non-training points). It is used for classification with an already trained neural network algorithm.

Train the RBFLN algorithm

7. Navigate to **C:\SDM\Scripts** in **Windows Explorer**, and double-click **geoxplore061106.exe** to run **GeoXplore**.
8. In the start-up screen, click the **RBFLN** button.
9. In the **RBFLN** dialog, click the **Train RBFLN** button.
10. In the **RBFLN Training** dialog:

    1) Click the **Load file** button to load the input training data file **unique_train.dta** from the workspace.
    2) Click the **Save as** button, navigate to the workspace, and enter **landslides.par** as the file name for storing the training parameters to be learnt during training. This file will hold information on the number of Gaussian radial basis functions, centres and radius values for the Gaussian radial basis functions, the weights on the lines from the hidden nodes to the output nodes, and the weights on the extra lines from the input nodes to the output nodes that bypass the hidden layer.
    3) Click **Run**. The **RBFLN Training Parameters** dialog pops up. The top two parameters (the number of the input feature vectors and the suggested number of radial basis functions) cannot be changed. The user can change the desired number of radial basis functions, but should not change them too much from the suggested number given in the second parameter value. For a larger number of radial basis functions, the neural network will be slower but more accurate. However, if the number of radial basis functions is too large, extraneous errors will build up. In practice, experiments can be conducted by using several smaller and larger numbers of radial basis functions so that an appropriate number of radial basis functions can be chosen based on the overall measures of training errors (see below). The number of Iterations also has an effect on the training results. If the training dataset is small, 100 iterations may be sufficient. But for larger training datasets, 400–600 iterations are better. The default of 200 is a trade-off between under-training and over-training. Click **OK** to accept the default parameter values.

After training is complete, the results and the overall measures of training errors are shown in the right-hand panel. The results are listed in three columns: Column 1, Vector No. – the ID number of each input vector (unique condition); Column 2, Fuzzy Target – the actual class membership value (1 for presence or 0 for absence) associated with each input vector; and Column 3, Training Fuzzy Beliefs – the output class membership value for each input vector, ranging from 0 to 1. A value of 1 indicates full class membership, with membership decreasing to 0 indicating it is not a member of the class. A higher fuzzy belief between 0 and 1 is better. A fuzzy belief value outside the range of 0–1 indicates a classification error.

(continued)

(continued)

The output errors are measured by the sum squared error (SSE) and the mean squared error (MSE). SSE is calculated as the sum of squared difference between the true class membership values and the output class membership values for each input vector. MSE is the SSE divided by the number of input vectors. The smaller the SSE and MSE, the better the predicted outcomes. The training may be repeated several times with different training parameters until the overall measures of goodness of training are near optimal.

4) Click the **Save Training Results** button to save the results.
5) Click **Return**.

Run RBFLN classification

11. In the **RBFLN** dialog, click the **Do RBFLN Classification** button.
12. In the **RBFLN Classification** dialog:

1) Click the **Load file** button for Step 2 (the second step of RBFLN classification) to load the input training result file **landslides.par** from the workspace.
2) Click the **Load file** button for Step 3 to load the input data file for the classification **unique_class.dta** from the workspace.
3) Click the **Save as** button, navigate to the workspace, and enter **landslides.rbn** as the file name for storing the classification results.
4) Click **Run**. After the operation is complete, **landslides.rbn** is created and saved to the workspace. This file is a text file, containing three columns of data: Column 1 – the ID number of each input vector; Column 2 – class (1 for presence); and Column 3 – the predicted class membership value (fuzzy belief) for the class in Column 2 for each input vector. The RBFLN classification overall measures, SSE and MSE are also displayed on the screen. Again, the smaller the SSE and MSE, the better the predicted outcomes.
5) Click **Return**.

13. Exit **GeoXplore**.

Read the RBFLN classification results, and create a response theme

14. In **ArcToolBox**, navigate to **Spatial Data Modeller Tools > Neural Network**, and double-click **Neural Network Output Files**. In the **Neural Network Output Files** dialog:

1) Select **unique** as the input unique conditions raster.
2) Add **landslides.rbn** from the workspace under **RBFLN filename**.
3) Name the output result table **landslidesNN.dbf**, and save it to the workspace.
4) Name the output raster **landslidesNN**, and save it to the workspace.
5) Click **OK**. **landslidesNN** is created and added to the table of contents. This layer is the response raster with all the unique combination of the evidence and the class membership values from **GeoXplore**.

15. Open the attribute table associated with **landslidesNN**. The field **RBFLN** stores the class membership (fuzzy belief) values for the presence of landslides for each unique condition. The higher the membership value, the more closely the unique condition resembles a pattern in which landslides are known to occur. In other words, landslides are more likely to occur under this condition. However, a number of unique conditions may get a membership value of greater than 1, which indicates a classification error. We define the areas with these conditions as uncertain. Use the field **RBFLN** to make a map of likelihood of landslides similar to that in Figure 9.21.

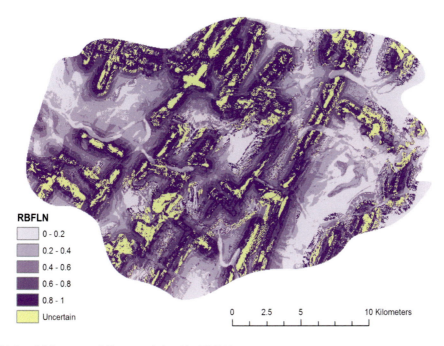

RBFLN

	0 - 0.2
	0.2 - 0.4
	0.4 - 0.6
	0.6 - 0.8
	0.8 - 1
	Uncertain

0   2.5   5   10 Kilometers

Figure 9.21 Landslide susceptibility map derived by RBFLN.

9.6 SUMMARY

- Multi-criteria decision analysis is concerned with prioritising decision alternatives based on a set of objectives and criteria. It involves establishing the decision context, structuring the problem, scoring the alternatives against each criterion, assigning weights to the criteria, evaluating the alternatives and conducting a sensitivity analysis. In GIS-based multi-criteria decision analysis, alternatives are typically locations. Criteria and scores associated with each alternative are represented in criterion map layers. They are standardised and combined via map algebra to derive a ranking of alternatives.
- Cellular automata use a grid of cells to represent the configuration of a self-organising and self-replication system, and simulate the evolution of the system by applying a set of transition rules based on the states of neighbouring cells through a number of discrete time steps.
- Agent-based modelling simulates complex systems composed of interacting, autonomous agents. Agents have their own goals, states and behaviours that are often described by simple rules. They interact with and influence each other, learn from their experiences, and adapt their behaviours so that they are better suited to their environment. This may

allow unanticipated behaviours to emerge, thus providing valuable information about the dynamics of the real-world system being emulated.
- Weights-of-evidence is a statistical method for combining evidence in support of a hypothesis. It involves the estimation of a response theme based on a set of evidence themes. It is mainly used to predict the likelihood of occurrence of particular geographical entities.
- An artificial neuron network is a computational model based on the structure and functions of biological neural systems. Individual elements of the network are artificial neurons. They read an input, process it and produce an output. A network of many neurons is an adaptive system, which can exhibit rich and intelligent behaviours. Artificial neuron networks have been used to extract patterns and detect trends.

REVIEW QUESTIONS

1. What are the key components of multi-criteria decision analysis?
2. Describe the general procedure of multi-criteria decision analysis.
3. Suppose you are going to identify potential sites for purchasing a new house based on land price

(in dollars), slope (in degree) and proximity to public transport (in metres). You will assign different weights to the three criteria, reflecting their importance in site selection. Use this example to describe a procedure for the use of the multi-criteria analysis approach in GIS.

4. What is a cellular automaton? How does it work?
5. What is the difference between cellular automata and agent-based modelling?
6. What are the four basic elements in an agent-based model?
7. How does weights-of-evidence measure the strength of spatial association of a particular environmental variable and the observed pattern of the modelled geographical entities?
8. Describe the general procedure of weights-of-evidence modelling with GIS.
9. Describe the basic elements of an artificial neural network, and discuss its capabilities.

REFERENCES

Agterberg, F.P. and Cheng, Q. (2002) 'Conditional independence test for weights-of-evidence modeling', *Natural Resources Research*, 11: 249–255.

Agterberg, F.P., Bonham-Carter, G.F. and Wright, D.F. (1990) 'Statistical pattern integration for mineral exploration', in G. Gaal and D.F. Merriam (eds) *Computer Applications in Resource Estimation*, Oxford: Pergamon Press, 1–21.

Batty, M. (2005) *Cities and Complexity: Understanding Cities with Cellular Automata, Agent-Based Models, and Fractals*, Cambridge, MA: MIT Press.

Bonham-Carter, G.F. (1994) *Geographic Information Systems for Geoscientists: Modelling with GIS*, Oxford: Pergamon.

Bonham-Carter, G.F., Agterberg, F.P. and Wright, D.F. (1988) 'Integration of geological data sets for gold exploration in Nova Scotia', *Photogrammetric Engineering and Remote Sensing*, 54: 1,585–1,592.

Carranza, E.J.M. and Hale, M. (2000) 'Geologically constrained probabilistic mapping of gold potential, Baguio District, Philippines', *Natural Resource Research*, 9: 237–253.

DeAngelis, D.L. and Mooij, W.M. (2005) 'Individual-based modeling of ecological and evolutionary processes', *Annual Reviews in Ecology, Evolution and Systematics*, 36: 147–168.

Drossel, B. and Schwabl, F. (1992) 'Self-organized critical forest-fire model', *Physical Review Letters*, 69: 1,629–1,632.

Edwards, W. and Barron, F.H. (1994) 'SMARTS and SMARTER: improved simple methods for multiattribute utility measurement', *Organizational Behavior and Human Decision Processes*, 60: 306–325.

Gardner, M. (1970) 'Mathematical games: the fantastic combinations of John Conway's new solitaire game "life"', *Scientific American*, 223: 120–123.

Graupe, G. (2013) *Principles of Artificial Neural Networks*, Hackensack, NJ: World Scientific Publishing.

Hagan, M.T., Demuth, H.B. and Beale, M.H. (1996) *Neural Network Design*, Boston, MA: PWS Publishing.

Haykin, S. (1999) *Neural Networks: A Comprehensive Foundation*, 2nd edn, Upper Saddle River, NJ: Prentice Hall.

Johnston, K.M. (2013) *Agent Analyst: Agent-Based Modeling in ArcGIS*, Redlands, CA: ESRI Press.

Linkov, I. and Moberg, E. (2012) *Multi-criteria Decision Analysis Environmental Applications and Case Studies*, Boca Raton, FL: CRC Press.

Looney, C.G. (2002) 'Radial basis functional link nets and fuzzy reasoning', *Neurocomputing* 48(1): 489–509.

Malczewski, J. (1999) *GIS and Multicriteria Decision Analysis*, New York: John Wiley & Sons.

Morgan, F.J. and Daigneault, A.J. (2015) 'Estimating impacts of climate change policy on land use: an agent-based modelling approach', *PLoS ONE*, 10(5), e0127317.

Parker, D.C., Manson, S.M., Janssen, M.A., Hoffmann, M.J. and Deadman, P. (2003) 'Multi-agent systems for the simulation of land-use and land-cover change: a review', *Annals of the Association of American Geographers*, 93(2): 314–337.

Rumelhart, D.E., Hinton, G.E. and Williams, R.J. (1986) 'Learning internal representations by error propagation', in D.E. Rumelhart and J.L. McClelland (eds) *Parallel Distributed Processing: Explorations in the Microstructure of Cognition*, Vol. 1, Cambridge, MA: MIT Press.

Saaty, T.L. (1980) *The Analytic Hierarchy Process*, New York: McGraw-Hill.

Santé, I., Garcia, A.M., Miranda, D. and Crecente, R. (2010) 'Cellular automata models for the simulation of real-world urban processes: a review and analysis', *Landscape and Urban Planning*, 96(2): 108–122.

Watkins, A., Noble, J. and Doncaster, C.P. (2011) 'An agent-based model of jaguar movement through conservation corridors', in T. Lenaerts, M. Giacobini, B. Mario, B. Hugues, P. Bourgine, M. Dorigo and R. Doursat (eds) *Advances in Artificial Life, ECAL 2011: Proceedings of the Eleventh European Conference on the Synthesis and Simulation of Living Systems*, Cambridge, MA: MIT Press, 846–853.

Zandbergen, P.A. (2013) *Python Scripting for ArcGIS*, Redlands, CA: ESRI Press.

Zhu, X. (2014) 'Computational intelligence techniques and applications', in T. Islam, P.K. Srivastava, M. Gupta, X. Zhu and S. Mukherjee (eds) *Computational Intelligence Techniques in Earth and Environmental Sciences*, Heidelberg, Germany: Springer, 3–26.

Environmental applications of GIS

10

This chapter provides a selection of case studies of representative environmental applications of GIS. It is by no means comprehensive, but is designed to illustrate how GIS is applied to improve our understanding of environmental systems and to showcase GIS solutions to some environmental issues. The case studies are selected from the literature, and cover the areas of hydrological modelling, land use analysis, atmospheric science research, ecological modelling and landscape valuation.

LEARNING OBJECTIVES

After studying this chapter, you should be able to:

- comprehend the breadth and versatility of GIS applications in environmental science;
- explain how value can be added to environmental studies with GIS and spatial data;
- understand how GIS can be applied in hydrology, ecology, land use science, atmospheric science and other representative areas of environmental science;
- appreciate the roles of GIS in environmental decision making;
- develop your own environmental applications of GIS.

10.1 HYDROLOGICAL MODELLING

Hydrology is concerned with study of the movement, distribution and quality of water through the hydrological cycle, as well as the transport of constituents such as sediment and pollutants in the water as it flows. Numerous models have been developed for simulating hydrological

systems in order to facilitate the understanding of hydrological processes. They range from full-catchment models dealing with all components of the hydrological cycle to smaller-scale models focusing on the processes in particular regions within a catchment, such as a hillslope. Some models describe the catchment as a single entity, assuming that rainfall falls evenly throughout the catchment, the hydrological parameters are uniform everywhere and the rainfall runoff processes are linear. These hydrological models are called lumped models – they lump the inherent spatial variability of environmental variables into catchment average characteristics. Other models incorporate a variety of spatially varying data from a proliferating set of spatial databases on land use, land cover and soil characteristics, and high-resolution precipitation, temperature and other environmental variables. These models are referred to as distributed models. They not only facilitate simulations and prediction with higher resolution than lumped models, but also improve hydrological predictions by using spatially distributed parameters of physical relevance. Many hydrological models were written as modelling programs (for example, the US Army Corps of Engineers' HEC-HMS and HEC-RAS), which can be coupled with GIS software systems, or as stand-alone software packages with built-in GIS functionalities (for example, MODFLOW, HEC-EFM and RiverCAD) or as an embedded GIS software module (such as ESRI's Arc Hydro). GIS, especially through its capacity for spatial representation of hydrological parameters by means of geographically referenced data and powerful capabilities for terrain analysis, provides an important platform for hydrological modelling. Coupled with hydrological models, GIS has been applied to a range of applications, from computation of hydrological variables, synthesis and characterisation of hydrological tendencies to prediction of response to hydrological events, such as storm water modelling, surface runoff estimation, flood risk assessment, non-point pollution modelling, erosion prediction and real-time hydrological process monitoring. Two case studies are presented below. One involves rainfall runoff modelling by Melesse and Shih (2002), and the other concerns non-point source pollution assessment by Mishra et al. (2010).

Case Study 1: rainfall runoff modelling

Runoff is the portion of the rainfall that flows over the land surface or through the soil towards surface water bodies (rivers, lakes, streams and so on). It occurs when the rainfall exceeds the capacities of interception, evaporation, infiltration and surface storage. Runoff is influenced by catchment morphology, soils and vegetation. For example, a larger catchment produces larger total runoff volumes and higher peak runoff rates than a smaller catchment. A longer and narrower catchment generally has lower peak runoff rates than a more compact catchment as it takes longer for runoff from the most remote areas of a long catchment to reach the outlet. A catchment with an extensive network of steep channels tends to produce greater runoff rates than a catchment with few channels or one with channels having a gentle slope. It is also not difficult to understand that a catchment with deep, permeable soils yields less runoff than a catchment with thin, less permeable soils. The amount and type of vegetation affect runoff too. Vegetation generally slows the movement of runoff water, and reduces the rate and volume of runoff. Runoff is usually modelled as a function of rainfall and parameters determined by catchment properties, including size, shape, topography, soils and vegetation. A rainfall runoff model is developed and used to predict how the surface runoff responds to a rainfall. There are literally thousands of conceptual rainfall runoff models in the literature. Many of them take into account both spatial and geomorphological variations. They are mainly used to extend stream flow records, to estimate catchment, regional or continental water availability, to predict or assess aspects of flooding, to aid in the prediction of transport of water-borne contaminants, and to estimate catchment inflows into river system models. One of the more popular rainfall runoff models is the SCS curve number model developed by the Soil Conservation Service (SCS) (now the Natural Resources Conservation Service, NRCS) of the US Department of Agriculture (USDA). It was developed from empirical analysis of runoff from small catchments and hillslope plots. The model estimates runoff volume for small catchments using the following equations (USDA 1986):

$$Q = \frac{(P - 0.2S)^2}{P + 0.8S} \qquad (10.1)$$

$$S = 254 \times \left(\frac{100}{CN} - 1\right) \qquad (10.2)$$

where Q is surface runoff depth (mm), P is storm rainfall (mm), S is catchment storage or maximum potential difference between P and Q, the constant 254 is a unit conversion factor for inches to millimetres and CN is the runoff curve number. CN is dimensionless, determined based on hydrological soil group, land use, hydrological conditions and antecedent moisture condition. It ranges between 1 and 100. The CN value of 100 implies a catchment with completely impermeable land cover, therefore S equals 0. A zero CN value means the catchment is completely permeable. The higher the CN value, the more impermeable the catchment and the higher the runoff. USDA (1986) provides the CN values for various land uses/covers, hydrological soil groups and hydrological conditions.

Melesse and Shih (2002) applied the SCS curve number model to estimate the runoff depth for the S-65A sub-basin of the Kissimmee River basin in south Florida. The basin suffered loss of a significant amount of river channels and flood plain wetlands due to construction of canals and reservoirs from 1962 to 1971. Begun in 1997, restoration work was undertaken to re-establish lost wetlands and flood plains. The objective of this research was to understand the impacts of these physical and ecological changes on the hydrology of the sub-basin by examining the runoff responses to land use changes in the sub-basin over twenty years from 1980 to 2000. They used satellite imagery to extract land cover information, and implemented the SCS curve number model in a GIS. Figure 10.1 outlines their solution process for mapping runoff depth. More specifically, the study involved:

1. acquiring Landsat images for 1980 (MSS images), 1990 (TM images) and 2000 (ETM+ images);
2. radiometrically and geometrically correcting all the images;
3. classifying the images into land cover classes using the unsupervised classification method ISODATA (see Section 6.5), and regrouping the results into seven land cover classes to create land cover layers for 1980, 1990 and 2000;
4. ground truthing the accuracy of the classifications according to Digital Orthophoto Quarter Quadrangle data for the same years;
5. acquiring the soil maps for the study area and converting them into 30m raster layers;

6.   overlaying the soil raster layers with the land cover layers to create soils-land cover layers;

7.   calculating *CN* values based on the information of soil and land cover for each grid cell in the soils-land cover layers according to the NRCS 1972 *National Engineering Handbook* (NEH 4) guidelines, and creating *CN* layers for 1980, 1990 and 2000;

8.   applying Equations 10.1 and 10.2 to the *CN* layers to calculate runoff depth for the case of 190.5mm of rainfall, which is a ten-year 48-hour rainfall for the S-65A sub-basin;

9.   comparing the spatial distributions of runoff depth for the three years, analysing the changes, and relating these changes to the land use changes occurring over the twenty-year period.

The change in the spatial distribution of *CN* and runoff depth is a sign of the change in the runoff response of the catchment due to land use change. The results from the study indicated that there was an increase in *CN* values and runoff volume in 2000 compared with those in 1980 and 1990, attributed to the increase in water- and wetland-covered areas. This suggested that the lost

wetlands and flood plains in the S-65A sub-basin were recovered as the result of the restoration work.

Case Study 2: non-point source pollution assessment

Surface waters can be contaminated by either point source pollution or non-point source pollution, or both. Point source pollution may come from any discernible, confined and discrete conveyances, such as pipes, ditches, channels, tunnels, conduits, wells, containers, rolling stocks and concentrated animal feeding operations. Non-point source pollution is derived from dispersive sources, including urban runoff, agriculture, industry, hydrological modification, solid waste disposal, atmospheric deposition, stream bank erosion and individual sewage disposal. It is mainly caused by rainfall or snowmelt moving over and through the ground and carrying away natural and human-made pollutants, finally washing them into different water bodies. Non-point source pollutants include nitrogen (N), phosphorous (P), heavy metals and other chemicals from different sources. Different approaches can be used to study and

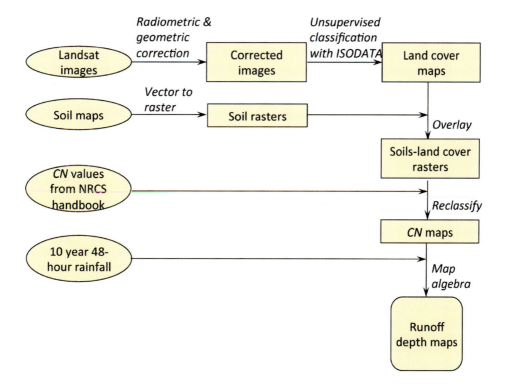

Figure 10.1 A GIS procedure for runoff modelling.

model non-point source pollution. Basically, non-point source pollution models represent the hydrological rainfall runoff transformation processes with attached quality components (Notovny and Chesters 1981). As the dispersion of non-point source pollutants is related to hydrological processes and land use patterns, non-point source pollution is often modelled by integrating hydrological and pollutant transport models with GIS. Mishra et al. (2010) provided such an example. They applied the SWAT model combined with ArcGIS to estimate non-point source N and P loads in a small catchment in India.

SWAT (Soil and Water Assessment Tool) is a catchment scale model jointly developed by USDA Agricultural Research Service and Texas A&M AgriLife Research to simulate the quality and quantity of surface and ground water, and predict the impact of land management practices and climate change on water, sediment and agricultural chemical yields in complex catchments with varying soils, land use and management conditions over long periods (Arnold and Fohrer 2005). The model is physically based, and has been widely used in assessing soil erosion, non-point source pollution and regional management in catchments (it is freely available at http://swat.tamu.edu/).

SWAT allows a number of different physical processes in a catchment to be simulated, including precipitation, runoff, erosion and nutrient processes (Neitsch et al. 2011). Non-point source pollution modelling with SWAT involves three specific models: a rainfall runoff model, an erosion model and a nutrient (N or P) loading model. SWAT uses the SCS curve number model described in Case Study 1 as the rainfall runoff model. Erosion caused by rainfall and runoff is calculated with the Modified Universal Soil Loss Equation (MUSLE) (Williams 1975), which is expressed as:

$$Y = 11.8 \times (V \times R)^{0.56} \times K \times LS \times C \times P \qquad (10.3)$$

where Y is the erosion and sediment yield (metric tons), V is the surface runoff volume (m^3), R is the peak runoff rate obtained during the erosion event (m^3/second), K is the soil erodibility factor, LS is the slope length-steepness factor, C is the cropping and management factor and P is the erosion control practice factor. This erosion model uses the amount of runoff to simulate erosion and sediment yield.

SWAT models the complete nutrient cycle for nitrogen and phosphorus in the soil. The amount of NO_3-N (nitrate nitrogen) contained in runoff, lateral flow and percolation are estimated by multiplying the volume of water and average concentration of nitrate in the layer. The amount of NO_3-N in runoff is estimated for the top 10mm soil layer only, using the following equation:

$$Y_{NO_3} = Q_t \times C_{NO_3} \qquad (10.4)$$

where Y_{NO_3} is the amount of NO_3-N lost by surface runoff from the top soil layer, Q_t is the total water lost from the layer and C_{NO_3} is the average concentration of NO_3-N in the layer. Leaching and lateral subsurface flows in lower soil layers are calculated in the same way as the top layer, but without considering surface runoff in water loss.

Organic N transport with sediment is estimated for individual runoff events using the following loading function, developed by McElroy et al. (1976) and modified by Williams and Hann (1978):

$$Y_{ON} = 0.001 \times \frac{Y}{A} \times C_{ON} \times R_e \qquad (10.5)$$

where Y_{ON} is the organic N runoff loss from an area (kg N/ha), Y is derived from Equation (10.3), A is the size of the area (ha), C_{ON} is the concentration of organic N in the top soil layer (g N/metric ton soil) and R_e is the enrichment ratio defined as the ratio of the concentration of organic N transported with the sediment to the concentration in the soil surface layer and calculated for each storm event as follows (Menzel 1980):

$$R_e = 0.78 \times C_s^{-0.2468} \qquad (10.6)$$

where C_s is the sediment concentration in surface runoff (Mg sediments/m^3 water), which is calculated as:

$$C_s = \frac{Y}{10 \times Q \times A} \qquad (10.7)$$

The main mechanism of P movement in the soil is by diffusion. Surface runoff only partially interacts with the soluble P stored in the top 10mm of soil. The amount of soluble P lost in surface runoff, Y_{SP} (kg P/ha), is calculated in SWAT as:

$$Y_{SP} = \frac{0.01 \times C_{SP} \times Q}{k_d} \qquad (10.8)$$

where C_{SP} is the concentration of soluble P in the top 10mm of soil (g/Mg), and k_d is the phosphorus soil partitioning coefficient (m^3/Mg), which is the P concentration in the sediment divided by that of the water.

Organic and mineral P attached to soil particles are associated with sediment, and may be transported by surface runoff. The amount of P transported with sediment, Y_{PSed} (kg P/ha), is estimated with the following loading function:

$$Y_{PSed} = 0.001 \times \frac{Y}{A} \times C_{PSed} \times R_e \qquad (10.9)$$

where C_{PSed} is the concentration of P attached to sediment in the top 10mm soil (g P/metric ton soil).

Mishra et al. (2010) applied the above SWAT models in the Banha catchment situated in the sub-humid subtropical region of northern India. It is a small catchment, covering an area of 1,695ha. About 40 per cent of the catchment is under crop (mainly rice) cultivation, 50 per cent under shrubs and forests and 10 per cent under barren land. Individual cattle and poultry farms occupy approximately 20ha. Fertilisers, animal wastes, crop and forest residues and individual housing effluent are the major sources of N and P, which can potentially contaminate surface and ground water. The objective of the study was to simulate runoff, sediment transport, nitrate nitrogen (NO_3-N), water-soluble P, organic N and organic P using the SWAT models, and to assess the loss of these non-point source pollutants to the downstream water bodies. The results were expected to be used in the development of catchment management strategies for mitigating the losses as nutrients as well as pollution of water resources. This study took the following steps:

1. Collect daily records of temperature, precipitation, stream flow and sediment transport from 1991 to 2001.
2. Monitor and measure NO_3-N and water-soluble P concentrations as well as sediment yields at the catchment outlet during the monsoon months of 2000 and 2001.
3. Measure runoff-associated NO_3-N and water-soluble P concentrations with ion analysers and standard methods.
4. Digitise contours and water bodies from topographic maps.
5. Take soil samples at twelve locations in the catchment, measure physical and chemical properties of surface and sub-surface soil layers (up to 100cm depth), and prepare a soil map.
6. Derive a land use/cover map from IRS-1D satellite imagery.
7. Use ArcGIS to build a DEM from the digitised contours, and derive slope, aspect, drainage network and sub-catchment boundary layers.
8. Use ArcGIS to generate other spatial data inputs (DEM, soils and land use/cover are the major input data) for the SWAT models.
9. Calibrate the SWAT hydrological and erosion models (that is, estimate the values of various constants and parameters in the models) with the time series of daily stream flow and sediment yield data collected in Step 1.
10. Implement the SWAT nutrient loading models with the calibrated hydrological and erosion models to estimate the loads of non-point source pollutants (NO_3-N, water soluble P, organic N and organic P) transported by runoff and percolated water from the catchment during the monsoon season.
11. Validate the modelling results with the measured nutrient concentration data of 2000 and 2001 collected in Steps 2 and 3.

Figures 10.2 and 10.3 show respectively the measured and SWAT-simulated NO_3-N and water-soluble P concentration at the catchment outlet in 2000 and 2001. The results revealed that the non-point pollutant load in runoff changed with various seasonal rainfall patterns. During the study period of 2000 and 2001, the maximum seasonal total loss of N from the catchment was about 14,984.14kg, whereas the P load was 50.85kg. Therefore, the major loss of N posed a serious threat to surface and ground water quality in the catchment.

The models were implemented in SWAT with some model input parameters generated from ArcGIS. These models can also be directly written and implemented in a GIS via map algebra. The most challenging part of such modelling is data collection. The model calibration and validation require longitudinal data records. In many cases, suitably long-term records may not be available. In addition, models developed in a non-spatial context, such as nutrient loading models, may require detailed input data that do not exist as spatial data. Moreover, some parameter values in the models such as those in MUSLE for a large catchment may be difficult to measure. In these circumstances, surrogate measures and interpolated data are often used. For example, the average concentration of a particular non-point source pollutant can be assumed to be directly related to land use. Therefore, it can be estimated based on previous field studies and land use information.

10.2 LAND USE ANALYSIS AND MODELLING

Land is a scarce resource, while almost every human activity uses land. As human requirements and economic activities continue to expand, pressures on land resources are ever-increasing, creating competition and conflicts, and leading to suboptimal use of both land and land resources. Therefore, there is a great need for the sound land use management and planning so that land resources can be utilised efficiently and effectively.

Land use is an assembly of different activities (such as agriculture, forestry, nature conservation and industry) performed within particular areas. Decisions on land use

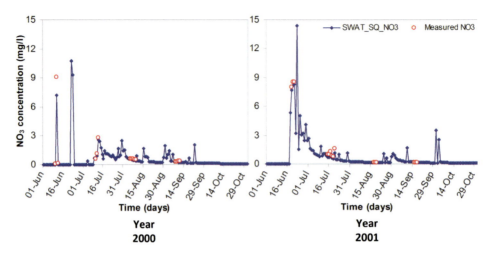

Figure 10.2 Measured and SWAT-simulated NO$_3$–N concentration at the catchment outlet in 2000 and 2001 (© Mishra, Singh and Singh, used under the terms of the Creative Commons Attribution License, which permits unrestricted use, distribution, and reproduction in any medium, provided the original author and source are credited).

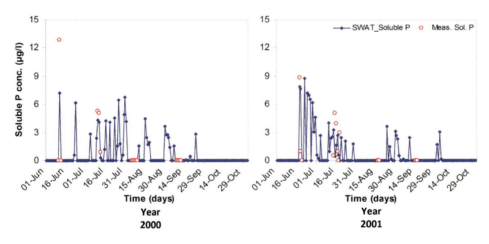

Figure 10.3 Measured and SWAT-simulated water-soluble P concentration at the catchment outlet in 2000 and 2001 (© Mishra, Singh and Singh, used under the terms of the Creative Commons Attribution License, which permits unrestricted use, distribution, and reproduction in any medium, provided the original author and source are credited).

may lead to great benefits or great losses, sometimes in economic terms, sometimes in less tangible environmental changes (McRae and Burnham 1981). Land use management and planning is a process of decision making as to how land resources should be managed and utilised. Its function is to guide management and planning decisions on land use in such a way that the land resources are put to the most beneficial use to meet the needs of the present, while at the same time conserving resources for the future (FAO 1976).

Land possesses both physical and economic characteristics. Land use management and planning must be based on an understanding of the natural and social environments as well as the kinds of land use envisaged. GIS can play three major roles in land use management and planning: (1) maintaining and managing general-purpose data, (2) generating specific-purpose information from such data and (3) using such information in land use decision making contexts. Typical applications of GIS include the identification of land use conflicts, the analysis and prediction of land use changes and the analysis of land use suitability.

Broadly speaking, land use suitability analysis involves evaluating and identifying suitable areas of land for a particular type or types of land use (for example, agriculture, forestry, nature conservation, recreation or urban development) according to specified objectives and criteria (Malczewski 2004). It aims to assist in the allocation of land to the uses that offer the greatest socio-economic and environmental benefits, and to support the transition to sustainable and integrated management of land resources. There are three types of land use suitability analysis: site selection, land evaluation and land allocation.

Site selection involves identifying the best or optimal site (or sites) for a particular land use activity among a set of the feasible sites, and ranking or rating the alternative sites based on their characteristics (such as location, size, shape and other relevant attributes). For instance, suppose we have a number of candidate sites for building a dam. Site selection analysis evaluates each site according to its potential environmental and socio-economic impacts, rates all the candidate sites and selects the best one.

Land evaluation involves assessing the fitness or suitability of a given tract of land for a defined use. The objective is to identify the most appropriate spatial pattern for future land uses. It aims to answer two questions: 'For any given type of land use, which areas of land are best suited?' and 'For any given area of land, for which type of use is it best suited?' (FAO 1976). Land evaluation is mainly based on biophysical characteristics of land, including topography, climate, hydrology and soils.

Land allocation allocates a certain amount of feasible land for particular types of land use. It is based on the comprehensive assessment of potentials and feasibility of land resources, and seeks to provide a compromise solution that attempts to maximise suitability of lands for multiple objectives.

In this section, three case studies are presented, each illustrating a type of land use suitability analysis with GIS.

Case Study 3: wind farm site selection

A wind farm is an area of land with a group of wind turbines used to produce energy. Wind farms can have as few as five wind turbines or as many as several hundred. As a potentially large source of renewable energy, wind farms are particularly popular in nations which are focusing on alternative energy. In the USA, twenty states plus the District of Columbia have renewable portfolio standards in place which require a certain percentage of energy

to come from renewable sources, including wind power, by a specific year. In 2009, 9,581MW of wind generation capacity was added in USA, which accounted for 41 per cent of the total added power capacity of the country that year (United States Energy Information Administration 2011). Obviously, the ideal place for a wind farm is a windy location. In some instances, a windy location may also be generally unusable or uninhabitable. In others, a wind farm may take up useful real estate which could be used for agriculture. In addition, wind farms may pose a severe threat to migratory birds. Therefore, the best location for a wind farm is not always where there are the greatest wind resources. Other socio-economic and environmental factors also play a role in the site selection of wind farms. Van Haaren and Fthenakis (2011) presented a case study of site selection of wind farms in New York State through multi-criteria analysis with ArcGIS.

Their case study considered several economic, planning, physical and ecological criteria in its three-stage siting process, including:

1. planning criteria – planning regulation, visual intrusion, noise and safety;
2. physical criteria – construction accessibility and foundation strength;
3. ecological criteria – wildlife habitat, hydrology and impacts on birds;
4. economic criteria – wind resources, road access and land clearing costs, and grid connection costs.

The three stages of the siting process were:

Stage 1 – identification of potentially feasible sites;

Stage 2 – economic evaluation;

Stage 3 – bird impact evaluation.

In Stage 1, infeasible sites were identified and excluded according to planning, physical and ecological constraints. They included (a) Federal and Indian land, such as national parks, military bases, wildlife refuges and other land serving specific functions (restricted by planning regulations), (b) urban areas, roads and their immediate vicinity (constrained by visual intrusion, noise and safety), (c) Karst porous grounds (constrained by foundation strength for wind turbine installation), (d) slopes greater than 10 per cent (limited by construction accessibility) and (e) lakes and their immediate vicinity (restricted by ecological significance). Seven map data layers, representing Federal land, Indian land, urban areas, roads, lakes, GTOPO30 DEM and Karst (above 100m depth), were obtained from the National Atlas GIS database created by the US Geological Survey; 500m buffers around roads, 1,000m buffers around towns, 2,000m buffers around cities and 3,000m buffers around lakes and Indian land were generated using the ArcGIS buffer function to represent the setbacks assigned to wind farm feasibility. The slope map was derived from the DEM using the ArcGIS slope function. The Federal land layer, the Karst area layer, the slope layer and the buffer layers were combined to create a map layer that excluded all the infeasible sites listed above. The remaining sites on the output map layer were potentially feasible sites. Figure 10.4 illustrates the Stage 1 process.

In Stage 2, the potentially feasible sites were evaluated and ranked based on their expected net present value calculated from costs for access roads, land clearing and grid connection, and revenue from generated electricity. Grid connection costs were calculated as a function of the cost required for building or upgrading a substation, the cost for a single cable per unit distance (for example, per km) and the distance to the nearest electricity line or substation. These were computed based on a map with substations and transmission lines. Land clearing costs depend on the type of vegetation present on site. The National Land Cover Database provided vegetation type data. Land clearing costs for different types of vegetation ($/acre) were obtained from Web references. Costs for access roads were set to $82,000/km to the nearest existing road. The revenue from generated electricity was determined based on local wind resources, which were measured in terms of annual energy production or electricity yield. Annual energy production data were offered by Associated Weather Services. They were estimated based on time series data on wind speed, direction, temperature and pressure at different locations. The total revenue from electricity production was converted into the net present value by annualising the electricity yield multiplied by the expected revenue. The result of this stage was a priority map showing the potential feasible sites ranked based on their net present values. Figure 10.5 shows the process.

In Stage 3, the potential feasible sites were evaluated for potential impacts on birds. A map of the important

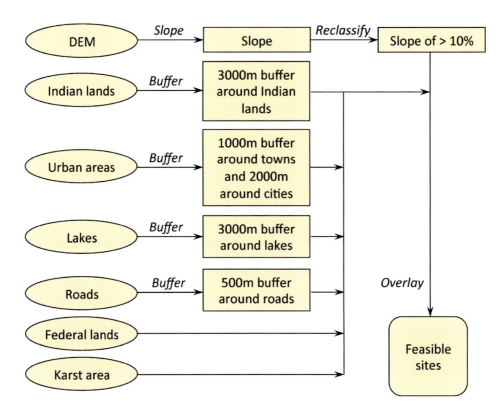

Figure 10.4 GIS analysis in Stage 1 of wind farm site selection.

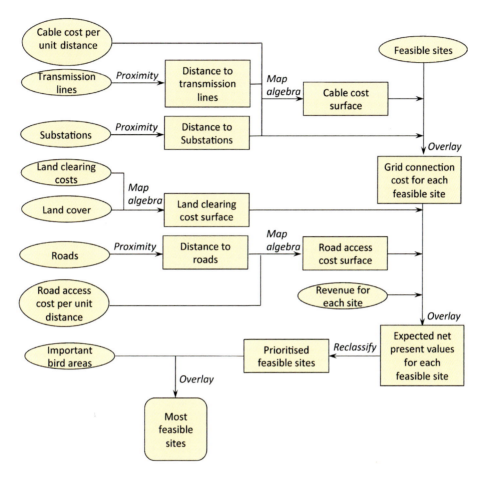

Figure 10.5 GIS analysis in Stages 2 and 3 of wind farm site selection.

bird areas produced by the New York State Important Bird Areas Program was intersected via an overlay operation with the priority map from Stage 2 (see Figure 10.5). The important bird areas represented potentially problematic sites in terms of bird mortalities. The most highly ranked sites located outside the important bird areas were identified as the most feasible sites. However, the authors suggested that individual site surveys and assessments were needed as the identified important bird areas were not a definitive statement of the presence or absence of all species at a given location, and a thorough environmental impact analysis was still required.

Case Study 4: sugarcane land allocation

This case study demonstrates the integration of GIS and multi-criteria analysis for sugarcane land allocation in the lower Herbert River catchment in Northern Queensland, Australia (Zhu et al. 2001).

The Herbert River Catchment covers an area of about 10,000km² (Figure 10.6). The sugar industry is the largest intensive agricultural industry in the catchment. Other important industries include forestry, beef and the small-scale cultivation of crops such as pineapples, melons and pumpkins. Land not used for arable agriculture is mainly under native vegetation or improved pasture for beef production. Small areas are utilised for mining and industrial activities. The study area is located in the southeast of the catchment (known as the 'Lower Herbert'), where sugarcane is the dominant land use.

This study aimed to assist industry and government users to explore and evaluate different land allocation scenarios for sugarcane production in the catchment. It used SMARTER (see Section 9.1) to help users identify

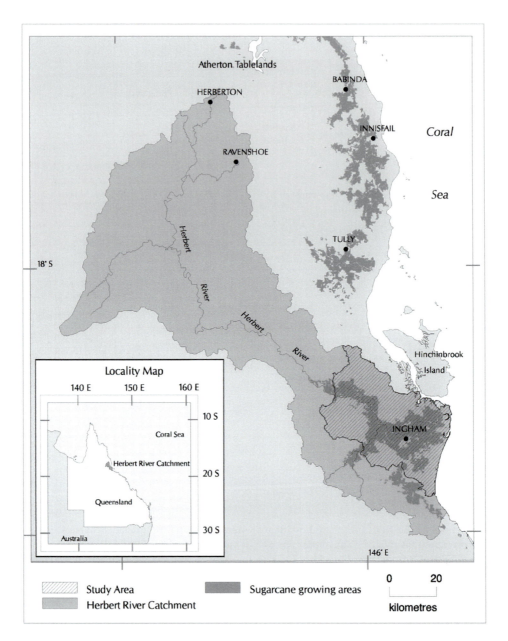

Figure 10.6 The Herbert River Catchment.

those criteria that were important for sugarcane land allocation, make subjective assessments of relative importance of those criteria, and to convert the assessments into a set of weights. GIS was used to develop map data layers to represent the identified allocation criteria and potential constraints for sugarcane land use, combine these map layers through spatial modelling, identify

the suitable land for sugarcane production by applying the SMARTER weights, and to generate map presentations of land allocation scenarios.

A detailed GIS database was collated, containing map data layers for land use/cover, slope, elevation, land suitability for sugarcane, roads, rivers, mill locations and flood-prone areas. All map data layers were in the

ARC/INFO GRID format, with a cell resolution of 100m × 100m. ArcView (the early version of ArcMap) was used for the analysis.

The following procedure was used:

1. Define land use constraints and allocation criteria.

Land use constraints represent restrictions imposed on areas for sugarcane production. They define regions of certain biophysical and environmental conditions which are unsuitable for sugarcane. These include areas where state forests, national parks, wetlands or flood-prone areas are located. The criteria for sugarcane land allocation include land suitability, slope, proximity to existing mills, proximity to roads, shape compactness of sugarcane land blocks and fragmentation of landscape.

The shape compactness of a sugarcane land block C_s is measured as:

$$C_s = 4 \times \frac{\sqrt{A}}{P} \qquad (10.10)$$

where A is the area (km²) and P is the perimeter (km) of a sugarcane land block (Bogaert et al. 2000). When C_s equals 1, the sugarcane land block is a square. A more compact sugarcane land block has a higher C_s and is easier to manage and maintain. The fragmentation of landscape F_g is calculated as:

$$F_g = 2 \times \frac{\log A_s}{\log P_s} \qquad (10.11)$$

where A_s is the total area of sugarcane land and P_s is the total perimeter of sugarcane land in the study area (Milne 1988). The higher F_g, the less fragmented the landscape. Both C_s and F_g can be calculated in GIS.

2. Prepare constraint data layers.

Each land use constraint is represented as a map data layer in the GIS. The constraint maps are created by eliminating regions characterised by attributes or certain values of attributes from consideration. They categorise the areas into two classes: feasible and infeasible. Feasible areas are assigned a value of 1, and infeasible areas are assigned a value of 0.

3. Prepare data for measuring the allocation criteria.

Each allocation criterion is also represented as a map data layer in the GIS.

4. Combine constraint and criterion maps to create a spatial unit map.

The spatial unit map excludes the restricted areas by applying the land use constraints. Each spatial unit represents an alternative area for sugarcane production, and contains the score values for all the allocation criteria.

5. Rank the allocation criteria in order of importance.

This is accomplished by asking decision makers:

> Imagine you have a sugarcane land block that has the worst possible performance on all criteria. You are selecting an alternative area. You can improve the value of one criterion to achieve the best possible attainment level of sugarcane production. Among all the n criteria, whose value would you improve to make the alternative most desirable?

The decision makers will then select one of the n criteria. This criterion will be removed from the list, and the decision makers would be asked to select one criterion from the remaining list whose value they would prefer to improve to make the alternative most desirable. This continues, with the outcome that a rank ordering of criteria is obtained. The most important criterion is the first selected in this operation, and the last selected is the least important.

6. Estimate weights.

This step estimates and assigns numerical weights to the allocation criteria based on their rank order of importance obtained in the previous step by applying SMARTER weighting method.

7. Calculate overall scores for each spatial unit.

This calculation is accomplished by applying the weighted summation method on all spatial units on the spatial unit map.

8. Allocate the most feasible land for sugarcane production.
9. Generate a future sugarcane land use plan.

The allocated sugarcane land is presented as a sugarcane allocation map.

Steps 1–8 can be repeated to develop different land allocation scenarios by defining different land use constraints and allocation criteria, and by providing different rank orders of the criteria. Three scenarios for sugarcane land allocation in the study area are shown in Figure 10.7.

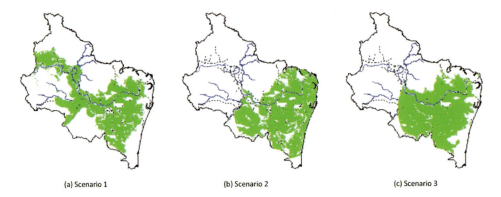

(a) Scenario 1 **(b) Scenario 2** **(c) Scenario 3**

Figure 10.7 Sugarcane land allocation scenarios: (a) scenario 1, (b) scenario 2 and (c) scenario 3.

In the first scenario, the plan is to allocate about 600km² of land to sugarcane production. The selected allocation criteria include land suitability for sugarcane, slope, distance to mills, distance to roads, shape compactness of a sugarcane land block and fragmentation of landscape. They are ranked in the order of importance: land suitability for sugarcane > slope > distance to mills > distance to roads > fragmentation of landscape > shape compactness of a sugarcane land block. National parks, state forests and those areas with a slope of 8 per cent or more are reserved. Figure 10.7a shows that the new sugarcane land allocation (mapped as green shaded areas) is distributed along the major rivers due to the importance placed on the criterion 'land suitability for sugarcane'.

In the second scenario, the same amount of land for sugarcane production, 600km², is planned. A different set of allocation criteria are selected, including land suitability for sugarcane, distance to mills and fragmentation of landscape. They are ranked in the following order: distance to mills > fragmentation of landscape > land suitability for sugarcane. The same land use constraints as in the first scenario are imposed. The result of sugarcane land allocation for this scenario is shown in Figure 10.7b. In this scenario, the new sugarcane land allocation is skewed towards the east due to the importance placed on the criterion 'distance to mills'.

With the same land use constraints, the third scenario allocates 600km² of land for sugarcane production according to the same set of criteria as used in the first scenario, but ranked in the following order: fragmentation of landscape > shape compactness of a sugarcane land block > distance to mills > distance to roads > land suitability for sugarcane > slope. Figure 10.7c shows this scenario.

Through the use of the method, sugarcane land allocation scenarios can be developed by: (a) using different sets of criteria, (b) imposing different sets of land use constraints, (c) providing different rank orders of importance of the selected criteria and (d) setting different targets for sugarcane production. By comparing land allocation scenarios based on deriving criteria and weightings, decision makers can explore the implications of different policies in allocation strategy to attempt to meet targets for expansion without violating constraints on expansion and while minimising conflicts with other land use aspirations.

Case Study 5: land evaluation

Land evaluation for rural land use planning was one of the earliest applications of GIS in the 1960s and 1970s. It is also known as land suitability assessment. It should be noted that land suitability is only meaningful in relation to specific kinds of land use. Different types of land use have different biophysical requirements. For example, an alluvial flood plain with impeded drainage caused by waterlogging may suit rice cultivation very well, but is unsuitable for many other forms of agriculture.

Land comprises the physical environment, including climate, topography, soils, vegetation and hydrology, while land use concerns how the land is used for human activities, such as agriculture, forestry, animal husbandry, nature conservation and recreation. Land evaluation involves comparing the characteristics and qualities of land with the requirements of specified types of land use, and determining how suitable the land is for those types of land use. Land characteristics refer to attributes of land that can be measured or estimated, such as slope, aspect, rainfall, air temperature, soil texture and rooting depth. Land qualities are complex attributes of land that have an influence on the suitability of land for a specific type of land use, such as moisture availability, capacity to retain fertilisers, land instability hazard, flood hazard, salinity hazard and waterlogging risk. The Food and Agriculture Organization of the United Nations developed a framework for land evaluation (FAO 1976) which has been widely used. The framework specifies a set of principles for land evaluation, the structure of a land suitability classification and the procedures for carrying out land evaluation. The basic procedure for land suitability assessment involves:

1. collecting and preparing land use, land characteristics and land quality data;
2. determining types of land use and their requirements (for example, for energy, water, nutrients, avoidance of erosion and salinity);
3. defining and creating land mapping units;
4. comparing the types of land use with the land present in each land mapping unit;
5. classifying the land suitability of each land mapping unit using land characteristics and land qualities as diagnostic criteria;
6. presenting the results.

A land mapping unit is a mapped area of land with certain land characteristics and land qualities. Traditionally, land mapping units were defined and mapped by natural resource surveys, such as soil surveys and forest inventories. Each land mapping unit is assumed to be homogenous or have little variation in soils, landforms and other physical properties. But their degree of homogeneity varies with the scale and intensity of the study. Each land mapping unit is compared with the requirements of the defined land use types, which leads to classification of land suitability. The degree of suitability is usually ranked into four classes: highly suitable, moderately suitable, marginally suitable and unsuitable. Table 10.1 lists their definitions.

In GIS, land evaluation can be based on either vector data or raster data. Vector-based analysis involves:

- preparing land characteristic and land quality data layers;
- creating land mapping units via map overlay of land characteristic and land quality data layers;
- determining the suitability class for each land mapping unit based on the combined attribute table;
- reclassifying the land mapping unit map layer to create a land suitability map.

Raster-based analysis is similar, but uses cells as land mapping units. The suitability class may be determined by using multi-criteria analysis or decision rules.

Chanhda et al. (2010) applied GIS to assess land suitability for tea in an area in northern Laos along the border with China, including six districts and covering an area of 10,323.67km². The area has a mountainous and hilly terrain with slash-and-burn cultivation practice (involving the cutting and burning of plants in forests to create agriculture fields). In order to promote sustainable land resource use and environmental management practices, prevent further expansion of slash-and-burn cultivation into forests, and at the same time improve the local economy, tea plantation was encouraged in this region. This study assessed land suitability for tea in the area using the FAO framework, and took the following steps:

1. mapping land uses and land use changes from 1992 to 2002 from satellite images;
2. determining land requirements for tea plantation by consulting experts (Table 10.2);

Table 10.1 Land suitability classes (adapted from FAO 1976)

Class	Definition
Class S1 Highly suitable	Land without significant limitations for a given land use, or with only minor limitations that will not significantly reduce productivity and will not raise inputs above an acceptable level.
Class S2 Moderately suitable	Land that is clearly suitable for a given use, but with limitations that will reduce productivity and increase the inputs required to sustain productivity.
Class S3 Marginally suitable	Land with limitations so severe for a given use that productivity will be reduced and the cost for the increased inputs necessary to sustain productivity will be only marginally justified.
Class N Unsuitable	Land that cannot support a given use on a sustained basis, or whose use will have unacceptable risk of damage to land resources.

3. preparing land characteristic and land quality data layers using ArcGIS;
4. creating a composite map of land mapping units by overlaying the land characteristic and land quality data layers in ArcGIS;
5. assigning weights obtained from experts to individual land characteristics and land qualities (Table 10.2);
6. using the land characteristics and land qualities as land suitability classification criteria, and applying AHP (see Section 9.1) to the land mapping units to derive their land suitability classes;
7. producing a land suitability map for tea plantation.

Ten land characteristics or qualities were used in the assessment, including elevation, slope, aspect, soil pH, soil depth, annual mean precipitation, ≥10°C accumulated temperature, current land use, distance from villages and distance from the Laos–China border checkpoint. Table 10.2 lists these criteria and their weights derived by AHP. Figure 10.8

shows the process of the GIS analysis. The results suggested that 3.11 per cent of the study area was highly suitable for tea plantation, 2.2 per cent of the area moderately suitable, 0.24 per cent marginally suitable and the rest unsuitable. The output land suitability map showed the spatial distribution of the four suitability classes.

10.3 ATMOSPHERIC MODELLING

Climatological and meteorological phenomena are spatially variable. GIS can play a dual role in atmospheric science research, providing an effective solution to the management, integration and dissemination of climate and meteorological data and offering a useful platform for atmospheric analysis and modelling (Chapman and Thornes 2003). This section focuses on GIS applications in climate interpolation, air pollution modelling and climate trend analysis.

Table 10.2 Land suitability classification criteria for tea cultivation (adapted from Chanhda et al. 2010)

Criteria	Land suitability				Weight
	Highly	Moderately	Marginally	Unsuitable	
Elevation (m)	1,000–1,400	500–1,000 or 1,400–1,600	250–500 or 1,600–1,800	< 250 or > 1,800	0.12
Slope (°)	15–25	25–35	6–15	> 35 or < 6	0.09
Aspect (°)	45–135	30–45 or 135–180	180–330	0–30 or 330–360	0.07
Soil pH	4.5–5.5	4–4.5 or 5.5–6	6–6.5	< 4 or > 6.5	0.13
Soil depth (m)	> 100	75–100	< 75		0.09
Annual mean precipitation (mm)	1,400–1,700	1,700–2,000	900–1,400 or 2,000–2,500	< 900	0.06
Accumulated temperature (°C)	> 4,500	4,000–4,500	3,700–4,000	< 3,700	0.04
Current land use	Potential forest	Permanent agriculture	Current forest, other on-forest, water		0.19
Distance from villages (km)	< 3	3–6	6–9	> 9	0.10
Distance from the border checkpoint (km)	< 20	20–40	40–60	> 60	0.11

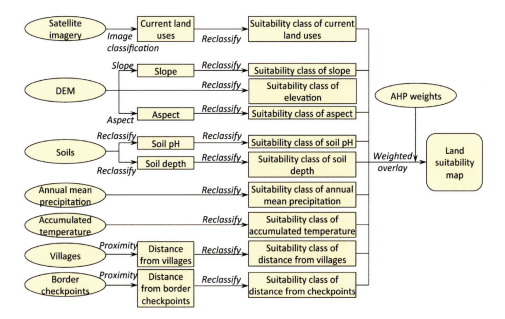

Figure 10.8 GIS analysis in land evaluation of tea plantation in Laos.

Case Study 6: climate interpolation

Climate and meteorological data are typically point station data. In order to infer climates from a sparse network of weather stations, spatial interpolation techniques are often used. In addition, interpolation of the point station data into surfaces (such as rainfall surfaces) facilitates the exploration of spatial patterns of climatological, meteorological and related phenomena. In particular, interpolation of climate and meteorological data based on digital terrain models (DTMs) enables good estimates of an area's baseline climatology without the need for extensive weather records (Chapman and Thornes 2003). Ustrnul and Czekierda (2005) provided a case study on spatial interpolation of air temperature in Poland and Central Europe using DTMs and meteorological data. This study adopted the residual kriging method.

In the residual kriging procedure, instead of kriging the dependent variable $Z(x)$ at point x directly, a multiple regression analysis is first performed between $Z(x)$ and k explanatory or independent variables $v_i(x)$ (i=1, 2, . . ., k) using point observation data, giving:

$$Z(x) = \hat{Z}(x) + \varepsilon(x) \tag{10.12}$$

$$\hat{Z}(x) = b_0 + b_1 v_1(x) + b_2 v_2(x) + \\ \ldots + b_k v_k(x) \tag{10.13}$$

where $Z(x)$ is the true value at point x, $\hat{Z}(x)$ is the estimated value at point x based on the observations at the same point, and $\varepsilon(x)$ is the residual or error term of the regression at point x. The coefficients b_i (i=0, 1, 2, . . ., k) are fitted using the ordinary least squares procedure. The residual $\varepsilon(x)$ preserves the spatial variability of $Z(x)$. The regression model is applied to every known point with observed or measured values. The residuals at these known points are calculated. Then the residuals are interpolated using ordinary kriging to generate a residual raster layer, which represents the corrections to apply to the regression model. The final estimates of the dependent variable $Z^{*}(x)$ at point x are given by:

$$Z^{*}(x) = \hat{Z}(x) + \hat{\varepsilon}(x) \tag{10.14}$$

where $\hat{\varepsilon}(x)$ is the estimated residual at point x interpolated by kriging. In GIS, residual kriging is implemented in the following steps:

1. representing point observations (for example, observations from meteorological stations) in a feature class;

2. representing independent variables in individual raster layers;

3. conducting a multiple regression analysis by combining the point observations with the independent variable rasters using Equation 10.13;

4. calculating the residuals at each observation point by subtracting the observed value from the estimated value by regression at the point;

5. implementing ordinary kriging to interpolate the residuals and creating the residual raster;

6. implementing the regression model created in Step 3 by combining all the independent variable rasters via map algebra to produce the regression raster;

7. adding the residual raster and the regression raster to produce the final map representing the final estimates of the dependent variable.

In the study by Ustrnul and Czekierda (2005), the dependent variable is air temperature, and the independent variables are elevation, latitude, longitude and distance to the Baltic coast (for stations located within 100km). Mean daily temperature data from 168 stations (synoptic and climatological) located across Poland and fifty-five stations located in bordering zones were used for residual kriging. A 250m DEM for Poland and a 1km DEM from GTOPO for the areas outside Poland were used to get elevation values. They mapped the mean annual, seasonal and monthly air temperatures for Poland and Central Europe with the residual kriging method.

Residual kriging assumes that a physical process, like the spatial distribution of air temperature, is composed of two components: deterministic and stochastic. The deterministic component is characterised by the spatial correlation of the phenomenon with a set of physical properties (such as elevation or distance to the sea), representing the large-scale trend. This trend is subtracted from the observations to obtain residuals. The interpolated residuals using kriging is regarded as stochastic. By adding the trend terms to the interpolated stochastic component, temperature can be estimated at any location. Residual kriging is widely used in climatology, and is one of the preferred methods for climatological mapping (Tveito et al. 2005). It is particularly valuable for spatial interpolation of climate and meteorological data in regions with complex topography such as mountainous areas, where climate elements show high variability caused by manifold interactions between relief and land use. In the case study presented above, elevation was the only terrain variable used in regression. Additional geographical variables, such as slope, aspect and land use, could also be included.

Case Study 7: air pollution modelling

Air pollution modelling aims to quantify the relationship between emissions and concentrations/depositions of a particular type of pollutant over an area. In addition to the diffusion properties of plumes emitted from the sources of pollutants, variations in meteorology, land use and terrain lead to complex patterns of air pollution. In recent years, GIS has been increasingly used to model and map the spatial and temporal patterns of air pollution (Maantay et al. 2009; Gulliver and Briggs 2011). Examples of GIS-based air pollution models include AERMOD (Kesarkar et al. 2007), DUSTRAN (Allwine et al. 2006), OSPM (Berkowicz et al. 2008) and STEMS-Air (Gulliver and Briggs 2011). Without coupling with any specific atmospheric dispersion model, Vienneau et al. (2009) developed a GIS-based moving window approach to model NO_2 pollution across Europe based on emissions and meteorological data using ArcGIS, which is presented as a case study below.

Basically, the moving window approach by Vienneau et al. (2009) uses focal functions (see Section 4.4) available in many GIS systems to estimate annual mean NO_2 concentrations by computing a measure of the emissions intensity at the focal cell as the distance-weighted sum of emissions in surrounding cells in the specified neighbourhood. It is based on the assumption that proximity to the source is one of the main determinants of pollutant concentrations. Figure 10.9 depicts their approach.

Three datasets were used in the modelling: emissions, concentrations and wind data. NO_2 emissions data were derived by disaggregating national data for 2001 from the inventory maintained by the United Nations Economic Commission for Europe under the Convention on Long-range Transboundary Air Pollution for eleven main source categories, including stationary and mobile combustion sources, industrial processes, fossil fuel extraction, solvent use, waste and agriculture. Different disaggregation methods were used for different source categories. For example, national totals of NO_2 emissions from residential sources were disaggregated by fuel types (such as electricity and gas) and population density, and those from industrial sources were disaggregated based on regional economy and land cover types. National NO_2 emissions from road transport were disaggregated using population and road density, traffic statistics and land cover as proxies. The disaggregated emissions data were mapped to a 1km grid, called the emissions grid.

NO_2 concentration data for 2001 were obtained from the Airbase database, a centralised database of measurements of monitored air pollution concentrations collated from the European Air Quality Monitoring Network. The dataset used in the study included 942 background NO_2 monitoring sites, among which the monitoring data from 714 sites were used for model building and those from the remaining 228 sites were used for model validation. Meteorological data were obtained from the 50km resolution European Centre for Medium-Range Weather

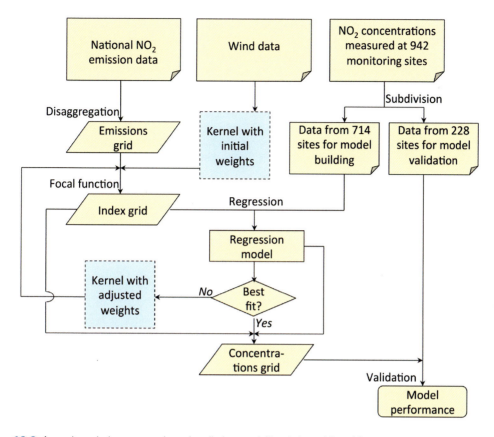

Figure 10.9 A moving window approach to air pollution modelling (adapted from Vienneau et al. 2009).

Forecasts operational archive, from which annual wind speed and direction were calculated for 2001 and then assigned to a 1km grid.

The focal function requires a specification of kernel (see Section 4.4). In this study, it was defined as a series of eleven concentric and non-overlapping annuli when wind effects were not considered. Each annulus had a width of 1km, allowing contributions from emission sources up to 11km from the focal cell to be modelled; 11km is the possible limit of detectable contributions according to the simulation results of a dispersion model. When wind effects were taken into account, wedge-shaped kernels were used. Wedges were created along the wind direction vector. The length and angular diameter of the wedges were set as a function of mean annual wind speed. Narrower and more elongated wedges represented higher wind speeds, and vice versa.

For the kernel of eleven annuli, each annulus was initially assigned a weight, calculated as a distance-decay function $(1/d^2)$, with the weight for the inner most annulus set as 1. The kernel then passed, as a moving window,

over the emissions grid to calculate the weighted sum of emissions within the window for every focal cell. The result was a grid representing the emission intensity for each cell. This grid was named the emission intensity index grid, or simply the index grid. A regression analysis was then carried out using the index grid and the NO_2 concentrations data measured at 714 monitoring sites, and the resulting regression equation was used to convert emission intensity values (tonnes/kg/year) into concentration values ($\mu g/m^3$). Note that before regression, the emission intensity index values and the observed concentration data were both log-transformed in order to correct for non-normality. Next, the weights for each annulus in the kernel were adjusted in a sequential, stepwise fashion while keeping the weight for the innermost annulus as 1 and the weight of an outer ring as less than or equal to that of the adjacent inner ring. For every round of adjustment, a new index grid was produced using a new set of weights in the kernel, followed by a regression analysis. The performance of the regression models was assessed by computing the coefficient of determination (r^2) and standard error of estimate (see Section 5.6). The best-fit model was used to create a concentration map.

When wedge-shaped kernels were applied to incorporate wind direction, eight index grids were created, one for each direction (north, northeast, northwest, east, southeast, south, southwest and west). The eight index grids were then summed to create a final emission intensity index grid. Similarly, regression analysis was conducted to create a concentration map based on the index grid derived using the wedge-shaped kernels and the monitored concentration data.

The NO_2 concentration maps produced using the above moving window approach were validated against the measured data from the 228 reserved monitoring sites in terms of r^2, RMSE, standard error and other measures. The validation results indicated that the moving window approach provided an acceptably accurate means of modelling long-term air pollutant concentrations across large areas. This study highlighted four major merits of the moving window approach: (1) it enables fast computation of large datasets at a high spatial resolution, as it is applied to raster data; (2) it is flexible, as users can define the shape and size of kernels used with focal functions; (3) it can easily accommodate new emission and monitored concentration data and take into consideration changes in emission source distribution such as those caused by land use/cover changes and (4) it facilitates the modelling of air pollution under different policy and emission scenarios. With these advantages, the moving window approach is especially suitable for high-resolution, continental-scale air pollution modelling and mapping.

Case Study 8: climate trend analysis

Most time series in climate exhibit various kinds of trends, cycles and seasonal patterns. Understanding climate trends or patterns is important for studying climate changes and their impacts on the environment, economy and public health. Historically, observations from a single location have been used to represent the climate for a large area. Accordingly, climate patterns have been investigated on a point-by-point basis. Temperature and rainfall trends from one location would be compared with other locations. However, such a point-based analysis cannot accurately portray the spatial variations of climate variables in a large region, particularly in complex climatic regions. Spatial analysis of climate variables on a local scale based on the full set of weather stations located in a region can overcome the above limitation and provide improved understanding of local climate

patterns over both space and time. Boyles and Raman (2003) presented a study on spatial and temporal analysis of climate patterns and trends in North Carolina, which has one of the most complex climates in the USA. In this case study area, complex topography in the west and warm waters off the east coast combined with weather patterns make it a region with locally variable weather and climate. Analysis of the climate trends and their local variability is crucial for agricultural and planning purposes in the region. In order to analyse local-scale climate and accurately represent the complex climate in the state, the authors applied the linear trend modelling technique with ArcView GIS.

Linear trend modelling is a statistical technique for trend analysis, often used to identify an underlying pattern of climate in a time series. A linear trend model is a trend line expressed as:

$$\hat{y}_t = \alpha + \beta \times t \qquad (10.15)$$

where t is the time index, α is the intercept and β is the slope of the trend line. α and β are usually estimated via a simple regression in which y_t is the dependent variable, and the time index t is the independent variable. A linear trend model tries to find the slope and intercept that give the best average fit to all the past data. In general, a positive slope indicates an increasing trend, and a negative slope represents a decreasing trend. Figure 10.10 shows a plot of winter rainfall trend over eighteen years in a hypothetical region. In this example, the slope of the trend line is −2.0226, suggesting that there was generally a decrease in rainfall during the winter over the eighteen-year period.

Boyles and Raman (2003) analysed the spatial and temporal trends of climate variability in North Carolina over the period from 1949 to 1998 through linear trend modelling with the following procedure:

1. collecting weather records – a complete record of monthly temperature and precipitation observations over the fifty years was collected from seventy-six weather stations located in the state;
2. calculating annual and seasonal averages of temperature and precipitation for each station;
3. constructing linear trend models to fit a linear trend to each time series at each station;
4. interpolating the slope values of the linear trend models using universal kriging (see Section 4.5), which produced a series of seasonal and annual precipitation, minimum and maximum temperature trend maps; interpolated slope values were represented using isolines on each map, allowing linear

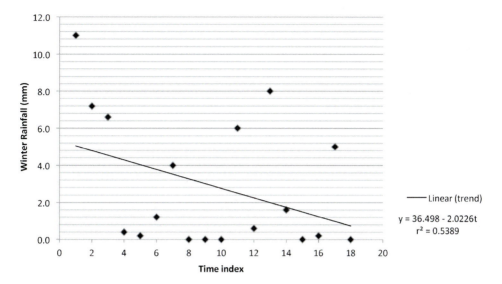

Figure 10.10 A time sequence plot for winter rainfall and the fitted linear trend model.

slopes – that is, climate trends at every location in the state – to be analysed; the maps effectively depicted the spatial patterns of climate variability across the region;

5. interpreting the resultant trend maps.

In addition, average precipitations and temperatures over subsequent non-overlapping ten-year periods were compared with the fifty-year average. By analysing the trend maps and ten-year period averages, this study concluded that during the period 1949–1998, precipitation in North Carolina had an increasing trend, especially during the autumn and winter seasons, and the last ten years were the warmest and wettest in half a century, but there was little change in maximum temperatures over the last thirty years.

Linear trend models fitting to the time series of climate data provide a simple picture of changes that have occurred at any location over a certain period. However, almost all naturally occurring climate time series do not behave as though there are straight lines fixed in space that they are trying to follow. As shown in Figure 10.10, deviations of some observations from the trend line may be quite large. Real trends change their slopes and/or their intercepts over time. Therefore, the best-fit trend model for a particular time series may be non-linear, and may not fit to another time series. Different time series may have different forms of best-fit non-linear trend models. Boyles and Raman (2003) argued that linear trend models are currently the best choice for multiple-station

spatial analysis of changes in climatic patterns, as only linear models can be used to compare changes in the time series of multiple stations across a region. With linear trend models, the slopes calculated at each weather station can be spatially analysed for regional changes. Nevertheless, we need to be aware that where the variations in observations are significantly larger than the straight-line trend, the choice of start and end points of time can significantly change the result.

10.4 ECOLOGICAL MODELLING

Ecological modelling involves the use of mathematical models, computer techniques and systems analysis combined with ecological theory and data for understanding and predicting ecological processes and for supporting the development of environmental and natural resource management strategies and policy options. Ecological modelling with GIS allows ecological modellers to integrate spatial and non-spatial data from various sources and deal with ecological processes in a spatially distributed manner, and provides them with new capabilities for analysing the space/time distribution of ecological phenomena. GIS and ecological modelling have been applied in studies of terrestrial and aquatic ecosystems. Examples include ecosystem dynamic modelling, habitat modelling, crop simulation modelling, landscape disturbance modelling, landscape pattern analysis and examination of the effects of changes in land use and environmental conditions. The case studies presented in this section are limited to landscape pattern analysis, characterisation of pest population dynamics and habitat potential mapping.

Case Study 9: landscape pattern analysis

Ecological phenomena are usually characterised by spatial structures, which involve multiple ecological processes operating at multiple spatial scales: trends at macro-scales, patches, gradients and patterns at meso-scales and local scales, and random patterns at local and micro-scales (Fortin et al. 2002). A disruption in landscape patterns may compromise the spatial structure's functional integrity by interfering with key ecological processes essential for population persistence of the organisms living in the system and the maintenance of biodiversity and ecosystem health. Therefore, landscape pattern analysis is considered a prerequisite to the study of pattern–process relationships, which constitutes the foundation of landscape ecology. While there are a variety of types of landscape patterns and purposes of landscape pattern analysis, the focus here is on using landscape metrics to quantify landscape patterns and the assessment of ecosystem condition.

Generally, landscape metrics are indices that characterise the geometric and spatial properties of landscape patterns represented at a particular scale. Here the term 'scale' refers to grain and extent. Grain is the spatial resolution of the data or map. Extent is the spatial area or boundaries of the landscape. There are literally hundreds of landscape metrics. Examples of landscape metrics include patch density, mean patch size, edge density, landscape shape index, contagion and Shannon's evenness index (Johnson and Patil 2007). Patch is an important concept in landscape ecology. From an ecological perspective, a patch is a discrete area of relatively homogeneous environmental conditions at a certain scale. Discrete land cover classes on a map are often treated as patches. Landscape metrics focus on the spatial properties and spatial distribution of patches. Therefore, GIS has been a useful and effective tool for landscape pattern analysis. Many landscape metrics have been integrated into GIS software, such as Patch Analyst for ArcView or for ArcGIS (http://www.cnfer.on.ca/SEP/patchanalyst/Patch5_2_Install.htm), the *r.le* programs in GRASS (Baker and Cai 1992) and FRAGSTATS for ArcGIS (http://www.umass.edu/landeco/research/fragstats/fragstats.html). However, it is important to understand that many landscape metrics are sensitive to changes in scale. Relationships between ecological processes and landscape patterns measured using a set of metrics at one scale may not exist or may not be the same at other scales. The case study by Mayer and Cameron (2003) highlighted the scale issue. They used Patch Analyst for ArcView to model relationships between landscape characteristics and bird diversity in Ohio, and evaluated the influence of scale.

Birds are often the most conspicuous animals within ecosystems, and are often held to be the most easily observed indicator of environmental change. Wormworth and Sekercioglu (2011, p. 4) described birds as 'flying data collectors' that can provide information about biodiversity and ecosystem health. Many studies have used birds to establish how taxonomic diversity is influenced by landscape characteristics by means of landscape metrics, and then used the information to assess the ecosystem condition. Mayer and Cameron (2003) conducted their study based on the breeding bird survey and land use/cover data using twelve landscape metrics. The Breeding Bird Survey (BBS) is an annual census coordinated by the USGS and carried out by volunteers in North America. The survey routes are mostly placed randomly along roads. Each route is 40km long, consisting of fifty stops. At each stop, the total number of spotted individuals per bird species is recorded by observers. Seventy-nine BBS routes in Ohio and their associated observation data from 1991 to 1997 were obtained, and fifty-eight of them were digitised into ArcView for use in the study. Land use/cover data for Ohio were also acquired from the USGS, which were classified from Landsat TM imagery with a 30m resolution.

The twelve landscape metrics were:

1. total core area of forest – the total number of pixels of forest located in the core area of the patch, which is the area defined with a core-buffer distance of 1 pixel (30m in this case);
2. per cent cover of forest – the number of pixels of forest divided by the total number of pixels in the landscape;
3. per cent cover of urban area – the number of pixels of urban area divided by the total number of pixels in the landscape;
4. per cent cover of agricultural land – the number of pixels of agricultural land divided by the total number of pixels in the landscape;
5. number of forest patches – the number of patches of forest cover;
6. number of urban patches – the number of patches of urban land cover;
7. mean patch size of forest – the total area of forest cover divided by the number of forest patches;

8. mean patch size of urban area – the total area of urban land cover divided by the number of urban patches;

9. total forest edge – the total length of edge around all forest patches in the landscape;

10. edge density of forest – the total length of edge around all forest patches in the landscape divided by the total area of the landscape;

11. mean patch fractal dimension of forest – calculated as:

$$MPFD = \frac{\sum_{i=1}^{m}\sum_{j=1}^{n_i} \dfrac{2 \times \ln\left(0.25 \times P_{ij}\right)}{\ln A_{ij}}}{N} \quad (10.16)$$

where N is the number of patches in the landscape, m is the number of land use/cover categories, n_i is the number of patches of land use/cover type i, P_{ij} is the perimeter of the jth patch of land use/cover type i and A_{ij} is the area of the jth patch of land use/cover type i;

12. mean nearest neighbour between forest patches – the sum of the smallest distances between the edge of each forest patch and its closest forest patch, divided by the number of forest patches in the landscape.

In this study, different landscape extents were defined in terms of buffers around each BBS route in order to evaluate the effect of changes in scale. This was done by first creating buffers of 50m, 100m, 500m, 1,000m, 2,500m and 5,000m around each BBS route on the digitised BBS route layer, then overlaying the buffer layer onto the land use/cover layer to create a new layer showing land use/cover classes (patches) within each buffer (extent). This layer was used to measure the above landscape metrics in each buffered landscape. The landscape metrics were calculated with Patch Analyst.

After the twelve landscape metrics were measured for each buffered landscape, correlation analyses were conducted at each buffer size (landscape extent) to determine how the landscape metrics were correlated with each other. The results showed that at the four smallest extents (50m, 100m, 500m and 1,000m buffers), mean patch size of forest, number of forest patches and per cent urban area were the least correlated, while mean patch fractal dimension of forest, total edge of forest and per cent urban area had the lowest correlations at the two largest extents (2,500m and 5,000m buffers). These least-correlated metrics were selected for subsequent analyses, and other highly correlated metrics were eliminated.

Bird diversity was measured in terms of species richness (number of species) and the Shannon diversity index. The Shannon diversity index is widely used to describe species diversity in a community. It accounts for both abundance and evenness of the species present. The index is calculated as:

$$I_d = -\sum_{i=1}^{n} p_i \ln p_i \quad (10.17)$$

where I_d is the Shannon diversity index, n is the total number of species in the community (richness) and p_i is the proportion of species i relative to the total number of species.

The species richness and Shannon diversity index were calculated for nine data groups in each BBS route in each year, including all bird species as one group, five habitat guilds (grassland, open water/wetland, successional/scrub, woodland and urban) and three migratory guilds (short-distance migrants, residents and neotropical migrants). An analysis of variance was performed, and confirmed that there was no significant among-year variation in richness and diversity.

The relationships between bird diversity and landscape characteristics were established by performing regression analyses at the six different scales (landscape extents), using the measured values of the selected least-correlated landscape metrics and the mean species richness and mean Shannon diversity index values from 1991 to 1997 for each habitat or migratory group in each route. All the statistic analyses were carried out using the SPSS statistical package. Figure 10.11 summarises the analysis process.

The study concluded that the diversity of woodland birds and neotropical migrants, and the richness of short-distance migrants were best explained by the landscape characteristics. However, scale effects were also observed. For example, the mean patch size of forest was positively related to diversity of successional/scrub birds at small landscape extents, while the total forest edge explained most of this guild's diversity at large landscape extents. This suggests that it is important to select an appropriate scale when using landscape characteristics to predict bird diversity and richness.

Case Study 10: characterisation of pest population dynamics

The case study by Beckler et al. (2004) demonstrated the use of GIS in the analysis of population dynamics of the western corn rootworm (WCR), *Diabrotica virgifera virgifera* Leconte (Coleoptera: Chrysomelidae), in relation to landscape characteristics in Brookings County, South Dakota, in the western portion of the US Corn Belt.

WCR is a devastating insect pest of maize in the US Corn Belt region, which can destroy significant

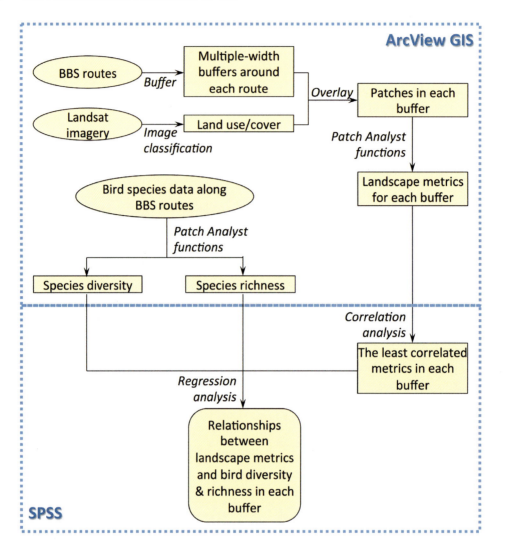

Figure 10.11 Analysis process of measuring landscape characteristics and their relationships with bird diversity at different scales in Ohio, USA.

percentages of corn if left untreated. Adult WCR prefer to feed on corn leaves. Once pollen shed is completed, they usually migrate in search of other pollen sources in later-planted fields. Therefore, the abundance and spatial distribution of adult WCR are influenced by cropping practices. In order to control and manage the WCR population, the USDA launched a corn rootworm area-wide pest management programme. The study aimed to support the population control tactics of the programme by investigating and documenting interactions of edaphic and landscape factors with WCR. More specifically, the study sought to characterise WCR population dynamics

in relation to the changes in landscape pattern over a five-year period (1997–2001), and to establish the relationships of the population dynamics with soil texture and elevation.

WCR population data were obtained through monitoring over the five years using emergence and post-emergence traps placed in fields classified as continuous maize, first-year maize, mixed maize, other maize and soybeans. Trap locations were recorded using a differential GPS. For each year, the number of WCR captured at each trap location was calculated. Vector map layers were created to represent locations of traps and associated yearly WCR numbers. These vector map layers were then used to produce WCR population raster maps of each year using the inverse distance weighted interpolation technique. The time series of WCR population rasters represents the WCR population dynamics.

In order to characterise the population dynamics, a number of class-level and patch-level landscape metrics were derived from the vegetation map of the study area using FRAGSTATS. The class-level metrics included area and percentage of landscape occupied by each patch type (indicators of landscape composition), number of patches, mean patch size, mean proximity index and mean nearest neighbour distance. The patch-level metrics included patch area size, proximity index of each patch and nearest neighbour distance of each patch. The proximity index and nearest neighbour distance are indicators of the spatial arrangement of patches, which may affect dispersal rates. Correlation analyses were then carried out to analyse relationships between patch-level metrics and mean numbers of post-emergence WCR captured, and between class-level metrics and the total and mean number of post-emergence WCR captured. Figure 10.12 outlines the analysis process.

Based on the above analysis, it was found that the total numbers of WCR captured correlated significantly with the area and per cent of landscape occupied by continuous maize, the number of continuous maize patches and their mean proximity. The total number of WCR captured decreased from 1997 to 2001, as the total class area, per cent of landscape and number of patches occupied by continuous maize decreased during this period. There were also significant correlations between the area and percentage of landscape occupied by each patch type and the number of patches with the total and mean numbers of WCR captured in all types of maize fields. In addition, in first-year maize, significant correlations were found between patch area, patch proximity and patch nearest neighbour with mean number of WCR captured in the post-emergence traps, while a negative correlation was

found between mean number of post-emergence WCR and nearest neighbour distance. Moreover, significant correlations existed between mean number of WCR post-emergence in all maize and patch areas and proximities.

The study also analysed the spatial associations of WCR with soil texture and elevation through contingency analysis using IDRISI. Both soil texture and DEM data were acquired from the USGS. This revealed that WCR abundance had strong spatial associations with both soil texture and elevation, with WCR most frequently occurring on loam and silty clay loam and in the elevation zone of 500–510m. The authors argued that the methodology used in this study could be extrapolated to larger areas, and could include climate, beneficial insect data and other relevant spatial data in the GIS analysis to find patterns in the landscape that might produce high insect population patches so that appropriate pest management strategies could be developed and implemented. The study also demonstrated that ecological systems are dynamic – the condition of their components changes through time. Landscape must be monitored through time to capture the system dynamics (including population demographics). GIS can play a useful role in this monitoring and analysis.

Case Study 11: habitat potential mapping

The term 'habitat' has been used in many ways in ecological studies. Spellerberg (1992) defined habitat as 'the locality or area used by a population of organisms and the place where they live'. Different combinations of light, air, water and soil, together with variations in topography and climate, create different habitats. For example, a shady area maintains more varieties of species of trees and shrubs than a sunny area, poor soil supports only certain plants, and rich soil produces greater growth. Most plants and animals are very specialised and can survive in only a very specific habitat where conditions are suited to

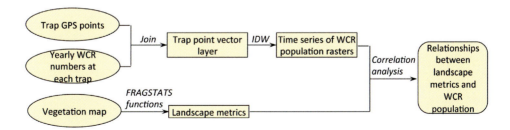

Figure 10.12 Analysis process of characterisation of population dynamics of western corn rootworm in South Dakota, USA.

them. For example, mangroves occur in dense, brackish swamps along coastal and tidally influenced low-energy shorelines, which often exclude other species. Reedbuck, bushbuck oribi and many other hoofed animals depend on grasslands that provide the abundance of grasses that make up their diet. Habitat potential mapping involves using statistical and spatial modelling techniques to establish empirical relationships between the spatial distribution of species and habitat conditions, and based on these, predicting and explaining geographical patterns and variability in species distribution, or mapping possible changes in species and biodiversity under changed environmental conditions (for example, climate change).

Habitat potential mapping is a kind of species distribution modelling. It involves generating potential species distribution maps by relating known species distributions to the spatial distribution of habitat or environmental variables. Many wildlife species are very difficult to census because of their low population density and our poor understanding of their ways of life. For such species, it can be more straightforward to infer possible distributions than to measure them directly. Habitat potential mapping has been widely recognised as a significant component of conservation planning. It is particularly useful in the domain of species and ecosystem management, where identification and protection of areas containing high biological diversity have become urgent priorities, but where species datasets are often limited or lacking.

The spatial distribution of species is governed by a range of biotic and abiotic factors. These factors can be classified into three types: resource, direct and indirect (Austin and Smith 1989). Resource variables constitute matter and energy consumed by plants or animals, including nutrients, water, light and food. Direct variables are environmental parameters which are not consumed, such as soil pH, temperature, precipitation and radiation. Indirect variables have no direct physiological relevance to a species, such as elevation, slope, aspect and geology. It has been recognised that indirect variables are most easily measurable in the field and have good correlation with observed patterns of species (Guisan and Zimmermann 2000), therefore these are often used for modelling and predicting species distributions. In particular, the use of indirect variables may provide better predictions for modelling the potential distribution of species at small spatial scales and in complex topography (Guisan and Zimmermann 2000). Topographic parameters have been found to have good prediction power in such cases (Guisan et al. 1999). Data related to resource, direct and indirect variables are best stored and manipulated in a GIS. With GIS, habitat potential mapping in general involves five steps:

1. conceptualisation – describing the potential habitat preferences of the target species based on scientific literature and experts' knowledge and expertise, formulating working hypotheses about species-environment relationships, and selecting the appropriate spatio-temporal resolution, spatial scale and geographical extent for the study;
2. data preparation – identifying data requirements and data sources, assessing available and missing data, assessing the relevance of environmental variables (predictors) for the target species at the given scale, identifying an appropriate sampling strategy for collecting species observation data and using GIS to process and manipulate the data;
3. model selection – identifying the most appropriate method(s) for modelling, and identifying methods or statistics needed for evaluating the predictive accuracy of the model(s);
4. spatial prediction – preparing input data (observations and environmental data), linking the selected modelling method(s) with GIS, parameterising the model(s), implementing the model(s) in the GIS environment and mapping the predictive results;
5. model validation – testing the model(s) with independently observed data.

Most modelling techniques for habitat potential mapping are statistical in nature and enable a probability of occurrence to be predicted in a location where no species information is available. The most commonly used statistical techniques include regression, classification techniques, environmental envelopes, Bayesian methods and neural networks (Guisan and Zimmermann 2000). Wright et al. (2011) presented a case study of applying weights-of-evidence with GIS (see Section 9.4) to predict the potential spatial distribution of toothed leionema (*Leionema bilobum* subsp. *serrulatum*) in the Strzelecki Ranges in Victoria, Australia.

Toothed leionema is one of four subspecies of *Leionema bilobum* from the Rutaceae family. It is a poorly investigated species that is considered rare in Victoria. There is no detailed recorded knowledge of the physiology and behaviour of this species (such as seed dispersal, migration and adaptation). Toothed leionema resembles either a dense shrub or a small tree (Figure 10.13). The average height observed within the Strzelecki Ranges was about 2.5m, and the maximum height approximately 3–3.5m. The leaves are in the shape of oblong or narrow elliptic-oblong, about 16–52mm in length and 4–8mm in width. The species is found primarily in the southern parts of the Eastern Highlands and the Gippsland Highlands of Victoria. The study area was limited to a small portion of the Strzelecki Ranges in West Gippsland (Figure 10.14), covering an area of 25 × 30km².

The species has been observed growing within wet schlerophyll forest or riparian open forest, often on steep slopes. Field observations indicated that the occurrence of toothed leionema in this region appeared to be dependent on disturbance, particularly near or within plantation forests, which are periodically harvested. The species also appeared to prefer open areas exposed to sunlight, where there was a low canopy cover near or within forested regions. Due to the lack of adequate understanding of the general botany and the physiology of toothed leionema, indirect environmental variables, particularly topographic variables, were used in the predictive modelling. Topography affects not only soil moisture, but also indirectly affects soil pH. Therefore, elevation, slope, aspect, wetness index, distance to water, distance to plantation areas and ecological vegetation

Figure 10.13 Toothed leionema (like a dense shrub or small tree) (courtesy of Mateusz Okurowski).

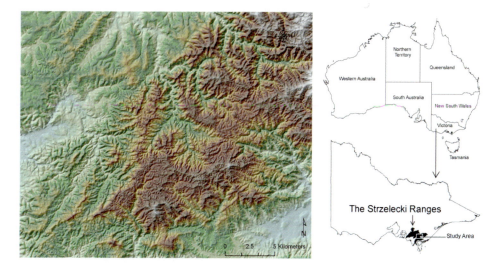

Figure 10.14 The study area of habitat potential mapping in the Strzelecki Ranges, Victoria.

classes were selected in this case study as major environmental factors that potentially influenced the growth and hence the distribution of toothed leionema.

Presence data for toothed leionema in the Strzelecki Ranges were obtained from HVP Plantations Pty Ltd and two field surveys. HVP provided thirty-three records of locations where toothed leionema was known to occur. These sites were verified during fieldwork, and three additional sites were identified during the first round of fieldwork. The boundaries of toothed leionema thickets were mapped using Trimble GeoXH GPS with an accuracy of 10cm. For the modelling purposes, the centre of each thicket colony was used to identify the location of each site. These thirty-six sites were used as training sites in the predictive modelling. The second field survey identified fourteen new sites for the occurrence of toothed leionema in the modelled area. In addition, HVP provided a further nine records. These twenty-three sites were used as control sites for model validation. The two surveys were guided by an HVP fieldworker who was knowledgeable with the region's vegetation and who selected the sites randomly for visit based on field knowledge of where the species were potentially located. Therefore, the training and control sites can be considered random samples.

In this study, a 10m DEM was used to derive topographic indices for modelling, including slope, aspect and wetness index (see Section 7.4). The ecological vegetation classes (EVC) data layer was acquired. Each EVC represents one or more plant communities that occur in similar types of environments and tend to show similar ecological responses to environmental factors such as disturbance

(for example, wildfire). EVCs can be used as a guide to the distribution of individual species or groups of species. A hydrology data layer was used to analyse the proximity to streams and other water bodies. A forest data layer supplied by HVP provided information on native and plantation forests. This information was used as a proxy for the disturbance associated with plantation forestry.

ArcGIS and ArcSDM (see Box 9.5) were used to establish spatial relationships between the species and the environmental variables listed above, and to predict the probabilities of the occurrence of the species at different locations through WofE modelling. Figure 10.15 shows the solution process. As a result, of the seven environmental factors considered, four environmental factors, including aspect, elevation, distance to water and proximity to plantation areas, were found to be correlated with the observed occurrence of toothed leionema, and were shown to have significant predictive power to estimate the spatial distribution of the species.

The potential spatial distribution of toothed leionema within the study area was determined using the posterior probabilities of species presence based on the four environmental factors that have predictive power, classified into four categories: high potential, moderate potential, low potential and uncertain, as defined in Section 9.4. Figure 10.16 shows the modelling results. Based on the analysis of the results, about 10 per cent of the study area was predicted as high potential habitat for toothed leionema, highlighting the rarity of the species in the region. These areas have a mean elevation of 395m (ranging from 336m to 555m), a mean distance to water of 983m, a mean distance to plantation of 120m and a

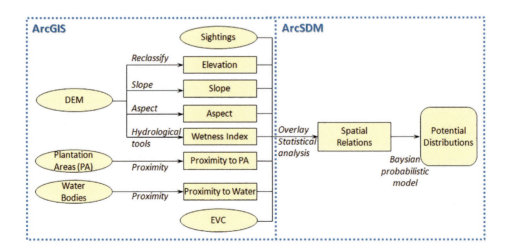

Figure 10.15 The solution process for predicting the potential spatial distribution of toothed leionema.

dominant southwestern aspect. About 35 per cent of the study area was predicted as moderate and low potential habitat for toothed leionema. Over 55 per cent of the study area was considered uncertain. The uncertain areas were generally located more than 680m away from plantation areas, with a northwestern or northern aspect. The areas with moderate potential tended to have an aspect of southwest, while those with low potential were likely to have an aspect of south or southeast.

Both model validation with training and control sites and ground truthing undertaken in the field suggested that the modelling results were valid, and the predicted spatial distribution provided a useful indication of potentially suitable habitat for toothed leionema. However, like any other predictive modelling methods, WofE cannot provide a perfect representation of real and accurate distributions of the species. Ecosystems are highly complex and heterogeneous, and are very difficult to model accurately in every aspect of time and space (Guisan and Zimmermann 2000). The WofE model for the distribution of toothed leionema developed in this case study was limited by a lack of sufficient environmental data and the limited number of known occurrences of the species. More understanding of the physiology and behaviour of the species is needed, and more datasets (such as soils, presence of ferns and upper storey species) are required to make the model more robust. It is also impossible to incorporate dynamic aspects of the species' succession into the model. In addition, the WofE model predicted a simulated distribution from field-derived observations (based on empirical field datasets) rather from theoretical physiological constraints. Therefore, the predicted potential spatial distribution is within the realised niche of toothed leionema, not a function of physiological performance and ecosystem constraints of the species, and cannot be applied in changing environmental situations. Nevertheless, the WofE model has been useful in identifying key habitat conditions of toothed leionema and predicting areas where the species is likely to be present in a way that has not been possible in the past. The results not only offered some understanding of the habitat requirements of the species, but also provided a new set of data about the species to develop strategies for its conservation and management.

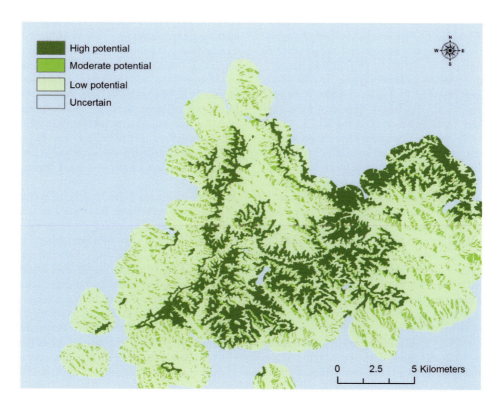

Figure 10.16 The potential spatial distribution of toothed leionema.

10.5 LANDSCAPE VALUATION

Landscapes are geographical areas that are spatially heterogeneous and characterised by diverse interacting patches or ecosystems. Intuitively, we distinguish between agricultural and urban landscapes, arid and humid landscapes, lowland and mountainous landscapes and so on. Landscapes are comprised of a variety of different elements, such as mountains, hills, forests, grassland, wetlands, farms, and other forms of land use and land cover. They are dynamic, and are shaped by ongoing physical, biological, social and spiritual processes. All landscapes have the capacity to enhance and regenerate the natural benefits and services provided by ecosystems in their natural state. The quality of the landscape underpins the viability of life on Earth. Regardless of scale, every landscape embodies a range of multidimensional and interdependent values, and manifests a living synthesis of people and place that is fundamental to local and national identity. Landscape values are increasingly recognised in environmental management, resource management, land use planning and cultural heritage conservation.

The value of landscape can be viewed and expressed differently by different disciplines, cultural conceptions, philosophical views and schools of thoughts. In the area of natural resource and environmental management, landscapes are often valued in terms of ecosystem services. Ecosystem services are the goods and services provided to humans by the ecological processes of healthy landscape systems – for example, clean air, water and food (Constanza et al. 1997). There are four types of ecosystem services: provisioning, supporting, regulating and cultural. Provisioning services are the products obtained from ecosystems, including food, water, timber, fibre, biochemical and genetic resources; regulating services are the benefits gained from the regulation of ecosystem processes that affect climate, floods, disease, wastes and water quality; cultural services are the non-material benefits obtained from ecosystems through spiritual enrichment, cognitive development, reflection, recreation and aesthetic experiences; and supporting services are those necessary for the production of all other ecosystem services, such as soil formation, photosynthesis and nutrient cycling (Millennium Ecosystem Assessment 2003). Landscape values can be assessed in terms of economic values of the ecosystem services they provide or according to the socio-cultural values of the services.

Case Study 12: economic valuation of ecosystem services

Ecosystem services have economic value because we derive utility from their actual or potential use, either directly or indirectly. Economic valuation of various ecosystem services determines their value to society and the economy in monetary terms, which allows them to be accounted for in natural resource management and sustainable development decision making. Various economic valuation methods focusing on utilitarian values have been used to quantify the benefits of ecosystem services, such as people's willingness to pay or willingness to accept hedonic pricing, travel cost, and benefit or value transfer methods (Millennium Ecosystem Assessment 2003). Troy and Wilson (2006) provided a case study of mapping ecosystem service values by combining GIS and value transfer.

Basically, the value transfer method involves estimating economic values of ecosystem services by adapting estimates obtained (by whatever method) from studies already completed in another location and/or context. For example, an estimate of the economic value for recreational fishing in one region might be used to estimate the benefit obtained from recreational fishing in another region. It has become a practical and important tool for informing environmental policy making when it is not feasible to collect primary data due to budget and time constraints yet some measure of benefits is needed. Troy and Wilson (2006) added the spatial dimension to value transfer by linking analyses of non-market economic valuation data and biophysical data, and developed a framework for carrying out spatially explicit value transfer. The framework is used to estimate and map ecosystem service values associated with each type of land cover. It can be outlined as a six-step procedure (Figure 10.17):

1. Defining the study area and its boundary – a small boundary adjustment may have a large impact on the final outcomes.
2. Acquiring GIS data and developing a land cover classification scheme – this is often done by examining and interpreting the GIS data available for the study area. The defined land cover classes should have significant differences in the flow and value of ecosystem services. This is accompanied by reviewing the valuation literature to collect empirical studies from a similar context, and to extract per unit area estimates of economic value for ecosystem services associated with each of the land cover types defined in this step.

3. Mapping land cover and associated ecosystem service flows – this involves the creation of a land cover map layer with associated ecosystem services using GIS overlay and other spatial analysis tools or through remote sensing image processing.

4. Calculating and mapping total ecosystem service values – the annual total ecosystem service value for each mapping unit on the land cover map is computed as:

$$V_{ij} = \Sigma_{k=1}^{n} A_i \times EV_{kj} \qquad (10.18)$$

where V_{ij} is the annual total ecosystem service value for mapping unit i of land cover type j, A_i is the area of mapping unit i, EV_{kj} is the annual value per unit area for ecosystem service type k provided by land cover type j and n is the number of ecosystem services.

5. Aggregating ecosystem service values in terms of management area units – this can include catchment areas and local government areas.

6. Conducting scenario analysis – by projecting land cover changes and repeating Steps 3 and 4 with the projected changes, future scenarios of ecosystem services and their economic values can be analysed.

This framework was used to estimate ecosystem service values in three study areas in the USA: (1) Massachusetts, (2) Maury Island, Washington State and (3) three counties in California. In the case of Massachusetts, a land cover map for Massachusetts was derived from 1:25,000 air photos including twenty-five land cover types, which were then combined into nine land cover classes for ecosystem service valuation. Forty-two empirical studies with similar land cover types and contexts were used to derive per unit area estimates of economic value for each ecosystem service associated with each type of land cover. The California case study used a typology of fourteen land cover classes derived from various sources of land cover information and eighty-four empirical valuation studies to get transferable values. In both the Massachusetts and California cases, the resultant total ecosystem service values were summarised and mapped by tributary basin. For the case of Maury Island, eleven land cover types were mapped mainly based on LANDSAT (30m resolution) and IKONOS (1m resolution) satellite imagery, and forty-three relevant empirical studies were used. The results were summarised by property parcel. Changes in ecosystem service values were also estimated under two alternative development scenarios in the case of Maury Island. Table 10.3 lists the ecosystem service values by land cover and ecosystem service type for Maury Island resulting from the study. Similar information was derived for the other two study areas. Maps of ecosystem service values and areal-aggregated totals were also produced. The maps show the spatial distribution of landscapes providing various ecosystem services and their economic values.

Table 10.3 also indicates that not all land cover types were valued for all possible associated ecosystem services due to the limited availability of economic valuation data, possibly leading to a significant undervaluation. To fill the gaps in the economic valuation data, new empirical studies need to be conducted. In addition, it is important to note that value transfers can only be as accurate as the initial empirical studies. Furthermore, valid and reliable estimates can only be obtained when the services being valued are identical in the empirical studies and the current application, and when the human population and biophysical conditions in the study areas have similar characteristics. The quality of land cover data affects the accuracy of the final results too.

Case Study 13: socio-cultural valuation of landscape

Troy and Wilson (2006) argued that economic valuation should be used as only one kind of evidence among many others in supporting management decisions. Indeed, in addition to economic values, many people attribute ecological, socio-cultural or intrinsic values to ecosystem services provided by landscapes (Millennium Ecosystem Assessment 2003). However, different people may attach different personal meanings and derive different values from the same landscape element and its surroundings. Landscape values are generally tied to an individual's personal experiences and purposes. Understanding how the public perceive landscape values is gaining recognition as an important component of decision making to determine appropriate nature conservation and development strategies and plans. Conservation and development decisions which neglect local knowledge and deeply embedded socio-cultural and ethical values may threaten individuals and their assets, and may undermine attempts by government agencies to obtain binding agreements between stakeholder groups or to manage conflicts.

Many approaches to gaining insight into perceptions of landscapes and their associated meaning and

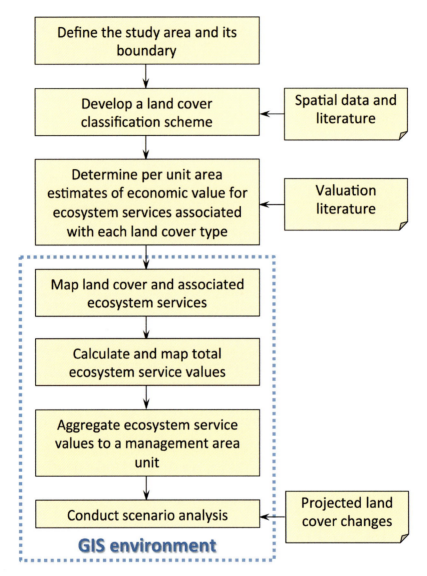

Figure 10.17 A value transfer method for estimating economic values of ecosystem services (based on Troy and Wilson 2006).

value have relied on participatory assessment or group valuation using interviews, focus groups or attitudinal questionnaires. Two methods have been applied for evaluation of landscape values perceived by the public or community obtained through participatory assessment: values suitability analysis and hot spot analysis.

Values suitability analysis involves measuring an inventory of place-specific social perceptions of landscape values (Reed and Brown 2003). Perceived landscape

values are measured and mapped by public surveys, and GIS-based density mapping techniques are then used to aggregate valuations and identify places with high densities. Zhu et al. (2010) provided a case study of values suitability analysis using ArcGIS for evaluation of landscape values in the Murray River region of Victoria, Australia.

The Murray River is Australia's second longest river. It stretches for over 1,500km from its headwaters in the Great Dividing Range in the northeast of Victoria

Table 10.3 Annual total ecosystem service values (in US$) by land cover and ecosystem service type for Maury Island (adapted from Troy and Wilson 2006)

Land cover	Average $/ha/year	Aesthetic and amenity	Climate and atmospheric regulation	Disturbance prevention	Food and raw materials	Habitat refuge	Recreation	Soil retention and formation	Waste assimilation	Water regulation and supply	Totals
Beach	88,204	—	—	—	—	—	2,371,006	—	—	—	2,371,006
Beach near dwelling	117,254	4,442,228	—	—	—	—	—	3,133,597	—	—	7,575,825
Coastal riparian	9,396	224,009	—	48,622	—	509,067	10,732	107,842	29,872	314,520	1,244,664
Forest	1,826	7,703	1,391,576	—	—	10,041	483,395	—	—	13,695	1,906,410
Freshwater stream	1,595	25	—	—	—	24,641	17,585	—	—	23,807	66,058
Freshwater wetland	72,787	17,866	—	56,893	—	85,466	4,203	—	104,642	20	269,090
Grassland/herbaceous	118	—	2 649	—	—	—	755	379	32,915	1,135	37,833
Near-shore aquatic habitat	16,283	—	—	—	2,080,557	—	3,605,238	—	—	—	9,204,633
Saltwater wetland	1,413	—	—	3,770	—	—	173	—	1,474	4,110	9,527
Disturbed and urban	—	—	—	—	—	—	—	—	—	—	—

to its mouth in South Australia. The river is one of the world's longest navigable rivers, and a major source of water for much of southeastern Australia. In addition to its importance and significance for agricultural production, the Murray River region has been valued greatly for its environmental and cultural qualities, which support a variety of unique river life, riparian habitats, historic towns and magnificent scenery. It has therefore become a major tourism destination. However, the health of the river and environmental quality of the region have declined significantly since European settlement, due to river regulation, logging, urban development and agricultural activities. In order to reconcile the conflicting needs of the environment, agricultural production and commercial development, it became necessary to assess whether the levels of protection of natural areas in the region should be changed to aid the conservation of important species, habitats and cultural assets. The project conducted by Zhu et al. (2010) aimed to provide some key insights into community-held landscape values of Murray River sites on the Victorian side, and to offer strategic information which could serve as a basis for benchmarking and prioritising sites for conservation, service provision and tourism development consistent with the community's values. Twelve values were used in landscape valuation in the project: aesthetic/scenic, biological diversity, economic, future, heritage, intrinsic,

learning, life-sustaining, recreation, spiritual, therapeutic and wilderness (Table 10.4). These landscape values reflect a wide range of land use potentials of the public land in the Murray River region.

Landscape valuation using the twelve values was conducted through postal surveys. Questionnaires, including a map of the study area and a set of sticker dots coded with the type of landscape value and the weight of importance, were sent to a random sample of 1,615 local residents within the study area, 508 visitors across the region and thirty-two tour operators. Surveys involved a mapping task to identify the locations or places with landscape values. Respondents were asked to nominate up to six locations for each of the twelve landscape values. This was done by placing coded sticker dots on a 1:294,000 greyscale map of the study area. Each dot had a code indicating the type of landscape value for places – for example, (a) stood for aesthetic/scenic value, (e) for economic value, (r) for recreation value and so on. Each dot also had an importance rating or weight of 5, 10, 20 or 50. The higher the weight, the more important the landscape value. All dots placed on the returned maps were digitised and stored in an ArcGIS shapefile. The distribution of the survey point data is shown in Figure 10.18. These point data were coded with the dot point values, the types of landscape value and the respondents' identification numbers. After eliminating outliers, a total

Table 10.4 Types of landscape values (adapted from Zhu et al. 2010 and Reed and Brown 2003)

Value	Description
Aesthetic	Having attractive scenery, sights, smells or sounds
Economic	Having economic benefits or potentials
Recreational	Having recreation opportunities
Life-sustaining	Helping produce, preserve and renew air, soil and water
Learning	Having opportunities for learning about the nature
Biological diversity	Having a variety of plants, wildlife, aquatic life or other living organisms
Spiritual	Being spiritually special
Intrinsic	Owning special values
Heritage	Possessing natural and human history
Future	Allowing future generations to know and experience them
Therapeutic	Making people feel better physically and/or mentally
Wilderness	Being wild

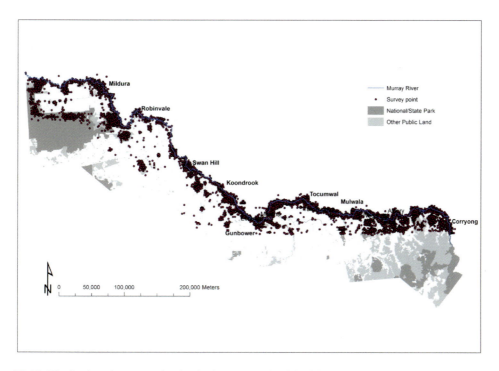

Figure 10.18 Distribution of survey point data in the case study of the Murray River region.

of 15,517 dot or point locations were obtained, each representing a place with a weighted landscape value perceived by the respondents.

Based on the weights of each landscape value recorded at each survey location, density maps of the twelve landscape values were produced using ArcGIS. Landscape value densities were measured using kernel estimation (Silverman 1986). Conceptually, a circular neighbourhood area is defined around each point with a weighted landscape value. A smoothly curved surface is fitted over each point, which is called a kernel surface (Figure 10.19). The surface value is highest at the survey point, and diminishes with increasing distance from it, reaching 0 at the edge of the neighbourhood area. The volume under the surface equals the weight of the landscape value for the point. The density of the landscape value at a certain location is calculated by adding the values of all the kernel surfaces where they overlay the location. Each density map illustrates how the intensity of a landscape value changes continuously over the study area. Figure 10.20 shows the resulting density map of aesthetic value.

The density maps reveal varying intensities of landscape values and show where the high values of a particular landscape value are concentrated. They provide a very useful starting point to identify areas with high landscape values. For example, by analysing the map in Figure 10.20, we may find that high aesthetic values are mainly concentrated around Mildura, Swan Hill, Echuca and Yarrawonga. The areas with especially high densities of a landscape value may be referred to as hot spots for that landscape value. However, at what density does an aggregation of a landscape value (represented by dot points) become a hot spot? It is not easy to subjectively determine a density threshold to identify hot spots. Because of this, it is difficult to identify hot spots with multiple values by simply overlaying the density maps. In addition, even if an analyst could do it, it would be difficult to establish or statistically test the significance of the individual hot spots identified.

Zhu et al. (2010) extended the values suitability analysis method by applying the Getis-Ord G_i^* statistic (see Section 5.5) to identify spatial clusters of statistically significant high landscape values – that is, hot spots of a landscape value. They divided the study area into 2,171 grid cells, as shown in Figure 10.21. Each cell covered an area of 5km × 5km. Each cell was identified with its centroid. In addition, each cell was associated with a density value of every landscape value. For a particular landscape value, the G_i^* statistic for each grid cell was calculated.

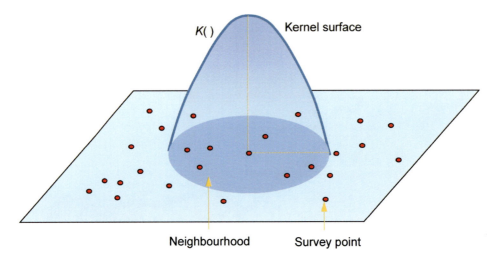

Figure 10.19 Kernel surface for density mapping.

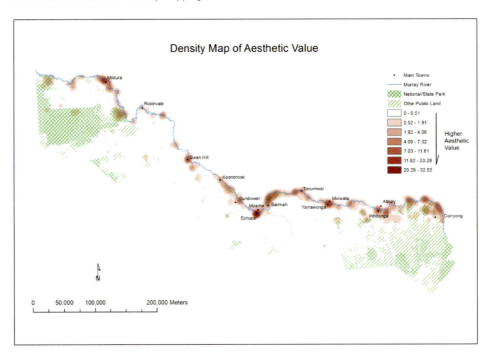

Figure 10.20 Density map of aesthetic value in the Murray River region.

The G_i^* statistic measures the concentration of the sum of density values of a particular landscape value in the study area. It indicates whether high density values or low density values (but not both) tend to cluster in the area. Given the set of grid cells with density values, the G_i^* statistic identifies those clusters of cells with density values of a particular landscape value higher in magnitude than one might expect to find by random chance. These clusters of cells are hot spots. The hot spots for each of the twelve landscape values were mapped. Figure 10.21 presents the hot spots of aesthetic value in the Murray River region. Fifty-nine hot spots (counted as the number of

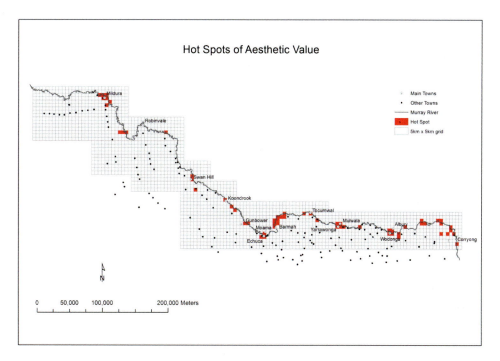

Figure 10.21 Hot spot map of aesthetic value in the Murray River region.

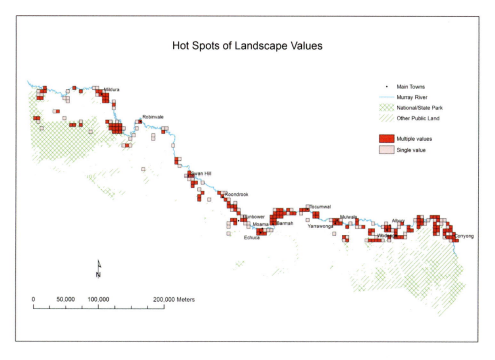

Figure 10.22 Hot spot map of multiple landscape values in the Murray River region.

grid cells) with aesthetic value were identified from the map, spread on or near the river bank.

The hot spot maps generally conformed to the corresponding density maps – that is, the hot spots identified corresponded to the areas with the highest densities on the density maps. The difference was that the hot spot maps precisely separated hot spots from areas without concentrations of high landscape values that were significant enough to be recognised as hot spots, as demonstrated in Figures 10.20 and 10.21.

By overlaying the hot spot maps of the twelve landscape values, the map in Figure 10.22 was produced. Overall, a total of 251 hot spots with one or more landscape values were identified. They were mainly distributed close to the Murray River, in national and state parks, state forests and on other public land, and near the major towns. These hot spots represented a specific group of sites in the region highly valued by the community for conservation or development. They were the sites prioritised for assessment of the existing conservation measures and tourism facilities, and for drawing up new plans for conservation, service provision and tourism development.

This case study demonstrated that hot spot analysis with GIS can effectively associate significant landscape values with specific places. Although landscape valuation using predefined values is based on subjective judgements and personal values, which are heavily influenced by respondents' understanding of the landscape values, their knowledge and past experiences, their familiarity with the study area, their desired activities and recreation requirements, their map literacy and the intrinsic appeal of the places themselves, the methods employed in this case study can be used to produce a useful inventory of the landscape values and special places for a region, which can then be easily integrated with biophysical data and information for conservation and resource planning to ensure that the conservation and development plans are both scientifically based and socially acceptable.

10.6 SUMMARY

- Environmental applications of GIS are very diverse. Examples include hydrological modelling, ecological modelling, land use analysis, water and air pollution modelling, and ecosystem service valuation. The thirteen case studies in this chapter provide some examples of these.
- Case Studies 1 and 2 are examples of GIS in hydrological modelling. In Case Study 1, GIS was used to implement the SCS curve number model to estimate

runoff for a sub-basin of the Kissimmee River basin in the USA. In Case Study 2, GIS coupled with the SWAT model was used to estimate the loads of non-point source pollutants, including NO_3-N, water-soluble P, organic N and organic P, transported by runoff and percolated water from the Banha catchment in northern India during the monsoon season. In both case studies, GIS was used to prepare model inputs.

- Case Studies 3, 4 and 5 are examples of GIS in land use analysis. Case Study 3 applied GIS and multi-criteria decision analysis to identify feasible sites for wind farms in New York State through three stages of the siting process. Case Study 4 integrated GIS and multi-criteria decision analysis for sugarcane land allocation in the lower Herbert River Catchment in Northern Queensland, Australia. It demonstrated that the integration of GIS and multi-criteria decision analysis allowed the development of different land allocation scenarios, which could help decision makers understand the implications of differing policies in land allocation strategy. Case Study 5 applied GIS to assess land suitability for tea cultivation in northern Laos, which implemented the FAO land evaluation framework.
- Case Studies 6, 7 and 8 are the examples of GIS in atmospheric science research. Case Study 6 is an application of spatial interpolation with residual kriging to estimate air temperature in Poland and Central Europe based on DEM. In Case Study 7, NO_2 pollution across Europe was modelled using GIS focal functions. Case Study 8 investigated spatial and temporal patterns and trends of climate variability in North Carolina through linear trend modelling and spatial interpolation.
- Case Studies 9, 10 and 11 are examples of GIS in ecological modelling. In Case Study 9, GIS was used to model relationships between landscape characteristics and bird diversity in Ohio. Landscape characteristics were measured using twelve landscape metrics. Case Study 10 used GIS to analyse the population dynamics of western corn rootworm in relation to changes in the landscape pattern in the western portion of the US Corn Belt. Case Study 11 modelled the potential spatial distribution of toothed leionema in the Strzelecki Ranges, Victoria, Australia using weights-of-evidence with GIS. It identified key habitat conditions of this rare tree species and predicted the areas where it is likely to be present.
- Case Studies 12 and 13 are examples of GIS in landscape valuation. Case Study 12 combined GIS and

the value transfer method to estimate economic values of ecosystem services in three areas in the USA. Case Study 13 applied values suitability analysis and hot spot analysis with GIS for socio-cultural valuation of landscape in the Murray River region of Victoria, Australia.

REVIEW QUESTIONS

1. What is the difference between lumped and distributed hydrological models? What are the roles of GIS in hydrological modelling?

2. For Case Study 1, what was the objective of the study? What does the result of runoff modelling indicate?

3. In Case Study 2, what was the most challenging aspect of the non-point source pollution modelling using GIS? Why?

4. What are the roles of GIS in land use management and planning?

5. For Case Study 3, discuss how GIS was used in each of the three stages of the siting process and what further studies are required to refine the results.

6. In Case Study 4, how could different scenarios of sugarcane land allocation be developed? How would the different scenarios help decision makers come to decisions?

7. Describe the general procedure of land evaluation using GIS. In Case Study 5, why was AHP used?

8. What is the dual role of GIS in atmospheric science research?

9. Referring to Case Study 6, describe how residual kriging was implemented in GIS for interpolation of meteorological variables.

10. In Case Study 7, what GIS functions were used to implement the moving window approach to modelling of NO_2 pollution?

11. In Case Study 8, how were precipitation and temperature trend maps created? What are the limitations of the approach the case study took to spatial and temporal analysis of climate variability?

12. What do the results of Case Study 9 tell us about the influence of scale on landscape metrics?

13. In Case Study 10, how were western corn rootworm population maps created?

14. What is habitat potential mapping? Referring to Case Study 11, describe the general procedure of habitat potential mapping with GIS.

15. Describe the value transfer method for economic valuation of ecosystem services used in Case Study 12, and discuss its limitations.

16. Referring to Case Study 13, compare values suitability analysis and hot spot analysis for socio-cultural valuation of landscape, and discuss their merits and limitations.

REFERENCES

Allwine, K.J., Rutz, F.C., Shaw, W.J., Rishel, J.P., Fritz, B.G., Chapman, E.G., Hoopes, B.L. and Seiple, T.E. (2006) *DUSTRAN 1.0 User's Guide: A GIS-Based Atmospheric Dust Dispersion Modelling System*, Technical Report PNNL-16055, Richland, WA: Pacific Northwest National Laboratory.

Arnold, J.G. and Fohrer, N. (2005) 'SWAT2000: current capabilities and research opportunities in applied watershed modeling', *Hydrological Processes*, 19(3): 563–572.

Austin, M.P. and Smith, T.M. (1989) 'A new model for the continuum concept', *Vegetation*, 83: 35–47.

Baker, W.L. and Cai, Y. (1992) 'The r.le programs for multiscale analysis of landscape structure using the GRASS geographical information system', *Landscape Ecology*, 7(4): 291–302.

Beckler, A.A., French, B.W. and Chandler, L.D. (2004) 'Characterization of western corn rootworm (Coleoptera: Chrysomelidae) population dynamics in relation to landscape attributes', *Agricultural and Forest Entomology*, 6(2): 129–139.

Berkowicz, R., Ketzel, M., Jensen, S.S., Hvidberg, M. and Raaschou-Nielsen, O. (2008) 'Evaluation and application of OSPM for traffic pollution assessment for a large number of street locations', *Environmental Modelling and Software*, 23(2): 296–303.

Bogaert, J., Rousseau, R., Hecke, P.V. and Impens, I. (2000) 'Alternative area-perimeter ratios for measurement of 2D shape compactness of habitats', *Applied Mathematics and Computation*, 111: 71–85.

Boyles, R.P. and Raman, S. (2003) 'Analysis of climate trends in North Carolina (1949–1998)', *Environment International*, 29: 263–275.

Chanhda, H., Wu, C., Ye, Y. and Ayumi, Y. (2010) 'GIS based land suitability assessment along Laos-China border', *Journal of Forestry Research*, 21(3): 343–349.

Chapman, L. and Thornes, J.E. (2003) 'The use of geographical information systems in climatology and meteorology', *Progress in Physical Geography*, 27: 313–330.

Costanza, R., d'Arge, R., De Groot, R., Farber, S., Grasso, M., Hannon, B., Limburg, K., Naeem, S., O'Neill, R.V., Paruelo, J., Raskin, R.G., Sutton, P. and Vandenbelt, M. (1997) 'The value of the world's ecosystem services and natural capital', *Nature*, 387: 253–260.

FAO (1976) *A Framework for Land Evaluation*, FAO Soils Bulletin no. 32, Rome, Italy: FAO.

Fortin, M.-J., Dale, M.R.T. and Ver Hoef, J. (2002) 'Spatial analysis in ecology', in A.H. El-Shaarawi and W.W. Piegorsch (eds) *Encyclopedia of Environmetrics*, Chichester, England: John Wiley & Sons, 2,051–2,058.

Guisan, A. and Zimmermann, N.E. (2000) 'Predictive habitat distribution models in ecology', *Ecological Modelling*, 135: 147–186.

Guisan, A., Weiss, S.B. and Weiss, A.D. (1999) 'GLM versus CCA spatial modelling of plant species distribution', *Plant Ecology*, 143: 107–122.

Gulliver, J. and Briggs, D. (2011) 'STEMS-Air: a simple GIS-based air pollution dispersion model for city-wide exposure assessment', *Science of the Total Environment*, 409(12): 2,419–2,429.

Johnson, G.D. and Patil, G.P. (2007) *Landscape Pattern Analysis for Assessing Ecosystem Condition*, New York: Springer.

Kesarkar, A.P., Dalvi, M., Kaginalkar, A. and Ojha, A. (2007) 'Coupling of the weather research and forecasting model with AERMOD for pollutant dispersion modeling: a case study for PM$_{10}$ dispersion over Pune, India', *Atmospheric Environment*, 41(9): 1,976–1,988.

Maantay, J.A., Tu, J. and Maroko, A.R. (2009) 'Loose-coupling an air dispersion model and a geographic information system (GIS) for studying air pollution and asthma in the Bronx, New York City', *International Journal of Environmental Health Research*, 19(1): 59–79.

Malczewski, J. (2004) 'GIS-based land-use suitability analysis: a critical overview', *Progress in Planning*, 62: 3–65.

Mayer, A.L. and Cameron, G.N. (2003) 'Landscape characteristics, spatial extent, and breeding bird diversity in Ohio, USA', *Diversity and Distributions*, 9: 297–311.

McElroy, A.D., Chiu, S.Y., Nebgen, J.W., Aleti, A. and Bennett, F.W. (1976) *Loading Functions for Assessment of Water Pollution from Non-point Sources*, Washington, DC: US Environmental Protection Agency.

McRae, S.G. and Burnham, C.P. (1981) *Land Evaluation*, Oxford: Clarendon Press.

Melesse, M.A. and Shih, S.F. (2002) 'Spatially distributed storm runoff depth estimation using Landsat images and GIS', *Computers and Electronics in Agriculture*, 37: 173–183.

Menzel, R.G. (1980) 'Enrichment ratios for water quality modeling', in W.G. Knisel (ed.) *CREAMS: A Field Scale Model for Chemicals, Runoff, and Erosion from Agricultural Management Systems*, USDA Conservation Research Report no. 26, 486–492.

Millennium Ecosystem Assessment (2003) *Ecosystems and Human Well-Being: A Framework for Assessment*, Washington, DC: Island Press.

Milne, B.T. (1988) 'Measuring the fractal geometry of landscape', *Applied Mathematics and Computation*, 27: 67–79.

Mishra, A., Singh, R. and Singh, V.P. (2010) 'Evaluation of non-point source N and P loads in a small mixed land use land cover watershed', *Journal of Water Resource and Protection*, 2(4): 362–372.

Neitsch, S.L., Arnold, J.G., Kiniry, J.R. and Williams, J.R. (2011) *Soil and Water Assessment Tool Theoretical Documentation Version 2009*, Texas Water Resources Institute Technical Report no. 406, College Station, TX: Texas A&M University System.

Notovny, V. and Chesters, G. (1981) *Handbook of Nonpoint Pollution Sources and Management*, New York: Litton Educational.

Reed, P. and Brown, G. (2003) 'Values suitability analysis: a methodology for identifying and integrating public perceptions of ecosystem values in forest planning', *Journal of Environmental Planning and Management*, 46: 643–658.

Silverman, B.W. (1986) *Density Estimation for Statistics and Data Analysis*, New York: Chapman & Hall.

Spellerberg, I.F. (1992) *Evaluation and Assessment for Conservation: Ecological Guidelines for Determining Priorities for Nature Conservation*, London: Chapman & Hall.

Troy, A. and Wilson, M.A. (2006) 'Mapping ecosystem services: practical challenges and opportunities in linking GIS and value transfer', *Ecological Economics*, 60: 435–449.

Tveito, O.E., Bjørdal, I., Skjelvag, A.O. and Aune, B. (2005) 'A GIS-based agro-ecological decision system based on gridded climatology', *Meteorological Applications*, 12: 57–68.

United States Energy Information Administration (2011) *Electric Power Annual 2009*, Washington, DC: United States Energy Information Administration.

USDA (1986) *Urban Hydrology for Small Watersheds, TR-55*, 2nd edn, Washington, DC: Natural Resources Conservation Service, Conservation Engineering Division, USDA.

Ustrnul, Z. and Czekierda, D. (2005) 'Application of GIS for the development of climatological air temperature maps: an example from Poland', *Meteorological. Applications*, 12: 43–50.

Van Haaren, R. and Fthenakis, V. (2011) 'GIS-based wind farm site selection using spatial multi-criteria analysis (SMCA): evaluating the case for New York State', *Renewable and Sustainable Energy Reviews*, 15: 3,332–3,340.

Vienneau, D., De Hoogh, K. and Briggs, D. (2009) 'A GIS-based method for modelling air pollution exposures across Europe', *Science of the Total Environment*, 408: 255–266.

Williams, J.R. (1975) 'Sediment-yield prediction with Universal Equation using runoff energy factor', in *Present and Prospective Technology for Predicting Sediment Yield and Sources: Proceedings of the Sediment-yield Workshop, ARS-S-40, November 28–30, 1972, Oxford, Miss.*, New Orleans, LA: US Department of Agriculture, Agricultural Research Service, Southern Region, 244–252.

Williams, J.R. and Hann, R.W. (1978) *Optimal Operation of Large Agricultural Watersheds with Water Quality Constraints*, Texas A&M University Technical Report no. 96, College Station, TX: Texas Water Resources Institute.

Wormworth, J. and Sekercioglu, C. (2011) *Winged Sentinels: Birds and Climate Change*, Cambridge: Cambridge University Press.

Wright, W., Zhu, X. and Okurowski, M. (2011) 'Identification of key environmental variables associated with the presence of toothed leionema (*Leionema bilobum serrulatum*) in the Strzelecki Ranges, Victoria, Australia', *Australian Journal of Botany*, 59: 207–214.

Zhu, X., Walker, D. and Mayocchi, C. (2001) 'Integrating multi-criteria modelling and GIS for sugarcane land allocation', in F. Ghassemi, M. McAleer, L. Oxley and M. Scoccimarro (eds) *Proceedings of MODSIM 2001 – International Congress on Modelling and Simulation*, Vol. 3, 10–13 December 2001, Canberra, Australia, 1,103–1,108.

Zhu, X., Pfueller, S., Whitelaw, P. and Winter, C. (2010) 'Spatial differentiation of landscape values in the Murray River Region of Victoria, Australia', *Environmental Management*, 45(5): 896–911.

Index